NOBEL LECTURES

CHEMISTRY

2001–2005

Nobel Lectures

Including Presentation Speeches and Laureates' Biographies

Physics

Chemistry

Physiology or Medicine

Literature

Peace

Economic Sciences

NOBEL LECTURES

INCLUDING PRESENTATION SPEECHES
AND LAUREATES' BIOGRAPHIES

CHEMISTRY

2001–2005

EDITOR

Per Ahlberg

Department of Chemistry
Göteborg University

World Scientific

NEW JERSEY • LONDON • SINGAPORE • BEIJING • SHANGHAI • HONG KONG • TAIPEI • CHENNAI

Published by

World Scientific Publishing Co. Pte. Ltd.
5 Toh Tuck Link, Singapore 596224
USA office: 27 Warren Street, Suite 401-402, Hackensack, NJ 07601
UK office: 57 Shelton Street, Covent Garden, London WC2H 9HE

NOBEL LECTURES IN CHEMISTRY (2001–2005)

ISBN-13 978-981-279-444-4
ISBN-10 981-279-444-1
ISBN-13 978-981-279-445-1 (pbk)
ISBN-10 981-279-445-X (pbk)

Printed in Singapore by Mainland Press Pte Ltd

FOREWORD

Since 1901 the Nobel Foundation has published annually *Les Prix Nobel* with reports from the Nobel award ceremonies in Stockholm and Oslo as well as the biographies and Nobel lectures of the Nobel laureates. In order to make the lectures available for people with special interests in the different fields the Nobel Foundation gave Elsevier Publishing Company the right to publish in English the lectures for 1901–1970, which were published in 1964–1972 through the following volumes:

Physics 1901–1970	4 vols.
Chemistry 1901–1970	4 vols.
Physiology or Medicine 1901–1970	4 vols.
Literature 1901–1967	1 vol.
Peace 1901–1970	3 vols.

Thereafter, and until the year 2000, the Nobel Foundation has given World Scientific Publishing Company the right to bring the series up to date and also publish the Prize Lectures in Economics from the year 1969. Between the years 2001–2005, the Nobel Foundation has given World Scientific Publishing Company a publishing right through an agreement between Nobel Media AB and World Scientific Publishing Company.

The Nobel Foundation is very pleased that the intellectual and spiritual message to the world laid down in the laureates' lectures, thanks to the efforts of World Scientific, will reach new readers all over the world.

Michael Sohlman
Executive Director
Stockholm, April 2008

PREFACE

In the present volume the 2001–2005 Nobel Lectures in Chemistry, together with autobiographies, portraits of the laureates and the presentation speeches delivered at the awarding ceremonies, are published. This information has previously been published in *Les Prix Nobel* together with lectures in Physics, Physiology or Medicine, Literature, Peace and Economic Sciences (Alfred Nobel Memorial Prize). The intention is that this collection of lectures, together with the previous ones, will give an impression of the spectacular discoveries and improvements that have been made in chemistry. It should also serve as a source of inspiration.

In 1901 the first Nobel Prize in chemistry was awarded to J. H. van 't Hoff for the discovery of the laws of chemical dynamics and osmotic pressure in solutions. Another of his important contributions is the concept of the tetrahedral carbon, which is a basis for chiral chemistry. One hundred years later, in 2001, the progress in chiral organic synthesis was recognized by jointly awarding **W. S. Knowles** and **R. Noyori** "for their work on chirally catalyzed hydrogenation reactions" and **K. B. Sharpless** "for his work on chirally catalyzed oxidation reactions". Many of today's drugs are based on our knowledge of fields, where these laureates have made important discoveries; they are based on mirror images of molecules.

In 2002 the development of mass spectroscopy and NMR spectroscopy was in focus. The award was shared by **J. B. Fenn** and **K. Tanaka** "for their soft desorption ionisation methods for mass spectrometric analyses of biological macromolecules" and **K. Wüthrich** "for his development of nuclear magnetic resonance spectroscopy for determining the three-dimensional structure of biological macromolecules in solution". These methods give us new important knowledge about structures and dynamics of the molecules of life.

The 2003 prize recognized insights into the transportation of water and ions through biological membranes. **P. Agre** was awarded "for the discovery of water channels" and **R. MacKinnon** "for structural and mechanistic studies of ion channels". Their discoveries are about the mechanisms of the central life processes that control where, when, and how often ions and water are let into or out of the cells in our body.

The target for the 2004 prize was progress in the understanding of protein degradation in the cells. **A. Ciechanover, A. Hershko** and **I. Rose** were jointly awarded "for their discovery of ubiquitin-mediated protein degradation". Their discovery of a system for controlled protein degradation in cells has fundamentally changed our way of thinking about protein degradation and is a basis for producing new medicines.

In 2005 advances of organic synthesis of carbon-carbon bonds were recognized. **Y. Chauvin, R. H. Grubbs** and **R. R. Schrock** were jointly awarded "for the development of the metathesis method in organic synthesis". Their discoveries have given us one of the most important methods for building organic molecules. It is among such molecules that we are finding new materials and new medicines.

Per Ahlberg

CONTENTS

Chemistry 2001

WILLIAM S. KNOWLES and RYOJI NOYORI

"for their work on chirally catalysed hydrogenation reactions"

and the other half to

BARRY K. SHARPLESS

"for his work on chirally catalysed oxidation reactions"

THE NOBEL PRIZE IN CHEMISTRY

Speech by Professor Per Ahlberg of the Royal Swedish Academy of Sciences.
Translation of the Swedish text.

Your Majesties, Your royal Highnesses, Honoured Nobel Laureates, Ladies and Gentlemen,

Science is exciting. At least I think so, along with everyone else here beside and behind me on this stage. As humans, we possess curiosity. With the aid of science we can discover answers – and surprises. That is exciting. But when I am going to explain something to a person who does not work with science, I hear: That may be interesting, but what is it good for? In the case of this year's Nobel Prize in Chemistry, I have no such problems. That is because it is very easy to answer the question. Many of today's drugs are based on our knowledge of fields where this year's Laureates have made important discoveries; they are based on mirror images of molecules.

This year's Nobel Prize in Chemistry concerns molecules that occur in two forms that are mirror images of each other. Such molecules are called *"chiral"* after the Greek word *cheir* meaning hand. Our hands are *chiral* – our right hand is a mirror image of our left hand – like most of life's molecules. In our cells we find only one of these forms. This is true, for example, of enzymes, antibodies, hormones and DNA.

Thus the enzymes in our cells are chiral, as are other receptors that play an important part in cell machinery. This means that they prefer to bind to one of the mirror image forms. The two forms of a chiral molecule often have totally different effects on cells. For example, the receptors in our nose are sensitive to mirror symmetry. One form of the substance *limonene smells like lemons,* while its mirror image smells like oranges. Most drugs consist of chiral molecules, and often only one of the mirror image forms is of interest. The other form may even be harmful. This was the case, for example, with the drug thalidomide, which was given in the 1960s to pregnant women. One mirror image form helped against nausea, while the other one – as it was discovered too late – could cause fetal damage.

It is obviously important to be able to produce each of the forms, as pure as possible. When producing compounds in the laboratory, ordinarily the result is equal quantities of both mirror image forms. This year's Nobel Laureates in Chemistry have developed chiral catalysts in order to produce only one of the forms. A catalyst is a substance that speeds up reactions without itself being consumed.

In 1968 William Knowles was the first who was able to demonstrate that chirally catalyzed hydrogenation was feasible. It was a timely discovery that immediately inspired many researchers. Knowles quickly applied his own basic research and that of others to create an industrial synthesis of the drug L-DOPA, which is used in the treatment of Parkinson's disease. Millions of patients have gained relief by using this medicine, including my own father, who suffered severely from this disease.

The man who led the further development resulting in today's more selective and general chiral hydrogenation catalysts was Ryoji Noyori. A single one of his catalyst molecules can yield millions of product molecules. His methods have gained very large practical importance, especially in industrial production of various antibiotics, for example. An important and, today, unfortunately highly newsworthy application.

Barry Sharpless has developed chiral catalysts for another important type of reactions – oxidations. His chirally catalyzed epoxidation and dihydroxylation reactions have given us new possibilities for also building complex molecules. They have gained very broad use, especially in industry, for example in producing medicines against ulcers and high blood pressure, two of our most serious health problems.

I have focused especially on the industrial role of the Laureates' discoveries, but they also constitute extremely important tools in academic research, thereby contributing to more rapid advances in research – not only in chemistry but also in materials science, biology and medicine.

Alfred Nobel himself focused on the utilitarian aspect of science when he wrote in his will that the prize should be awarded to those who "shall have conferred the greatest benefit on mankind." Thus this year's Nobel Prize in Chemistry not only makes it easy to answer what science is good for, but is also a prize entirely in the spirit of Alfred Nobel.

Dear Dr. Knowles, Dr. Noyori and Dr. Sharpless,

I have briefly described your discoveries and improvements. The beneficial consequences of your achievements to mankind are already plentiful and will certainly be amplified by the work of your numerous followers in the field of chiral catalysis.

On behalf of the Royal Swedish Academy of Sciences, I wish to convey our warmest congratulations and ask you to step forward to receive the Nobel Prize from the hands of His Majesty the King.

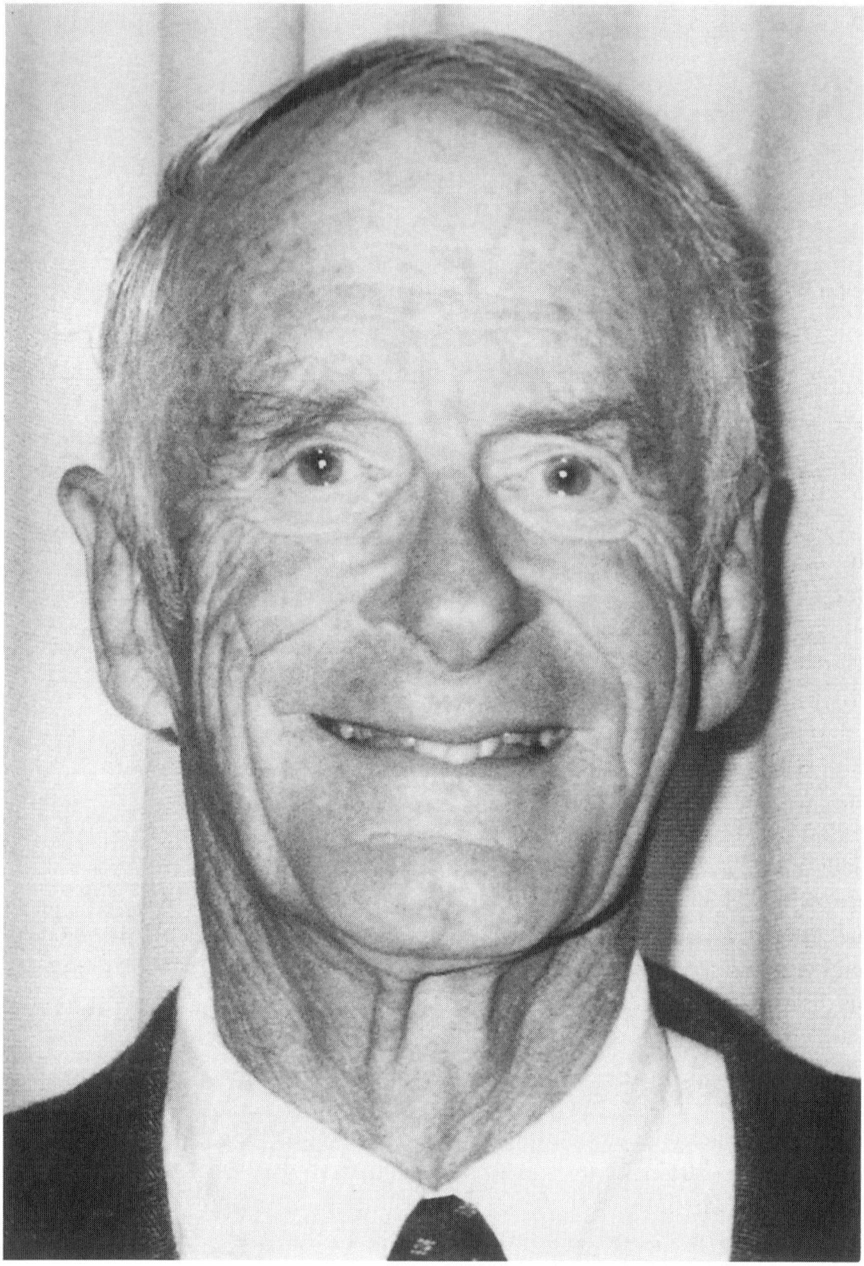

William S. Knowles

WILLIAM S. KNOWLES

I was born in Taunton, Massachusetts on June 1, 1917, but I actually grew up in nearby New Bedford. My family background was heavily slanted toward business and seafaring matters. I can't think of any relatives that ever went into science. My family gave me the best in education. To my father, business was the highest calling, but to my mother, medicine was the top profession. She would probably have gone to medical school if she had been born in a more enlightened era.

I went to boarding school at Berkshire in western Massachusetts, definitely the most beautiful part of the state. I'll never forget the fall colors on the Berkshires. In those days I was terrible at athletics and never made a team, but quite easily led my class in academics. I was particularly good at math and science. I also got a good lesson in New England thrift. To get free ice for our physics experiments we had to wait until it snowed.

On graduating, I was easily admitted to Harvard. In that era all one had to do was pass the College Board exams. If any one in my family went to college that was where he went. My father spent a year there and quit to go into the textile business. At this point I was strongly advised that I was too young socially to go to college so I took a second senior year at Andover, another boarding school. At that time many students did this. At Andover I took my first chemistry course from a teacher named Bushy Graham and was fascinated by the subject. I remember him trying to explain Avagadro's number and his discussion of the dangers of hydrogen and oxygen. At the end of the year, I took a competitive exam and won my first prize, the $50 Boylston prize in chemistry.

That summer I took a cruise on a 75-foot schooner with no engine, sailing from Gloucester, Massachusetts to Norway. We sailed around the Baltic ending up at Stockholm. I didn't think of it at the time, but we spent most of three weeks on the north Atlantic with no contact at all with the outside world. Today one is always in touch with home base even if you go to the South Pole or the Moon. Memories of this sailing trip have always been vivid. On one instance we were mistakenly arrested in Tallin, Estonia and got a ride in the paddy wagon. Later we were released without comment. Little did I think that one day years later I would be returning to Stockholm to share the Nobel Prize in Chemistry.

At Harvard I majored in chemistry with a strong inclination toward math. I took the minimum of humanities. I was told I'd be a natural for physical chemistry but taking organic with Louis Fieser changed my mind. It was there I got my introduction to optical isomerism and the tetrahedral carbon atom. At

Harvard competition was fierce and I always got a solid B, but not the straight A's of many of my class mates. These were the days when most got a gentleman's C.

On graduating in 1939 I was strongly advised to go elsewhere to graduate school. I went to Columbia with Professor Elderfield and worked on making simple analogs of the cardiac aglucones. These were tested at Eli Lilly for cardiac activity. Bob Elderfield was at his best when he talked steroids when he was at Rockefeller Institute. Paralleling Nobel's experience I too had an explosion. Mine came when distilling diazomethane. No one was hurt, but a bottle of intermediate that I had labored on for months was destroyed.

In those days Professor Elderfield spent a lot of time away on the anti-malarial project in the military, and we were on our own a lot. Professor Nelson Leonard, long at the University of Illinois in Champagne, was in our research group. Later he consulted at Monsanto.

New York was an exciting place to be in those war years and my draft board forced Columbia to push me out sooner than would ordinarily happen. In those days industry would hire any chemist that could breathe. In 1942, I started in Dayton, Ohio at the Thomas and Hochwalt laboratories, which had recently joined Monsanto. Most of my assignments were pretty mundane, like making super pure hexamethylenetetramine to be used for making the explosive cyclonite.

In 1944, I was transferred to St. Louis to work on plasticizers and intermediates. We did make a lot of benzyl benzoate as a mite repellant for soldiers clothing. We later had a DDT project which never got into production until the war was over. More interesting, we had a synthetic process for vanillin but lost out to lignin as a way to get that desirable molecule. In those days we did get involved with the custom manufacture of the antibiotic chloramphenicol and made 10–15000 lbs. before it was taken off the market because a very small percentage of patients developed aplastic anemia. At the time my dog had a fungus on her chest that wouldn't heal and resisted treatment. I made an ointment with our product and it cleared up in two days. She lived to 17 years.

Shortly after the war the discovery that cortisone might become a large volume pharmaceutical caused Monsanto to engage Professor Woodward with the hope of commercializing his synthetic approach. I was selected to join this effort since I had a steroid background.

Actually I got to spend nine months in his lab at Cambridge on this total synthesis. The experience working with the "great man" is one I'll never forget. For the first three months in his lab he would come in at noon and say, "Let's go to Schrafts." We would spend an hour or more scribbling chemical structures on the menu or placemats. His phenomenal memory was beyond anything I'd ever seen. In those days he never kept a file or wrote a reference. He'd just say look on page so and so in Beilstein and you'll find something on that. He lost some of this ability as he grew older and it bothered him. He really hoped Monsanto would commercialize his steroid synthesis, but the Mexican yam with its high content of diosgenin eventually killed our effort.

Our program for cortisone got fairly well along. We made a few milligrams of racemic cortisone and we had resolved an early intermediate which we intended to carry through to the real thing. It was made too complex to compete with the lowly yam.

Later in the fifties I got involved in kinetic studies using my long forgotten math background. These studies led to improvements in several of our processes by doubling production with little or no additional capital. In those days, industry was hungry for chemicals and much effort was spent to get more out in the same equipment.

Monsanto had developed a separate line of advancement for those who wished to stay in technology and I rose to the top of that ladder before I even thought of asymmetric hydrogenations. I was one who liked to work with my hands as well as my brain. Chemical research in the lab was ideal for filling this need. The work on the asymmetric project, which started in the mid-sixties, is the subject of my lecture. Obviously, I kept active in this area until I retired in 1986, and continued in a consulting capacity for several years after.

On the home front we had purchased a cabin in Jackson Hole, Wyoming 25 years ago and have spent summers and some winter skiing time there ever since. It is there that our four children and four grand children often meet. On several occasions, Professor Kagan has visited us there and we've been able to talk asymmetric hydrogenations. I have always loved doing things outdoors, including fly-fishing, hiking and biking. When things are going wrong I find splitting wood quite therapeutic.

I received a number of awards for our work.

1974 IR 100 Awards for Asymmetric Hydrogenation
1978 St. Louis ACS Section Award
1981 Monsanto Thomas and Hochwalt Award
1982 ACS Award for Creative Invention
1996 The Organic Reactions Catalysis Society – Paul N. Rylander Award

ASYMMETRIC HYDROGENATIONS

Nobel Lecture, December 8, 2001

by

William S. Knowles

Monsanto Co., St. Louis, MO 63167, USA.

Actually this account is the story of the genesis of an invention[1]. The inventative process is not clearly understood, but one factor that seems to be important is to have a heavy infusion of naivety. That is why so frequently it is not the experts that do the inventing, but they are the ones who, that once the lead is established, come in and exploit the area. Our work is an excellent illustration of this phenomenon.

In the study of any of the life sciences, chiral compounds are important. In the past when chiral compounds were needed, chemists have had to use biochemical processes or make racemic mixtures followed by laborious resolutions. In industry the problem is particularly severe since resolution with its numerous recycle loops and fractional crystallizations, is an inherently expensive process. Thus, large volume products like monosodium L-glutamate, L-lysine and L-menthol have been made traditionally by biochemical routes even though efficient procedures are available to make their racemic forms.

In the early 1960's we became aware of this problem, when we made a paper evaluation of a monosodium glutamate process. The racemic mixture was easy to obtain, but by the time we had resolved, the projected costs doubled, even though we racemized and recycled back the unwanted D-isomer. It looked as though if one wanted to beat "the bug" it would be necessary to have a catalyst which, when an asymmetric center was formed, would direct the reaction to give a predominance of the desired isomer. For this purpose the 100% efficiency of enzymes would not be needed to have something of real value.

At this point in time I was aware of the extensive studies by Akabori, starting in the mid fifties, modifying heterogeneous catalysts like Raney Ni and palladium with a chiral agent. Always the asymmetric bias was too small to be of preparative interest. All these thoughts remained fallow for several years.

In the interim, I became part of a program for doing exploratory research. I was given a new Ph.D. to train for a year before going into more pressing things. Industrial labs are always wrestling with the problem of how much undirected research they should do, and this was just one of many ways to achieve this goal. I had been thru several new employees on a number of projects when I became aware of Professor Wilkinson's discovery of chlorotris(triphenyl-phosphine) Rhodium, $[RhCl(PPh_3)_3]$, and its amazing properties as a soluble hydrogenation catalyst for unhindered olefins. Homogeneous catalysts had been reported before but this was the first one that compared in rates with the well known heterogeneous counterparts.

A second development in the mid-sixties was the development of methods for making chiral phosphines by Mislow and also by Horner. Phosphorus like carbon is tetrahedral, and when four different substituents are attached, can exist in D and L forms. In the case of phosphines the lone pair of electrons counts as a substituent. Earlier it was thought that phosphines might pyramidally invert like their nitrogen analogs but Mislow and also Horner showed they were stable at room temperature. They turned out to have a half-life of a couple of hours at 115°. For our contemplated hydrogenations this stability would be quite adequate. Then the basic strategy was to replace the triphenylphosphine of Wilkinson's catalyst with a chiral counterpart and hydrogenate a prochiral olefin. This experiment was performed on α-phenylacrylic acid using the known chiral methylpropylphenylphosphine, giving an enantomeric excess (ee) of 15% (Figure 1).

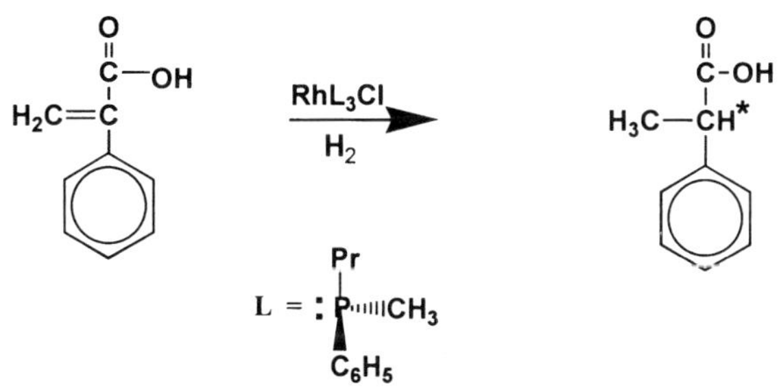

ee (enantiomeric excess) = 15%

Figure 1

This modest result of course was of no preparative value, but it did establish that the hydrogenation technique gave a definite asymmetric bias. In order to achieve this bias the hydrogen, the ligand, and the substrate all had to be on the metal at the same time. Furthermore we established that the hydrogenation was accomplished in solution and not from some extraneous rhodium plating out in our reactor. The inherent generality of the method offered almost unlimited opportunities for matching substrate and catalyst for moving toward the goal of achieving efficient results.

We were not alone in having this idea, but were the first to report on it. I think it was discussed in the question session after Wilkinson's lecture on his soluble hydrogenation catalyst at a Welch Foundation conference. Horner, shortly after our paper, reported even more modest results with the same methylpropylphenylphosphine on a substituted styrene. There were others using other phosphines with uninteresting results. Seemingly we may have been the only ones naïve enough to pursue this lead in depth. Actually there was definitely nothing in the literature to encourage us to proceed further. A mechanistic study showed that just two ligands were all that were needed and not the three as in Wilkinson's structure. α-Phenylacrylic acid worked better as a triethylamine salt, but even so we never got good results with this system.

While groping in this area another seemingly unrelated development appeared, which played an important role in our project. This was the discovery that a fairly massive dose of L-DOPA was useful in treating Parkinson's disease. It created a sizeable demand for this rare amino acid. Because of Monsanto's position in vanillin, which provided the 3,4-dihydroxyphenyl moiety, we found that they were custom manufacturing a racemic intermediate, which Hoffman-LaRoche was resolving and deblocking to L-DOPA. The synthesis, which followed closely the Erlenmeyer azlactone procedure described in Organic Syntheses, went by way of a prochiral enamide, which was hydrogenated to blocked DL-DOPA (Figure 2).

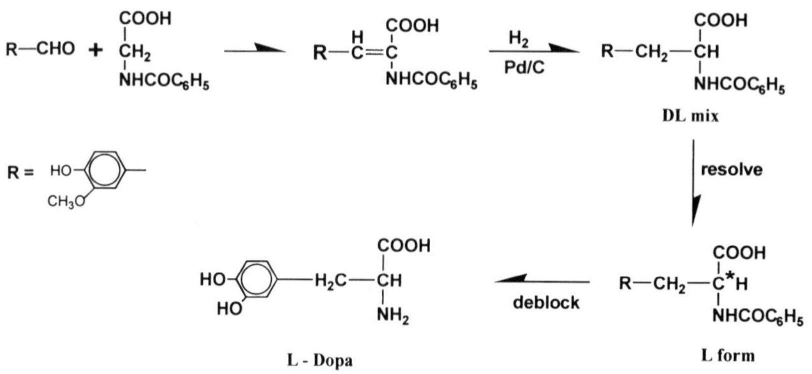

Hoffman - LaRoche L-Dopa Process

Figure 2

This enamide offered a golden opportunity for commercializing this burgeoning technology. We soon found out that these prochiral enamide precursors of α-amino acids hydrogenated much faster than one would expect for such a highly substituted olefin. Even so the chiral results were only 28%, but the stage was nicely set to make a structure vs. activity study. We had a good test reaction in Figure 3 where we used the simple phenylalanine intermediate.

We also had a good test for efficiency since all we had to do was run rotations on a properly diluted reaction mixture. Our job was to find a phosphine

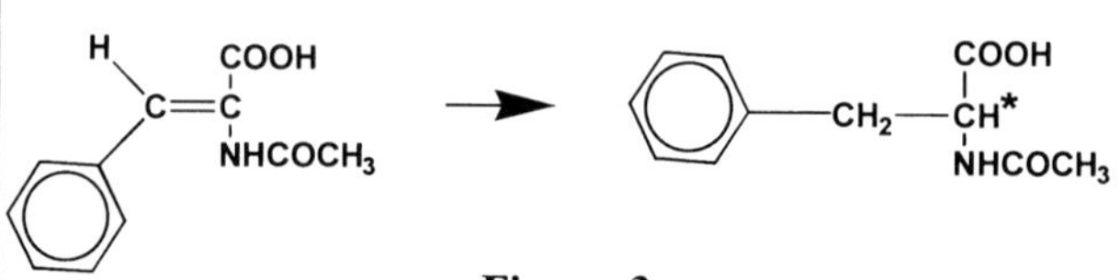

Figure 3

of the proper structure. Early on we tried phosphines with a chiral alkyl side chain, and the asymmetric bias was barely detectible. We felt strongly that if one wanted to get high ee's the asymmetry would have to be directly on the phosphorus. That is where the action is.

Initially we varied the alkyl groups on the phosphorus converting the normal propyl to the more hindered isopropyl or cyclohexyl but ee's still remained in the range of 28–32%.

Our first real variation was to introduce the *o*-anisyl group. This should provide some steric hindrance as well as a possible hydrogen bonding site. Furthermore, the ether linkage would be stable enough to survive the rigors of a phosphine synthesis. In those days our small group was in continuous contact and what we decided to do was arrived at by informal consensus. I hate to admit it but it's much easier to invent in a small under-funded group. Being lean and hungry is conducive to invention.

We made methylphenyl-*o*-anisylphosphine (PAMP) and got ee's after playing with hydrogenation conditions up to 58%. Further modification of this molecule gave us methylcycohexyl-*o*-anisylphosphine (CAMP), this change gave up to 88%. These results are summarized in Figure 4.

It all seems too easy and simple but this was the first time *ever* that anyone had obtained enzyme like selectivity with a man-made catalyst! Never in our wildest imagination did we think a structure vs. activity study would converge so quickly to a product with commercial potential. CAMP was our sixth candidate. As I look back from this perspective I don't think that even we were

Figure 4 Ligands for Asymmetric Hydrogenation

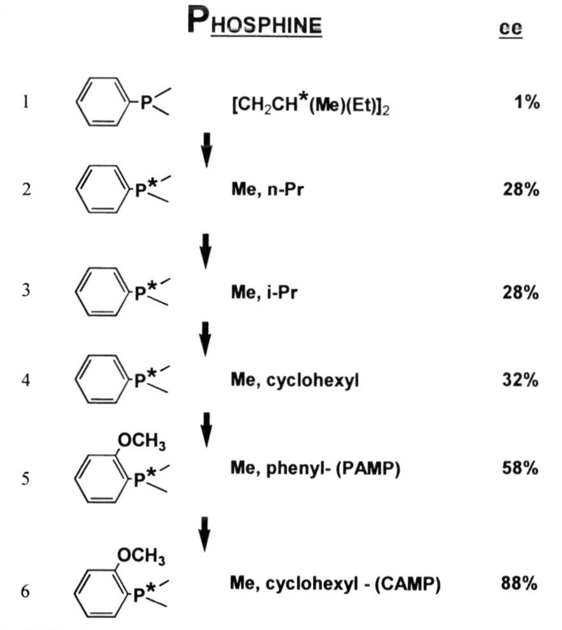

emotionally equipped to realize what we had done. Here with this simplest of molecules (CAMP), we had solved one of the toughest of synthetic problems. For the past hundred years it had been almost axiomatic among chemists that only nature's enzymes could ever do this job.

Our patent department always considered our invention was the use of chiral phosphines with rhodium, but of course without finding PAMP and CAMP we would have had only a new way of doing what had been done before. The lawyers felt that this first result wasn't much but that we could very rapidly come up with an improved embodiment and in this they were unusually prescient.

We have called these catalysts man-made but this is not strictly true. We have not violated the general principal that if you want chiral molecules you will have to get them with the assistance of previously formed natural products. Our asymmetry was obtained from the menthol used in the chiral phosphine synthesis, but being a catalyst a small amount of L-menthol could lead to a large amount of chiral product.

CAMP worked equally well in the L-DOPA precursor, Figure 2, and it made no difference whether the amine-blocking group was benzoyl or acetyl.

At this point we were strongly motivated to develop a commercial L-DOPA process. It is a rare thing that the emergence of a substantial demand for a chemical is so closely timed with an invention for a new way of making it. Our management reluctantly increased our manpower but didn't really believe we could do it until the hydrogenation was done on a 50-gallon scale without incident.

Since CAMP was already good enough we stopped exploring phosphines and concentrated on converting this unique hydrogenation to large-scale production. This process was helped when another fortuitous event occurred. Monsanto decided to get out of its first product, saccharin, and an idle plant was now available for this kind of fine chemical manufacture. These things were then put together to give our simplified L-DOPA process starting with vanillin (Figure 5).

The chiral hydrogenation was the simplest step in the sequence. We started with a slurry of prochiral olefin in an alcohol-water mix and ended with a slurry of chiral product, which could be filtered leaving the catalyst and residual racemate in the mother liquor. We could use an *in-situ* prepared catalyst but it was more convenient to use a solid, air stable complex of the type $[Rh(1,5\text{-}COD)L_2]^+BF_4^-$. These catalysts were fast so that mole ratios of substrate to catalyst were about 20,000 to 1. Thus even this super expensive complex was used at close to throwaway levels.

Even in the best case some racemic product is made and must be separated. This separation is easy or hard depending on the nature of the racemate. If the racemic modification has a different crystalline form than pure D or L then separation of the pure excess enantiomer will be inefficient. If one achieves a 90% ee then it is possible to get out easily, only 75 to 80% pure enantiomer. With lower ee's the losses become prohibitive. For such a system a catalyst of very high efficiency must be used. Unfortunately, most com-

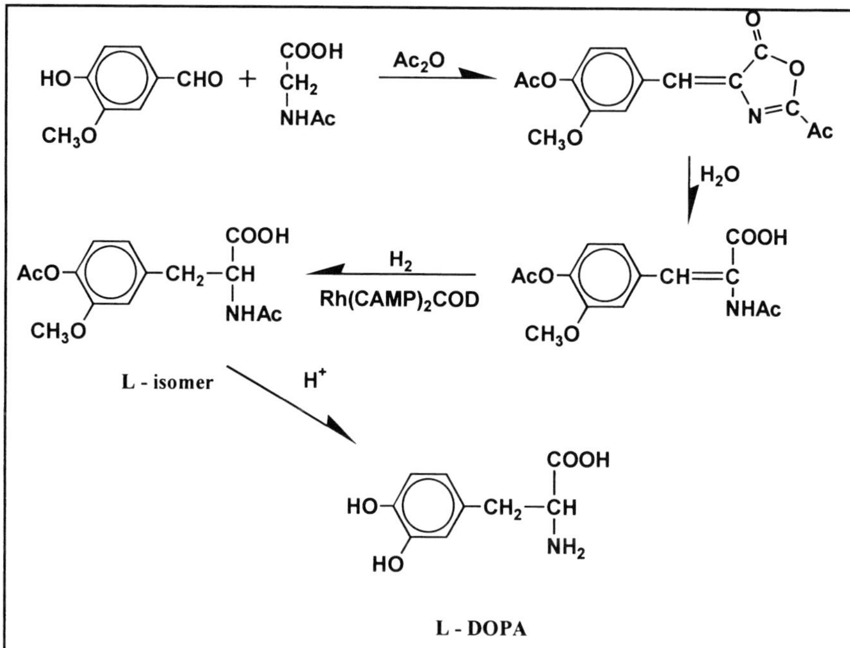

Figure 5 Monsanto L-DOPA Process

pounds are of this type. If on the other hand the racemic modification is a conglomerate or an equal mix of D and L crystals, then recovery of excess L can be achieved with no losses. Since the L and DL are not independently soluble a 90% ee easily gives a 90% recovery of pure isomer. In our L-DOPA process the intermediate is such a conglomerate and separations are efficient. This lucky break was most welcome. If one thinks back ours was the same luck that Pasteur encountered in his classical tartaric acid separations 150 years ago.

At the time of our initial commercialization we learned of a new efficient ligand invented by Professor Kagan, which he called DIOP (Figure 6).

This was a chelating bisphosphine made from tartaric acid with chirality on

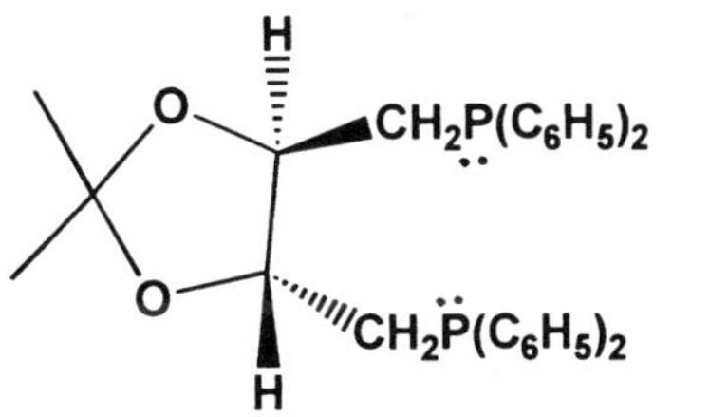

83% ee

Figure 6 R,R-DIOP, Kagan

the carbon backbone and it gave results comparable to CAMP. We had hypothesized that to get good results one needed chirality directly on the phosphorous atom. It made sense, but Professor Kagan showed us to be totally wrong. It is most appropriate that this invention using tartaric acid should have come from a Frenchman in the land of Louis Pasteur, who of course was the one who got it all started. Kagan's discovery was the wave of the future for a whole series of bisphosphine ligands with asymmetry on the chiral backbone.

Shortly after we came up with our own chelating bisphosphine by dimerizing PAMP with another Mislow procedure. We called it DiPAMP, where chirality resided on the phosphorus (Figure 7).

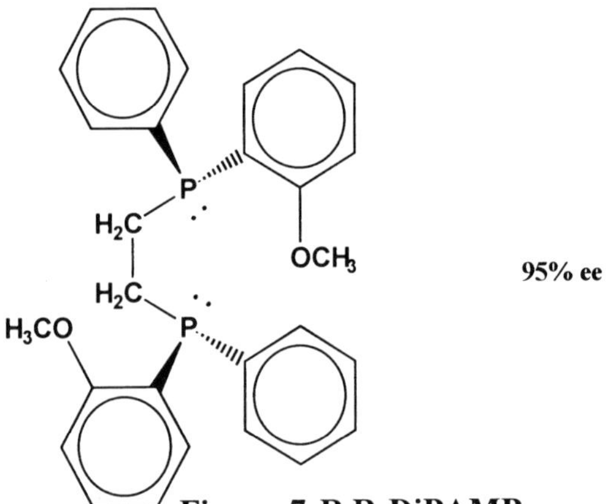

Figure 7 R,R-DiPAMP

It worked at about 95% ee in our L-DOPA system and we quickly converted our commercial process to use it. Part of our motivation to make a quick change was that DiPAMP was easier to make than CAMP, and in addition was a nice crystalline air-stable solid.

When we started this work we expected these man made systems to have a highly specific match between substrate and ligand just like enzymes. Generally in our hands and in the hands of those that followed us a good candidate has been useful for quite a range of applications. This feature has substantially enhanced their value in synthesis. It turned out these chiral hydrogenations as applied to enamides were entirely general, especially with DiPAMP. Here it should be pointed out that these prochiral enamides can exist in both E and Z forms. The Z hydrogenates efficiently whereas the E is less so. Both give the same product. Fortunately for us the base condensation used in their preparation gives us only the desirable Z form.[3c]

In Figure 8, R can be just about anything except – COOH. It's easy to see how our rhodium catalyst could become confused with two carboxyls so close together. Thus aspartic acid is best made by an enzyme process.

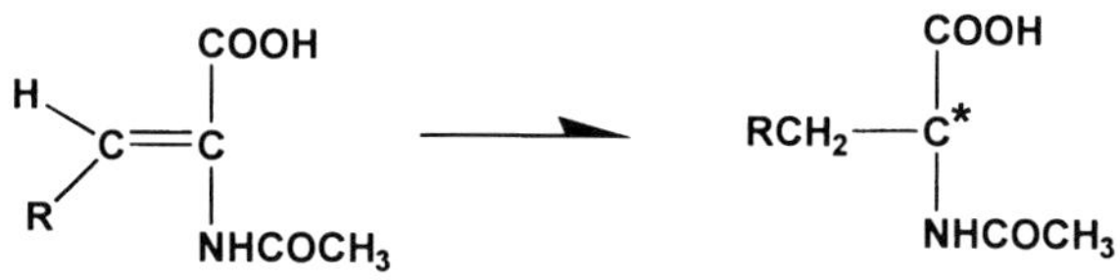

Figure 8 Generality for Z-Enamides

However, almost all the known familiar α-amino acids can be prepared this way since at least in principle an enamide precursor is possible. Evidently the polar carboxyl and amide overwhelm any variation in the R-group. Also the carboxyl and the nitrogen blocking group can be varied extensively. Once again lady luck was with us, since if we had a choice where the catalyst would be useful we couldn't have selected a more important area than the α-amino acids, the building blocks of the proteins.

A few of the more important ones are listed in Figure 9. Our colleagues at Hoffman LaRoche have added about a dozen more non-aromatic members to this list using DiPAMP.

This generality can be extended to a variety of enol-esters and itaconic derivatives. Evidently what is required is the ability to chelate with the metal. Thus the nitrogen can be replaced with oxygen or methylene (Figure 10).[3c]

One system which did not work well in our system was our original model, α-phenylacrylic acid. A number of these aryl propionic acids have value as non-steroidal anti-arthritics. Here as is the usual case only one enantiomer is active and thus a process to make one isomer directly was needed. We tried hard to solve this problem even using ruthenium-ligand systems but without success. It took Professor Noyori with his BiNAP-Ruthenium complex to solve this problem. I'm afraid this is just another example in the history of inven-

Figure 9 Amino Acids

PRODUCT	%ee
L - DOPA	94
L - Phenylalanine	96
L - Tryptophane	93
L - Alanine	90
L - Lysine	85

I

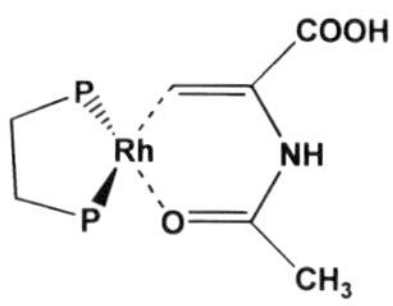

enamides

II

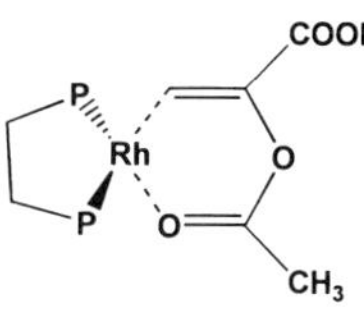

enol-esters

III

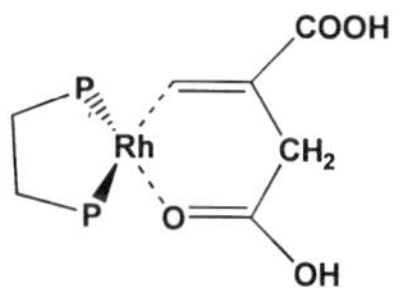

carboxylic acids

Figure 10

tion. The one who makes the first discovery seldom makes the second. On a grander stage this may explain why there are so few Nobel Laureate repeats.

Soon after the appearance of DiOP and DiPAMP a considerable number of bisphosphines with chiral carbon backbones were found. All of these worked well with the same enamides and on related oxygen analogs. A few of these are shown in Figure 11.[3a]

It is interesting that over the years we made a lot of chiral phosphines but never got a good one without our beloved *o*-anisyl group. Others have used it in connection with their bisphosphines but it gives them no particular advantage. Thus, choosing suitable structures is still pretty much guess work.

In our hands DiPAMP was by far the most versatile and remained supreme for enamides for many years. Later in the nineties an improved bisphosphine was reported by Burk, then working at DuPont, which he gave the name DuPHOS (Figure 12).[2]

This bisphosphine complexed with rhodium gave fast hydrogenations of enamides with efficiencies of 99%. Once again the next invention was made by someone else. These high ee's can be important where the racemate is not a conglomerate.

PHOSPHINES

The key to asymmetric hydrogenation is the structure of the chiral ligand. The phosphines are prepared by a multi-step route and are quite expensive, but fortunately one mole will make many thousands of moles of product.

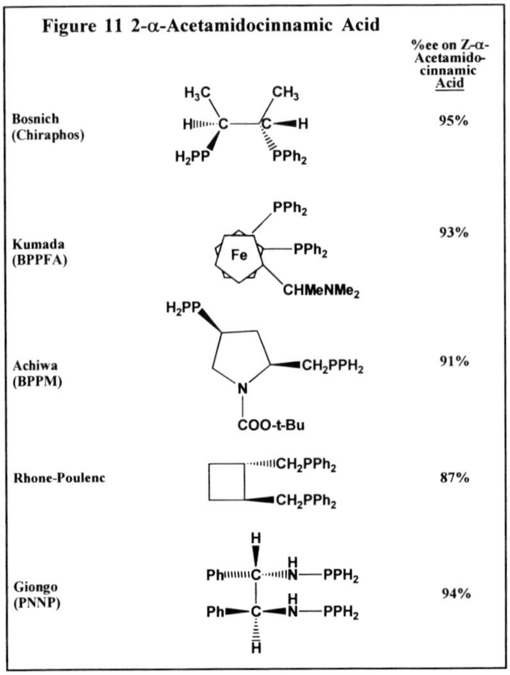

Figure 11 2-α-Acetamidocinnamic Acid

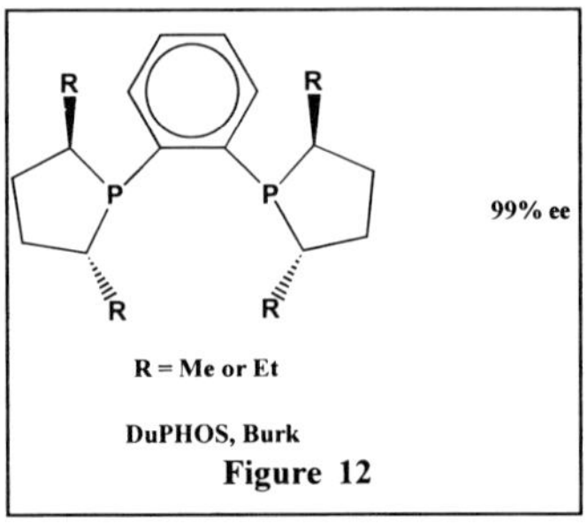

Figure 12

Even so, the ligand must be made from cheap starting materials. Some economy of scale is achieved by making a 10 year supply in a few plant-size batches. At first CAMP was prepared from phenyldichlorophosphine via Mislow's menthyl ester, introducing the *o*-anisyl group last. Unfortunately, the desired isomer was produced in minor amount and, to correct this situation, it was necessary to reverse the order of addition of aryl groups. The sequence starting with trimethylphosphite is outlined in Figures 13 and 14.

A large excess of trimethylphosphite was needed to get good yields of monosubstitution product I. In the sequence I–III only the nicely crystalline phosphinic acid III was isolated. The fact that the acid chloride IV can be converted to an 80/20 mix of (*S*)*p* and (*R*)*p* isomers means that the menthol preferentially reacts with one form while the other isomer rapidly racemizes. Thus, the catalyst preparation was greatly facilitated by an asymmetric synthesis of its own directed by *l*-menthol.

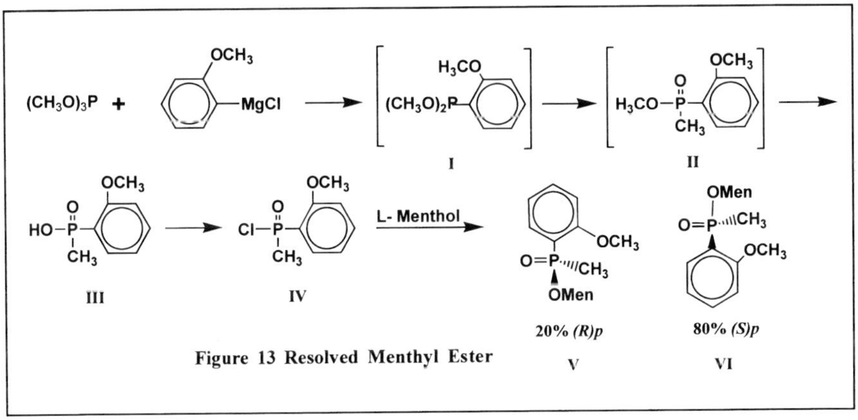

Figure 13 Resolved Menthyl Ester

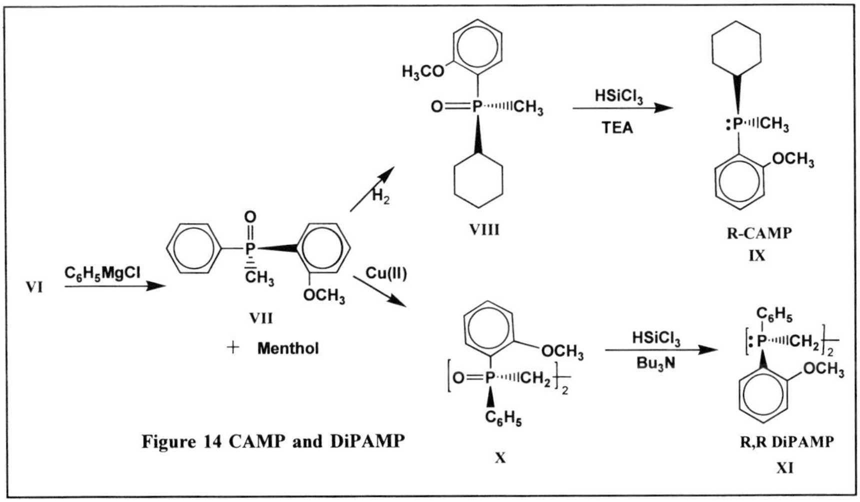

Figure 14 CAMP and DiPAMP

Another advantage of the sequence in Figures 13 and 14 was that CAMP and DiPAMP were prepared from a common intermediate VII and no new resolution procedure needed to be worked out. Thus, the change to an improved ligand could be done with minimum dislocation both at the synthesis and the utilization end. It is a clear advantage of catalytic processes that it is often easy to shift from the old to the new.

CAMP was prepared by a selective hydrogenation of VII, Figure 14, using a rhodium on carbon heterogeneous catalyst. It was important to monitor the reaction closely and stop before the anisyl ring started to hydrogenate. Reduction with trichlorosilane and TEA gave *R*-CAMP, IX, with inversion. The (*R*)*p* menthyl ester V could also have been used in this sequence if the last step was run with pure $HSiCl_3$ which goes with retention.

In the case of DiPAMP the copper coupling step run with lithium diisopropyl amide and CuCl did not affect the stereochemistry. However, only the base promoted trichlorosilane to give a double inversion was applicable. In this case an empirical study showed that use of tributylamine minimized meso formation.

In principle, the menthol recovered in Figure 4 could be recycled, making the usage of chiral agent derived from nature truly minimal, but in practice it has not been worth the effort. More useful is the recovery by hydrolysis of the phosphinic acid from the *(R)p* menthyl ester (Figure 13).

In contrast to CAMP, DiPAMP is a stable solid melting at 102°. Heated at 100° it has a half-life of 3–5 hrs. This racemization was somewhat faster than Mislow's phosphines, which did not invert appreciably until 10–15° higher. The rate was reasonable if one considers that inversion at either end destroys chirality. Use of DiPAMP complexed with rhodium must invert much more slowly because efficient asymmetric hydrogenations have been obtained at 95–100°.

For the sake of convenience, particularly on a large scale, a solid complex was made by reacting two equivalents of phosphine with one equivalent of $[Rh(COD)Cl]_2$ in alcohol. This air-stable orange solid $[Rh(Bisligand)(COD)]^+BF_4^-$ made a most suitable catalyst precursor.

We have used the resolved menthyl ester VII to make a variety of phosphines. The first and most obvious is to convert DiPAMP to DiCAMP. You will recall that in the monophosphine series converting a phenyl to a cyclohexyl gave an enormous increase in selectivity. Not so with DiCAMP which gave only 60–65% ee on our enamide systems. It was, however, our best candidate for preparing the more hindered amino acid valine, where the other systems were very poor (Figure 15).

In the monophosphine series we only found one that was marginally better than CAMP. This rarity of good monophosphines shows how lucky we were to find an efficient one on almost the first try (Figure 16).

We never found a good candidate without the *o*-anisyl group. This is contrasted with all our colleagues in other labs who never found much benefit from it.

We could sulfonate DiPAMP and make it water-soluble. It worked fine but

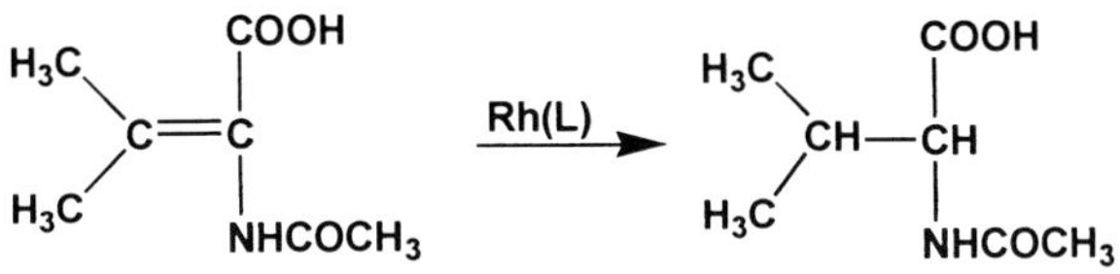

DiCAMP ee 70%

Figure 15 Valine

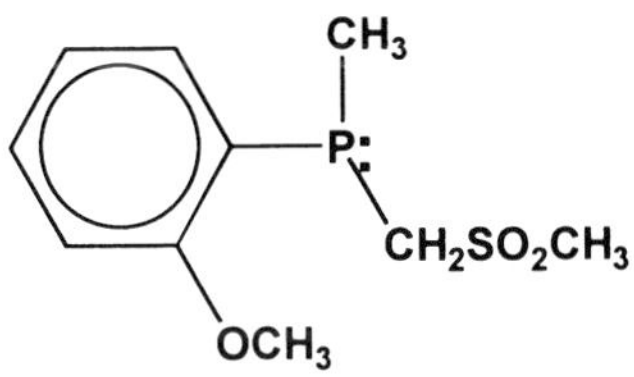

ee 92%

Figure 16

gave only 85% ee which by current standards is too low. I winced when I came in one morning and saw our valuable DiPAMPO being treated with concentrated sulfuric acid, but it worked. This exploratory effort suffices to show that as one might expect the catalysis continues to be a very sensitive function of ligand structure and that our ability to predict or proceed in a rational manner is severely limited.

MECHANISM[3b]

Now that we have these catalysts and have the ability to use them commercially, we would like to know how they work. When we look at energy calculations and realize that to get 90% ee we are talking only a 2 kg cal difference, and this is just about the same as the rotation barrier in ethane. Thus, the asymmetric bias may be caused by very subtle effects.

Using the ball and stick models in Figure 17 to illustrate a typical pro-chiral olefin we can see that attack at the si face gives the D isomer and at the re face the L, corresponding to R and S in more modern nomenclature. These of course are mirror images and our catalyst must discriminate between them.

We did an x-ray crystal structure of the catalyst precursor Rh(DiPAMP) $(COD)^+BF_4^-$ and we noticed that this system presented an array of four aryl groups arranged in an edge-face manner. The phenyl groups present an edge and the *o*-anisyl group a face. This is depicted in Figure 18 where for clarity's sake we have omitted the cyclooctadiene and the counter ion as well as over simplifying the x-ray chart.

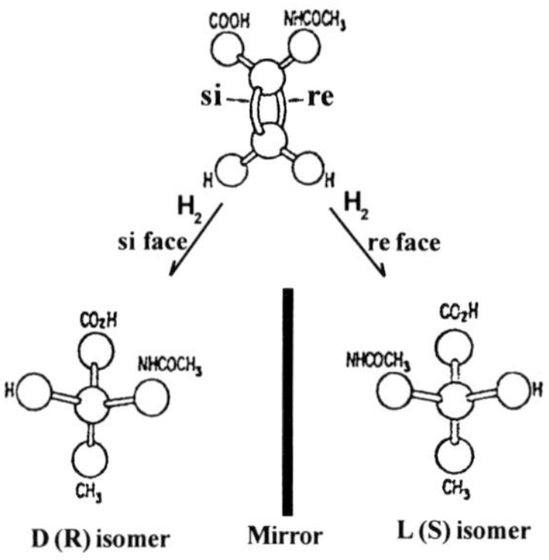

Figure 17 Ball and Stick Models

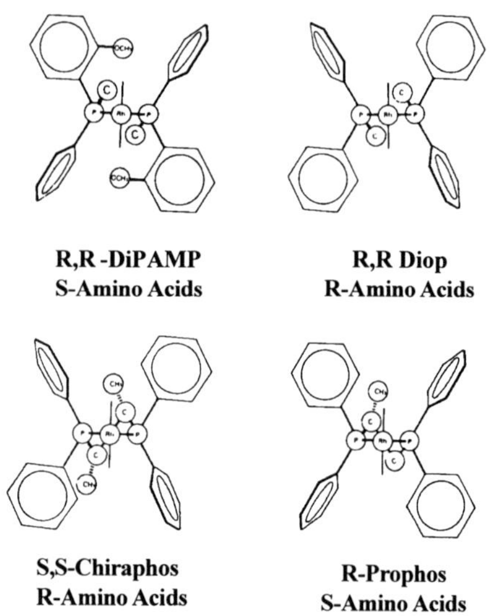

Figure 18 Edge-Face Diagrams

In this picture we are looking along the phosphorus-rhodium-phosphorus plane. One could speculate that an approaching substrate might prefer to lie on the flat face rather than on the hindered edge. We can more easily show this by a quadrant diagram, Figure 19, where the shaded quadrants represent the edge or hindered side. We speculated that a prochiral olefin might prefer to lie in the unhindered quadrant.

You will note that all the other bisphosphines in Figure 11 also present in their x-ray structures a similar array of four phenyl groups, though not quite as convincingly, but always there was a face-exposed ring next to the skewed methylene.

It makes no difference whether one attributes the bias to the edge-face configuration, as I prefer, or the skewed methylene. Using this quadrant interpretation we could predict from the chirality of the phosphine what the chirality of the product would be. Since for any single case where there is a 50% chance of being right, such a prediction has no significance, but having predicted correctly for five cases where x-ray structures were available one gains credibility. We felt pretty good with how things were fitting into place.

Then along came Halpern's studies.[3b] He had been able to isolate a more advanced intermediate where the enamide substrate actually formed a complex with the metal ligand system. He got it crystalline and it was with considerable eagerness we awaited the x-ray results. It turned out that the enamide was lying nicely in the hindered quadrant.

So much for our theory. As so often happens in science, one comes up with an explanation where everything seems to be fitting in nicely, and someone else shows your whole interpretation may well be wrong. We were stranded with the argument that at the square planar stage Halpern reported these steric factors may not be important. However, to get asymmetric bias we know

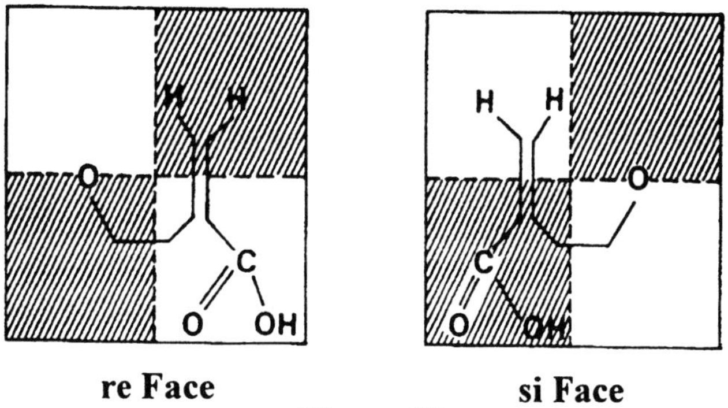

Figure 19

that the hydrogen, the ligand, and the substrate must all be on the metal at the same time. Such a configuration requires an octahedral structure. Perhaps then these quadrant constraints are important. So far as I know there is no evidence either to support or reject this contention. Our theory, though possibly wrong, does predict correctly.

All this thinking does not explain CAMP, unless we argue that during the hydrogenation step this monodentate ligand prefers to occupy adjacent sights on the metal and acts like a bi-dentate species.[3b]

In any case this unique catalysis has enabled chemists to study mechanistic details not previously possible. When one thinks of it, it is quite remarkable that we are even in a position to debate such subtle features.

CONCLUDING COMMENTS

These soluble hydrogenation catalysts have started a new era in catalytic processes. Since we are now dealing with pure complexes we can design something to do just the job you want.

This catalysis will continue to find many uses in industry whenever an efficient route to the unsaturated precursor is available. They can be a nice alternative but by no means will replace biochemical processes. Here the problems with dilute solutions and difficult isolations are often less then the problems of a multi-step synthesis.

One area where these catalysts will reign supreme is in the preparation of D-amino acids or other unnatural isomers. Here biochemical alternatives will not be available.

Perhaps the most important use of these catalysts will be to provide an easy way to make a large number of chiral compounds. In the past research chemists have been reluctant to run laborious resolutions and have done so only when necessary. Now they can get chiral compounds for their life sciences research with very little effort. We can look on these catalysts as a labor saving device for the laboratory. For this they will have an impact for as long as chemists run reactions.

For an invention to succeed, Paul Ehrlich, the father of chemotherapy, has stated that four Gs are required, Geist, Geld, Geduld and Gluck. The first of these is axiomatic. You have got to have a good idea. The second is essential. One needs financial support but I would suggest a proper balance. Too much or too little is inhibitory. For the third you must have patience. Things never move as fast as you would have them. Finally luck is all-important. I suspect that no invention has ever been made without some fortuitous help.

I have pointed out and will continue to do so that ours has been very much a joint effort. It would not have been possible without my associates, Jerry Sabacky and Billy Vineyard.

In closing, I would like to add a couple of Gs to Ehrlich's list. We are extremely Grateful to have so Great an honor bestowed on us by your committee.

REFERENCES

1) Accounts of Chemical Research, **1983**, 16, 106. William S. Knowles and references cited therein.
2) M. J. Burk, J. E. Feaster, W. A. Nugent, and R. L. Harlow, *J. Am. Chem. Soc.* 115, 10125 **(1993)**.
   M. J. Burk, M. F. Gross, and J. P. Martinez, *J. Am. Chem. Soc.* 117, 9375 **(1995)**.
3) *Asymmetric Synthesis,* Vol. 5, Chiral Catalysis, James D. Morrison, Editor, Academic Press, **1985.**
   a) Chapter 1. *Chiral Ligands for Asymmetric Catalysis* – Henri B. Kagan.
   b) Chapter 2. *Asymmetric Catalytic Hydrogenation: Mechanism and Origin of Enantioselection* – Jack Halpern.
   c) Chapter 3. *The Applicability of Asymmetric Homogeneous Catalytic Hydrogenation* – Karl E. Koenig.

RYOJI NOYORI

I was born on September 3, 1938 in a suburb of Kobe (now Ashiya), Japan, the first son of Kaneki and Suzuko Noyori. Our family moved to Kobe soon afterwards. I grew up with two younger brothers and a sister in a pleasant city blessed by beautiful natural surroundings. Except for a short period at the end of World War II, I attended an elementary school affiliated to Kobe University from ages six to twelve, and then moved on to Nada Middle and High School from ages twelve to eighteen. I enjoyed many out-door activities in my youth.

My father, Kaneki, was a gifted research director of a chemical company, and his profession strongly influenced the path of my life. At home, we were surrounded by his scientific journals and books and various samples of plastics and synthetic fibers, and were frequently asked to test the quality of products which were under development for commercialization. When I entered middle school, my father took me to a public conference, the topic of which was "nylon". The lecturer explained proudly that this new fiber could be synthesized from coal, air, and water (the then famous catchphrase of DuPont company). Although I knew nothing about industrial technology, I was deeply impressed by the power of chemistry. Chemistry can create important things from almost nothing! The event had an enormous impact on this 12-year-old schoolboy, because it was in 1951, shortly after World War II when Japan was so poor. We were very hungry. It was at this point that it became my dream to be a leading chemist to contribute to the society by inventing beneficial products.

My appetite for chemistry was further wetted through class work led by enthusiastic teachers in middle/high school including Dr. Kazuo Nakamoto (then Osaka University and afterward Illinois Institute of Technology and Marquette University) who gave me my first chemistry lesson. I also liked other sciences and mathematics. Together with regular schoolwork, "judo" (one of Japan's traditional sports) was a major passion at this time. It was very popular amongst us because Nada School and Kodokan Judo School were founded by the same family. I highly appreciate the educational efforts of many schoolteachers as well as the warm friendship of classmates in those days, which strongly influenced the formation of my personal character.

In 1957, at the age of 18, I entered Kyoto University, which was known to be the most active institution in the research of polymer chemistry. Incidentally, this was the year when the USSR launched into space for the first time an artificial satellite, the Sputnik, thereby demonstrating the power of science-based technology. I recall that this success substantially shocked young science students in Japan. After three years, I started to study organic chemistry, rather than polymer chemistry, under the guidance of Professor Keiiti Sisido. The laboratory environment was very hospitable and I obtained my Bachelor degree in 1961. Upon completion of my Master's degree in 1963, I was immediately appointed Instructor of Professor Hitosi Nozaki's laboratories at Kyoto University and, in 1967, received my doctorate (DEng). My career path, that is the appointment to Instructor without a doctorate, is a little unusual but this is partly due to the difference in Japanese and Western education/teaching systems. Professor Nozaki strongly encouraged us to pursue new, original chemistry rather than tracing traditional subjects, while I served as a leader of his subgroup working on flourishing physical organic chemistry. It was under such conditions that in 1966 we discovered an interesting asymmetric catalysis that later became a life-long interest. This finding emerged during the course of an investigation of the transition metal effects in carbene reactions. Reaction of styrene and ethyl diazoacetate in the presence of a small amount of a chiral Schiff base–Cu(II) complex gave optically active cyclopropane derivatives, albeit with <10 % ee. Although the enantioselectivity was not synthetically meaningful, this was probably the first example of asymmetric catalysis using structurally well-defined organometallic molecular complexes. In the early 1960s, homogeneous catalysis (typically Reppe chemistry) was already well known, however, the notion of "molecular catalysis" that utilizes the structural and electronic characteristics of molecules more positively, was not clearly documented. This discovery opened my research perspective. In any event, I intended first to expand my scientific background under the supervision of an eminent chemist abroad, and Professor E. J. Corey at Harvard kindly agreed to accommodate me in his laboratories as a postdoctoral fellow. This plan, however, was postponed for reasons outlined below.

The situation changed drastically in the fall of 1967, when I received a totally unexpected offer from Nagoya University. I was asked to chair a newly created organic chemistry laboratory. This invitation surprised me. I was a mere 29-year-old Instructor at Kyoto enjoying daily research work with some young students. Nothing had prepared me to be a Professor at a major national university. Being too young and inexperienced to be a Full Professor, I was first appointed Associate Professor of Chemistry. In February, 1968, when I launched my own research group, Professor Yoshimasa Hirata, a senior faculty known for his outstanding accomplishments in natural products of organic chemistry, asked me to create a new stream of organic chemistry at Nagoya, different from his own field, thereby making the Chemistry Department more visible. I immediately decided to focus on organic synthesis using organometallic chemistry, which then comprised a branch of inorganic chemistry. Although not many researchers were aware of the high utility in organic synthesis, I was intuitively confident of the bright future of this scientific field. Professor Hirata consistently helped me in many aspects during his time at Nagoya University.

In 1969, as planned earlier, I went to Harvard. I was amazed by the enormous difference in the standard of living and science between the US and my mother country. Professor Corey was then already a leading organic chemist and I learned much from him. In addition, I became acquainted with many promising students and postdoctoral fellows including K. Barry Sharpless who was working with Professor Konrad Bloch. Later many of these reliable friends, together with their scientific relatives, grew to become eminent researchers in the scientific community and helped me in many ways. Synthesis of prostaglandins (PGs) was my research theme in the Corey group. After completing several works, I was asked to selectively hydrogenate a $PGF_{2\alpha}$ derivative that has two C=C bonds to a $PGF_{1\alpha}$ compound possessing a single C=C bond. This was the start of my three-decade-long work on hydrogenation. My interest in homogenous hydrogenation was enhanced by reading almost all available literature on this very new topic and also through personal interaction with Assistant Professor John A. Osborn, who had joined Harvard Chemistry Department from Geoffrey Wilkinson's laboratory at Imperial College, London. Osborn, an authority of Rh-catalyzed homogeneous hydrogenation, taught me many aspects of organometallic chemistry. It was in 1968, when W. S. Knowles and L. Horner reported independently the first homogeneous asymmetric hydrogenation using chiral phosphine–Rh catalyts, albeit in low optical yields. The fruitful Harvard experience, coupled with our earlier asymmetric cyclopropanation in 1966, led to my life-long research on asymmetric hydrogenation.

After returning to Nagoya in 1970, I began to study organic synthesis and homogeneous catalysis via organometallic chemistry, while in August 1972, at the age of 33, I was promoted to Full Professor. In the hope of development of efficient asymmetric hydrogenation and other reactions, we became interested in BINAP [2,2'-bis(diphenylphosphino)-1,1'-binaphthyl], a novel C_2 chiral diphosphine possessing a beautiful molecular shape. Synthesis of the optically pure diphosphine was unexpectedly difficult. It was in 1974, that I started stereospecific synthesis from optically pure 2,2'-diamino-1,1'-binaphthyl with my long-term collaborator, the late Professor Hidemasa Takaya, who was with me at Nagoya and afterwards moved to the Institute of Molecular Science and Kyoto University. After two years, we managed to obtain optically active BINAP, however, the result was disappointingly irreproducible. In 1978, we reached a reliable method for resolution of racemic BINAP with a chiral amine–Pd complex. Unfortunately, the results of BINAP–Rh(I) catalyzed asymmetric hydrogenation of dehydro amino acids were highly variable depending on the reaction conditions. Eventually, in 1980 after a six-year endeavour, thanks to the unswerving efforts of my young colleagues and students, we were able to publish our first work on asymmetric synthesis of amino acids via this BINAP chemistry.

The success in our asymmetric hydrogenation largely relies on the invention of BINAP and the use of Ru element, which behaves differently from conventional Rh. A major breakthrough in asymmetric hydrogenation came in 1986, when we developed BINAP–Ru(II) dicarboxylate complexes that enjoy a much greater scope of olefinic substrates. Furthermore, in 1987–1988, we developed a versatile general asymmetric hydrogenation of functionalized ketones with BINAP–Ru(II) dihalide complexes. The scope of this method is far reaching. These asymmetric hydrogenation methods allow for the synthesis of a wide array of terpenes, vitamins, β-lactam antibiotics, α- and β-amino acids, alkaloids, prostaglandins, and other compounds of biological and physiological interest. BINAP chemistry has been applied to the large-scale production of the synthetic intermediates of antibiotic carbapenems (Takasago International Co.) and levofloxacin, a quinolone antibacterial agent (Takasago International Co./Daiichi Pharmaceutical Co.). The efficiency of BINAP chemistry rivals or in certain cases even exceeds that of enzymes. In addition, a team of the Noyori Molecular Catalysis Project (ERATO, 1991–1996) discovered the catalysts of type $RuCl_2$(diphosphine)(diamine) leading to another major breakthrough in hydrogenation. The reaction of unsaturated ketones occurs preferentially the C=O function leaving the olefinic linkage intact. The combined use of the BINAP ligand and a chiral diamine effects asymmetric hydrogenation of a range of aromatic, hetero-aromatic, and olefinic ketones. The reaction is very rapid, productive and stereoselective, providing the most practical method for converting simple ketones to chiral secondary alcohols.

BINAP–Rh(I) complexes are useful for asymmetric isomerization of allylic amines to enamines of high enantiomeric purity. In the early 1980s, a fruitful academic/industry collaboration was made between the groups at Osaka University (S. Otsuka and H. Tani), Nagoya University, Institute of Molecular Science (H. Takaya), Shizuoka University (J. Tanaka and K. Takabe), and Takasago International Co., realizing the industrial production of (-)-menthol and other optically active terpenes.

In 1995–1996, we invented a range of Ru(II) catalysts modified with a chiral β-amino alcohol or 1,2-diamine derivative that effects asymmetric transfer hydrogenation of ketones and imines using 2-propanol or formic acid as hydrogen donors. More recently, the reaction has proven to proceed via a nonclassical metal – ligand bifunctional mechanism. My interest in asymmetric chemistry is broad. In 1986, we found a highly enantioselecive addition of dialkylzincs to aldehydes using a small quantity of a camphor-derived chiral amino alcohol, where the alkylation products with high enantiomeric excesses are accessible with a partially resolved chiral ancillary. We could fully elucidate the origin of this striking chiral amplification phenomenon at the molecular structure level. My stay at Harvard in 1969–1970 spurred me to develop an efficient way to synthesize prostaglandins (PGs). In this connection, a series of selective synthetic methods was explored in our laboratories. Our binaphthol-modified lithium aluminum hydride reagent (1979) was applied to the commercial Corey PG synthesis (Ono Pharmaceutical Co.). Furthermore, we realized the long-sought three-component PG synthesis in 1985, which now plays an important role in biochemical and physiological studies of PGs.

Chemical synthesis provides a logical basis for molecular science and related technologies which require a high degree of structural precision. I have tried to select general and fundamental research subjects in this important field. A clear-cut solution to a long-persistent problem, when accomplished, often results in an enormous scientific or technological impact. Asymmetric hydrogenation is a typical example. BINAP chemistry is now utilized worldwide in research laboratories and also at the industrial level. In fact, the selective synthesis of single enantiomers using well-designed chiral molecular catalysts has now become common practice. This fascinating field is still growing rapidly, and recent advances have dramatically changed the way of chemical synthesis, opening tremendous potential for molecular technologies.

Our broad research activity goes beyond asymmetric synthesis. In 1994, we discovered the remarkable utility of supercritical carbon dioxide as a medium for homogeneous catalysis. Thus Ru-catalyzed hydrogenation produces formic acid, methyl formate, and dimethylformamide with an extremely high turnover number. More recently, we devised practical, environmentally sound methods for olefin epoxidation and alcohol oxidation using aqueous H_2O_2, whose utility is highlighted by the direct conversion of cyclohexene to adipic acid (1996–1998). The stereospecific living polymerization of phenylacetylenes was achieved by using a structurally defined tetracoordinate Rh complex (1994). We also developed an efficient synthesis of solid-anchored DNA oligomers using organopalladium chemistry (1990). In my early days at Nagoya, we invented the iron carbonyl–polybromo ketone reaction which allows the construction of five- and seven-member carbocycles in a 3 + 2 and 3 + 4 manner, respectively. In the late 1970s, we exercised initiative in the catalytic use of organosilicon compounds for organic synthesis.

Organometallic chemistry is a scientific space leading to an enormous technical impact and even more general social benefits. I am very pleased to be involved in contributing to the progress of this significant scientific realm. The above described scientific accomplishments are not my own, but the credit in fact belongs to my research family at Nagoya University and many collaborators at other institutions. My initial ideas in solving problems were not always appropriate and, sometimes even nonsensical. However, my serendipitous collaborators incubated such research themes through careful experiments and much deliberation, and eventually reached new chemical concepts and useful methodologies. Their intellect, sense, and skills are highly appreciated. In addition, through my frequent traveling abroad as a visiting professor or invited lecturer at research institutions and conferences, I have met many superb colleagues from the international scientific community. Their encouragement as well as the exposure to different cultures and environments have deeply affected my way of thinking. Furthermore, for over three decades, my scientific work has been supported generously and consistently by the Ministry of Education, Culture, Sports, Science and Technology of Japan as well as the Research Development Corporation of Japan, various private foundations, and numerous industrial companies.

My activities are not limited to education and research. I have served on the editorial boards of some thirty international journals including the editorship of *Advanced Synthesis & Catalysis* (Wiley/VCH) which emphasizes the "practical elegance" of chemical synthesis. Furthermore, I have been involved in much administrative work, for example, as Science Advisor (1992–1996) and Member of the Scientific Council (1996–present) of the Ministry of Education, Culture, Sports, Science and Technology; Dean of the Graduate School of Science at Nagoya University (1997–1999); and President of the Society of Synthetic Organic Chemistry, Japan (1997–1999). Such official duties significantly hamper my research activity but are unavoidable for a senior scientist.

In 1972, I married Hiroko Oshima (a daughter of a Professor of Medicine at Tokyo University) who was studying the immunology of cancer at a research institute in Tokyo. Since then, she has played the most important part in our private life at Nagoya. We have two children. Our first son, Eiji (born in 1973), is an active staff writer of a newspaper company, and our second son, Koji (born in 1978), studies painting at an art university in Tokyo.

APPOINTMENTS

Kyoto University: Instructor, 1963–1968.

Nagoya University: Associate Professor, 1968–1972. Professor, 1972–present. Director Chemical Instrument Center, 1979–1991. Dean, Graduate School of Science, 1997–1999. Director, Research Center for Materials Science, 2000–present.

Kyushu University: Professor (adjunct), 1993–1996.

Ministry of Education, Science, Sports and Culture: Science Advisor, 1992–1996. Member of Scientific Council, 1996–2001.

Ministry of Education, Culture, Sports, Science and Technology: Member of Scientific Council, 2001–present.

Japan Society for the Promotion of Science: Committee Chairman, Research for the Future Program on "Advanced Processes", 1996–present. Science Advisor, 2001–present.

The Research Development Corporation of Japan: Director of the ERATO Molecular Catalysis Project, 1991–1996.

The Society of Synthetic Organic Chemistry, Japan: Vice President, 1994–1996. President, 1997–1999.

The Chemical Society of Japan: President-Elect, 2001.

HONORARY DEGREES

Technische Universität München, Germany, 1995.

Universite de Rennes 1, France, 2000.

HONORARY PROFESSORSHIPS

Shanghai Institute of Organic Chemistry, China, 2001.

FELLOWSHIPS AND MEMBERSHIPS

Fellow, American Association for the Advancement of Science, 1996.
Honorary Member, Chemical Society of Japan, 1998.
Honorary Fellow, Royal Society of Chemistry, UK, 2000.
Foreign Honorary Member, American Academy of Arts and Sciences, 2001.
Honorary Member, European Academy of Sciences and Arts, 2001.

AWARDS

The Chemical Society of Japan Award for Young Chemists, 1972.
The Matsunaga Prize (Matsunaga Memorial Foundation, Japan), 1978.
The Chunichi Cultural Prize (Chunichi Newspaper Co., Japan), 1982.
The Chemical Society of Japan Award, 1985.
The Naito Foundation Research Prize (Naito Science Foundation, Japan), 1988.
The Fluka Prize, Reagent of the Year (Fluka Chemie AG, Switzerland), 1989.
The Centenary Medal (The Royal Chemical Society, UK), 1990.
The Toray Science & Technology Prize (Toray Science Foundation, Japan), 1990.
The Merck-Schuchardt Chair (BOSS Symposium, Belgium), 1990.
The J. G. Kirkwood Award (Yale University, USA), 1991.
The Asahi Prize (Asahi Culture Foundation, Japan), 1992.
Tetrahedron Prize for Creativity in Organic Chemistry (Pergamon Press, UK), 1993.
The Keimei Life Science Prize (Keimei Foundation, Japan), 1994.
The Japan Academy Prize, 1995.
The Arthur C. Cope Scholar Award (American Chemical Society), 1996.
Bonn Chemistry Award (University of Bonn and Pinguin Foundation, Germany), 1996.
The Arthur C. Cope Award (American Chemical Society), 1997.
The Chirality Medal (International Symposium on Chiral Discrimination), 1997.
The George Kenner Award (University of Liverpool, UK), 1997.
Person of Cultural Merit (Japanese Government), 1998.
The King Faisal International Prize for Science (King Faisal Foundation, Saudi Arabia), 1999.
The Cliff S. Hamilton Award (University of Nebraska, USA), 1999.
ISI Citation Laureate Award (ISI/Thomson Scientific Inc., USA/Japan), 2000.
The Order of Culture (Japanese Emperor/Government), 2000.
The Special Award (The Society of Synthetic Organic Chemistry, Japan), 2001.
The Wolf Prize in Chemistry (Wolf Foundation, Israel), 2001.

The Roger Adams Award in Organic Chemistry (American Chemical Society), 2001.
The Nobel Prize in Chemistry (Royal Swedish Academy of Sciences), 2001.

PUBLICATIONS

Over 400 publications in scientific journals.

PATENTS

Over 160.

SELECTED REFERENCES

Monograph: "Asymmetric Catalysis in Organic Synthesis." R. Noyori, John Wiley & Sons, New York, 1994.

Asymmetric Induction in Carbenoid Reaction by Means of a Dissymmetric Copper Chelate. H. Nozaki, S. Moriuti, H. Takaya, and R. Noyori, *Tetrahedron Lett.*, 5239 (1966).

Organic Syntheses via the Polybromo Ketone–Iron Carbonyl Reaction. R. Noyori, *Acc. Chem. Res.*, **12**, 61 (1979).

Synthesis of 2,2'-Bis(diphenylphosphino)-1,1'-binaphthyl (BINAP), an Atropisomeric Chiral Bis(triaryl)phosphine, and Its Use in the Rhodium(I)-Catalyzed Asymmetric Hydrogenation of α-(Acylamino)acrylic Acids. A. Miyashita, A. Yasuda, H. Takaya, K. Toriumi, T. Ito, T. Souchi, and R. Noyori, *J. Am. Chem. Soc.*, **102**, 7932 (1980).

Prostaglandin Syntheses by Three-Component Coupling. R. Noyori and M. Suzuki, *Angew. Chem., Int. Ed. Engl.*, **23**, 847 (1984).

The Allylic Protection Method in Solid-Phase Oligonucleotide Synthesis. An Efficient Preparation of Solid-Anchored DNA Oligomers. Y. Hayakawa, S. Wakabayashi, H. Kato, and R. Noyori, *J. Am. Chem. Soc.*, **112**, 1691 (1990).

An Organometallic Way to Prostaglandins: The Three-Component Coupling Synthesis. R. Noyori and M. Suzuki, *Chemtracts–Org. Chem.*, **3**, 173 (1990).

Chiral Metal Complexes as Discriminating Molecular Catalysts. R. Noyori, *Science*, **248**, 1194 (1990).

BINAP: An Efficient Chiral Element for Asymmetric Catalysis. R. Noyori and H. Takaya, *Acc. Chem. Res.*, **23**, 345 (1990).

Enantioselective Addition of Organometallic Reagents to Carbonyl Compounds: Chirality Transfer, Multiplication, and Amplification. R. Noyori and M. Kitamura, *Angew. Chem., Int. Ed. Engl.*, **30**, 49 (1991).

Stereoselective Organic Synthesis via Dynamic Kinetic Resolution. R. Noyori, M. Tokunaga, and M. Kitamura, *Bull. Chem. Soc. Jpn.*, **68**, 36 (1995).

Homogeneous Catalysis in Supercritical Fluids. P. G. Jessop, T. Ikariya, and R. Noyori, *Science*, **269**, 1065 (1995).

Asymmetric Hydrogenation. R. Noyori, *Acta Chem. Scand.*, **50**, 380 (1996).

The Catalyst Precursor, Catalyst, and Intermediate in the Ru^{II}-Promoted

Asymmetric Hydrogen Transfer between Alcohols and Ketones. K.-J. Haack, S. Hashiguchi, A. Fujii, T. Ikariya, and R. Noyori, *Angew. Chem., Int. Ed. Engl.*, **36**, 285 (1997).

Asymmetric Transfer Hydrogenation Catalyzed by Chiral Ruthenium Complexes. R. Noyori and S. Hashiguchi, *Acc. Chem. Res.*, **30**, 97 (1997).

A "Green" Route to Adipic Acid: Direct Oxidation of Cyclohexenes with 30% Hydrogen Peroxide. K. Sato, M. Aoki, and R. Noyori, *Science*, **281**, 1646 (1998).

Asymmetric Catalysis by Architectural and Functional Molecular Engineering: Practical Chemo- and Stereoselective Hydrogenation of Ketones. R. Noyori and T. Ohkuma, *Angew. Chem., Int. Ed.*, **40**, 40 (2001).

Self and Nonself Recognition of Chiral Catalysts: The Origin of Nonlinear Effects in the Amino-Alcohol Catalyzed Asymmetric Addition of Diorganozincs to Aldehydes. R. Noyori, S. Suga, H. Oka, and M. Kitamura, *Chem. Rec.*, **1**, 85 (2001).

Metal–Ligand Bifunctional Catalysis: A Nonclassical Mechanism for Asymmetric Hydrogen Transfer between Alcohols and Carbonyl Compounds. R. Noyori, M. Yamakawa, and S. Hashiguchi, *J. Org. Chem.*, **66** 7931, (2001).

ASYMMETRIC CATALYSIS: SCIENCE AND OPPORTUNITIES

Nobel Lecture, December 8, 2001

by

Ryoji Noyori

Department of Chemistry, Graduate School of Science, and Research Center of Materials Science, Nagoya University, Chikusa, Nagoya 464-8602, Japan.

PROLOGUE

Chirality (handedness; left or right) is an intrinsic universal feature of various levels of matter[1]. Molecular chirality plays a key role in science and technology. In particular, life depends on molecular chirality, in that many biological functions are inherently dissymmetric. Most physiological phenomena arise from highly precise molecular interactions in which chiral host molecules recognize two enantiomeric guest molecules in different ways. There are numerous examples of enantiomer effects which are frequently dramatic. Enantiomers often smell and taste differently. The structural difference between enanatiomers can be serious with respect to the actions of synthetic drugs. Chiral receptor sites in the human body interact only with drug molecules having the proper absolute configuration, resulting in marked differences in the pharmacological activities of enantiomers. A compelling example of the relationship between pharmacological activity and molecular chirality was provided by the tragic administration of thalidomide to pregnant women in the 1960s. (*R*)-Thalidomide has desirable sedative properties, while its *S* enantiomer is teratogenic and induces fetal malformations[2,3]. Such problems arising from inappropriate molecular recognition should be avoided at all costs. Nevertheless, even in the early 1990s, about 90 % of synthetic chiral drugs were still racemic – that is, equimolar mixtures of both enantiomers, reflecting the difficulty in the practical synthesis of single enantiomeric compounds[4]. In 1992, the Food and Drug Administration in the US introduced a guideline regarding "racemic switches", in order to encourage the commercialization of clinical drugs consisting of single enantiomers[5]. Such marketing regulations for synthetic drugs, coupled with recent progress in stereoselective organic synthesis, resulted in a significant increase in the proportion of single-enantiomer drugs. In 2000, the worldwide sales of single enantiomer compounds reached 123 billion US dollars[6]. Thus, gaining access to enantiomerically pure compounds in the development of pharmaceuticals, agrochemicals, flavors, is a very significant endeavor.

Discovery of truly efficient methods to achieve this has been a substantial challenge for chemists in both academia and industry. Earlier, enantiomerically pure compounds were obtained by the classical resolution of a racemate or transformation of readily accessible naturally occurring chiral compounds such as amino acids, tartaric and lactic acids, carbohydrates, terpenes, or alkaloids. Even though stereoselective conversion of a prochiral compound to a chiral product, namely, through an asymmetric reaction is the most attractive approach, practical access to pure enantiomers relied largely on biochemical or biological methods. However, the scope of such methods using enzymes, cell cultures, or whole microorganisms is limited because of the inherent single-handed, lock-and-key specificity of biocatalysts. On the other hand, a chemical approach allows for the flexible synthesis of a wide array of enantiopure organic substances from achiral precursors. The requirements for practical asymmetric synthesis include high stereoselectivity, high rate and productivity, atom economy, cost efficiency, operational simplicity, environmental friendliness, and low energy consumption. Traditional asymmetric synthesis using a stoichiometric amount of a chiral compound, though convenient for small to medium scale reactions, is practical only if the expensive chiral auxiliary deliberately attached to a substrate or reagent is readily recyclable; otherwise it is a wasteful procedure.

Figure 1 illustrates a general principle of asymmetric catalysis which provides an ideal way for multiplying molecular chirality[7]. A small amount of a well-designed chiral catalyst can combine A and B, producing the chiral AB compound stereoselectively in a large quantity. Of various possibilities, the use of chiral organometallic molecular catalysts would be the most powerful strategy for this purpose. Asymmetric catalysis is an integrated chemical approach where the maximum chiral efficiency can be obtained only by a combination of suitable molecular designing with proper reaction conditions. The reaction must proceed with a high turnover number (TON) and a high turnover frequency (TOF), while the enantioselectivity ranges from 50:50

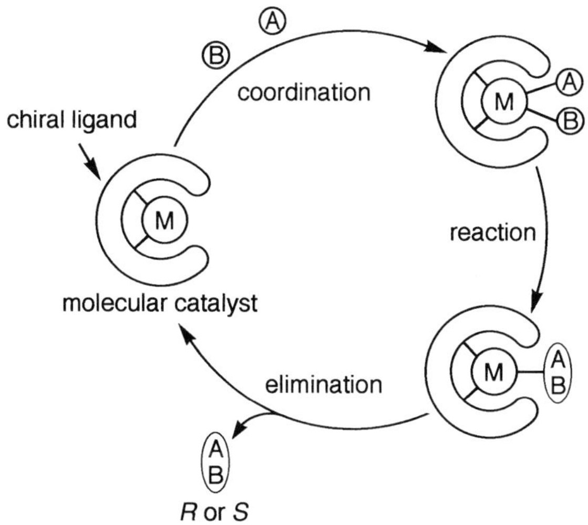

Figure 1. A general principle of asymmetric catalysis with chiral organometallic molecular catalysts. M = metal; A, B = reactant and substrate.

(nonselective) to 100:0 (perfectly selective). The chiral ligands that modify intrinsically achiral metal atoms must possess suitable three-dimensional structures and functionality to generate sufficient reactivity and the desired stereoselectivity. Sometimes the properties of achiral ligands are also important. The chiral catalyst can permit *kinetically* precise discrimination among enantiotopic atoms, groups, or faces in achiral molecules. Similarly, enantiomeric molecules can also be discriminated. Certain well-designed chiral metal catalysts not only accelerate the chemical reactions repeatedly but also differentiate between diastereomeric transition states (TSs) with an accuracy of 10 kJ/mol. In this way such compact molecular catalysts with a molecular weight less than 1000, or <20 Å in length or diameter, allow for an ideal method for synthesizing enantiomeric compounds. The diverse catalytic activities of metallic species as well as the virtually unlimited structural variation of the organic ligand provides enormous opportunities for asymmetric catalysis.

DISCOVERY OF ASYMMETRIC CATALYSIS VIA CHIRAL ORGANOMETALLIC COMPLEXES

In 1966, when I was in H. Nozaki's laboratory at Kyoto, we discovered the first example of asymmetric catalysis using a structurally well-defined chiral transition metal complex[8]. This finding resulted from research done for an entirely different purpose which was to elucidate the mechanism of carbene reactions. As illustrated in Figure 2, when a small amount (1 mol%) of a chiral Schiff base–Cu(II) complex was used as a molecular catalyst in the reac-

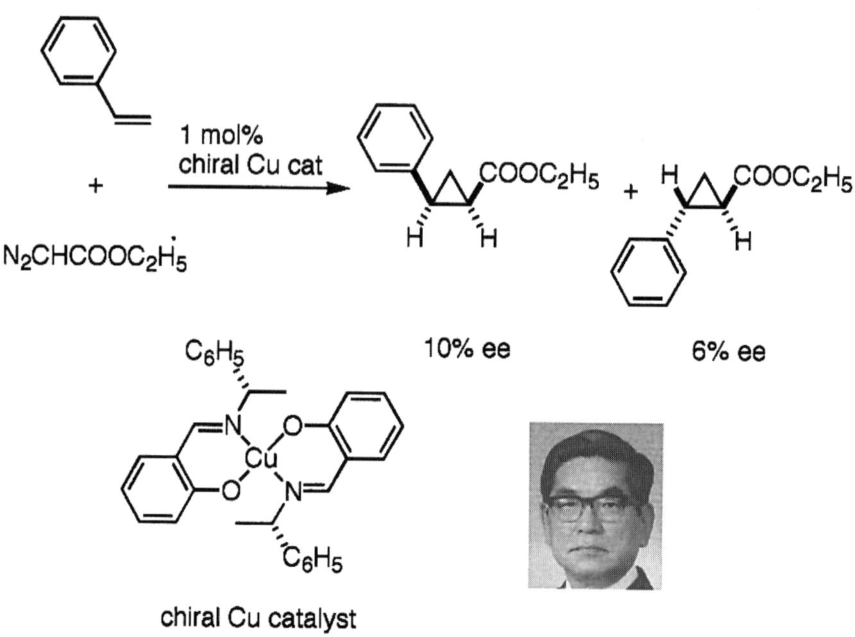

Figure 2. Discovery of asymmetric reaction by means of a chiral organometallic molecular catalyst. Photo: Professor H. Nozaki (1985).

tion of styrene and ethyl diazoacetate, the *cis*- and *trans*-cyclopropanecarboxylates were obtained in 10 and 6% enantiomeric excess (ee), respectively. We also observed asymmetric induction in carbene insertion to a C–O bond of 2-phenyloxetane giving optically active 2,3-substituted tetrahydrofuran derivatives. At that time, the finding was synthetically primitive since the degree of enantioselection was meaningless practically. Later, T. Aratani, a Kyoto student, went to Sumitomo Chemical Co., where he invented an excellent chiral Cu catalyst for asymmetric cyclopropanation, achieving the industrial synthesis of chrysanthemates (efficient insecticides) and (*S*)-2,2-cyclopropanecarboxylic acid. The latter compound is a building block of cilastatin, an in vivo stabilizer of the carbapenem antibiotic, imipenem (Merck Co.) (Figure 3) [9].

ASYMMETRIC HYDROGENATION IN THE EARLY DAYS

At present the asymmetric cyclopropanation is important practically, but in the late 1960s, it was just a special reaction in organic synthesis. I decided to pursue hydrogenation, which is a core technology in chemistry. It is the

Figure 3. Sumitomo asymmetric cyclopropanation. a) Useful chiral cyclopropanes and b) reaction apparatus.

simplest but most powerful way to produce a wide array of important compounds in large quantities using inexpensive, clean hydrogen gas without forming any waste. Hydrogenation was initiated at the end of the 19th century by P. Sabatier (1912 Nobel laureate) who used fine metal particles as heterogeneous catalysts. Some notable achievements that attracted me, before doing research in this area, include: activation of H_2 by a transition metal complex in the late 1930s (M. Calvin, 1961 Nobel laureate)[10]; homogeneous hydrogenation of olefinic substrates with $RuCl_3$ in 1961 (J. Halpern, J. Harrod, and B. R. James)[11]; and hydrogenation of olefinic compounds using $RhCl[P(C_6H_5)_3]_3$ in 1965 (G. Wilkinson, 1973 Nobel laureate)[12]. Most importantly, in 1956, S. Akabori at Osaka reported that metallic Pd drawn on silk catalyzes asymmetric (heterogeneous) hydrogenation of oximes and oxazolones[13]. This pioneering work, though not effective synthetically, was already well known throughout Japan. In 1968, two years after our asymmetric cyclopropanation in 1966, W. S. Knowles (fellow Nobel laureate in 2001)[14] and L. Horner[15] reported independently the first homogeneously catalyzed asymmetric hydrogenation of olefins with chiral monodentate tertiary phosphine–Rh complexes, albeit in 3–15% optical yield[16]. H. B. Kagan provided a major breakthrough in this area in 1971, when he devised DIOP, a C_2 chiral diphosphine derived from tartaric acid. He used its Rh complex for asymmetric hydrogenation of dehydro amino acids leading to phenylalanine in about 80% ee, then recorded as 72% ee[17]. The Knowles group at Monsanto established a method for the industrial synthesis of L-DOPA, a drug for treating Parkinson's disease, using his DIPAMP–Rh catalyzed asymmetric hydrogenation as a key step[18]. These achievements significantly stimulated the subsequent investigation of this important subject.

Shortly after moving from Kyoto to Nagoya in 1969–70, I spent a postdoctoral year at Harvard with E. J. Corey (1990 Nobel laureate). He asked me to hydrogenate selectively one of the two C = C bonds in a prostaglandin $F_{2\alpha}$ derivative to the $F_{1\alpha}$ form having only one C = C bond[19]. This research was helped by K. B. Sharpless (another fellow Nobel laureate in 2001) who was then a postdoctoral fellow with K. Bloch (1964 Nobel laureate in Physiology or Medicine) and who suggested a convenient TLC technique for analyzing the structurally very similar olefinic compounds. In addition to this background, my personal interaction with J. A. Osborn, a former Wilkinson student and co-inventor of $RhCl[P(C_6H_5)_3]_3$[12] who was then an Assistant Professor at Harvard, greatly enhanced my interest in asymmetric hydrogenation, which later became my life-long research interest. My desire was to develop a truly efficient asymmetric hydrogenation which would have a wide scope of applications. In the early 1970s, chiral phosphine–Rh complexes could hydrogenate satisfactorily only dehydro amino acids but not many other olefins. Asymmetric hydrogenation of ketones was totally unexplored[20].

BINAP, A BEAUTIFUL CHIRAL MOLECULE

H_2 is the simplest molecule but it has enormous potential from both a scientific and technical point of view. To discover high-performance asymmetric catalysts, developing an excellent chiral ligand is crucial. Attracted by its molecular beauty[21], we initiated the synthesis of BINAP (2,2'-bis(diphenylphosphino)-1,1'-binaphthyl)[22] in 1974 at Nagoya with the help of H. Takaya, my respected long-term collaborator. BINAP was a new fully aromatic, axially dissymmetric C_2 chiral diphosphine that would exert strong steric and electronic influences on transition metal complexes. Its properties could be fine-tuned by substitutions on the aromatic rings. However, synthesis of this optically pure C_2 chiral diphosphine was unexpectedly difficult. In 1976, for the first time, we managed to obtain optically active BINAP starting from optically pure 2,2'-diamino-1,1'-binaphthyl (Figure 4a). However, this seemingly straightforward synthetic pathway was not reproducible due to the tendency of the chiral intermediates to cause racemization[23]. In 1978, we found a reliable method for resolving racemic BINAP with an optically active dimethyl(1-phenylethyl)amine–Pd(II) chloride complex[22], while later optically pure BINAP became available more conveniently by resolution of BINAP dioxide with camphorsulfonic acid or 2,3-*O*-dibenzoyltartaric acid (Figure 4b)[24,25]. Although the elusive BINAP was available, achieving our goal was

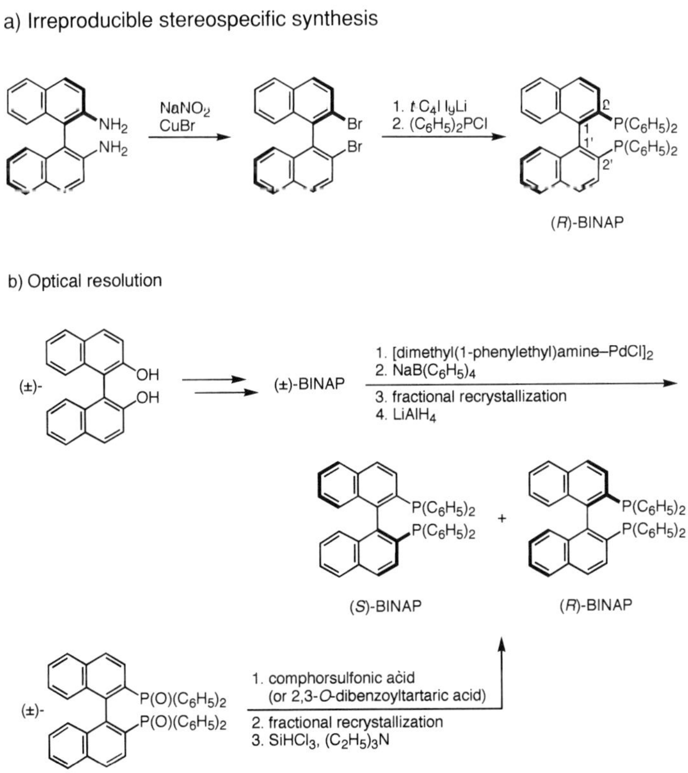

Figure 4. Access to enantiomerically pure BINAP.

still in the distance. Enantioselectivity of BINAP–Rh(I) catalyzed asymmetric hydrogenation of α-(acylamino)acrylic acids was highly variable and not satisfactory at that time, ee of the chiral products being at most about 80%. However, we remained patient. In 1980, six years after the start, thanks to the unswerving efforts of my young associates, we published our first work on asymmetric synthesis of amino acids of high enantiomeric purity, up to 100% ee, together with the X-ray crystalline structure of a cationic BINAP–Rh(norbornadiene) complex[22,26].

BINAP, a conformationally flexible atropisomeric C_2 diphosphine, can accommodate a range of transition metals by rotating about the binaphthyl C(1)–C(1') pivot and C(2 or 2')–P bonds without seriously increasing torsional strain, while the resulting seven-membered chelate rings containing only sp^2 carbon atoms are in turn skeletally unambiguous. The chirality of BINAP is transmitted to other metal coordination sites through the chelate structure[22,26]. The δ or λ geometry is highly skewed and determines the chiral disposition of the *P*-phenyl rings that play a key role in generating outstanding chirality-discriminating ability at the reactive coordination sites. Thus BINAP-based metal complexes were expected to exhibit high chiral recognition ability in various catalytic reactions in addition to hydrogenation.

ASYMMETRIC SYNTHESIS OF MENTHOL

The cationic BINAP–Rh complex was best used in asymmetric isomerization of allylic amines[27], realizing an industrial synthesis of (–)-menthol from myrcene (Figure 5)[28]. This resulted from a fruitful academic/industrial

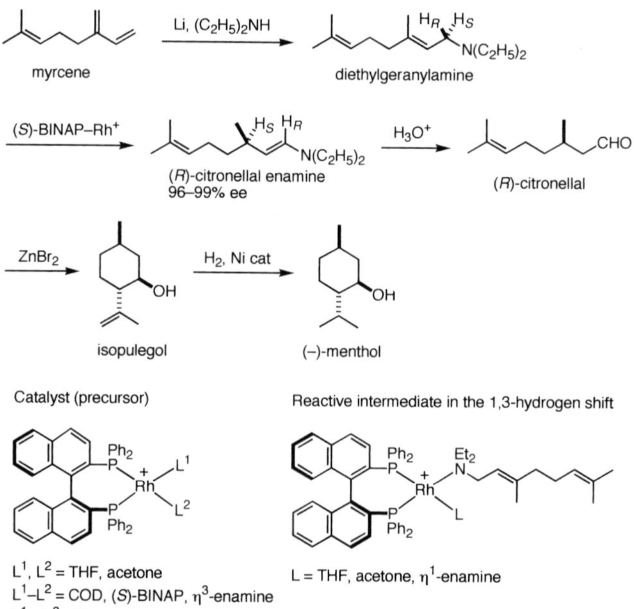

Figure 5. Takasago menthol synthesis.

collaboration between groups at Osaka University (S. Otsuka and H. Tani)[29], Nagoya University (R. Noyori), Institute for Molecular Science (H. Takaya), Sizuoka University (J. Tanaka and K. Takabe)[30], and Takasago International Co.(Figure 6). The key step was the asymmetric isomerization of geranyldiethylamine catalyzed by an (*S*)-BINAP–Rh complex in THF forming (*R*)-citronellal enamine, which upon hydrolysis gives (*R*)-citronellal in 96–99% ee. This is far superior to the 80% ee of the naturally occurring product available from rose oil. Among various Rh and other catalysts examined, the BINAP-based cationic Rh complex was the most reactive and the most stereoselective. The BINAP–Rh catalyst clearly differentiates between the pro-*S* and pro-*R* hydrogens on the flexible allylic amine skeleton during the 1,3-suprafacial shift that occurs via a nitrogen-triggered mechanism[31]. The asymmetric reaction is performed on a nine-ton scale. The full technical refinements of the position- and stereoselective addition of diethylamine to myrcene, giving the starting geranylamine, and the $ZnBr_2$-catalyzed intramolecular ene reaction of (*R*)-citronellal, forming isopulegol with the three correct stereogenic centers, allowed for the production of terpenic substrates totaling about 1500 tons per year at Takasago International Co. Most of the (*R*)-citronellal is converted to 1000 tons per year of (–)-menthol, one-third of the world demand. (*R*)-7-Hydroxydihydrocitronellal thus prepared is a perfumery agent that smells like lily of the valley. Its methyl ether is an intermediate in the synthesis of methoprene, a growth regulator of the yellow fever mosquito[28,29].

Figure 6. At the Takasago plant for (–)-menthol synthesis (February, 1984). From the left, K. Tani, H. Takaya, R. Noyori, S. Otsuka, S. Akutagawa, and H. Kumobayashi.

ASYMMETRIC HYDROGENATION OF OLEFINS BY BINAP–RUTHENIUM COMPLEXES

Returning to the topic of asymmetric hydrogenation, our success resulted from the invention of the BINAP ligand[32] and also from the use of Ru element that behaves differently from conventional Rh[33,34]. The cationic BINAP–Rh complexes catalyze hydrogenation of α-(acylamino)acrylic acids or esters to give the corresponding amino acid derivatives of high ee (Figure 7)[22,23]. However, the reaction is relatively slow, and high enantioselectivity is obtained only under special conditions probably due to operation of the unsaturate/dihydride mechanism. J. Halpern[35] and J. M. Brown[36] showed that hydrogenation of enamides in the presence of a C_2 chiral diphosphine –Rh complex proceeds via oxidative addition of H_2 to diastereomeric Rh–substrate chelate complexes, followed by stepwise transfer of the two hydrides to the coordinated olefin. Most significantly, the minor diastereomer of these complexes is the more reactive one[37]. Because of the excellent chiral recognition ability of BINAP, the reactive species, leading to the desired hydrogenation product, is present in a very small quantity and is even NMR-invisible in the equilibrium mixture[23a]. Therefore, conditions such as hydrogen pressure, temperature, and concentration must be chosen carefully to obtain high enantioselectivity. Furthermore, asymmetric hydrogenation was limited to the synthesis of amino acids.

A major breakthrough occurred when we devised the BINAP–Ru(II) dicarboxylate complexes in 1986 (Figure 8)[38,39]. The Ru complexes are excellent catalysts for asymmetric hydrogenation of various functionalized olefins, as summarized in Figure 9. The reaction proceeds via a Ru monohydride intermediate formed by heterolysis of H_2 by the Ru complex. The Ru center remains in the +2 oxidation state throughout the catalytic cycle[40] in contrast to the Rh complex which involves a +1/+3 redox process. Heteroatoms in the functional groups serve as a binding tether to the catalytic Ru center. This hydrogenation has a very wide scope. Hydrogenation of α,β- and β,γ-unsaturated carboxylic acids takes place in alcoholic media, where the sense and de-

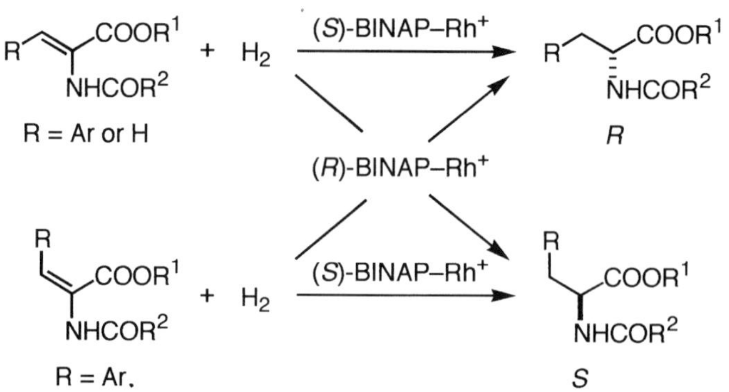

Figure 7. Asymmetric hydrogenation of α-(acylamino)acrylic acids catalyzed by BINAP–Rh complexes.

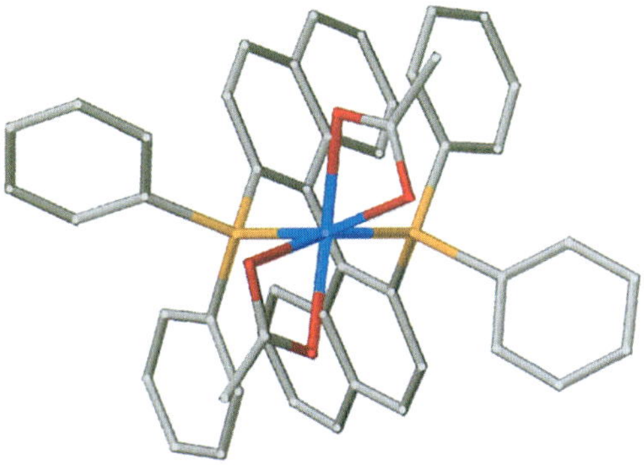

(*R*)-BINAP–Ru(II) diacetate

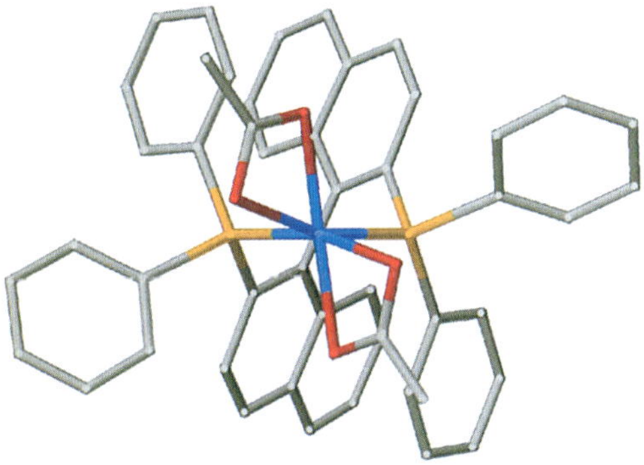

(*S*)-BINAP–Ru(II) diacetate

Figure 8. Structures of BINAP–Ru diacetate complexes.

gree of the enantioselection are highly dependent on the substitution pattern and hydrogen pressure[41]. Allylic and homoallylic alcohols are also hydrogenated with high enantioselection[42]. Certain racemic allylic alcohols can be resolved by the BINAP–Ru catalyzed hydrogenation[43]. The chiral Ru complexes effect highly enantioselective hydrogenation of (*Z*)-2-acyl-1-benzylidene-1,2,3,4-tetrahydroisoquinolines[38,44]. In a similar manner, enantio-enriched α- and β-amino acids[45] as well as α-amino phosphonic acids[46] are obtainable from suitably amido-substituted olefins. Notably, the Ru(II) and Rh(I) complexes possessing the same BINAP chirality form antipodal amino acids as the predominant products[47].

Figure 10 illustrates some chiral compounds that can be obtained by this asymmetric hydrogenation. An important application is the synthesis of the anti-inflammatory drug, naproxen, in 97% ee from an α-aryl-acrylic acid[41,46]. Natural and unnatural citronellol with up to 99% ee are obtainable from geraniol or nerol without saturation of the C(6)–C(7) double bond with a high substrate to catalyst (S/C) ratio. The hydrogenation of (*R*,*E*)-

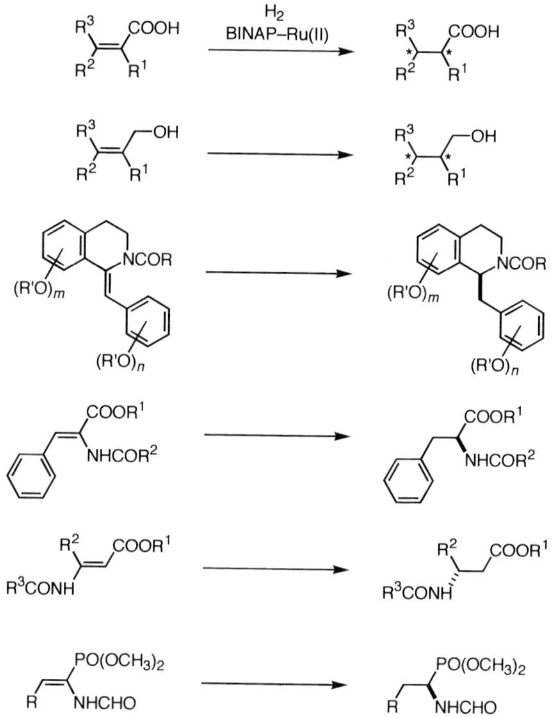

Figure 9. Asymmetric hydrogenation of functionalized olefins catalyzed by (*S*)-BINAP–Ru dicarboxylates.

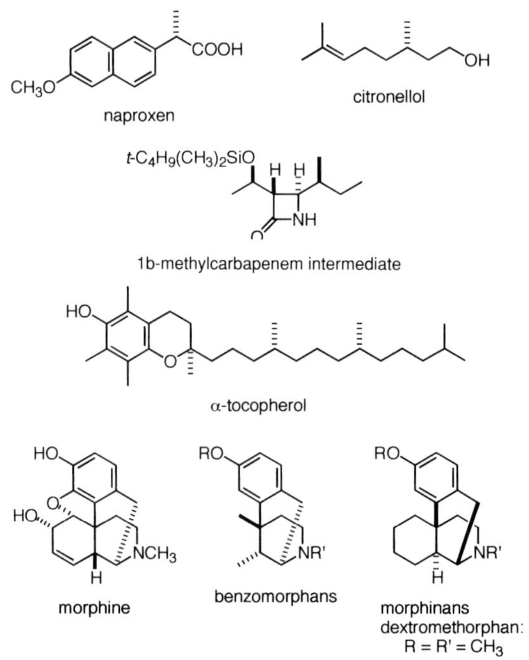

Figure 10. Applications of BINAP–Ru catalyzed hydrogenation of olefins.

6,7,10,11-tetrahydrofarnesol produces (3*R*,7*R*)-hexahydrofarnesol, a C_{15} side chain of α-tocopherol (vitamin E) and a part of vitamin K_1. The hydrogenation of an allylic alcohol possessing a chiral azetidinone unit gives a 1β-methylcarbapenem synthetic intermediate diastereoselectively[48]. The discovery of this asymmetric hydrogenation made possible the general asymmetric synthesis of isoquinoline alkaloids including morphine, benzomorphans, and morphinans such as the antitussive dextromethorphan[43,49].

Importantly, the list of substrates can be extended to include various ketones, as generalized in Figure 11. The halogen-containing BINAP–Ru(II) complexes (oligomers)[50], but not the diacetate complexes, are efficient cata-

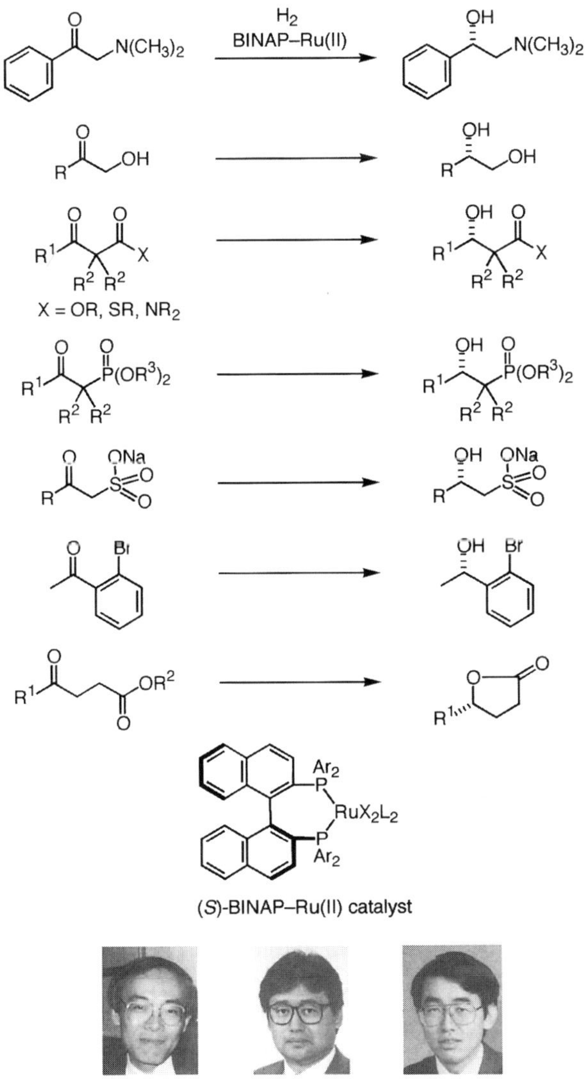

Figure 11. Asymmetric hydrogenation of functionalized ketones catalyzed by (*S*)-BINAP–Ru dihalide complexes (X = halogen). Photo: major contributors (from the left, H. Takaya, M. Kitamura, and T. Ohkuma).

lysts for the asymmetric hydrogenation of a range of functionalized ketones, wherein coordinative nitrogen, oxygen, and halogen atoms near C=O functions direct the reactivity and stereochemical outcome in an absolute sense[51]. A wide variety of achiral ketones are hydrogenated enantioselectively to the corresponding chiral alcohols in 90–100% ee in a predictable manner. The reaction can be performed normally in alcohols with up to 50% substrate concentration under 4–100 atm at room temperature with an S/C ratio of up to 10 000 on any scale even using >100 kg of the substrate. Figure 12 shows some synthetic applications of this asymmetric hydrogenation. (*R*)-1,2-Propanediol thus obtained from hydroxyacetone is used for industrial synthesis of the antibacterial levofloxacin (Takasago Co./Daiichi Pharmaceutical Co.). In addition, γ-amino-β-hydroxybutyric acid (GABOB) and a compactin intermediate can be prepared with high enantiomeric purity[49,52]. Pre-existing stereogenic centers in the ketonic substrate significantly affect the steric course. Statines can be obtained with a high diastereo- and enantioselectivity[53]. The double hydrogenation of 1,3-diones via chiral hydroxy ketones leads to the anti 1,3-diols in close to 100% ee[51a].

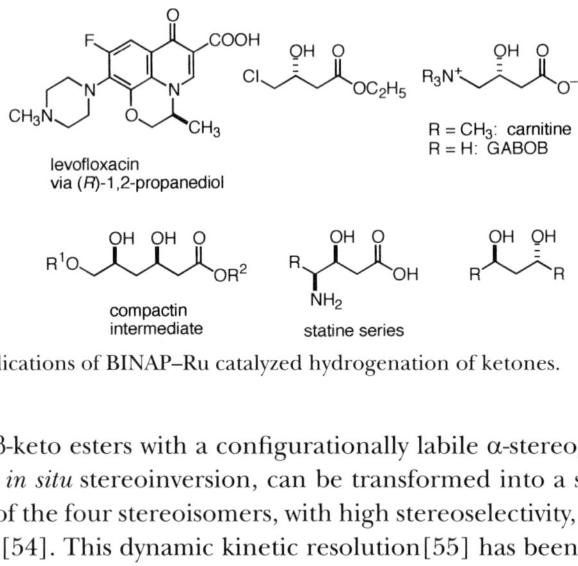

Figure 12. Applications of BINAP–Ru catalyzed hydrogenation of ketones.

Racemic β-keto esters with a configurationally labile α-stereogenic center, undergoing *in situ* stereoinversion, can be transformed into a single stereoisomer, out of the four stereoisomers, with high stereoselectivity, as illustrated in Figure 13[54]. This dynamic kinetic resolution[55] has been used for the synthesis of various biologically important compounds such as threonine, L-DOPS[52], phosphothreonine[56], and fosfomycin[57]. Its utility was highlighted by the industrial synthesis of carbapenem antibiotics at Takasago International Co. (Figure 14). The requisite chiral 4-acetoxyazetidinone is prepared by the (*R*)-BINAP–Ru catalyzed hydrogenation of racemic methyl α-(benzamidomethyl)acetoacetate in dichloromethane giving the 2*S*,3*R* hydroxy ester with 94:6 erythro:threo diastereoselectivity[58] and 99.5:0.5 enantioselectivity[54a]. Quantitative analysis[54] indicates that the 2*S* substrate is hydrogenated 15 times faster than the *R* enantiomer, and the slow-reacting *R* isomer is inverted to the 2*S* enantiomer 92 times easier than it is hydrogenated. The extent of the BINAP catalyst-based asymmetric induction is calcu-

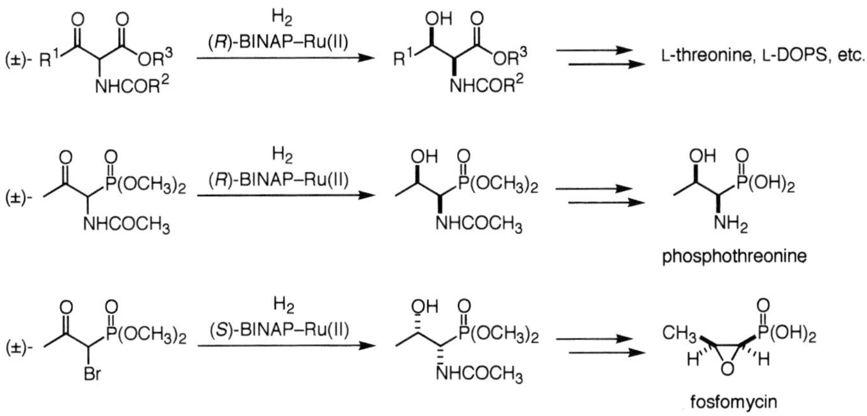

Figure 13. Asymmetric hydrogenation via dynamic kinetic resolution.

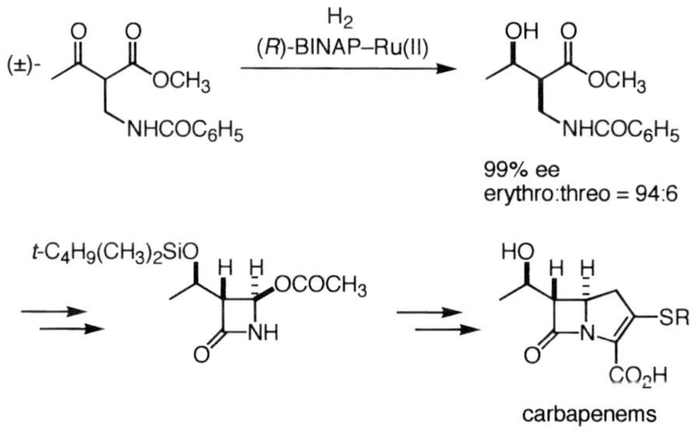

Figure 14. Stereoselective synthesis of carbapenem antibiotics.

lated to be 104 in favor of the 3*R* isomer, whereas the substrate-based asymmetric induction is 9 in favor of the C(2)/C(3) erythro stereochemistry. The volume of the hydrogenation reactor shown in Figure 15 is 13 m^3.

β-Keto esters are the best substrates for the Ru catalyzed asymmetric hydrogenation leading to the β-hydroxy esters in >98% ee[59]. Figure 16 illustrates the mechanistic model. The halide ligand in the Ru complex, generating a strong acid and a RuHCl species by the action of H_2, is important to facilitate the hydride transfer from Ru to the carbonyl carbon[55]. In addition, the presence of the ester moiety interacting with the Ru center is crucial for both high reactivity and enantioselectivity. Because of the excellent chiral recognition ability of BINAP, the two stereo-determining diastereomeric TSs are well differentiated with the assistance of the oxygen–Ru interaction. The *R*-directing TS is highly favored over the *S*-generating diastereomer, which suffers from substantial R/*P*-phenyl repulsive interaction. The oxygen–Ru dative bond (and related interaction in the reactions in Figure 11) exerts a pivotal function in the acceleration of hydrogenation as well. Thus, β-keto

Figure 15. A large-scale BINAP–Ru catalyzed hydrogenation at Takasago International Co.

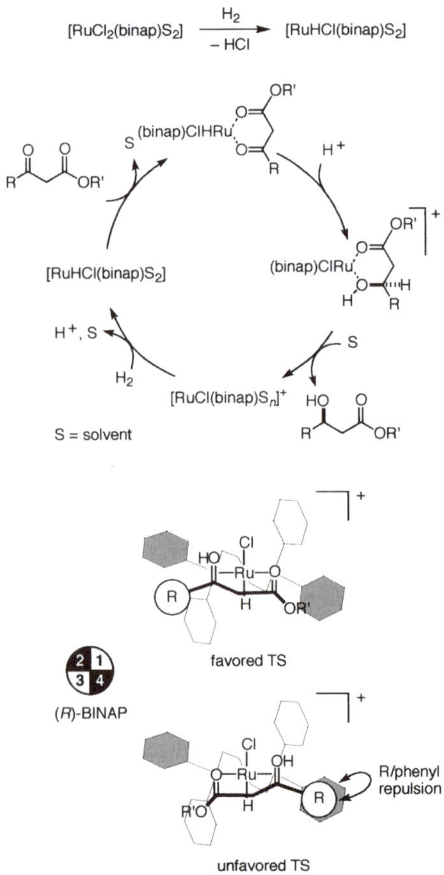

Figure 16. Mechanism of (*R*)-BINAP–Ru catalyzed hydrogenation of β-keto esters.

esters are hydrogenated smoothly even in the simplest ketone, acetone, containing a small amount of water. Thus, although BINAP–Ru dihalide catalysts have a very wide scope, they are unable to hydrogenate simple unfunctionalized ketones.

ASYMMETRIC HYDROGENATION OF SIMPLE KETONES BY BINAP/DIAMINE RUTHENIUM COMPLEXES

For more than half a century, selective reduction of simple ketones relied heavily on the metal hydride chemistry developed largely by H. C. Brown (1979 Nobel laureate). Chemoselective reduction of a C=O function in the presence of a C=C group has been best effected by the stoichiometric $NaBH_4$ reagent[60]. Diastereoselective reduction of ketones has frequently been achieved by Selectrides[61]. Enantioselective reduction of achiral ketones are effected by chiral stoichiometric reagents including BINAL-H[62], DIP chloride[63], and Alpine-borane[64] or by the Corey–Bakshi–Shibata (CBS) method combining B_2H_6 or catecholborane and a chiral oxazaborolidine catalyst[65]. Until very recently, these types of selective C=O reductions were not generally achievable by catalytic hydrogenation[49d,66].

In 1995 when I was the director of the ERATO Molecular Catalysis Project, we found that hydrogenation catalyzed by a $RuCl_2$(phosphine)$_2$(diamine) complex and an alkaline base provided a general solution to this long-standing problem[67]. The use of appropriate chiral diphosphines and chiral diamines allows asymmetric hydrogenation of simple ketones which lack any Lewis-basic functionality capable of interacting with the metal center. The reactivity and stereoselectivity are fine-tuned by changing the steric (bulkiness and chirality) and electronic properties of the auxiliaries. As generalized in Figure 17, the newly devised BINAP/diamine complex catalyzes rapid, productive and highly enantioselective hydrogenation of a range of aromatic, hetero-aromatic, and olefinic ketones in 2-propanol containing KO-*t*-C_4H_9 or KOH[68–70]. Among various complexes, $RuCl_2$(xylbinap)(daipen)[71] is particularly effective. For example, acetophenone and its derivatives are hydrogenated with S/C of up to 100 000 giving the secondary alcohols quantitatively in 99% ee[72], although the diamine free BINAP–Ru complexes are totally ineffective. Normally C=C bonds are much more reactive than C=O in catalytic hydrogenation, but this system allows for the preferential saturation of a C=O function over a coexisting C=C linkage[73,74]. Olefinic ketones, either conjugated or nonconjugated, can be converted to olefinic alcohols selectively. The hydrogenation tolerates various functionalities including F, Cl, Br, I, CF_3, OCH_3, $OCH_2C_6H_5$, $COOCH(CH_3)_2$, NO_2, NH_2, and NRCOR. Both electron-rich (furan, thiophene, thiazole, etc.) and -deficient rings (pyridine and pyrimidine) are left intact[75]. The simple $RuCl_2(PAr_3)_2(NH_2CH_2CH_2NH_2)$ complex hydrogenates various substituted cyclic and acyclic ketones with high diastereoselectivity, where the RuH intermediate acts as a bulky hydride species[76]. Because of the basic and protic nature of the reaction environment, hydrogenation of configurationally labile ketones allows for the

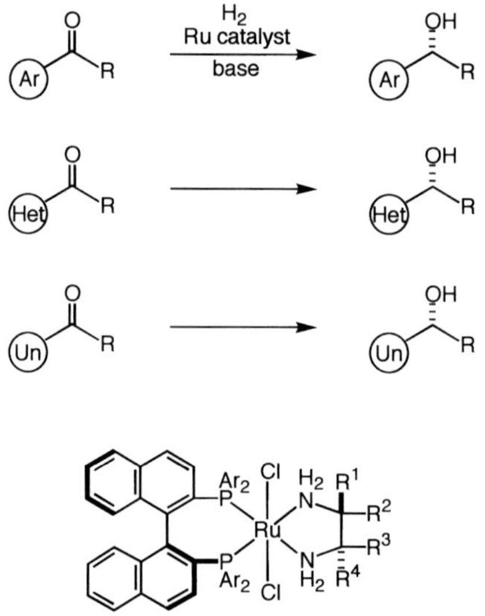

(*S*)-BINAP/(*S*)-diamine–Ru(II) catalyst

(*S*)-BINAP: Ar = C_6H_5

(*S*)-TolBINAP: Ar = 4-$CH_3C_6H_4$

(*S*)-XylBINAP: Ar = 3,5-$(CH_3)_2C_6H_3$

(*S,S*)-DPEN: $R^1 = R^4 = C_6H_5$; $R^2 = R^3 = H$

(*S*)-DAIPEN: $R^1 = R^2 =$ 4-$CH_3OC_6H_4$; $R^3 = H$; $R^4 = (CH_3)_2CH$

Figure 17. General asymmetric hydrogenation of simple ketones. Ar = aryl, Het = heteroaryl, Un = alkenyl.

dynamic kinetic discrimination of diastereomers, epimers, and enantiomers[76–78], effecting a new type of stereoselective reductions of ketones which are not possible with stoichiometric hydride reagents.

This asymmetric hydrogenation shows promise for the practical synthesis of a wide variety of chiral alcohols. The chiral diphosphine/diamine Ru complexes effect enantioselective hydrogenation of certain amino or amido ketones via a non-chelate mechanism without interaction between Ru and nitrogen or oxygen[78]. This method has been applied to the asymmetric synthesis of various important pharmaceuticals, including (*R*)-denopamine, a β_1-receptor agonist, the antidepressant (*R*)-fluoxetine, the antipsychotic BMS 181100, and (*S*)-duloxetine which is a potent inhibitor of serotonin and norepinephrine uptake carriers (Figure 18). Benzophenones can be hydrogenated to benzhydrols with an S/C ratio of up to 20 000 without over-reduction[79]. Enantioselective hydrogenation of certain *ortho*-substituted benzophenones leads to the unsymmetrically substituted benzhydrols with high ee, allowing convenient synthesis of the anticholinergic and antihistaminic (*S*)-orphenadrine. The antihistaminic (*R*)-neobenodine can be synthesized by using asymmetric hydrogenation of *o*-bromo-*p*′-methylbenzophenone.

This approach is the first example of general and efficient asymmetric hydrogenation of α,β-unsaturated ketones to chiral allylic alcohols of high

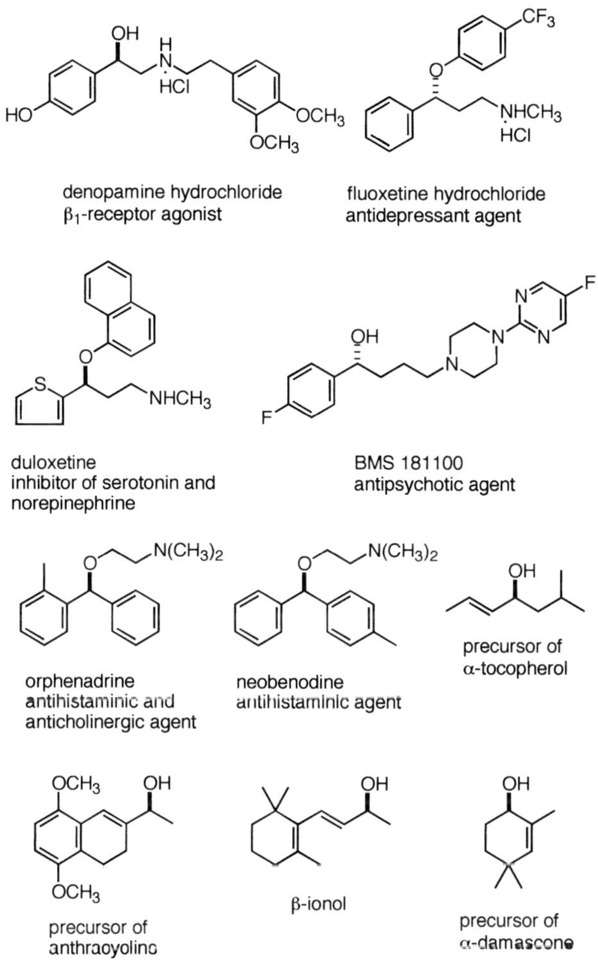

Figure 18. Application of asymmetric hydrogenation of simple ketones.

enantiomeric purity[72–74]. The selectivity profile is in sharp contrast to that observed with the diamine-free BINAP–Ru complex, efficiently catalyzing asymmetric hydrogenation of allylic alcohols (Figure 9). Its utility has been demonstrated by the synthesis of intermediates of an α-tocopherol side chain and anthracyclines as well as β-ionol (Figure 18)[72,73]. The asymmetric hydrogenation shown in Figure 17 is generally achieved by the combined use of an (*S*)-BINAP ligand and an (*S*)-1,2-diamine (or *R* and *R*). This is also the case for the reaction of *s*-cis exocyclic enones, such as (*R*)-pulegone. However, asymmetric hydrogenation of 2,4,4-trimethyl-2-cyclohexenone was effected best with $RuCl_2$[(*S*)-tolbinap][(*R*,*R*)-dpen][74,80]. The cyclic allyl alcohol obtained in 96% ee (Figure 18) can be converted to a series of carotenoid-derived odorants and bioactive terpenes, such as α-damascone. The *R* or *S* alcohols with ee as high as 95% can be obtained even with a racemic TolBINAP–$RuCl_2$ complex in the presence of (*R*,*R*)- or (*S*,*S*)-DPEN via asymmetric activation[80–82]. In this case, the highly enantioselective hydrogenation catalyzed by the (*S*)-diphosphine/(*R,R*)-diamine complex (or *R/S,S* combination) turns over 121 times faster than the less stereoselective reaction promoted by the diastereomeric *S/S,S* (or *R/R,R*) complex[83].

The reaction is rapid and highly productive. For example, when a mixture of acetophenone (601 g), the (*S*)-TolBINAP/(*S,S*)-DPEN Ru complex (2.2 mg), and KO-*t*-C_4H_9 (5.6 g) in 2-propanol (1.5 L) (30% w/v substrate concentration) was stirred under 45 atm H_2 at 30 °C for 48 h, the *R* alcohol was obtained with 80% ee and 100% yield[71,84]. Under such conditions, the turnover number was greater than 2 400 000, while the turnover frequency at 30% conversion was 228 000 h^{-1} or 63 s^{-1}.

This high rate and chemoselectivity for the C=O function are due to the nonclassical metal–ligand bifunctional mechanism (Figure 19)[68,70]. The hydrogenation involves a metal hydride intermediate. Hydride transfer from the metal to the carbonyl carbon has been considered to occur via a 2 + 2 mechanism. This reaction involves a Ru hydride species possessing an NH_2 li-

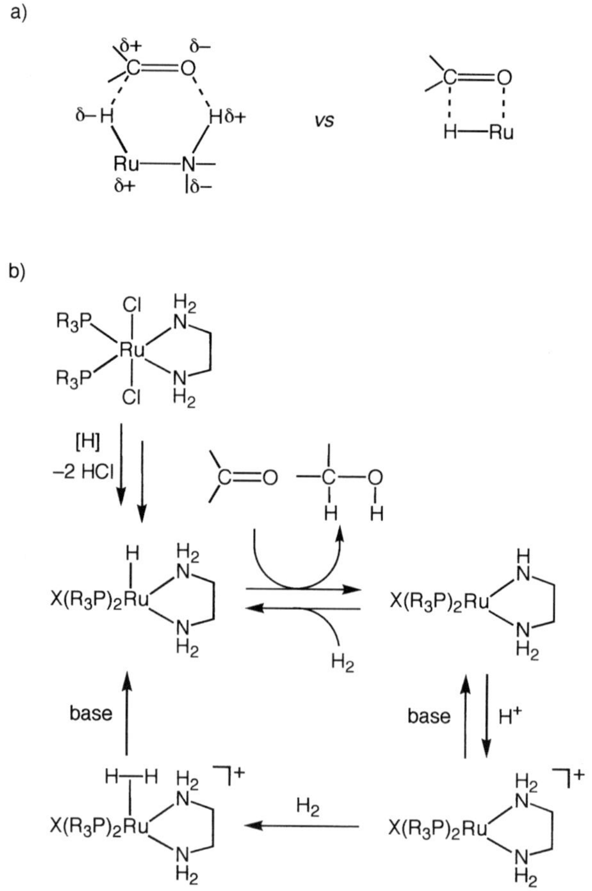

Figure 19. a) Nonclassical metal–ligand bifunctional mechanism and conventional 2 + 2 mechanism. b) Catalytic cycle of hydrogenation of ketones with a $RuCl_2(PR_3)_2(NH_2CH_2CH_2NH_2)$/strong base combined system in 2-propanol. X = H, OR, etc.

gand, whose hydridic Ru–H and protic N–H are simultaneously transferred to the C=O linkage via a six-membered pericyclic TS, thereby forming an alcoholic product directly, without forming a metal alkoxide (Figure 19a). In this hydrogenation, the metal and the ligand participate cooperatively in the bond-forming and -breaking processes. A more detailed mechanistic model is given in Figure 19b. The 18e RuH species reduces the ketone substrate via the pericyclic mechanism and the formal 16e Ru amide reacts directly with H_2 in a 2 + 2 manner or by a stepwise mechanism assisted by an alcohol and a base, giving back the reducing RuH complex[85]. The reducing activity of the RuH species is generated by the hydrogen-bonding NH_2 end in the diamine ligand that forms a *fac* relationship with the hydride ligand in the octahedral geometry. Neither ketone substrate nor alcoholic product interacts with the metallic center throughout the hydrogenation. The enantiofaces of prochiral ketones are differentiated on the molecular surface of the coordinatively saturated RuH intermediate. This notion is in contrast to the conventional mechanism for hydrogenation of unsaturated bonds that requires the metal–substrate π complexation.

This NH effect is common to the mechanism of Ru–catalyzed asymmetric transfer hydrogenation[86]. Recently we found that RuCl[(*S,S*)-YCH-$(C_6H_5)CH(C_6H_5)NH_2](\eta^6$-arene) (Y = O, NTs) or their analogues catalyze asymmetric transfer hydrogenation of aromatic and acetylenic carbonyl compounds using a 2-propanol/alkaline base system to give the corresponding *S* chiral alcohols of high enantiomeric purity, as generalized in Figure 20[87,88]. A formic acid/triethylamine mixture often serves as a better reducing agent. Certain imines are also reduced enantioselectively by this method. The detailed experimental[89] and theoretical analyses[90] revealed that the transfer hydrogenation of carbonyl compounds with 2-propanol proceeds via a coordinatively saturated 18e complex, RuH[(*S,S*)-YCH(C_6H_5)CH-$(C_6H_5)NH_2](\eta^6$-arene), as illustrated in Figure 21. The metal–ligand bifunctional mechanism allows for simultaneous delivery of the Ru–H and N–H to the C=O function via a six-membered pericyclic TS, giving an *S* alcohol and Ru[(*S,S*)-YCH$(C_6H_5)CH(C_6H_5)NH](\eta^6$-arene). The latter 16e Ru amide complex dehydrogenates 2-propanol to regenerate the Ru hydride species[86,91].

TOWARD CEREBRAL MOLECULAR SCIENCE

The major goals of synthetic chemists and the chemical industry have been the efficient synthesis of known valuable compounds. Another and perhaps more important pursuit is the creation of new valuable substances and materials through chemical synthesis. Toward this end, mere chemical knowledge or technology is often insufficient and basic research through interdisciplinary collaboration with scientists in other fields is needed. The recent progress in asymmetric synthesis has, in fact, spurred such endeavors which are directed toward the creation of molecularly engineered novel functions.

In the mid-1980s, we established the long-sought after three-component

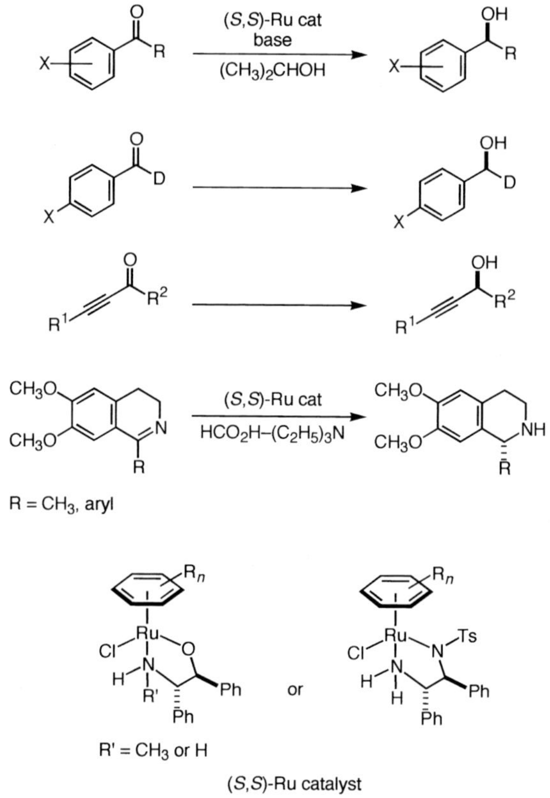

Figure 20. Asymmetric transfer hydrogenation of carbonyl compounds and imines catalyzed by chiral Ru complexes.

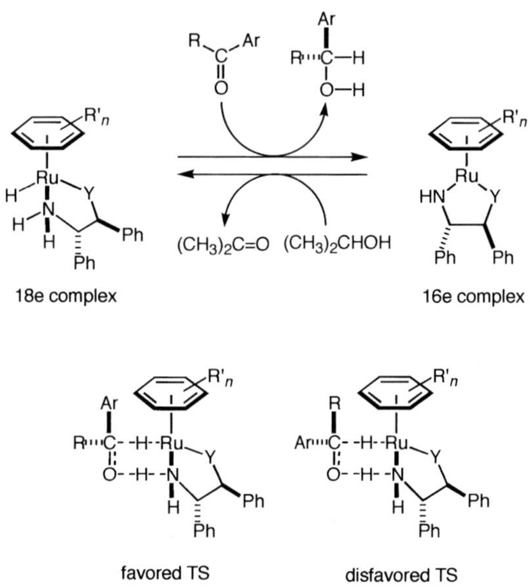

Figure 21. Metal–ligand bifunctional mechanism in asymmetric transfer hydrogenation catalyzed by RuH[(*S,S*)-YCH(C_6H_5)CH(C_6H_5)NH_2](η^6-arene). R = alkyl or D; Y = O or NTs.

coupling synthesis of prostaglandins (PGs) illustrated in Figure 22[92]. The five-membered unit could be combined with the two C_7 and C_8 side chain units by organometallic methodologies. Our asymmetric methods play a key role in controlling the C(11) and C(15) OH-bearing stereogenic centers. The requisite (*R*)-4-hydroxy-2-cyclopentenone is conveniently prepared on a multi-kilogram scale by kinetic resolution of the racemate by BINAP–Ru catalyzed hydrogenation[43]. The BINAL–H reagent is useful for asymmetric synthesis of the lower side-chain block[93,94]. This straightforward procedure is useful for the synthesis of not only naturally occurring PGs but also their artificial analogues[95].

To explore applications to the science of the human brain, we collaborated with the research groups led by M. Suzuki (my long-term collaborator at Nagoya and now at Gifu University), Y. Watanabe (Osaka City University), and B. Långström (Uppsala University)[96,97]. After a long investigation, (15*R*)-TIC, a PGI_2-type carboxylic acid, was found to show strong, selective binding in the central nervous system, thereby identifying the novel IP_2 receptor (Figure 23). Interestingly, this compound has the unnatural 15*R* configuration, although most biologically active PG derivatives have the natural 15*S* configuration. This discovery was made by an in vitro study using frozen sections of rat brain and (15*R*)-[^{3}H]TIC as a probe[98]. However, this radioactive probe is not appropriate for studies on living monkey or human brain, since β^- particles emitted from ^{3}H can not penetrate tissues. Incorporation of ^{11}C, a positron emitter having a short half-life of about 20 min and a high specific radioactivity, as a radioactive nuclide is essential for noninvasive studies using positron emission tomography (PET). Positrons (β^+) interact with free electrons in biological materials producing γ rays that can penetrate tissues

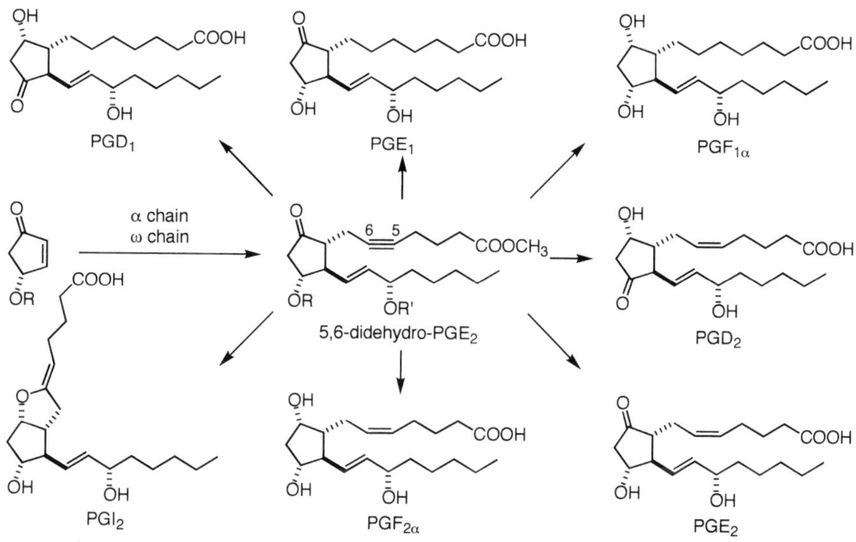

Figure 22. Three-component synthesis of prostaglandins. α chain = $ICH_2C \equiv C(CH_2)_3$-$COOCH_3$; ω chain = (*E,S*)-LiCH=CHCH(OR')$(CH_2)_4CH_3$.

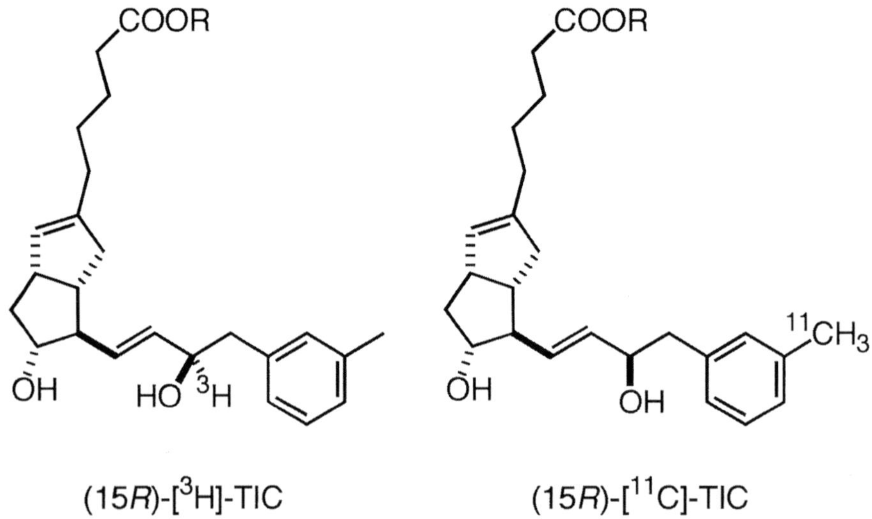

Figure 23. (15*R*)-TIC (R = H) and the methyl esters (R = CH_3) labeled by radioactive nuclides and an interdisciplinary collaborative team (from the left, M. Suzuki, R. Noyori, Y. Watanabe, and B. Långström).

and are detectable outside the human body. However, this presents a new chemical problem. The $^{11}CH_3$ group must be incorporated in the final step of the synthesis of (15*R*)-TIC methyl ester, and the total time for synthesis, work-up, purification, and sterilization should be less than 40 min because of the short half-life time of ^{11}C. A student in my group at Nagoya made a concerted effort to achieve this and, eventually succeeded with a rapid Pd-mediated coupling of methyl iodide and tributyl(aryl)stannane (excess) which is applicable to the synthesis of (15*R*)-[^{11}C]TIC methyl ester[99].

This technology was then transferred to the PET Center at Uppsala. A very dedicated colleague in our team, Suzuki, volunteered to test this new artificial compound on himself. After being carefully examined, (15*R*)-[^{11}C]TIC methyl ester was injected into his right arm. The methyl ester was carried through his blood stream, passed through the blood-brain barrier, reached his brain, and was hydrolyzed to the free carboxylic acid, which was bound to IP_2 receptors in his central nervous system. Figure 24 shows the PET images of horizontal slices of his brain, from the lower to the upper portions. From this trial, a new receptor, IP_2, was found in various important structures of the human brain. Thus, (15*R*)-TIC and its analogues are expected to have effects

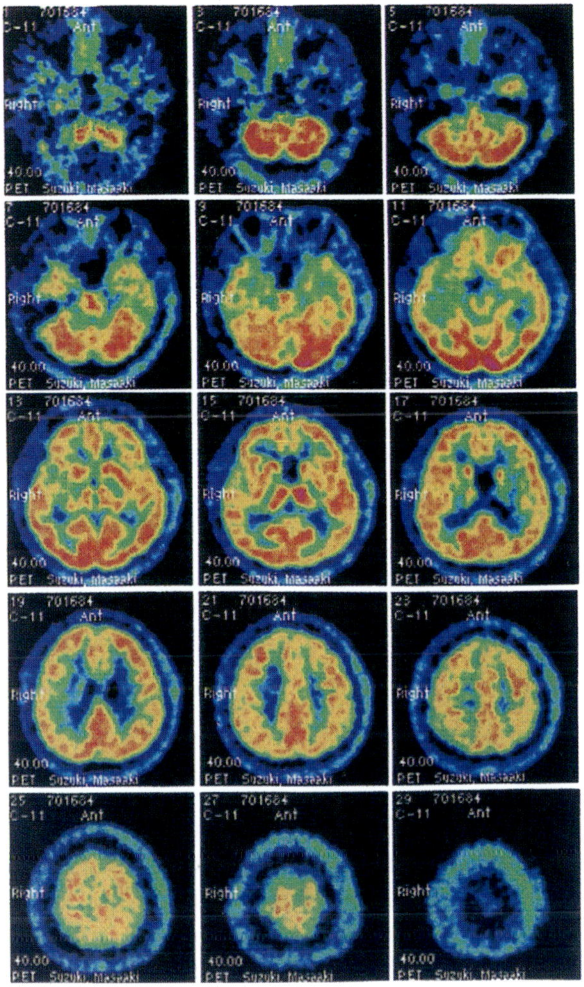

Figure 24. The uptake of (15*R*)-[^{11}C]-TIC in the human brain. The PET images of 18 horizontal slices from the lower to the upper portions of the brain (volunteer: M. Suzuki; June 13, 2000. PET Center of Uppsala University).

on the brain and, in fact, do show a unique neuroprotective effect, which may be of clinical benefit. Primary cultured hippocampal neurons exposed to a high oxygen concentration display the morphological features of apoptotic cell death and (15*R*)-TIC effectively protects them against such oxygen toxicity[100]. Similar neuroprotective effects were demonstrated in other *in vitro* experiments using serum deprivation and in *in vivo* studies of ischemic insults with gerbils, rats, and monkeys. Thus, the IP_2 receptor is a novel target for developing drugs which may be neuroprotective in brain disorders and neutrodegenerative diseases.

Studies of molecular chirality have the promise to yield significant clinical, scientific, and industrial benefits in the future. A structurally diverse array of molecular substances exist. All molecules possess common characteristics, viz., fixed elemental composition, definite atomic connectivity, a defined relative and absolute stereochemistry, and some conformation. From such precise nanometer-scale structures, certain significant properties and functions emerge. Chemists can design and synthesize molecules at will, based on accumulated scientific knowledge. The practical synthesis of enantiomers with a defined absolute stereochemistry is one of the most significant areas of research. This endeavor is not only an intellectual pursuit but is also a fertile area for the development of beneficial technologies[101]. Its utility is obvious, ranging from basic scientific research at a sub-femtomol scale, as in the case of brain research described above, to the industrial production of high-value compounds in multi-thousand tons per annum quantities. Louis Pasteur stated in 1851 that, "Dissymmetry is the only and distinct boundary between biological and nonbiological chemistry. Symmetrical physical or chemical force cannot generate molecular dissymmetry." This notion is no longer true. The recent revolutionary development in asymmetric catalysis has totally changed the approach to chemical synthesis. This field is still growing rapidly and I am certain that it will play a pivotal role in the development of the life sciences and nano-technology in the 21st century.

The highest honor for me is to be recognized with the prestigious 2001 Nobel Prize in Chemistry. This honor must be shared with my research family at Nagoya and with many collaborators at other institutions. Asymmetric hydrogenation has been the life-long focus of my research, and my studies have relied largely on BINAP chemistry which I initiated with the late Professor Hidemasa Takaya. Subsequently BINAP chemistry was developed further in our laboratories at Nagoya, where Professors Masato Kitamura and Takeshi Ohkuma made major contributions. Other asymmetric hydrogenation methods were discovered during my directorship of the ERATO Molecular Catalysis Project (1991–1996), which was managed by Professor Takao Ikariya (now Tokyo Institute of Technology) and Dr. Shohei Hashiguchi (Takeda Chemical Industry). Our laboratory at Nagoya is small. To realize the utilization of our scientific achievements, it was important to collaborate with other institutions. In this regard, I appreciate the cooperation of the groups led by Professors Sei Otsuka and Kazuhide Tani (Osaka University), and Professors Masaaki Suzuki (Gifu University), Yasuyoshi Watanabe (Osaka City University), and Bengt Långström (Uppsala University). These are just the names of the leaders of the research groups, although many young associates and students also contributed significantly. I had opportunities to have fruitful collaborations with many other scientists whose names are cited in the references. We have been supported by many companies, particularly Takasago International Corporation and Teijin Company. The generous and consistent support from the Ministry of Education, Culture, Sports, Science and Technology was essential for the success of my research. I am also grateful to the Japan Science and Technology Corporation and many private foundations for their support. Last, but not least, I acknowledge Professor Hitosi Nozaki at Kyoto University, my mentor who first introduced me to this fascinating and rewarding field of research.

REFERENCES

[1] For example, a) M. Gardner, *The New Ambidextrous Universe, 3rd Rev. Ed.*, W. H. Freeman & Co., New York, **1990**; b) E. Heilbronner, J. D. Dunitz, *Reflections on Symmetry*, VHCA–VCH, Basel, Weinheim, **1993**; c) R. Hoffmann, *The Same and Not the Same*, Columbia University Press, New York, **1995**; d) H. Brunner. *Rechts oder links in der Natur und anderswo*, Wiley–VCH, Weinheim, **1999**.

[2] G. Blaschke, H. P. Kraft, K. Fickentscher, F. Köhler, *Arzneim.-Forsch./Drug. Res.* **1979**, *29(II)*, 1640.

[3] This interpretation must be considered carefully, because the *R* enantiomer racemizes *in vivo*.

[4] a) *Chem. Eng. News* **1990**, *68*(12), 26; b) S. Borman, *Chem. Eng. News* **1990**, *68*(28), 9.

[5] S. C. Stinson, *Chem. Eng. News* **1992**. *70*(39), 46.

[6] S. C. Stinson, *Chem. Eng. News* **2001**, *79*(20), 45.

[7] R. Noyori, *Asymmetric Catalysis in Organic Synthesis*; Wiley, New York, **1994**.

[8] a) H. Nozaki, S. Moriuti, H. Takaya, R. Noyori, *Tetrahedron Lett.* **1966**, 5239; b) H. Nozaki, H. Takaya, S. Moriuti, R. Noyori, *Tetrahedron* **1968**, *24*, 3655.

[9] a) T. Aratani, *Pure Appl. Chem.* **1985**, *57*, 1839; b) T. Aratani in *Comprehensive Asymmetric Catalysis, Vol. 3* (Eds.: E. N. Jacobsen, A. Pfaltz, H. Yamamoto), Springer, Berlin, **1999**, p. 1451.

[10] a) M. Calvin, *Trans. Faraday Soc.* **1938**, *34*, 1181; b) M. Calvin, *J. Am. Chem. Soc.* **1939**, *61*, 2230.

[11] J. Halpern, J. F. Harrod, B. R. James, *J. Am. Chem. Soc.* **1961**, *83*, 753.

[12] J. F. Young, J. A. Osborn, F. H. Jardine, G. Wilkinson, *Chem. Commun.* **1965**, 131.

[13] S. Akabori, S. Sakurai, Y. Izumi, Y. Fujii, *Nature* **1956**, *178*, 323.

[14] W. S. Knowles, M. J. Sabacky, *Chem. Commun.*, **1968**, 1445.

[15] L. Horner, H. Siegel, H. Büthe, *Angew. Chem. Int. Ed. Engl.* **1968**, *7*, 942.

[16] These achievements were based on the classic work of Mislow and Horner on resolution of chiral tertiary phosphines: a) O. Korpiun, R. A. Lewis, J. Chickos, K. Mislow, *J. Am. Chem. Soc.* **1968**, *90*, 4842; b) L. Horner, *Pure Appl. Chem.* **1964**, *9*, 225,

[17] T. P. Dang, H. B. Kagan, *J. Chem. Soc. Chem. Commun.* **1971**, 481.

[18] a) W. S. Knowles, M. J. Sabacky, B. D. Vineyard, *J. Chem. Soc., Chem. Commun.* **1972**, 10; b) B. D. Vineyard, W. S. Knowles, M. J. Sabacky, G. L. Bachman, D. J. Weinkauff, *J. Am. Chem. Soc.* **1977**, *99*, 5946; c) W. S. Knowles, *Acc. Chem. Res.* **1983**, *16*, 106; d) J. Crosby in *Chirality in Industry: The Commercial Manufacture and Applications of Optically Active Compounds* (Eds.: A. N. Collins, G. N. Sheldrake, J. Crosby), Wiley, Chichester, **1992**, chap 1.

[19] E. J. Corey, R. Noyori, T. K. Schaaf, *J. Am. Chem. Soc.* **1970**, *92*, 2586.

[20] One of the early examples of successful asymmetric catalysis was the Ni-catalyzed hydrovinylation of norbornene: See: B. Bogdanovic, B. Henc, A. Lösler, B.Meister, H. Pauling, G. Wilke, *Angew. Chem. Int. Ed. Engl.* **1973**, *12*, 954.

[21] R. Noyori, H. Takaya, *Chemica Scripta* **1985**, *25*, 83.

[22] A. Miyashita, A. Yasuda, H. Takaya, K. Toriumi, T. Ito, T. Souchi, R. Noyori, *J. Am. Chem. Soc.* **1980**, *102*, 7932.

[23] a) A. Miyashita, H. Takaya, T. Souchi, R. Noyori, *Tetrahedron* **1984**, *40*, 1245; b) K. J. Brown, M. S. Berry, K. C. Waterman, D. Lingenfelter, J. R. Murdoch, *J. Am. Chem. Soc.* **1984**, *106*, 4717.

[24] H. Takaya, K. Mashima, K. Koyano, M. Yagi, H. Kumobayashi, T. Taketomi, S. Akutagawa, R. Noyori, *J. Org. Chem.* **1986**, *51*, 629.

[25] For the more recent stereospecific synthesis by Ni-catalyzed coupling between optically pure binaphthol ditriflate and diphenylphosphine, see: D. Cai, J. F. Payack, D. R. Bender, D. L. Hughes, T. R. Verhoeven, P. J. Reider, *Org. Synth.* **1998**, *76*, 6.

[26] K. Toriumi, T. Ito, H. Takaya, T. Souchi, R. Noyori, *Acta Crystallogr. Sect. B*, **1982**, *38*, 807.

[27] a) K. Tani, T. Yamagata, S. Otsuka, S. Akutagawa, H. Kumobayashi, T. Taketomi, H.

Takaya, A. Miyashita, R. Noyori, *J. Chem. Soc. Chem. Commun.* **1982**, 600; b) K. Tani, T. Yamagata, S. Akutagawa, H. Kumobayashi, T. Taketomi, H. Takaya, A. Miyashita, R. Noyori, S. Otsuka, *J. Am. Chem. Soc.* **1984**, *106*, 5208.

[28] a) S. Akutagawa in *Organic Synthesis in Japan: Past, Present, and Future* (Eds.: R. Noyori, T. Hiraoka, K. Mori, S. Murahashi, T. Onoda, K. Suzuki, O. Yonemitsu), Tokyo Kagaku Dozin, Tokyo, **1992**, p. 75; b) S. Akutagawa in *Chirality in Industry: The Commercial Manufacture and Applications of Optically Active Compounds* (Eds.: A. N. Collins, G. N. Sheldrake, J. Crosby), Wiley, Chichester, **1992**, chap 16.

[29] a) S. Otsuka, K. Tani in *Asymmetric Synthesis Vol. 5* (Ed.: J. D. Morrison), Academic Press, Orlando, **1985**, chap 6; b) S. Otsuka, K. Tani, *Synthesis* **1991**, 665; c) S. Otsuka, *Acta Chem. Scand.* **1996**, *50*, 353.

[30] K. Takabe, T. Yamada, T. Katagiri, J. Tanaka, *Org. Synth.* **1989**. *67*, 48.

[31] M. Yamakawa, R. Noyori, *Organomet.* **1992**, *11*, 3167.

[32] a) R. Noyori, *Chem. Soc. Rev.* **1989**, *18*, 187; b) R. Noyori, *Science* **1990**, *248*, 1194; c) R. Noyori, H. Takaya, *Acc. Chem. Res.* **1990**, *23*, 345; d) R. Noyori, *CHEMTECH* **1992**, *22*, 360; e) R. Noyori, *Tetrahedron* **1994**, *50*, 4259; f) R. Noyori in *(IUPAC) Stereocontrolled Organic Synthesis* (Ed.: B. M. Trost), Blackwell Scientific Publications, Oxford, **1994**, p. 1; g) R. Noyori, *Acta Chem. Scand.* **1996**, *50*, 380.

[33] An early example of phosphine–Ru complex catalyzed hydrogenation: P.S. Hallman, B. R. McGarvey, G. Wilkinson, *J. Chem. Soc. A* **1968**, 3143.

[34] Ru-catalyzed asymmetric hydrogenation of olefins was first achieved using a DIOP–Ru complex: B. R. James, D. K. W. Wang, R. F. Voigt, *J. Chem. Soc., Chem. Commun.* **1975**, 574.

[35] C. R. Landis, J. Halpern, *J. Am. Chem. Soc.* **1987**, *109*, 1746.

[36] J. M. Brown, P. A. Chaloner, *J. Am. Chem. Soc.* **1980**, *102*, 3040.

[37] For an exmple of a solution to this problem, see: I. D. Gridnev, N. Higashi, K. Asakura, T. Imamoto, *J. Am. Chem. Soc.* **2000**, *122*, 7183.

[38] R. Noyori, M. Ohta, Y. Hsiao, M. Kitamura, T. Ohta, H. Takaya, *J. Am. Chem. Soc.* **1986**, *108*, 7117.

[39] a) T. Ohta, H. Takaya, R. Noyori, *Inorg. Chem.* **1988**, *27*, 566; b) M. Kitamura, M. Tokunaga, R. Noyori, *J. Org. Chem.* **1992**, *57*, 4053; c) H. Takaya, T. Ohta, S. Inoue, M. Tokunaga, M. Kitamura, R. Noyori, *Org. Synth.* **1993**, *72*, 74.

[40] a) T. Ohta, H. Takaya, R. Noyori, *Tetrahedron Lett.* **1990**, *31*, 7189; b) M. T. Ashby, J. Halpern, *J. Am. Chem. Soc.* **1991**, *113*, 589.

[41] T. Ohta, H. Takaya, M. Kitamura, K. Nagai, R. Noyori, *J. Org. Chem.* **1987**, *52*, 3174.

[42] H. Takaya, T. Ohta, N. Sayo, H. Kumobayashi, S. Akutagawa, S. Inoue, I. Kasahara, R. Noyori, *J. Am. Chem. Soc.* **1987**, *109*, 1596 and 4129.

[43] M. Kitamura, I. Kasahara, K. Manabe, R. Noyori, H. Takaya, *J. Org. Chem.* **1988**, *53*, 708.

[44] a) M. Kitamura, Y. Hsiao, R. Noyori, H. Takaya, *Tetrahedron Lett.* **1987**, *28*, 4829; b) M. Kitamura, Y. Hsiao, M. Ohta, M. Tsukamoto, T. Ohta, H. Takaya, R. Noyori, *J. Org. Chem.* **1994**, *59*, 297.

[45] W. D. Lubell, M. Kitamura, R. Noyori, *Tetrahedron: Asymmetry* **1991**, *2*, 543.

[46] M. Kitamura, M. Yoshimura, M. Tsukamoto, R. Noyori, *Enantiomer* **1996**, *1*, 281.

[47] T. Ikariya, Y. Ishii, H. Kawano, T. Arai, M. Saburi, S. Yoshikawa, S. Akutagawa, *J. Chem. Soc. Chem. Commun.* **1985**, 922. This was the first use of BINAP in Ru-catalyzed asymmetric hydrogenation.

[48] M. Kitamura, K. Nagai, Y. Hsiao, R. Noyori, *Tetrahedron Lett.* **1990**, *31*, 549.

[49] For other synthetic applications, see reviews: a) G. M. Ramos Tombo, G. Belluš, *Angew. Chem. Int. Ed. Engl.* **1991**, *30*, 1193; b) A. Börner, J. Holz in *Transition Metals for Organic Synthesis, Vol. 2* (Eds.: M. Beller, C. Bolm), Wiley–VCH, Weinheim, **1998**, p. 3; c) J. M. Brown in *Comprehensive Asymmetric Catalysis, Vol. 1* (Eds.: E. N. Jacobsen, A. Pfaltz, H. Yamamoto), Springer, Berlin, **1999**, p. 121; d) T. Ohkuma, M. Kitamura, R. Noyori in *Catalytic Asymmetric Synthesis, 2nd Ed.* (Ed.: I. Ojima), Wiley–VCH, New York, **2000**, p. 1.

[50] a) M. Kitamura, M. Tokunaga, T. Ohkuma, R. Noyori, *Tetrahedron Lett.* **1991**, *32*, 4163; b) M. Kitamura, M. Tokunaga, T. Ohkuma, R. Noyori, *Org. Synth.* **1993**, *71*, 1; c) K. Mashima, K. Kusano, N. Sato, Y. Matsumura, K. Nozaki, H. Kumobayashi, N. Sayo, Y. Hori, T. Ishizaki, S. Akutagawa, H. Takaya, *J. Org. Chem.* **1994**, *59*, 3064. The halogen-containing Ru complexes are also effective for asymmetric hydrogenation of various functionalized olefins.

[51] a) M. Kitamura, T. Ohkuma, S. Inoue, N. Sayo, H. Kumobayashi, S. Akutagawa, T. Ohta, H. Takaya, R. Noyori, *J. Am. Chem. Soc.* **1988**, *110*, 629; b) K. Mashima, K. Kusano, T. Ohta, R. Noyori, H. Takaya, *J. Chem. Soc. Chem. Commun.* **1989**, 1208; c) T. Ohkuma, M. Kitamura, R. Noyori, *Tetrahedron Lett.* **1990**, *31*, 5509.

[52] M. Kitamura, T. Ohkuma, H. Takaya, R. Noyori, *Tetrahedron Lett.* **1988**, *29*, 1555.

[53] T. Nishi, M. Kitamura, T. Ohkuma, R. Noyori, *Tetrahedron Lett.* **1988**, *29*, 6327.

[54] a) R. Noyori, T. Ikeda, T. Ohkuma, M. Widhalm, M. Kitamura, H. Takaya, S. Akutagawa, N. Sayo, T. Saito, T. Taketomi, H. Kumobayashi, *J. Am. Chem. Soc.* **1989**, *111*, 9134; b) M. Kitamura, T. Ohkuma, M. Tokunaga, R. Noyori, *Tetrahedron: Asymmetry* **1990**, *1*, 1.

[55] a) M. Kitamura, M. Tokunaga, R. Noyori, *J. Am. Chem. Soc.* **1993**, *115*, 144; b) M. Kitamura, M. Tokunaga, R. Noyori, *Tetrahedron* **1993**, *49*, 1853; c) R. Noyori, M. Tokunaga, M. Kitamura, *Bull. Chem. Soc. Jpn.* **1995**, *68*, 36.

[56] M. Kitamura, M. Tokunaga, T. Pham, W. D. Lubell, R. Noyori, *Tetrahedron Lett.* **1995**, *36*, 5769.

[57] M. Kitamura, M. Tokunaga, R. Noyori, *J. Am. Chem. Soc.* **1995**, *117*, 2931.

[58] For the threo/erythro nomenclature, see: R. Noyori, I. Nishida, J. Sakata, *J. Am. Chem. Soc.* **1981**, *103*, 2106.

[59] R. Noyori, T. Ohkuma, M. Kitamura, H. Takaya, N. Sayo, H. Kumobayashi, S. Akutagawa, *J. Am. Chem. Soc.* **1987**, *109*, 5856.

[60] a) H. C. Brown, H. I. Schlesinger, I. Sheft, D. M. Ritter, *J. Am. Chem. Soc.* **1953**, *75*, 192; b) H. I. Schlesinger, H. C. Brown, A. E. Finholt, *J. Am. Chem. Soc.* **1953**, *75*, 205. Reviews: c) H. C. Brown, S. Krishnamurthy, *Tetrahedron* **1979**, *35*, 567; d) H. C. Brown, P. V. Ramachandran in *Reductions in Organic Synthesis: Recent Advances and Practical Applications (ACS symposium series 641)* (Ed.: A. F. Abdel-Magid), American Chemical Society, Washington, DC, **1996**, chap 1.

[61] a) H. C. Brown, S. Krishnamurthy, *J. Am. Chem. Soc.* **1972**, *94*, 7159; b) S. Krishnamurthy, H. C. Brown, *J. Am. Chem. Soc.* **1976**, *98*, 3383.

[62] a) R. Noyori, I. Tomino, Y. Tanimoto, M. Nishizawa, *J. Am. Chem. Soc.* **1984**, *106*, 6709; b) R. Noyori, I. Tomino, M. Yamada, M. Nishizawa, *J. Am. Chem. Soc.* **1984**, *106*, 6717.

[63] P. V. Ramachandran, H. C. Brown in *Reductions in Organic Synthesis: Recent Advances and Practical Applications (ACS symposium series 641)* (Ed.: A. F. Abdel-Magid), American Chemical Society, Washington, DC, **1996**, chap 5.

[64] M. M. Midland, *Chem. Rev.* **1989**, *89*, 1553.

[65] E. J. Corey, C. J. Helal, *Angew. Chem. Int. Ed.* **1998**, *37*, 1986.

[66] For early efforts, see: a) J. Solodar, *CHEMTECH* **1975**, *5*, 421; b) T. Hayashi, T. Mise, M. Kumada, *Tetrahedron Lett.* **1976**, 4351; c) S. Törös, B. Heil, L. Kollár, L. Markó, *J. Organomet. Chem.* **1980**, *197*, 85; d) J. Bakos, I. Tóth, B. Heil, L. Markó, *J. Organomet. Chem.* **1985**, *279*, 23; e) X. Zhang, T. Taketomi, T. Yoshizumi, H. Kumobayashi, S. Akutagawa, K. Mashima, H. Takaya, *J. Am. Chem. Soc.* **1993**, *115*, 3318. See also: f) F. Robin, F. Mercier, L. Ricard, F. Mathey, M. Spagnol, *Chem. Eur. J.* **1997**, *3*, 1365; g) U. Nagel, C. Roller, *Z. Naturforsch., Ser. B* **1998**, *53*, 267; h) Q. Jiang, Y. Jiang, D. Xiao, P. Cao, X. Zhang, *Angew. Chem. Int. Ed.* **1998**, *37*, 1100; i) P. Cao, X. Zhang, *J. Org. Chem.* **1999**, *64*, 2127. Heterogeneous hydrogenation, see: j) T. Osawa, A. Tai, Y. Imachi, S. Takasaki in *Chiral Reactions in Heterogeneous Catalysis* (Eds.: G. Jannes, V. Dubois), Plenum Press, New York, **1995**, p. 75; k) T. Harada, T. Osawa in *Chiral Reactions in Heterogeneous Catalysis* (Eds.: G. Jannes, V. Dubois), Plenum Press, New York, **1995**, p. 83.

[67] T. Ohkuma, H. Ooka, S. Hashiguchi, T. Ikariya, R. Noyori, *J. Am. Chem. Soc.* **1995**, *117*, 2675.
[68] R. Noyori, T. Ohkuma, *Angew. Chem. Int. Ed.* **2001**, *40*, 40.
[69] R. Noyori, T. Ohkuma, *Pure Appl. Chem.* **1999**, *71*, 1493.
[70] R. Noyori, M. Koizumi, D. Ishii, T. Ohkuma, *Pure Appl. Chem.* **2001**, *73*, 227.
[71] Abbreviations: TolBINAP = 2,2'-Bis(di-4-tolylphosphino)-1,1'-binaphthyl. XylBINAP = 2,2'-Bis(di-3,5-xylylphosphino)-1,1'-binaphthyl. DPEN = 1,2-Diphenylethylenediamine. DAIPEN = 1,1-Di-4-anisyl-2-isopropyl-1,2-ethylenediamine.
[72] T. Ohkuma, M. Koizumi, H. Doucet, T. Pham, M. Kozawa, K. Murata, E. Katayama, T. Yokozawa, T. Ikariya, R. Noyori, *J. Am. Chem. Soc.* **1998**, *120*, 13529.
[73] T. Ohkuma, H. Ooka, T. Ikariya, R. Noyori, *J. Am. Chem. Soc.* **1995**, *117*, 10417.
[74] T. Ohkuma, H. Ikehira, T. Ikariya, R. Noyori, *Synlett* **1997**, 467.
[75] T. Ohkuma, M. Koizumi, M. Yoshida, R. Noyori, *Org. Lett.* **2000**, *2*, 1749.
[76] T. Ohkuma, H. Ooka, M. Yamakawa, T. Ikariya, R. Noyori, *J. Org. Chem.* **1996**, *61*, 4872.
[77] T. Matsumoto, T. Murayama, S. Mitsuhashi, T. Miura, *Tetrahedron Lett.* **1999**, *40*, 5043.
[78] T. Ohkuma, D. Ishii, H. Takeno, R. Noyori, *J. Am. Chem. Soc.* **2000**, *122*, 6510.
[79] T. Ohkuma, M. Koizumi, H. Ikehira, T. Yokozawa, R. Noyori, *Org. Lett.* **2000**, *2*, 659.
[80] T. Ohkuma, H. Doucet, T. Pham, K. Mikami, T. Korenaga, M. Terada, R. Noyori, *J. Am. Chem. Soc.* **1998**, *120*, 1086.
[81] T. Ohkuma, H. Takeno, Y. Honda, R. Noyori, *Adv. Synth. Catal.* **2001**, *343*, 369.
[82] Review: K. Mikami, M. Terada, T. Korenaga, Y. Matsumoto, M. Ueki, R. Angelaud, *Angew. Chem. Int. Ed.* **2000**, *39*, 3532. See also: K. Mikami, T. Korenaga, M. Terada, T. Ohkuma, T. Pham, R. Noyori, *Angew. Chem. Int. Ed.* **1999**, *38*, 495; K. Mikami, T. Korenaga, T. Ohkuma, R. Noyori, *Angew. Chem. Int. Ed.* **2000**, *39*, 3707.
[83] Non-linear effects in catalysis should be investigated carefully because of the homochiral and heterochiral interaction of the chiral species present in the catalytic system. For reviews, see: a) R. Noyori, S. Suga, H. Oka, M. Kitamura, *Chem. Rec.* **2001**, *1*, 85; b) C. Girard, H. B. Kagan, *Angew. Chem. Int. Ed.* **1998**, *37*, 2922; c) D. R. Fenwick, H. B. Kagan, *Top. Stereochem.* **1999**, *22*, 257.
[84] H. Doucet, T. Ohkuma, K. Murata, T. Yokozawa, M. Kozawa, E. Katayama, A. F. England, T. Ikariya, R. Noyori, *Angew. Chem. Int. Ed.* **1998**, *37*, 1703.
[85] The ketone substrate reacts in an outer sphere of the coordinatively saturated 18e complex without interaction with the metallic center. Thus the general scheme given in Figure 1 is to be modified to some extent. For other mechanistic investigations, see: a) K. Abdur-Rashid, M. Faatz, A. J. Lough, R. H. Morris, *J. Am. Chem. Soc.* **2001**, *123*, 7473; b) R. Hartmann, P. Chen, *Angew. Chem. Int. Ed.* **2001**, *40*, 3581.
[86] R. Noyori, M. Yamakawa, S. Hashiguchi, *J. Org. Chem.* **2001**, *66*, 7931.
[87] R. Noyori, S. Hashiguchi, *Acc. Chem. Res.* **1997**, *30*, 97.
[88] a) K. Matsumura, S. Hashiguchi, T. Ikariya, R. Noyori, *J. Am. Chem. Soc.* **1997**, *119*, 8738; b) I. Yamada, R. Noyori, *Org. Lett.* **2000**, *2*, 3425.
[89] K.-J. Haack, S. Hashiguchi, A. Fujii, T. Ikariya, R. Noyori, *Angew. Chem. Int. Ed. Engl.* **1997**, *36*, 285.
[90] a) M. Yamakawa, H. Ito, R. Noyori, *J. Am. Chem. Soc.* **2000**, *122*, 1466; b) D. A. Alonso, P. Brandt, S. J. M. Nordin, P. G. Andersson, *J. Am. Chem. Soc.* **1999**, *121*, 9580.
[91] M. Yamakawa, I. Yamada, R. Noyori, *Angew. Chem. Int. Ed.* **2001**, *40*, 2818.
[92] a) M. Suzuki, A. Yanagisawa, R. Noyori, *J. Am. Chem. Soc.* **1985**, *107*, 3348; b) M. Suzuki, A. Yanagisawa, R. Noyori, *J. Am. Chem. Soc.* **1988**, *110*, 4718; c) M. Suzuki, Y. Morita, H. Koyano, M. Koga, R. Noyori, *Tetrahedron* **1990**, *46*, 4809.
[93] M. Suzuki, H. Koyano, Y. Morita, R. Noyori, *Synlett* **1989**, 22.
[94] This has been used at Ono Pharmaceutical Co. for the industrial synthesis of PGs using the Corey method.
[95] a) R. Noyori, M. Suzuki, *Angew. Chem. Int. Ed. Engl.* **1984**, *23*, 847; b) R. Noyori, M. Suzuki, *Chemtracts—Org. Chem.* **1990**, *3*, 173; c) R. Noyori, M. Suzuki, *Science* **1993**, *259*, 44.

[96] a) M. Suzuki, K. Kato, R. Noyori, Yu. Watanabe, H. Takechi, K. Matsumura, B. Långström, Y. Watanabe, *Angew. Chem. Int. Ed. Engl.* **1996**, *35*, 334; b) H. Takechi, K. Matsumura, Yu. Watanabe, K. Kato, R. Noyori, M. Suzuki, Y. Watanabe, *J. Biol. Chem.* **1996**, *271*, 5901.

[97] Account: M. Suzuki, R. Noyori, B. Långström, Y. Watanabe, *Bull. Chem. Soc. Jpn.* **2000**, *73*, 1053.

[98] Yu. Watanabe, K. Matsumura, H. Takechi, K. Kato, H. Morii, M. Björkman, B. Långström, R. Noyori, M. Suzuki, Y. Watanabe, *J. Neurochem.* **1999**, *72*, 2583.

[99] a) M. Suzuki, H. Doi, M. Björkman, Y. Andersson, B. Långström, Y. Watanabe, R. Noyori, *Chem. Eur. J.* **1997**, *3*, 2039; b) M. Suzuki, H. Doi, K. Kato, M. Björkman, B. Långström, Y. Watanabe, R. Noyori, *Tetrahedron* **2000**, *56*, 8263.

[100] T. Satoh, Y. Ishikawa, Y. Kataoka, Y. Cui, H. Yanase, K. Kato, Yu. Watanabe, K. Nakadate, K. Matsumura, H. Hatanaka, K. Kataoka, R. Noyori, M. Suzuki, Y. Watanabe, *Eur. J. Neurosci.* **1999**, *11*, 3115.

[101] R. Noyori, S. Hashiguchi, T. Yamano in *Applied Homogeneous Catalysis by Organometallic Complexes, 2nd ed.* (Eds.: B. Cornils, W. A. Hermann), VCH–Wiley, Weinheim, in press.

K. BARRY SHARPLESS

From 6th through 12th grades I attended a Quaker school on the Philadelphia city line. Twice a week the entire school attended Quaker Meeting, silent gatherings except when someone received a personal call to speak.

I never got a call, but nonetheless my head was full: I thought about fishing and boats. Or else I thought about when next I could get from Philadelphia to our cottage on the New Jersey Shore in order to go out fishing in a boat. Beneath my picture in one high school yearbook it says, "I'm going to the Shore".

While I had an overwhelming passion for fishing, school I merely enjoyed and I never planned to be a scientist. In fact, passion, not planning, is the engine driving all my thought and action. The almost unimaginably good fortune of my youth was that other people made such very, very good plans and choices for me.

My parents selected the excellent Friends Central School where, fortuitously, Clayton Farraday was both a science teacher and the school's beloved Mr. Chips. The counselors there decided, wisely, that I should attend a college rather than a large university, and I departed Philadelphia for Dartmouth College in the fall of 1959. Though literature courses there were my favorites, I was a pre-medical student solely because my parents always hoped that I'd become an MD like my father. Pre-meds majored in chemistry or biology, and between the two I leaned toward chemistry. I didn't get really interested, however, until I had two semesters of organic my sophomore year from a young chemistry professor who chose me to do research in his lab. When I graduated Dartmouth a few years later, in 1963, the same prof called my next move, a PhD in organic chemistry instead of medical school. He even chose the graduate school I attended and my research supervisor there. Such a strong intervention in a student's life is no doubt unusual, but the precipitating events were unusual, too.

Generally speaking, colleges have the best undergraduate teaching, and universities, whose labs are filled by graduate and post-graduate students, have the best research. When I arrived at Dartmouth College in 1959, the chemistry department had a graduate program, which meant great teachers who were just as good at research. However, the program was small, and only a master's degree was awarded, so consequently professors were perpetually hungry for more manpower for their labs, more "hands". Undergraduates who performed well in lab courses were actively recruited to do "real" graduate-level research.

Thomas A. Spencer, a brand-new assistant professor of chemistry, arrived at

Harbo Jensen (left) and Rudy Lauer (right), my first "full-term" PhD students at MIT graduation, 1974 (Tom Flood received his doctorate earlier, but he had transferred from another group).

My family: (clockwise from upper left) youngest son Ike, wife Jan, daughter Hannah, middle kid Will, Jan's mother Lillian Duder.

Dartmouth when I did, and I was part of his research group for three years. Because Tom was (and still is) so smart and such a good chemist, he could recognize not just talent, but the potential to do something significant; because Tom was also born a great teacher, he was obliged to give a swift kick to my comfortable obliviousness. Fishing, now in the form of working all summer on charter boats, continued as my abiding passion, which meant I continued to need a wise person to make good decisions for me. If some variables in my adult life were changed, I might still have made it onto these pages, but it never would have happened without Tom Spencer.

Since some family background and professional activities (and lots more about fishing) are in the Nobel lecture that follows, and since the standard biographical folderol is most easily found online at www.scripps.edu/chem/sharpless/, I hope to provide a more interesting read with the highly subjective and largely unorthodox personal information that follows.

I met my future wife, Jan Dueser, at a beach party at San Gregorio, west across the foothills from Stanford University. I was a first-year graduate stu-

My parents, Evelyn Anderson Sharpless and Edwin Dallett Sharpless, M.D.

Asymmetric Dihydroxylation pioneers, MIT, 1988: (l to r) John Wai, Istvan Markó, John Svendsen.

Valery Fokin, an Assistant Professor in the group, who, along with Erik Rubin, Wall Pringle, Allen Thomas, Robert Epple, and Malin Andersson tirelessly championed the so-called "second cycle" ligand strategies which have led to a greatly improved processes for osmium-catalyzed dihydroxylation and aminohydroxylation of olefins.

Jan Dueser, my future wife, on the day I met her at San Gregorio Beach, California, Fall 1963.

Hiking in the High Sierra with Papagena and Hartmuth Kolb, Summer 2000: one of the best dogs of a lifetime and an equally valued coworker.

dent, and she was a Stanford sophomore and, on that day, my roommate's date. I admired her touch football form, and she entrusted me with her delicate wristwatch, which I lost in the sand. We were married about a year-and-a-half later, on April 28, 1965, my 24th birthday, at the Palo Alto courthouse. David Schooley, a fellow chemistry grad student and now a professor at University of Nevada, Reno, was our best man.

Jan and I practiced with dogs before we had children; chemists still ask about our first, the black and enormous Lionel, a regular laboratory and classroom visitor at MIT. Our daughter Hannah (whose nickname "Pippi", comes from "pipette", not from Miss Longstocking) was born in 1976, and is a middle

The Asymmetric Epoxidation team and my family at MIT in 1982:(front, l to r) Will & Jan Sharpless, Larry Reed, Sotaro Miyano, Ike Sharpless, Fred Walker, Joel Hawkins, Janice Klunder, Mary Jean Schweiter, Marie Martin; (standing, l to r) Ken Kirshenbaum, Soo Ko, Steve Viti, KBS with Hannah, Maurice Caron, Carl Behrens, M. G. Finn, Mamoru Uchiyama, Gordon Hill, John Ellman.

Eric Jacobsen (left) did the early Asymmetric Dihydroxylation mechanism studies and documented the ligand acceleration phenomenon; Yun Gao's cheery countenance was as uplifting as his chemistry.

An exemplary mentor during his productive year as a visiting scientist from Upjohn Pharmaceuticals, Roy Johnson remained a contributor to the group for decades through review articles and book chapters.

My passion for chemistry was preceded by a passion for fishing.

Dr. Alfred Bader and his wife (and companion in all things) Isabel. Like many academic chemists throughout the world, I was privileged to receive the Baders' longtime support and friendship. With the founding of Aldrich Chemical, Dr. Bader created the reliable and diverse catalog of research chemicals that launched modern organic chemistry.

school teacher in Boston. To chemists who've attended my seminars, she is permanently six years old, the familiar Alice in Wonderland who gazes at the huge book of Looking Glass Sugars. William ("Will") and Isaac ("Ike") followed Hannah at two-year intervals. Both of our sons are still college undergraduates. None of our children has much interest in science, and I'm sorry, but not disappointed, that that is so.

A 1983 reunion with the extraordinary cadre of Japanese coworkers who made such a huge contribution to my research in the previous decade: (seated, l to r) Kiyoshi Kondo, Koichiro Oshima, KBS, Tetsuo Hori Onami; (rear, l to r) Tsutomu Katsuki, Yasuhiro Yamada, Masanori Ikeda, Ichiro Takagi, Muneo Takatani, Susumu Tsushima.

Maybe my Scandinavian heritage contributed to the immediate personal and professional bond created when Björn Åkermark (left) came to van Tamelen's Stanford group as a visiting professor in the mid-1960s. The Scandinavian Connection took off when Björn's doctoral student Jan-Erling Bäckvall joined my group. Scandinavian co-workers helping Jan and me celebrate a 1995 honorary doctorate from Stockholm's Royal Institute of Technology are: (r to l) Torbjörn Frejd, Pher Andersson and Per-Ola Norrby.

With no children at home any more, dogs are, once again, our companions of choice – for play, for exercise and for hanging out with in bed. I haven't gone fishing for probably over thirty years, but the ocean is still programmed into me like the birth stream of a salmon. One of the glories of moving to Scripps in 1990 has been seeing the Pacific Ocean every day, and, when its temperature reaches 70° (July or August), swimming in it every day as well. In windy New England I wind-surfed and we loved our little catamaran; San Diego's sail-less ocean vistas still seem weird.

My most important award, the greatest honor I've ever received, and the grandest and most memorable occasions I've ever witnessed, are, of course, benefits of sharing the 2001 Nobel Prize in Chemistry. But other honors have peerless characteristics as well, notably:

The heaviest object in our bank deposit box is the 1995 King Faisal Prize for Science medal; the most beautiful one is the 1988 Prelog Medal from the Swiss Federal Institute of Technology (ETH). Its exquisite relief rendering of Old Vlado's profile rivals the most beautiful portraits found on coins from antiquity, and the gold has a gorgeous, pliant, velvety warmth that has to be seen to be believed (by appointment only). A friend once asked, quite appropriately, if the portrait was of Alexander the Great.

Three unique objects, and I treasure each one, celebrate the day in 1995 when I received an honorary doctorate from Stockholm's Royal Institute of Technology. My only top hat, frequently brought out for guests to admire, bears the Institute's seal; my only ring, always admired when I wear it, is a heavy gold band surrounded by a garland of leaves and acorns in deep relief. These two I share with all the Institute's doctoral recipients, but I also

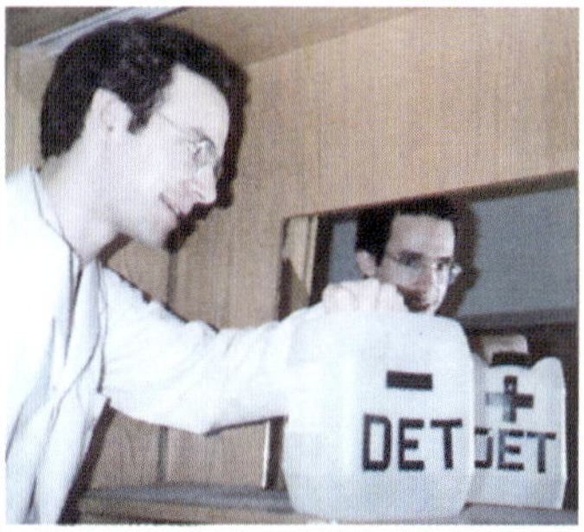

M.G. Finn, now a Scripps Research Institute professor and colleague, unraveled the mechanism of the Asymmetric Epoxidation, assisted by Scott Woodard and Steve Pedersen. This 1984 photo of M. G. showed how, unlike Nature's protein catalysts, the means to make either right- and left-handed catalytic Asymmetric Epoxidation products were both available off-the-shelf and in large quantities.

Stockholm guests in December 2001 included (right) Dartmouth College Prof. Thomas A. Spencer, my undergraduate research supervisor, and Stanford University Emeritus Prof. Eugene E. van Tamelen, who was both Tom Spencer's and my doctoral research supervisor.

have a large brass cannon shell casing, fired during the cannon salute that accompanied the conferring of the degree and the ceremonial placing of the hat on my head. The shell sits on my desk at home.

Only one award commemoration of mine is lettered on real vellum, and it is the largest one, as well: in both English and Hebrew the 1998 Harvey

In 1996 the Sharpless Group at Scripps was investigating the recently discovered Asymmetric Aminohydroxylation and continuing to develop the Asymmetric Dihydroxylation: (front, l to r) Hubert Angert, Andreas Gypser, Administrative Assistant Laureen Stav, KBS, Paul Richardson; (middle, l to r) Pui Tong "Joe" Ho, Bryan Hauger, Erik Rubin, SonBinh Nguyen, Ahmad Houri, Guigen Li, Dominique Michel; (rear, l to r) Peter Sennhenn, Milan Bruncko, David Nirschl, Joachim Rudolph, Cornelis Vlaar, Hans Henniges, Gunther Schlingloff, Han-Ting Chang.

Tsutomu Katsuki (front, 3rd from left), discoverer of the catalytic asymmetric epoxidation and some Sharpless Group contemporaries at Stanford University, 1980: (front, l to r) Yasuhiro Yamada, Judith Zilanovsky, Tsutomu Katsuki, Katherine Takaki; (standing, l to r) Victor Martin, Carl Behrens, Bryant Rossiter, Scott Woodard.

Alice (daughter Hannah, age 6) contemplates the looking-glass symmetry of the hexose sugars. Mirror-image symmetry fascinated both Lewis Carroll and his contemporary Emil Fischer, the German chemist who determined the sugars' structures. This 1985 photo celebrates the syntheses of the hexoses by the Sharpless and Masamune groups at MIT, and was taken by photographer Tito Simboli (wife of 2000 Nobel Laureate in Economics Daniel McFadden).

Science and Technology Prize of the Israel Institute of Technology, Haifa's Technion, is proclaimed.

In the category "news received most delightfully ", the winner is... an April, 1984, telephone call Jan took in a Jacksonville, Florida, hotel room all five of us were sharing while I attended a meeting. On the line from Washington, D. C. was my MIT colleague George Büchi, the most generous and thoughtful colleague I have ever known. George said he was calling because it was announced just minutes before that I had been elected to membership in the National Academy of Sciences. When Jan replied that she'd pass the message on, George said, no, she must go immediately to the meeting room and give me the message. Our children were too young to be left alone, especially since the meeting room was next to a swimming pool. I was giving my talk when I saw Jan and the three children, and all of them on tiptoes, enter the room and move along the back in the semi-darkness. She was looking for a familiar face, and she whispered the message when she found one. I stopped talking as Jan's informant walked to the front of the room and asked for the lights to be brought up so he could make an important announcement. Why the audience was so enthusiastic I wasn't quite sure. Not only did I not know I'd been nominated, I didn't even know one had to be nominated. I thought the National Academy was something like a high-level appointed government advisory committee. Learning otherwise was a wonderful surprise.

Inaugural events always have special significance and vivid memories; these "firsts" mean a lot to me:

Receiving the first Paul Janssen Prize for Creativity in Organic Synthesis, presented by HRH Prince (now King) Albert of Belgium. Security forces were everywhere that day in 1986, and I asked Prince Albert if having to travel with such a large group wasn't inconvenient. No, he replied, all those soldiers were required because I was an American – he didn't need them.

Being Texas A & M's first Barton Lecturer, 1997. Nothing is dearer me than having been selected by Sir Derek, my career-long scientific role model and mentor, to deliver the first edition of a lectureship endowed in his honor, the only Barton Lecture to take place before his death in 1998.

Launching the University of Sydney's Conforth Foundation for Chemistry (which honors both Rita and Kappa) with the inaugural Cornforth lecture in 2002. Like Sir Derek, Sir John is one of our gods; I stand awed at having participated in these events honoring them.

And, finally, if I had a crown, its jewels would be the 75-or-so former Sharpless Group members who are research professors. The training received in the group is neither predictable nor quantifiable; likewise, it is not intended to produce a product that, for example, industry wants. Since nothing original is intentionally discovered by scientists who cannot tolerate (indeed, they should welcome it) a high degree of uncertainty, group membership does not guarantee results. Because of the nature of our research, however, group members preselect themselves and possess a remarkably high degree of independence of thought as well as scientific motives tilted toward discovery, not reward. As a group, they hold superior standards for judging the significance of research, and I share with all them all of the glory that is a Nobel Prize.

SEARCHING FOR NEW REACTIVITY*

Nobel Lecture, December 8, 2001

by

K. Barry Sharpless

The Scripps Research Institute, 10550 N. Torrey Pines Rd, La Jolla, CA 02037, USA.

In 1938, three years before I was born, a live coelacanth was taken from the waters off the eastern coast of South Africa. Previously known only in the fossil record from some hundred million years ago, the coelacanth and the implications of its discovery remained big news for years, fueling an enthusiasm for "creatures" that persisted for decades. Those of us born in the Forties grew up on photos of eminent scientists setting off on expeditions, their sunburnt faces dwarfed by mountain explorer's garb, or making thumbs-up signs as they entered the water in scuba gear. We shared their confident expectation that the Loch Ness Monster, Sasquatch, the Yeti – even a dinosaur – soon would be taken alive.

I grew up loving the sea and loving fishing in particular, but unlike most fishermen I cared less for the size or quantity of the catch than for its rarity. Nothing could be more exciting than pulling (if not this time, surely the next!) a mysterious and hitherto unknown creature from the water. As a kid, I passionately wanted to be one who caught the next coelacanth, the first to see something that was beyond reasoning, even beyond imagining.

Opening a Nobel Lecture with a fishing expedition may seem frivolous, even indecorous, but I assure you no disrespect is intended. These are the circumstances that shaped my professional life: my first laboratory was New Jersey's Manasquan River, whose astonishingly rich variety addicted me to discovery; a few years later, when I was as comfortable at sea as I'd been on the river, my laboratory became the Atlantic Ocean; later when I started doing chemistry, I did it the way I fished – for the excitement, the discovery, the adventure, for going after the most elusive catch imaginable in uncharted seas.

Chemists usually write about their chemical careers in terms of the different areas and the discrete projects in those areas on which they have worked. Essentially all my chemical investigations, however, are in only one area, and I tend to view my research not with respect to projects, but with respect to where I've been driven by two passions which I acquired in graduate school: I am passionate about the Periodic Table (and selenium, titanium and osmium are absolutely thrilling), and I am passionate about catalysis.

* Adapted with the permission of the editors from "Coelacanth and Catalysis": K. B. Sharpless, *Tetrahedron* **1994**, 50, 4235.

What the ocean was to the child, the Periodic Table is to the chemist; new catalytic reactivity is, of course, my personal coelacanth.

Even though I grew up in Philadelphia, if someone asks me where I'm from, I usually say "the Jersey Shore," because that's where my family spent summers, as well as many weekends and holidays, with my father joining us whenever he could. My father had a flourishing one-man general surgery practice which meant he was perpetually on call. With him at home so little and practically guaranteed to be called away when he was, my mother liked being near family and friends at the Shore, where her parents had settled and established a fishery after emigrating from Norway. When I was a baby, my parents bought land on a bluff overlooking the Manasquan River about four miles up from where it enters the Atlantic.

Like many scientists I was a very shy child, happier and more confident on my own, and my interest was totally absorbed by the river. In those days, the incoming tide transformed our part of the river from a channel flanked by broad mud flats to a quarter-mile basin that exploded with life of myriad variety – about a dozen kinds of fish big enough to make it to the dinner table, plus blue crab, eel, and a bounty of fry and fingerlings that would graduate downstream to the ocean. I was obsessed with finding and observing everything that lived in the river and knowing everyone who worked on it.

My most delicious childhood memory is the excitement I experienced at the instant of awakening almost every summer morning. The sound I associate with that feeling is the distant whine of my first scientific mentor's outboard motor. That was my wake-up call, and within minutes I was at river's edge, waiting in the pre-dawn stillness for Elmer Havens and his father Ollie to make their way across the river from Herbertsville to pick me up to "help" them seine for crabs. Amused by my regularly walking along the bank to watch them haul their seine, Elmer eventually installed me in the boat, which he used for transportation as well as for steadying himself as he dragged the seine's deep-water end. Ollie walked one end of the seine along the shore, alarming the crabs gathered at the river's edge, and frightening them toward deeper water and so into the net's pocket. Chest deep in water and mud, Elmer walked parallel to his father, one arm clasping the seine, the other hooked over the boat's gunwale. Elmer and I, our heads close together, would speculate about the catch, taking into account all the variables – the weather, the season, the tide. Every hundred yards or so, Elmer doubled ahead toward the shore to draw the purse. I liked it best if a big eel or a snapping turtle got caught up in the net, making the water boil and the net flop into the air. I always hoped we'd catch something new.

I had a little dinghy, and my realm of exploration expanded in direct proportion to my rowing ability. But the same tide that created this abundant estuary also was my nemesis, forever stranding me upriver in the narrows or perhaps at Chapman's Boat Yard, a mile downstream and on the opposite bank. Since my parents couldn't keep me off the water, they opted for increasing the likelihood of my getting home unaided by giving me a boat with an outboard. It wasn't long before I went down river all the way to the inlet (absolutely forbidden, of course), and, soon after, the prospect of new creatures to pull from the water lured me out through the rock jetty and into the ocean; at the time I was only seven or perhaps eight years old.

By the time I was ten, I ran crab and eel traps and supplied everyone we knew with fish as well; at fourteen I started working during the summer as the first (and only) mate on a charter boat. My parents allowed me to go to sea when I was so young and small even for my age because I was offered a job on a relative's boat – little did my parents or I know that Uncle Dink, a cousin actually, offered me the job so he wouldn't have to pay a "full-sized" helper. I so wanted to keep working on the boats that it was years before I dared tell my parents what went on aboard the Teepee, like how the Coast Guard refused assistance to Dink because his boat was in chronic disrepair. (Consequently, some of our adventures at sea were memorable indeed – grappling hooks and guns have their place in the canon – and I mention this trove of Uncle Dink stories because for years my MIT colleagues begged me to tell them over and over again.)

On a charter boat the captain pilots the ship and finds the fish the customers reel in. Meanwhile, the mate is over the boat like a dervish, skillfully arraying the water with fishly temptations – adjusting outriggers, finding the perfect combination of lure or bait and tackle, always mindful of the action on nearby boats competing for the same fish.* Since my friends were all mates we naturally agreed that enticing fish to bite was the greatest challenge, but I alone felt that getting the strike was the most fun, even more exciting than landing the fish. I worked as a mate almost daily every summer, right up until the day before I set out from New Jersey headed toward the biggest ocean and graduate school at Stanford University.

That was in 1963. In the spring of that year my inspiring Dartmouth College chemistry professor and first research director, Tom Spencer, talked me into delaying entering medical school to try a year of graduate school. He sent me to Stanford specifically to work for E. E. van Tamelen, Tom's own mentor at Wisconsin. The appeal of fishing was such that Tom, to my later regret, never succeeded in getting me to spend any summers working in his lab. In fact even in graduate school I expressed my ambivalence by continuing to fantasize about finding a boat out of Manasquan to skipper and by failing –

*This diversion into fishing-as-metaphor-for-research could go on for pages: consider how when a boat was hooking tuna – the catch of choice – word spread by radio and the competition converged from every compass point. The hot boat's captain greeted this acknowledgment of his success with some anxiety: while he liked setting the other captains' agendas and pleasurably speculating that the parties on the other boats were considering chartering him next time, the secrets of his success nonetheless required protection, so trolling speeds were lowered to sink the lures and prevent rubberneckers from identifying them, and red herrings (literally, on occasion!) were casually displayed on the fish box. Isaak Walton and John Hersey devoted whole books to this metaphor, so indulge me for a few more sentences. The handy process vs. product dichotomy that applies so neatly to much of human endeavor illuminates this fisherman-chemist comparison, too. Conventional wisdom places fly-fishing at the "process" end of the scale, while a "product" fisherman uses sonar to find a school before he bothers to get his line wet. Process person though I am, only the Manasquan River ran through my fishing days: trolling for the unknown always had more appeal than hooking a trout I already knew was there.

this did not please v.T. – to do the simple paperwork required to renew my NSF predoctoral fellowship.

However, toward the end of my first year at Stanford, a serendipitous misunderstanding catalyzed the complete transfer of my passion (some would say my monomania) from one great science to another, from fishing to chemistry. Before leaving for a lengthy European visiting professorship, v.T. sent me to the library to look for reactive inorganic species that might produce interesting transformations of organic compounds. My first projects with v.T. were selective oxidation of polyolefins and titanium-mediated deoxygenative coupling of alcohols, and I was already primed to appreciate useful chemistry employing "strange" elements after selecting the Wittig Reaction from a list of suggested topics for my student seminar. The Wittig Reaction really engaged my enthusiasm, and I ingenuously concluded that finding new reactions other chemists could use looked like a lot of fun.

In any event, upon v.T.'s return, I discovered he had not intended for me to spend all those months immersed in the literature. While I had no research results to report, I did have a notebook filled with ideas and an eagerness to drop my line throughout the vastness of the Periodic Table. I don't think I've gone fishing in the literal sense a dozen times since then!

From van Tamelen, a Gilbert Stork protégé, I inherited enthusiastic disdain for "safe" problems, deep admiration for traditional multistep organic synthesis, and awe before selective biological catalysis: studying the squalene oxide/lanosterol cyclase enzyme left me impressed by enzymic selectivity but depressed by the difficulty of using enzymes for synthetic transformations. After getting a double dose of him in the classroom, Derek Barton became my model. At Dartmouth Tom Spencer taught a course on conformational analysis based on one he took at Wisconsin from William S. Johnson (Tom's uncle, in fact), then I experienced the original at Stanford.* Being wet behind the ears, I took conformational analysis for granted: it was Sir Derek's search for new reactivity that electrified me. A postdoc with Jim Collman (the only person, I concede, who gets more excited about chemistry than I do) ignited my interest in using simple metal complexes to develop catalysts (in the Collman lab, incidentally, I had the privilege of many hours at the blackboard with labmate Bob Grubbs). Then, before taking up my job at MIT, a postdoc with Konrad Bloch confirmed my hunch that impatience rendered me incompetent around enzymes. Konrad graciously let me start working on my own ideas when his proved much too frustrating for me.

One other part of my background seems to have contributed to my chemistry. The first American Sharpless ("Sharples" then) came to Pennsylvania in the Seventeenth Century, not long after William Penn. My father was a prac-

* When teaching MIT undergraduates I always said, "The lights came on with conformational analysis," without thinking where I picked up the phrase, but now I know: the previous Tetrahedron Prize article states, "Just as chemists of the Robinson generation worked without concern for stereochemical factors so we, in the early days, were working in ignorance of conformational considerations until Derek Barton showed us the light in 1950." The author is, of course, Bill Johnson (see reference 1).

ticing Quaker only as a child, but the values in our home were Quaker values, and I was educated in a Quaker school. The Quakers encourage modesty, thrift, initiative, and enterprise, but the greatest good is being a responsible member of the community – being useful. "Elegant" and "clever" were the chemical accolades of choice when I started doing research, just as "novel" is high praise now. Perhaps the Quakers are responsible for me valuing "useful" most.

So that is my background as a chemist. I've been accused of going too far when I speculate that chirality fascinates me because I handled my umbilical cord in utero, but I'm quite sincere in proposing that the extraordinary training I received as a young chemist transformed an existing passion for discovering the unknown into the search for new reactivity, and that Quaker utilitarianism made the selective oxidation of olefins so appealing.

With respect to chemical reactions, "useful" implies wide scope, simplicity to run, and an essential transformation of readily available starting materials. Clearly, if useful new reactivity is the goal, investigating the transformations chemists rely on is the obvious strategy. The processes for the selective oxidation of olefins have long been among the most useful tools for day-to-day organic synthesis because of these appealing characteristics of olefins:

they are among the cheapest functionalized organic starting materials,
they can be carried "hidden" through conventional acid/base-catalyzed transformations, then "revealed" at will by adding heteroatoms through selective oxidations,
most simple olefins are prochiral, providing a prominent portal to the chiral world.

The trisubstituted olefin geraniol, in addition to being one of my favorite smells, provides an excellent case study both for laying out the challenges of selective olefin oxidation as well as for noting some benchmarks in meeting those challenges.

As shown in Scheme 1, geraniol (**1**) has two trisubstituted olefinic units, one of which has a hydroxyl in the allylic position. Four monoepoxides are possible: making either racemic **2** or racemic **3** requires regio- (or chemo-) selectivity, while making each of the individual enantiomers requires enantioselectivity. When Henbest showed that the electronic deactivation by the oxygen substituent at C-1 causes peracids to prefer the 6,7-double bond (especially on the ester derivatives), making racemic 3 became possible.[2] When I started doing research in the Sixties, neither racemic 2 nor any of the enantiomers could be synthesized directly. Solving the other half of the regioselectivity problem was an obvious challenge, but enantioselectivity was considered well-nigh impossible.

In 1973, Bob Michaelson cracked the other half of the regioselectivity problem presented by geraniol.[3] Since early-transition-metal-catalyzed epoxidations with alkyl hydroperoxides proved highly selective for the 2,3-position, racemic **2** could be prepared as well.

In 1980, Tsutomu Katsuki discovered the titanium-catalyzed asymmetric

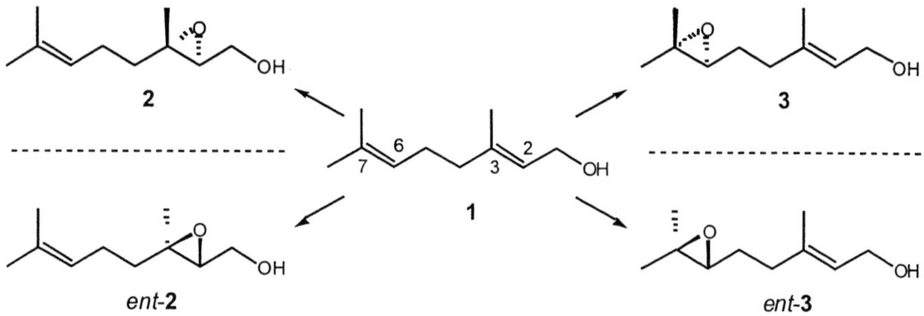

Scheme 1. Regio- and enantioselective monoepoxidations of geraniol.

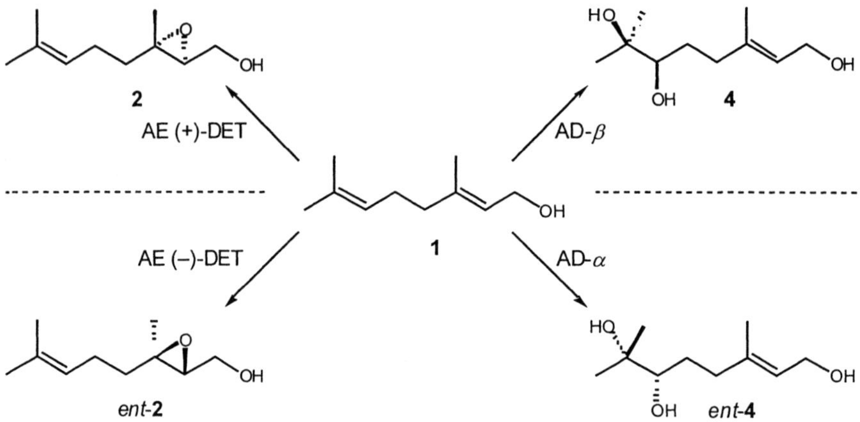

Scheme 2. Asymmetric epoxidation and dihydroxylation reactions of geraniol

epoxidation (AE); the enantioselective oxidation of olefins bearing allylic hydroxyl groups made it possible to make either **2** or *ent*-**2** thus solving one side of the enantioselectivity problem.[4]

The osmium-catalyzed asymmetric dihydroxylation (AD), discovered in 1987, subsequently was improved to the point that either **3** or *ent*-**3** could be made by way of the diol, an indirect solution to enantioselective epoxidation at the 6,7-position (Scheme 2).[5]

In 1990, came the breakthrough introduction of enantioselectivity into existing manganese salen ligand catalysts for the epoxidation of isolated-olefins.[6] Developed independently by the groups of Jacobsen[6a, 10] and Katsuki[6b], these epoxidation catalysts work best on only a few of the six olefin-substitution classes. Nonetheless, their very existence is tantalizing, encouraging the hope that a general, off-the-shelf solution will be found for the direct asymmetric epoxidation across the full range of isolated-olefin substitution patterns.

The greater generality of man-made catalysts, such as these, compared with enzymes was noted first by Knowles[7a,c] and Kagan.[7b] During the lean times in the first decade of my career, their pioneering development of man's first highly enantioselective catalysts (the L-dopa synthesis that came out of

Knowles' Monsanto lab was the asymmetric hydrogenation's first commercial application) sustained my faith that a catalyst for asymmetric oxidation could be found. Jack Halpern's mechanistic studies[7d] on asymmetric hydrogenation catalysis likewise inspired me. Several Japanese chemists, chief among them Ryoji Noyori,[7e] hugely extended both the scope and application of the asymmetric hydrogenation process[8].

This focused search has frustrated but never bored me even after so many years, and the geraniol paradigm illustrates why. My own investigations into the oxidation of olefins commenced at MIT in 1970, but, fittingly, I was back at Stanford on January 18, 1980, for Tsutomu Katsuki's dramatic discovery of the titanium-catalyzed asymmetric epoxidation.[4, 9a] Two years later the most scientifically stimulating and professionally gratifying collaboration of my career, the total syntheses of the eight L-hexoses with my MIT colleague Sat Masamune, capped the AE's discovery.[11] Previous articles[12] in a vein similar to this one describe that chemistry; understanding the AE's significance and putting that understanding to work are the purview here.

After the euphoria of completing the hexose syntheses, three years were spent developing, refining, and finding more applications for the AE. During this time I returned to the search for new reactivity, but it was clear that my random, scattershot attempts were going nowhere,* so I was grateful for the

* I have enormous admiration for colleagues who can keep multiple research projects alive and large groups humming, but the "monomania" that prevents me from being able to do that is my long suit as well, making it possible to concentrate – for years, actually – on a single topic. I know some chemists call my approach "intuitive," a term I've always thought underestimates the rigor that frames my method; perhaps "unstructured" or "contemplative" is more accurate. Many of my cohorts are quick and facile and can jump on a few interesting bits of data and start building tentative edifices that get taken apart and reassembled to suit new data. I, on the other hand, am ruminative: my training after all consisted of busily poking and perturbing the Manasquan River, a curriculum both urgent and leisurely, one that permitted exploration without assumptions and without the structure imposed by deadlines or competition or by knowing too little or too much. Since I was compelled by shyness to learn to do much on my own, there was (and is) no right or wrong way, only many ways, some more or less suited to a given endeavor. The discipline, nonetheless, is exacting: everything that can be observed should be observed, even if it is only recalled as the bland background from which the intriguing bits pop out like Venus in the evening sky. The goal is always finding something new, hopefully unimagined and, better still, hitherto unimaginable. When I became a bench- and desk-bound explorer the method stayed the same. I try to imagine away the packaging information arrives in, then let bits and pieces move around lazily, rather like objects tumbling slowly in zero gravity, but eventually, over time, exploring every possible relationship with other information that's previously arrived. Since joining the faculty of The Scripps Research Institute, I've discovered that ocean swimming and running on the beach provide an excellent medium for this kind of activity; however, in any climate the best catalyst is generous, stimulating conversation. This slow but endlessly fascinating method is like an exotic ritual courtship, full of displays of bright feathers or offerings of shiny metal or towers of sticks – what does it all, what does any of it mean? Enormous concentration is required to remember it all in a way that causes little sparks when certain conjunctions appear, making a connection with something noted previously, perhaps decades ago. Sadly, as I grow older, the connections become harder to summon up, so the sparks, though seeming as bright as ever, are less frequent. I describe this process at length because it's not the way most scientists approach their work, nor is it well suited to the demands of funding agencies that are railroaded into answering questions posed for political rather than scientific reasons, nor to the needs of graduate students who require publications to compete for jobs. Academic chemistry is much harder now, and I'm glad I was born when I was.

opportunity to spend the first three months of 1987 as a Sherman Fairchild Scholar at Caltech.

Many universities and institutions have handsome Fairchild buildings, but Caltech, ever the bastion of collegiality and camaraderie, used its Fairchild grant to endow a program that brings scientists from many fields to be housed graciously in the sunshine for as long as a year. Since my research group's investigation of the AE had reached the point of diminishing returns, I left for Pasadena hoping to renew my mission.

I love reading journals, and I love mountains, so the Caltech library with its panoramic view of Mt. Wilson became my thinking place of choice. Every day Mt. Wilson offered new vistas, especially on those occasions when snow fell during the night. One morning the mountain was completely cloaked (the first time a freezing temperature was recorded in downtown LA, I recall), and the melting snow receded at such a clip I was sure I saw it happening. Mt. Wilson was the perfect backdrop for bringing my own big picture back into focus, and I returned to MIT eager to renew my search for new reactivity. Meditating on the AE yielded this lesson to guide that search:

> ligand-accelerated catalysis (the significance of which is documented in M. G. Finn's fine MIT thesis on the mechanism of the AE[13]), is crucial to the AE and not merely a feature of it; despite its rarity this phenomenon might be the agent for uncovering more catalytic processes.

Of course, the first and best-known example of ligand acceleration is found in Criegee's papers from the Thirties.[14] He observed that pyridine accelerates the reaction in his classic study of osmium tetroxide and olefins. Ironically, the lesson from the AE was directing me back toward Criegee, whose discoveries in olefin oxidation and osmylation were, in large measure, the jumping off point for my own research career.

I first looked into Criegee's process shortly after becoming an assistant professor at MIT. My attraction to the reaction of OsO_4 with olefins was inevitable. Osmium tetroxide not only accomplishes an important synthetic transformation, but it does so with a scope and reliability unique among reactions used for organic synthesis. It reacts only with olefins and it reacts with all olefins (slight poetic license here). Even R. B. Woodward valued Criegee's stoichiometric transformation so much he was willing to use 100 g of OsO_4 in one shot. Osmium's expense was not compatible with "useful," however and, since the existing catalytic variants were not very effective, I started searching for a reliable catalytic method. In 1975 Kagayasu Akashi found a good process for us based on a hydroperoxide as oxidant, *tertiary*-butyl hydroperoxide (TBHP) [15], but the brass ring was ultimately captured that same year with the publication of the famous Upjohn process based on *N*-methyl morpholine-*N*-oxide (NMO).[16]

Throughout the rest of the Seventies osmium remained our primary tool for looking for new reactivity: we discovered that imido osmium(VIII) species effected stoichiometric *cis*-oxyamination of olefins in direct analogy to the *cis*-dihydroxylation of olefins by osmium tetroxide; even more effective catalytic versions of those transformations came shortly thereafter.

In 1977, I left MIT, where I had been a contented member of a wonderful chemistry faculty since 1970, for Stanford University, where I previously spent six contented years as a graduate student and postdoc, surrounded by a wonderful chemistry faculty. I never made the transition back to contentment at Stanford, probably because my research wasn't churning up much. This frustrated me and scared off potential graduate students who wanted publications, not a fishing expedition. In addition, at Stanford I remained awed by a faculty I worshiped when a graduate student, and I lacked the confidence to stand firm on issues, particularly faculty appointments, that meant a lot to me. In 1979, at about the same time I made the decision to return to MIT, Steve Hentges, who worked in our well-developed osmium imido area and already had the material for a good Ph.D. thesis in hand, decided to take on one more project before writing up.

The notion of an asymmetric ligand for osmium tetroxide had been knocking around the lab for years, and Steve first approached the idea by making several pyridines with chiral substituents at the 2-position; these gave diols with essentially 0% *ee*![17] Pyridine is only a modest ligand for osmium tetroxide, and, as we discovered, any *ortho* substituent is lethal to binding. But since William Griffith at Imperial College had shown that quinuclidine binds much more strongly to OsO_4, I suggested trying the cinchona alkaloids, essentially substituted quinuclidines.[18] (Many chemists have expressed surprise at how quickly we arrived at what is now the best ligand framework for the AD: anyone with a natural products background and who is also a fan of Hans Wynberg's chemistry recognizes the cinchona alkaloids as the obvious next step.) The results were spectacular, even without taking into account a measurement error (discovered years later) that caused most of the *ee*'s to be underreported by 5 to 15%![17]

Steve had a dramatic story to cap his thesis work, so he started writing; my attention was taken up by the decision to return to MIT. Then, a couple of months later, Katsuki discovered an asymmetric process with ingredients so cheap it made working with osmium look like Rolls-Royce chemistry. Although the AE was only weakly catalytic in the early days,[19] its uniformly high *ee*'s and nontoxic, inexpensive reagents were enough to completely divert our attention from its promising but stoichiometric predecessor, the OsO_4/cinchona asymmetric dihydroxylation.

The preceding paragraph has no doubt failed to deflect your attention from the obvious question: Why didn't I try the Hentges ligands in the Upjohn system in 1979? Indeed, why did I propose the experiment in my NIH grant renewal in January, 1984, but not follow up on it? "As for the ligand," I wrote in the proposal, "it is probably best to stay with the cinchona derivatives because the quinuclidine moiety is the best ligand we know of for Os(VIII) complexes. The substrate will be stilbene... the osmium catalyst will be recycled using an amine *N*-oxide. Ideally, both the osmium and the chiral alkaloid could be used in catalytic quantities. A successful system of this type could be of great practical importance."

Instead of poking and perturbing, the Jersey Shore School of Thinking's

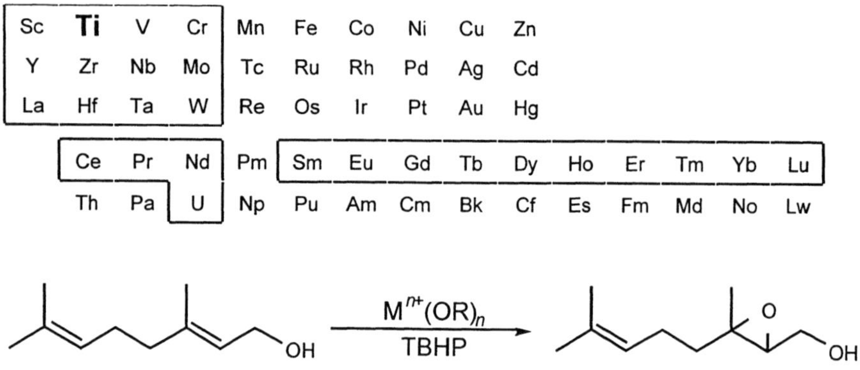

1 *rac*-**2**

Figure 1. Metals Catalyzing the Epoxidation of Allylic Alcohols by TBHP. Adding tartrate ligand always affects reactivity: the titanium system is accelerated.

cardinal rule, I stuck with the odds logic suggested: ligands accelerate the reaction of OsO_4 with olefins, but they also bind avidly to the resulting osmate ester, lethally effecting catalyst turnover. This ability of ligands such as pyridine and quinuclidine to kill turnover in catalytic osmylation systems had been often observed in my laboratory. What I did not, nor could not, anticipate is the perfect balance cinchona alkaloids achieve in ligating ability, binding well enough to accelerate the key step, but weakly enough to slip off allowing the hydrolysis/reoxidation steps of the catalytic cycle to proceed. At the time, however, the precedents seemed clear, so the AD languished until 1987.

Unraveling the mechanism of the AE was largely the work of M. G. Finn.[13] His persistent exploration during the early- to mid-eighties of the AE's titanium-tartrate catalyst system exposed a complex mixture of species in dynamic equilibrium with one other.[20] M. G. discovered the main species $[Ti(DIPT)(O\text{-}i\text{-}Pr)_2]_2$ is substantially more active than the many other species present (significantly, it is five to ten times more active than $Ti(OR)_4$, a catalyst for the formation of racemic epoxy alcohol) and this rate advantage funnels catalysis through the appropriate tartrate-bearing species.

If the tartrate-induced acceleration of the titanium-catalyzed epoxidation reaction came as a surprise, investigating that phenomenon brought even more surprising results. We ultimately found twenty-four metals other than Ti that catalyze the epoxidation of allylic alcohols by TBHP (Figure 1), but all these systems were strongly inhibited or killed by adding tartrate![21] Ligand-decelerated catalysis was clearly the rule while ligand acceleration was the extraordinarily valuable exception.

Shortly before I left for Caltech, Chris Burns, encouraged by Pui Tong Ho, presciently lobbied to resurrect the OsO_4/cinchona asymmetric dihydroxylation, and, without any encouragement from me, I must admit, he embarked on the synthesis of a stoichiometric C3-symmetric ligand for the AD.[22] A few months later, I, too, was recommitted to osmium, and when Bill Mungall and Georg Schröder reexamined the work from 1979 they uncovered *ee*'s even better than previously reported. Meanwhile Eric Jacobsen attacked the problem from the mechanistic side, discovering that the ligand-dependent rate accelerations could be enormous.[23]

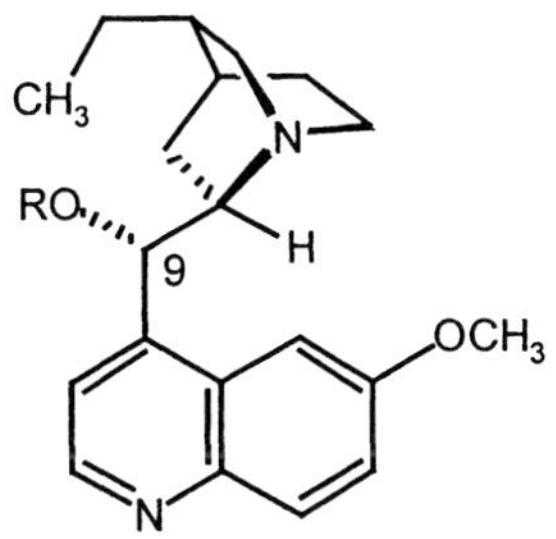

dihydroquinidine derivatives

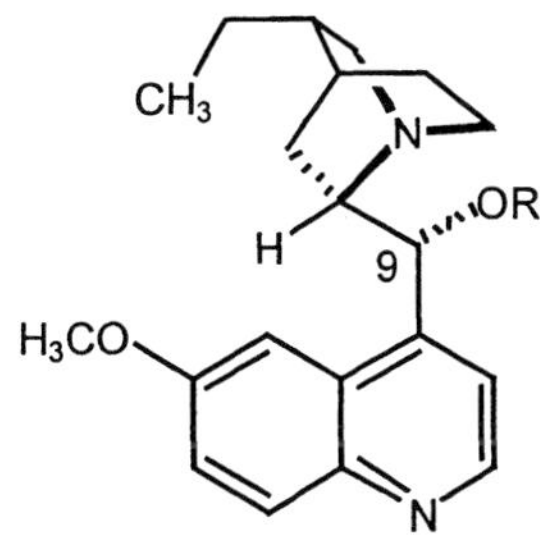

dihydroquinine derivatives

With these very encouraging results on the stoichiometric reaction just in, Istvan Markó joined the project. I was travelling at the time, and on his own initiative, unaware of the NIH proposal, he combined Hentges' system[17] with the reliable Upjohn NMO-based catalytic osmylation system,[16] immediately getting results indicating the reaction was catalytic.[24] However, unlike the dramatic "Eureka!" that accompanied the discovery of the AE, cautious optimism was the response to the catalytic AD and its initially modest *ee*'s. Now, however, after years of research since Markó's first experiments in October of 1987, the AD's utility rivals and often surpasses the AE's.[9]

Unlike the AE, for which Katsuki's initial tartrate ester ligands have yet to be eclipsed, the ligands for the AD have evolved substantially in effectiveness and scope through substitution at the C-9 hydroxyl.

The simple ester derivatives (e.g. the acetate and *para*-chlorobenzoate esters) gave way in 1990 and 1991 to aryl ether derivatives, first proposed by Yun Gao during a late night group meeting to address the mechanistic question of a possible ligating role of the ester carbonyl. Brent Blackburn made the phenyl ether which, to our surprise, gave good *ee*'s, but was too hard to make to be competitive with the then dominant *para*-chlorobenzoate (CLB) ligand.

Almost a year later Declan Gilheany correctly predicted that aryl ethers should be better for aliphatic olefins than the CLB ligand,[25] and these results laid the foundation for a steady expansion of this ligand class, culminating in the phenanthryl ether ligand.[26] Another big jump in effectiveness came with the dimeric alkaloid ligands having a phthalazine core, first made by Jens Hartung in 1990.[27] Along with the pyrimidine ligands[28] whose development they inspired, they remain the best general ligands for the AD reaction.

The search for better ligands was paralleled by advances in catalyst turnover efficiency:

John Wai found both the second-catalytic-cycle problem and its partial remedy, slow addition of the olefin;[29]

Since ferricyanide in *tert*-butanol/water provides an excellent two-phase system for catalytic osmylation,[30] Hoi-Lun Kwong applied it to the AD, solving the second-cycle problem and the need for slow addition;[31]

Willi Amberg found that adding organic sulfonamides greatly facilitates the rate of catalyst turnover for olefins whose osmate esters resist hydrolysis.[27]

As the practicality (it has been scaled up to run in 4000 liter reactors with no ill effects on yield or *ee*[32]) and scope of the AD process grew, so did the pressure to understand the origin of its enantioselectivity. Mechanistic studies dating from the early Seventies by Alan Teranishi and Jan Bäckvall[33] were rekindled by Eric Jacobsen in 1987 and continued into the mid-90's.[34]

While a complete and general solution to the geraniol paradigm's final challenge is clearly within reach, comparing selectivity at the bench with selectivity in living systems remains striking. For example, the squalene monooxygenase in our livers unerringly deposits a single oxygen atom on the squalene molecule and, in so doing, further chooses only the si-enantioface of the terminal double bond (Scheme 3).[35] On the other hand, the attempted AD of a single double bond of squalene does give the terminal diol in 96% *ee*. The preference for the terminal double bond is slight, however, and internal diols as well as tetraols also can be isolated from the reaction.[36] Thus, while the AD catalyst cannot match the exquisite selectivity of the enzymic system, this very inability to discriminate between the six trisubstituted double bonds of squalene allows the exhaustive AD of squalene (Scheme 4) in an overall yield of 79.8% for the AD-β reaction.[37]

Serial multistep reactions such as these are generally stymied by Bob Ireland's "arithmetic demon" – the geometric fall in yield in sequential chemical reactions. The AD of each double bond is one step in a procession of six dihydroxylations, each with a chemical and an optical yield, twelve yields in all. Thus the average yield of each step is $(0.798)^{1/12}$ or 98%, translating to 98% for each chemical yield, 96% *ee* for the single enantioselective reaction and 96% *de* for each of the five diastereoselective reactions. The high yield of a single enantiomer from the multiple hydroxylation events required to com-

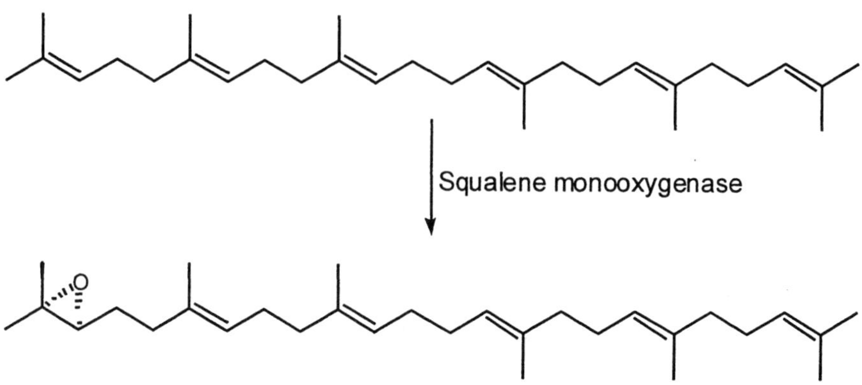

Scheme 3. Enzymatic epoxidation of squalene.

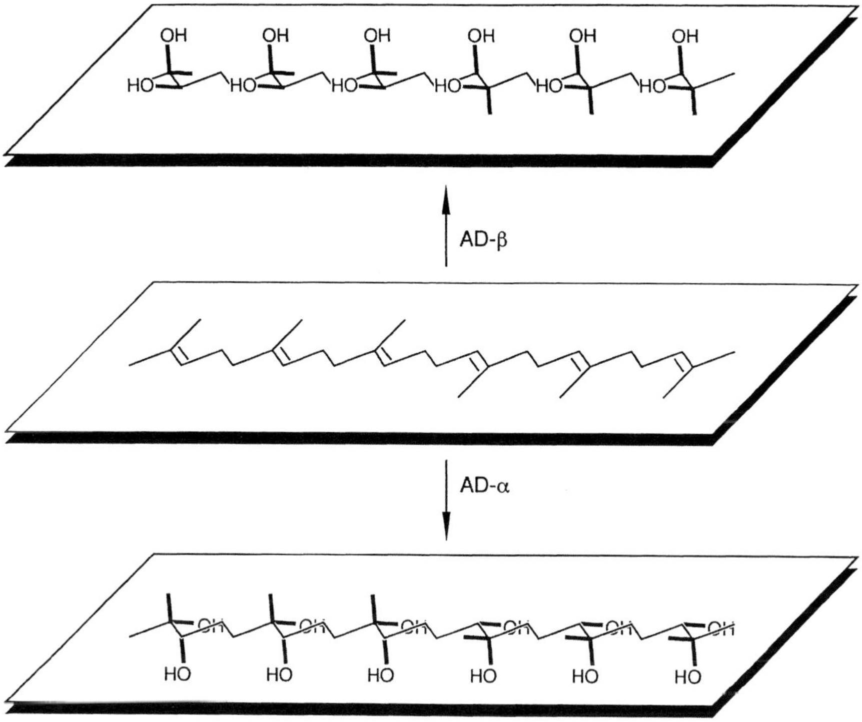

Scheme 4. Exhaustive, stereoselective dihydroxylations of squalene.

pletely oxidize squalene reflects the reliability and selectivity of the AD process. Joel Hawkins' Berkeley lab kinetically resolved the chiral fullerene C^{76} resulting in the first enantiomerically pure allotrope of carbon, the AD's most intriguing use to date.[38]

My decision, nearly twenty-five years ago, to study the selective oxidation of olefins produced an unexpected bonus, one that gave me an opportunity to investigate uncharted territory on a scale that is more associated with the previous half-century than with our own. Selenium, titanium/alkyl peroxides, and osmium, my three most successful olefin oxidation catalysts, all had phobias associated with them, with the result that much of their chemistry remained *terra incognito*. Selenium and osmium were considered highly toxic, and the peroxide oxidants used with titanium had a nasty reputation. Rarely did I find myself in another chemist's territory; likewise, few wanted to cast a line in mine.

Tracking these elements offers a rather curious way to view my research. Figure 2a plots the time course of their dominance (as measured by publications for want of a more qualitative ruler) during the years 1970–1993. Selenium came first, flourished, then ended abruptly. Osmium research came next, co-existing with selenium until both were eclipsed by titanium, the descendant of molybdenum and vanadium. Osmium made a strong comeback, knocking off titanium.

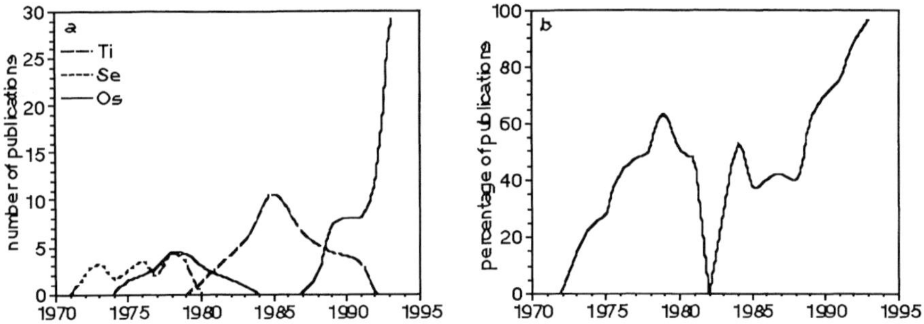

Figure 2. (a) Selenium, titanium, and osmium chemistry; note the osmium line's bimodality. (b) Growth of Catalysis in my laboratory.

Figure 2b, charting my research with respect to catalytic transformations, looks quite unlike Figure 2a, but relates directly to it. As my involvement with catalysis grew, the largely stoichiometric selenium reagents lost their appeal; titanium fell because the effectiveness of the titanium catalyst for the AE is modest, with about only twenty turnovers per titanium center before all activity is lost. Osmium, despite a bimodal presentation, was never actually out of the picture, merely quiescent until the discovery of the highly catalytic AD (it has been run to completion with as little as 1/50,000 of osmium catalyst).

In Figure 2b the only real defection from the steady growth of catalysis to dominion in my research was the 1982 trough caused by the hexose synthesis collaboration with Sat Masamune. Stepping out of the realm of catalysis is almost unimaginable to me now.

Because of its unique potential for channeling a reaction sequence along one of myriad possible pathways, selective catalysis lies at the heart of both pure and applied chemistry, not to mention life chemistry. In addition to the selectivity benefits of catalysis, the phenomenon of turnover (which equals amplification), implicit in the definition, highly leverages its potential impact. For all these reasons, catalysis was and continues to be the engine driving my research.

Nature's enzymes made it possible to imagine simpler asymmetric catalysts. What we found, however, was unimaginable on two scores: small, highly enantioselective catalysts that were not only not fettered by nature's "lock-and-key" modus operandi, but tolerant as well of substrates throughout the entire range of olefin substitution patterns. Now, going on four decades later, I am still plumbing the vastness of the Periodic Table in search of new catalytic reactivity. The unpredictability and rarity of what I seek are not deterrents since I am, after all, the product of optimistic times. There are other coelacanths to be found!

ACKNOWLEDGEMENT

Above all I thank and express my deep gratitude to my past and present coworkers at MIT, Stanford and The Scripps Research Institute. Many of you learned to tolerate my style of directing research (an oxymoron perhaps?); indeed, some of you flourished. Others were not well served, and to you I sincerely apologize. I'm exceedingly proud of the MIT undergraduates who got their feet wet in my lab and now hold leading academic and industrial positions: remember you got your opportunities because Tom Spencer gave me mine and I expect you to do the same. Mentioning Tom brings me back, as so many things do, to E. E. van Tamelen: the bright flashes of his career remain of the first magnitude and still inspire me. And finally, my scientific career would have been unthinkable without the constant support and counsel of my wife, best friend – and ghost writer – Jan.

I also thank the National Institute of General Medical Sciences, National Institutes of Health (GM-28384) and the National Science Foundation for their continuous financial support over the past twentyfive years, and, more recently, the W. M. Keck Foundation and the Skaggs Institute for Chemical Biology for helping to make possible my present tenure at The Scripps Research Institute in La Jolla.

REFERENCES AND NOTES

1. Johnson, W.S. *Tetrahedron* **1991**, 47, xi-1.
2. Henbest, H.B.; Wilson, R.A.L. *J. Chem. Soc.* **1957**, 1958.
3. Sharpless, K.B.; Michaelson, R. C. *J. Am. Chem. Soc.* **1973**, 95, 6136.
4. Katsuki, T.; Sharpless, K.B. *J. Am. Chem. Soc.* **1980**, 102, 5974.
5. (a) David Xu synthesized 4 and ent-4 in 1991 during his work on the AD mechanism, from which synthesis of 3 and ent-3 was implicit; two other groups reported this reaction sequence: (b) Corey, E.J.; Noe, M.C.; Shieh, W.C. *Tetrahedron Lett.* **1993**, 34, 5995–5998; (c) Vidari, G.; Dapiaggi, A; Zanoni, G.; Garlaschelli, L. *Tetrahedron Lett.* **1993**, 34, 6485–6488; (d) Vidari, G.; Giori, A.; Dapiaggi, A; Lanfranchi, G. *Tetrahedron Lett.* **1993**, 34, 6925–6928. Gerry Crispino's juvenile hormone iii synthesis demonstrates the same conversion: (e) Crispino, G.A.; Sharpless, K.B. *Synthesis* **1993**, 8, 777.
6. (a) Zhang, W.; Loebach, J.L.; Wilson, S.R.; Jacobsen, E.N. *J. Am. Chem. Soc.* **1990**, 112, 2801; (b) Irie, R.; Noda, K.; Ito, Y.; Matsumoto, N.; Katsuki, T. *Tetrahedron Lett.* **1990**, 31, 7345.
7. (a) Knowles, W.S.; Sabacky, M.J. *J. Chem. Soc. Chem. Commun.* **1968**, 1445; (b) Dang, T.P.; Kagan, H.B. *J. Chem. Soc. Chem. Commun.* **1971**, 481 (c) Knowles, W.S.; Sabacky, M.J.; Vineyard, B.D.; WeinkauV, D.J. *J. Am. Chem. Soc.* **1975**, 97, 2567–2568. (d) See his excellent review: Halpern, J. *Science* **1982**, 217, 401. (e) Miyashita, A.; Yasuda, A.; Takaya, H.; Toriumi, K.; Ito, T.; Souchi, T.; Noyori, R. *J. Am. Chem. Soc.* **1980**, 102, 7932–7934.
8. (a) Takaya, H.; Ohta, T. "Asymmetric Hydrogenation" In Catalytic Asymmetric Synthesis, Ojima, I. Ed.; VCH Publishers: New York, **1993;** pp. 1–39; (b) Kumobayashi, H.; Akutagawa, S.; Otsuka, S. *J. Am. Chem. Soc.* **1978**, 100, 3949; (c) Akutagawa, S.; Tani, K. "Asymmetric Isomerization of Allylamines" In Catalytic Asymmetric Synthesis; Ojima, I. Ed.; VCH Publishers, Inc.: New York, 1993; pp. 41–61. A history and review of the development of the asymmetric hydrogen migration of allylamines can be found in: (d) Otsuka, S.; Tani, K. *Synthesis* **1991**, 665–680.
9. For review chapters on both the AE and AD written largely by Roy Johnson despite being co-authored by both of us, see: (a) Johnson, R.A.; Sharpless, K.B. "Catalytic Asymmetric Epoxidation of Allylic Alcohols" In Catalytic Asymmetric Synthesis, Ojima, I. Ed.; VCH Publishers: New York, **1993**; pp. 101–158; (b) Johnson, R.A.; Sharpless, K.B. "Catalytic Asymmetric Dihydroxylation" In Catalytic Asymmetric Synthesis, Ojima, I. Ed.; VCH Publishers: New York, **1993**; pp. 227–272.
10. Jacobsen, E.N. "Asymmetric Catalytic Epoxidation of Unfunctionalized Olefins" In

Catalytic Asymmetric Synthesis, Ojima, I. Ed.; VCH Publishers: New York, **1993**; pp. 159–202.

11. (a) Ko, S.Y.; Lee, A.W.M.; Masamune, S.; Reed, L.A.; Sharpless, K.B.; Walker, F.J. *Science* **1983**, 220, 949; (b) Ko, S.Y.; Lee, A.W.M.; Masamune, S.; Reed, L.A.; Sharpless, K.B.; Walker, F.J. *Tetrahedron* **1990**, 46, 245.
12. (a) Sharpless, K.B. *CHEMTECH* **1985**, 692; (b) Sharpless, K.B. *Chem. Brit.* **1986**, 22, 1.
13. M.G. Finn's thesis work has appeared in two full papers: (a) Woodard, S.S.; Finn, M.G.; Sharpless, K.B. *J. Am. Chem. Soc.* **1991**, 113, 106; (b) Finn, M.G.; Sharpless, K.B. *J. Am. Chem. Soc.* **1991**, 113, 113.
14. (a) Criegee, R. *Justus Liebigs Ann. Chem.* **1936**, 522, 75; (b) Criegee, R. *Angew. Chem.* **1937**, 50, 153; (c) Criegee, R. *Angew. Chem.* **1938**, 51, 519; (d) Criegee, R.; Marchand, B.; Wannowias, H. *Justus Liebigs Ann. Chem.* **1942**, 550, 99.
15. Sharpless, K.B.; Akashi, K. *J. Am. Chem. Soc.* **1976**, 98, 1986.
16. van Rheenen, V.; Kelly, R.C.; Cha, P.Y. *Tetrahedron Lett.* **1976**, 1973.
17. Hentges, S.G.; Sharpless, K.B. *J. Am. Chem. Soc.* **1980**, 102, 4263.
18. (a) Cleare, M.J.; Hydes, P.C.; Griffith, W.P.; Wright, M.J. *J. Chem. Soc., Dalton Trans.* **1977**, 941; (b) Griffith, W.P.; Skapski, A.C.; Woode, K.A.; Wright, M.J. *Inorg. Chim. Acta* **1978**, 31, l413.
19. Five years later Bob Hanson made the simple but wonderful discovery that adding molecular sieves to the reaction dramatically increases catalyst turnover so that the AE process became truly catalytic: Hanson, R.M.; Sharpless, K.B. J. *Org. Chem.* **1986**, 51, 1922.
20. Steve Pedersen provided invaluable structural results to aid M.G. Finn's studies during this period. Steve's work gave us our first glimpse into the dynamic world of complex mixtures in rapid equilibrium with each other. His contributions are summarized in: Sharpless, K.B. *Chemica Scripta (Sweden)* **1987**, 27, 521. These studies led to no new catalysts, but they drove home our appreciation for the absolute need for a ligand-acceleration effect in catalyst systems where ligand exchange is rapid.
21. Kirshenbaum, K.S. Ph.D. Thesis, Massachusetts Institute of Technology, **1985.**
22. Burns, C.J. Ph.D. Thesis, Massachusetts Institute of Technology, **1989**.
23. Jacobsen, E.N.; Markó, E.I.; France, M.B.; Svendsen, J.S.; Sharpless, K.B. *J. Am. Chem. Soc.* **1989**, 111, 737.
24. Jacobsen, E.N.; Markó, E.I.; Mungall, W.S.; Schröder, G.; Sharpless, K.B. *J. Am. Chem. Soc.* **1988**, 110, 1968.
25. Shibata, T.; Gilheany, D.G.; Blackburn, B.K.; Sharpless, K.B. *Tetrahedron Lett.* **1990**, 31, 3817.
26. Sharpless, K.B.; Amberg, W.; Beller, M.; Chen, H,; Hartung, J.; Kawanami, Y.; Lübben, D.; Manoury, E.; Ogino, Y.; Shibata, T.; Ukita, T. *J. Org. Chem.* **1991**, 56, 4585.
27. Sharpless, K.B.; Amberg, W.; Bennani, Y.; Crispino, G.; Hartung, J.; Jeong, K.; Kwong, H.L.; Morikawa, K.; Wang, Z.M.; Xu, D.; Zhang, X.L. *J. Org. Chem.* **1992**, 57, 2768.
28. Crispino, G.; Jeong, K.S.; Kolb, H.; Wang, Z.M.; Xu, D.; Sharpless, K.B. *J. Org. Chem.* **1993,** 58, 1958.
29. Wai, J.S.M.; Markó, I.; Svendsen, J.S.; Finn, M.G.; Jacobsen E.N.; Sharpless, K.B. *J. Am. Chem. Soc.* **1989**, 111, 1123.
30. Minato, M.; Yamamoto, K.; Tsuji, J. *J. Org. Chem.* **1990**, 55, 766.
31. Kwong, H.; Sorato, C.; Ogino, Y.; Chen, H.; Sharpless, K.B. *Tetrahedron Lett.* **1990**, 31, 2999.
32. Private communication from Dr. Yun Gao, formerly with Sepracor, Inc., Marlborough, MA.
33. Sharpless, K.B.; Teranishi, A.Y.; Bäckvall, J.E. *J. Am. Chem. Soc.* **1977**, 99, 3120.
34. (a) Jacobsen, E.N.; Markó, E.I.; France, M.B.; Svendsen, J.S.; Sharpless, K.B. *J. Am. Chem. Soc.* **1989**, 111, 737; (b) Göbel, T.; Sharpless, K.B. *Angew. Chem., Int. Ed. Eng.* **1993**; 32, 1329; (c) Norrby, P.O.; Kolb, H., Sharpless, K.B. *Organometallics,* **1994,** 13, 44; (d) Kolb, H.C.; Andersson, P.G.; Bennani, Y.L.; Crispino, G.A.; Jeong, K.S.; Kwong, H.L.; Sharpless, K.B. *J. Am. Chem. Soc.* **1993**, 115, 12226; (e) Kolb, H.C; Andersen, P.G.;

Sharpless, K.B. *J. Am. Chem. Soc.* **1994**, 116, 1278; (f) Kwong, H.L. Ph.D. Thesis, Massachusetts Institute of Technology, **1993**.
35. (a) Boar, R.B.; Damps, K. *Tetrahedron Lett.* **1974**, 3731; (b) Barton, D.H.R.; Jarman, T.R.; Watson, K.G.; Widdowson, D.A.; Boar, R.B.; Damps, K. *J. Chem. Soc., Chem. Commun.* **1974**, 861; (c) Barton, D.H.R.; Jarman, T.R.; Watson, K.G.; Widdowson, D.A.; Boar, R.B.; Damps, K. *J. Chem. Soc., Perkin I* **1975**, 1134.
36. Crispino, G.; Sharpless, K.B. *Tetrahedron Lett.* **1992**, 33, 4273.
37. Crispino, G.A.; Ho, P.T.; Sharpless, K.B. *Science* **1993**, 259, 64.
38. Hawkins, J.M.; Meyer, A. *Science* **1993**, 260, 1918.

Chemistry 2002

JOHN B. FENN and KOICHI TANAKA

"for their development of soft desorption ionisation methods for mass spectrometric analyses of biological macromolecules"

and the other half to

KURT WÜTHRICH

"for his development of nuclear magnetic resonance spectroscopy for determining the three-dimensional structure of biological macromolecules in solution"

THE NOBEL PRIZE IN CHEMISTRY

Speech by Professor Astrid Gräslund of the Royal Swedish Academy of Sciences.
Translation of the Swedish text.

Your Majesties, Your Royal Highnesses, Ladies and Gentlemen,

What would life be without proteins! Proteins are large molecules that do almost all the work in our cells. Every living organism, including a human being, has a large array of various kinds of proteins. Like diligent worker bees, they take care of what needs to be done. In principle, it works the same in this flower as in me or you.

To learn more about the diligent workers inside of cells, we want to know how they look, in order to understand what they do. This year's Nobel Laureates in Chemistry have developed methods that enable us to weigh and create pictures of giant molecules like proteins in new ways that we could hardly believe were possible. I am convinced that biochemistry is now standing on the threshold of a new era – we are beginning to become acquainted with the complete genetic code of many organisms. Soon we will be able to survey all the thousands of protein varieties that work simultaneously in a given cell. It is in this new era that the discoveries of the 2002 Nobel Laureates are so important.

Mass spectometry has been part of the chemist's toolkit for identifying small molecules since the beginning of the 20th century. But for many years, being able to make accurate measurements of the molecular masses of large proteins was a dream for chemists. For this reason, it caused a minor revolution in the field when John Fenn and Koichi Tanaka, each in his own way, succeeded to making intact proteins fly through the mass spectrometer.

Fenn discovered that it was possible to spray a water solution of the protein, in the presence of an electrical field, in order to obtain hovering electrically charged drops. The water evaporates and the drops are scattered by their electrical charge, becoming smaller and smaller. Finally only pure protein molecules are left. Then their mass is determined by measuring the time it takes them to accelerate across a given distance – the principle being that the heavier the molecule is, the more slowly it moves.

Tanaka's special method was to fire a laser pulse toward the sample. With the right wavelength in the laser, he could make the proteins be released from their surroundings without falling apart, so that they hovered freely as charged particles. Their mass could then also be determined by measuring their time of flight.

This was one half of the year's Chemistry Prize – now I will move to the second half. This time it is not a matter of flying proteins, but swimming proteins. Using nuclear magnetic resonance, or NMR, a method that Kurt Wüthrich has further refined, it is now possible to determine the three-dimensional structure of protein molecules in a water solution. NMR is one of the chemist's best methods for examining molecules, and it has been used extensively for small molecules since the mid-20th century. But large molecules like proteins involve special problems. One of the fine points of NMR is that it enables us to see individual signals, for example, from each hydrogen nucleus in a molecule. But because a protein can contain thousands of hydrogen nuclei, how do you know which signal belongs to which nucleus?

Wüthrich devised a way of systematically determining how each signal fits together with its special hydrogen nucleus. In the bargain, he was also able to determine a large number of pairwise distances between hydrogen nuclei. This enabled him to calculate a three-dimensional structure for the protein molecule. It is something like drawing a picture of a house if you know a large number of distances in the house. So thanks to Wüthrich's discovery, we can now use NMR to examine and depict proteins in their natural environment, surrounded by water like in a cell.

So, what would life be without proteins? Since I view the world through the eyes of a biochemist, my answer is: nothing at all! Next question: How would life be as a biochemistry researcher without the tools that this year's Nobel Laureates have given us? My answer to this question is: much more difficult, and also more dull! So I would like to conclude by saying to the 2002 Nobel Laureates in Chemistry: Thank you for your fantastic contributions, which help us to better understand the chemical miracles that constantly occur in our cells – what we call life.

Dr. Fenn, Mr. Tanaka and Dr. Wüthrich,

You have made pioneering contributions to the development of methods for identification and structure analyses of biological macromolecules. Your work to make mass spectrometry and NMR applicable for detailed studies of large molecules like proteins has given us new tools for investigations of the processes that constitute life.

In recognition of your services to chemistry, the Royal Swedish Academy of Sciences has decided to confer upon you this year's Nobel Prize in Chemistry.

On behalf of the Academy, I convey to you our warmest congratulations and I now ask you to receive the Prize from the hands of His Majesty the King.

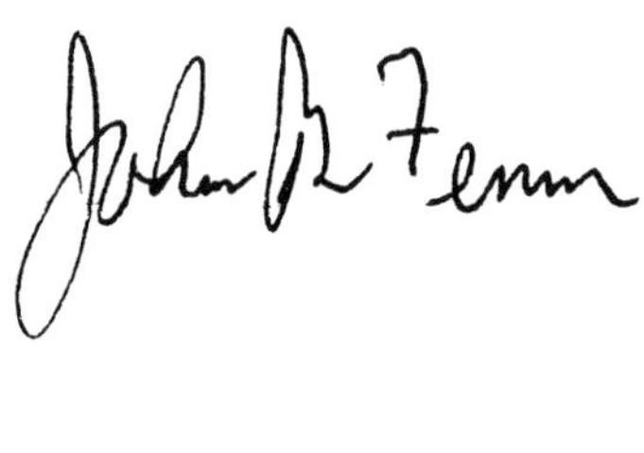

JOHN B. FENN

My father, Herbert Bennett Fenn, the eldest of three children was born and raised on a farm in northern Delaware which his father operated but did not own. I never saw that farm but I vividly remember my Grandmother's frequent reference to a single chestnut tree in the front yard, so large and prolific that the nuts from that one tree paid the taxes on the farm every year! My mother was the sixth of ten children in the family of John Clarence Dingman, a country doctor in Spring Valley, N.Y. whose three surviving sons also became physicians. Dad worked his way through college, graduating in Electrical Engineering from Rutgers in 1910, the same year that mother received a degree in Home Economics and Nutrition from Columbia. They were both hired by the Presbyterian Mission Board to teach at the Sheldon Jackson Mission School in Sitka, originally settled by the Russians and sometime capital (until replaced by Juneau in 1900). The Chairman of that Board told each one about the other and warned them both not to fall in love, but if they did and got married he would provide them with a house for as long as they stayed in Sitka! They did and he did, so after mother's death in 1990 (Dad had died in 1944) we found in her papers a clipping from the Sitka paper with a photograph of the "Honeymoon Cottage".

During a cruise in 1996 I spent a day in Sitka. The Mission School is now Sheldon-Jackson University. I visited the library, discovered a complete file of student newspapers, and found several articles about my parents! Moreover, the present Librarian was then living in the "Honeymoon Cottage" and gave me a tour so I was actually able to look out through its front windows and see the magnificent view of Sitka Bay that my parents used to rave about.

The newlyweds were very happy in Alaska and no doubt would have remained there indefinitely had it not emerged that mother would be unable to bear children except by Caesarian Section, a procedure not then available in Sitka. Determined to have a family she persuaded Dad to move back to the States where he became manager of the Metacloth Company, a small enterprise in Lodi, N.J. whose main product was cotton duck treated by immersion in a concentrated ammoniacal solution of copper hydroxide followed by a wash in acid. The deep blue "cuprammonium" solution dissolved some of the cellulose which reprecipitated during passage through the succeeding acid bath, filling the pores of the cloth enough to make it "water resistant" but not "water-proof". Residual copper provided a characteristic blue-green color along with a high resistance to attack by micro-organisms and white ants. For these reasons, "Metacloth" was in fairly steady demand for tents and tarpaulins in tropical latitudes. My abrupt introduction to "chemistry" occurred on one of the treasured Saturdays when Dad took me to the plant. Nosing around outside I lifted a small flap on the cover of a large tank half full of cuprammonium solution. I still recoil at the memory of that ammoniacal, and demoniacal, assault on my eyes and nose. It was a startling revelation on how and why a whiff of smelling salts can often revive people from a dead faint!

Our home was in Hackensack, N.J., next door to Lodi and County Seat of Bergen County. I was born in New York City in 1917 and three plus years later my brother Norman arrived in Paterson, N.J. where two of mother's brothers were surgeons. The Metacloth company was sold in 1926 and Dad was unceremoniously dumped by the new owners. He was not a vindictive man but he got no little satisfaction out of the several occasions in the next two years when his help was needed in overcoming mistakes made by the new management. Meanwhile, approaching fifty and finding equivalent jobs scarce, he was paying the bills by working as a draftsman at the Fokker Aircraft Company in Teterboro, N.J. The value of having a trade as back-up for a profession was an enduring object lesson for a youngster on the eve of the Great Depression but there was for him a far more exciting consequence of the Fokker connection. When Lindbergh's "Spirit of St. Louis" was shipped back from Paris after its famous flight across the Atlantic, it was parked for a time in a hangar at the Teterboro airport. The thrill of a lifetime for a ten year old boy was when his Dad took him into that hangar where he was allowed to sit in the cockpit and move the controls, pretending to be pilot of that famous plane!

Meanwhile the family fortunes were going down hill, much further than my brother and I were aware. Just before Dad lost his job, he and mother had invested their life savings in a new (for us) house that everyone insisted was worth "every nickel" of the $15,000 that it cost! Unfortunately, after we had to move and before the house could be sold the great depression had begun. No prospective buyers had enough nickels so those savings disappeared in the meltdown of foreclosure. Meanwhile, a door of opportunity had opened in Berea, Kentucky, a small community of 3500 or so inhabitants at the edge of bluegrass country some 40 miles south of Lexington and roughly halfway between Cincinnati, Ohio and Knoxville, Tennessee. The town of Berea was home for a remarkable institution known by the same name but officially entitled "Berea College and Allied Schools". Both the community and the school had their roots in a non-sectarian Union Church founded in 1848 by John G. Fee, on a ridge of land donated by Cassius Clay, a brother of Henry Clay, the famous orator and statesman from Lexington, Ky.

Fee was a Congregational Minister from Massachusetts and a fervent abolitionist determined to provide educational opportunities for needy students, regardless of their race or creed. After decades of effort by himself and his followers his dream grew into what in 1928 comprised a coeducational student body of about 1700 divided among four schools: (1) The *Foundation-Junior-High School* with an ungraded program into which students with as little as two years of formal schooling could enroll and progress at their own rate up through the equivalent of 8th grade into a standard and accredited 9th grade curriculum; (2)The *Academy*, with an accredited curriculum for grades 10 through 12, (3) the *Normal School* with a two year program leading to a Teaching Certificate, and (4) the *College* which offered accredited curricula leading to BA degrees in the liberal arts and sciences as well as BS degrees in Home Economics and Agriculture. Mother's sister was teaching in the College and knew that the Industrial Arts Department of the Foundation School and the Academy needed someone to teach Auto Mechanics and Practical Electricity. Dad was eminently qualified and got the job. We moved to Kentucky in time for me to enter the eighth grade in the "Training School" of Berea's Normal School in the fall of 1928. The next fall I entered ninth grade in the Foundation School and continued on through the Academy and College.

A lot of misgivings, pain and confusion attended our passage through the newly opened door, including an automobile accident on the way to Kentucky, but life on the other side turned out to be rich and rewarding beyond what any of us had dreamed. In later years Dad and Mother both said, time and again, that losing his "good" job in New Jersey turned out to be the greatest blessing that we could have received! To this day my brother and I share those sentiments and count ourselves especially privileged to have been reared in what was a truly remarkable community. Its soul was its President, William J. Hutchins, father of Robert Maynard Hutchins, the "boy wonder" of the American education scene who became Secretary of Yale at the age of 24, Dean of its Law School at 26 and President of the University of Chicago at 30! "William J", as the father was fondly known at his own institution, was truly one of nature's noblemen. A striking man of vision and patrician to the core, his Berea was a singular stage on which the play was always provocative and the message meaningful. Name an outstanding man or woman of letters, the arts, science, or religion, of those days and the odds are high that he or she came to Berea to talk at one of the thrice weekly "United Chapels" for students from all schools. Attendance was required of all, and resented by most, but at my 50th class reunion there was a remarkable consensus among my surviving classmates that the community experiences of those United Chapel services were by far the most memorable and valuable components of their educational. Alas, that tradition has long since disappeared, one more victim of "students' rights" and television. Moreover, as the public education system improved in surrounding "Appalachia", the need for Berea's Foundation School, Academy and Normal School faded over the years, so that the only survivor is a now a substantially larger Berea College.

A unique feature of Berea was and is a student labor program that provided much of the manpower required to run the institution. Under the supervision of permanent staff, the dining halls, dormitories, offices, and yard work were all maintained and operated by student labor. In addition were several "Industries" including a bakery, broom factory, dairy, large truck garden, sheep farm, piggery, a "woodwork department" that made fine furniture, and a weaving establishment. The normal work load was two hours a day by which a student could earn as much as half or two thirds of his or her out of pocket expenses that in the 1930's averaged only $250 to $300 per year because tuition was free! The system was very democratic because every student was required to work at least two hours a day. A fair fraction of the student body comprised so-called half-day students who worked four hours a day, earning enough to pay all expenses. I knew students who arrived at Berea with nothing but the clothes on their backs, having dropped out of school after the second grade, and worked their way through college! It is not surprising that competition to enter Berea, especially the college, was extremely keen. In order to avoid having to reject such a large fraction of applicants, the Trustees had established a policy requiring that 85 per cent of the student body come from a region of Appalachia comprising some 500 or so counties in Virginia, West Virginia, Kentucky, Tennessee, and North Carolina. The remaining 15 percent of the students came from the rest of the world. Today's Berea is quite a different place because the growth of alternative educational opportunities in Berea's "Territory", has decreased demand to the point that in 1987, at my 50th reunion, I learned there were only two applicants for every opening in the College compared to 20 or 30 in my day. Part of that decrease in demand is due to a higher level of average income in that "territory" so that fewer students can show that they cannot afford to go elsewhere, a requisite for admission.

When I graduated from the Academy in 1932 I was only 15, too young, my parents thought, to go to college, so I stayed in the Academy an extra year taking courses in mechanical drawing and shorthand. I also continued the piano lessons that I had started the year before and spent enough time practicing so that I actually gave a recital during my last year in college! Alas, whatever piano skills I had have faded until now I do well to play chopsticks with my grandchildren! When I entered College with the class of '38 I had not decided on a major but I leaned toward a science, probably because of my long affair with the Book of Knowledge, an encyclopedia for young people which my parents bought when I was around eight or nine. During the hours that I pored over them, its 20 volumes became a well worn magic carpet to new and fascinating worlds. I've often quipped that "I got through college on the Book of Knowledge", a bit of rhyming hyperbole that contains an appreciable kernel of truth.

The College required at least one course in science for all students no matter what their major. Having enjoyed my chemistry course in the Academy I enrolled in Introductory Chemistry during my freshman year. It was taught by Professor Julian Capps, the senior of the two member chemistry faculty. He was a wonderful teacher who made his subject live. I was so seduced that Chemistry became my major even though Gravimetric Analysis gave me fits in my sophomore year. I repeated the lengthy phosphorous determination on three samples each of ten unknowns before I got one set of three results with acceptable agreement! It seems ironic that mass spectrometry, the center of my scientific life for the last 20 years, is really just an exercise in gravimetric analysis!

During that sophomore year an experiment in another kind of chemistry turned out to have a big effect on my chemical education. I fell in love with a girl in the junior class. Determined to catch up with her and graduate when she did, I went to summer school in 1936 at the University of Iowa because it had a 10 week session that allowed me to earn 12 semester hours of credit in Organic Chemistry under "Uncle Charlie" Raiford, and Inorganic Chemistry under Professors Jacob Cornog and Perry Bond. All in all it was a very hot and grueling summer. The romance that had blossomed in the spring and sent me off to summer school, withered the following fall. However, because I would be within two courses of finishing my degree requirements by end of that year, my third in college, I managed to get classified as a Senior and became a provisional member of Class of '37 at the June commencement, the objective inspired by a young man's fancy of the previous spring. My courses in chemistry thus far had required only some facility in simple algebra. Straight A's in all Prof. Peck's courses in the Academy (plane geometry, advanced algebra, solid geometry and trigonometry) had excused me from freshman math in college, all that was then required for a major in chemistry.

Fortunately for my future our next door neighbor, George Bent, was visited by his brother Henry, then an assistant professor of chemistry at Harvard. When Henry discovered that I was contemplating graduate work in chemistry, having had no college math, he was aghast and spent a couple of hours impressing me with the importance of mathematics in chemistry. As a result I spent much of my third and last year of college immersed in that subject. Fortune had tossed me another favor in the form of Berea's new curriculum based on four 9 week quarters instead of two 18 week semesters. So-called "intensive" courses met 6 hours a week for nine weeks while "running" courses met 3 hours a week for eighteen. Thus, a standard full load of 15 class hours per week comprised two intensive courses and one running course, an arrangement which I heartily endorse. One major advantage is that a student takes only three courses at any one time instead of five. It is much less distracting to be studying three subjects at a time rather than five. The particular advantage for me was that the calendar time for a three credit-hour course was cut in half so that both Course B, and its prerequisie Course A, could be completed in two nine week quarters instead of two eighteen week semesters. Thus, in my senior year I was able to take Analytic Geometry, Differential Calculus, Integral Calculus, and Complex variable, all in that same year! It was pretty heavy going, but it enabled me to survive in graduate school where I took one more undergraduate course in Advanced Calculus. The record seems to say that I have also survived after graduate school but a lack of mathematical skills has been a great handicap throughout my life.

Though I had taken part in the June Commencement exercises at Berea in 1937, I still had course requirements to complete before I could get my degree. I fulfilled those requirements at Purdue University that summer by taking Physical Chemistry with lectures by Roy Newton and laboratory under Hershel Hunt, along with a course in Chemical Microscopy under Ed F. Degering.

During that hectic last year in college I had applied to several universities for graduate study in chemistry. Yale and Northwestern responded with offers of teaching assistantships that would meet most of my expenses. I first leaned toward Northwestern but, for reasons I won't take space to enlarge on, I finally decided on Yale. That decision delayed by half a century my meeting Malcolm Dole who in 1937 was a young assistant professor of chemistry at Northwestern. It was the experiments he reported much later (in 1968) that changed the course of my scientific life.

At Yale I found myself in a new and different world. I had done time in two much larger universities and one much smaller college but I was completely unprepared for, and awed by, the splendor of Yale's campus. Its architectural theme has been disdainfully dubbed "Fifth Avenue Gothic" by the self-styled experts who sneer at such expensive imitation of the old world. But a wide eyed small town youth from Kentucky found it all fascinating and loved to wander around and through the gardens, gargoyles and gates that were scattered throughout the campus. To me it was rewarding recreation to discover and explore the little nooks and courtyards where one could sit in the shade of a tree completely oblivious to the cacaphony of a busy city, only a few yards away but completely shut out by the thickest walls and heaviest masonry I had ever seen. Sterling Library, then said to be the largest in the world with open stacks, became a favorite haunt where I spent many winter weekend afternoons with a good book and an apple or a candy bar. My room in the Hall of Graduate Studies looked out on a forest of chimneys, slate roofs, and Gothic towers that resembled a Hollywood version of medieval London. I ate in an elegant dining hall, complete with menus and waiters — twenty one meals per week for a now incredible eight dollars! Even so, to save money I took advantage of an option in the board contract and "signed out" every week for dinner on Friday night along with breakfast and lunch on Saturday. This "fasting" saved me $1.15 every week and probably was good for my figure! My assistantship paid a total of $850 for nine months, of which $350 went for tuition and fees. By such frugalities as the weekly "fast" I could make the remaining $500 cover most of my other expenses. My parents were horrified when they later learned of my miserly existence at Yale but I was determined to get along without their support.

On the appointed day in September the six new graduate students in physical chemistry met in Sterling Chemistry Laboratory with Professor Herbert Harned, the ranking faculty member in that subject whose colleagues included Lars Onsager, Benton B. Owen, Rodney Smith, George Murphy, John Vance, and Gosta "Gus" Akerlof. With little or no ceremony, no introductions and no opportunity to ask questions or indicate preferences, we were assigned to advisors. I recall "Herbie" as he was known but not addressed, saying to me: "Fenn, you might as well go with Gus." Thus, began what blossomed into a close friendship that lasted until the day Gus died, many years later, in Princeton.

The Yale curriculum for chemistry graduate students in those days was pretty cut and dried. During the first year we took four courses, spent 12 to 14 hours per week as assistants in undergraduate labs and the rest of the time getting our feet wet in the laboratory on what would become our dissertation research. At the end of that first year the faculty, on the basis of course grades and day to day observations, decided whether or not each student was Ph.D. material. Those for whom thumbs were down were given a Master's degree and dismissed. The others continued, almost always receiving a Ph.D. after two more years. The second year was much like the first except that we took only a couple of courses and spent much more time on our research projects. At the beginning of the third year we had to take a two day written comprehensive exam. While I was a student, all who took that exam passed it so I don't know what the consequences of failing would have been. One very big difference in the third year was that we could no longer be teaching assistants but were expected to spend full time on research. Thus, unless we were lucky enough to win one of the two fellowships available to third year students, we were on our own financially. My solution to this problem was to get married, not to my sophomore flame but to Margaret Wilson, who became supervisor of my student labor at Berea while I was working in the Registrar's Office. She was a beautiful woman by whom I (then in the tenth grade!) was smitten the first day she started work as Assistant Registrar. Always ten years older than I she seemed to be so far beyond my reach that I never dreamed she could ever be more than the close friend and confidante that she became. I had escorted her to several events while I was in college and when I was home after I graduated. Romance bloomed while I was at Yale and she finally agreed to become my bride at end of my second year. She also became my "fellowship" during that third year at Yale, supporting us both on the 50 cents per hour she earned at various odd jobs. To this day I marvel that the daughter of a very conservative minister had the courage to ignore the disapproval of her parents and the inevitable elevation of eyebrows by everybody, everywhere, at our flagrant departure from the social norm. One of my closest friends told us decades later that he had been sure our marriage would not last more than a year or so. In fact it was not until 1992 that I lost my bride of 53 years in a New Zealand car crash. She was then 85 but looked and acted ten years younger than the husband she had "snatched from the cradle."

The plain truth is that in matters of matrimony this society's norms are genetically out of joint. Women on average live about ten years longer than men and yet become the wives of husbands who are several years older. That is why widows greatly outnumber widowers. Some say that this imbalance in life expectancy will be redressed as women increasingly become exposed to the stresses of employment in the workplace. I am much more inclined to believe the geneticist who told me that the female of the species lived longer than the male because she is fundamentally tougher and more resistant to the ravages of age. In choosing a mate men should follow the advice of Ben Franklin in his letter of advice to a young man on choosing a mistress. His message? "be sure she is older than you are!"

Only in choosing a graduate school did I exert any influence on the research I would do for my thesis. That influence was unwitting because I knew nothing of the faculty research interests at either Yale or Northwestern. Indeed I had only the vaguest idea of what graduate study and research were all about. Thus, I felt neither joy nor apprehension when I was summarily assigned to Gus Akerlof's group. Most of the research by Harned and his colleagues was on the properties of electrolyte solutions, the more dilute the better. So narrow was this focus that when I left Yale I halfway believed the old cliche that "Physical Chemistry was the study of slightly contaminated water." Gus was a bit of a maverick because he had an interest in concentrated solutions. In essence what I had to do was measure the potential difference between electrodes of silver-silver chloride and platinum-hydrogen in solutions of HCl with molalities from 0.01 to 10.0 in solvents comprising methanol in water at concentration intervals of 10 per cent from zero to ninety, at temperatures every ten degrees from zero to 50 Celsius. As I remember the total number of acceptable emf measurements was around 3000, taking into account that each measurement was made in duplicate. Thus, my routine every afternoon for most of two years was to prepare three pairs of cells with both cells of each pair containing the same solution. The cells were suspended in a large water bath or "thermostat" and allowed to equilibrate over night (i.e. with hydrogen bubbling over the Pt electrodes.) If by the next morning the emfs in each pair of cells agreed to within some forgotten small fraction of a millivolt, then I would haul some 80 pounds or so of ice from the basement to lower the bath temperature to near 273 K. The rest of the day would be spent recording emf values for all six cells at every ten K as the thermostat was heated to 313 or 323 K depending on the vapor pressure (i.e. the methanol content) of the cell solvent. All the results were fitted by least squares to a quadratic curve for the dependence of emf on concentration. The least square calculations were all carried out on a hand-cranked Monroe Calculator, except when I was lucky enough to grab the one electrically driven machine in the calculating room. Clearly, I was less than inspired by this introduction to "original research". The experiments were a boring chore with few redeeming features. The results contained no surprises and nothing of much interest to anyone, least of all me. It was some years later before any of them found their way into the literature and then only as part of a table in a review article that Gus had been asked to write. My dissertation attested to the sterility of that project, consisting as it did of 45 pages of tables with only three pages of text! Although this first experience in "original research" shattered some illusions, my three years at Yale were a rewarding experience. I made many long lasting friendships with both fellow students and faculty. The Physical Chemists were a very congenial group and Herbie Harned was really a very interesting man, except in the class room. He came by almost every lab almost every day to chat with us students about almost everything. Sometimes his visits were inconveniently long and would interfere with experiments. One of the older graduate students had told me that Herbie would run for cover if he were asked to help, for example by holding a light or a tool during some adjustment. That ploy worked like a charm and saved me many hours.

Actually, the Harneds were very kind and warm hearted people. They held an open house every Friday night which both students and faculty felt obliged to attend more or less regularly. More often than not we ended up around the piano in a group sing-along with Roger Bates, then a Sterling Fellow and accomplished musician, at the keyboard.

I also learned a lot about life and lore outside of chemistry, e.g. to play bridge, to drink beer and to smoke a pipe during the mostly dull seminars every Tuesday night. I had many interactions with many interesting people. The weekly university calendar was filled with provocative events ranging from seminars to sermons and concerts to contests. My perspectives were stretched by memorable if abstruse encounters with great minds, including lectures on resonance by Linus Pauling, and two required courses in statistical mechanics under Lars Onsager (known to the students as Norwegian I and II!). Gus and his wife Rosalie, who was the sister of Joe Hirschfelder, looked after his students with great care and affection, making their house our home away from home. Rosalie took my new bride under her wing, showering us newlyweds with almost too much attention. I was extremely fond of Gus and most grateful for his always calm demeanor and seeming imperturbability. He strongly disapproved of marriage for graduate students but when I told him I wanted to take a wife he protested not at all and simply said "OK, we take it from here." He and "Magee" as my wife was later known, became extremely fond of each other and he later told me many times that marrying her was the smartest thing I ever did, an opinion shared by everyone who has known us both. All in all, my time at Yale was happy and rewarding but I left with a much diminished interest in "scholarly research", and no desire to return. I often think about how surprised and pleased Gus would have been to know that six decades later his son George Akerlof would share the 2001 Nobel Prize in economics and his student John Fenn would share the 2002 Prize in Chemistry!

My immediate and exciting prospect after graduation from Yale was a job in the research department of the Phosphate Division of Monsanto Chemical Company in Anniston, Alabama. The starting salary was $2700 per year, then the going rate for new PhD's in industry and a mind-boggling jump from the $1000 or so a year Magee and I had been living on. I couldn't understand why my mother laughed when I once remarked that I didn't know how we could spend so much money!

Anniston, the county seat of Calhoun county and "Cast Iron Pipe Capital of the World", was located halfway between Atlanta and Birmingham. Living there was my introduction to the deep south where cotton still grew and the Confederate Flag still flew. It was the home of Fort McClellan, a long time army base that was already gearing up for America's forthcoming participation in World War II. Two of my mother's brothers had been medical officers at McClellan during World War I and two of our closest friends from New Haven passed through there after being drafted in the wake of Pearl Harbor. We had the chance to compensate them a wee bit with an occasional home-cooked meal. I was more fortunate because the army was not yet drafting husbands.

Monsanto had recently acquired the Anniston Plant when it bought Swann Chemical Company started by a legendary character named Theodore Swann. He had pioneered the production of phosphoric acid from elemental phosphorous shipped in by tank car from a plant in Columbia, Tennessee where it was produced by smelting "phos rock" in an electric furnace. At Anniston the phosphorous was burned in air to produce its pentoxide which could be hydrated to form acid of any desired concentration. Major customers for the acid per se included the soft drink industry, especially the Coca Cola Company in nearby Atlanta. According to local folk lore, the first tank car of food grade acid to be delivered was filtered through laboratory Buechner funnels because the filtration plant was not ready in time to meet the deadline on the sales contract! Much of the phosphoric acid left the plant in the form of sodium and potassium phosphates that among other applications were widely used as soap builders, water conditioners and in food products such as baking powders.

The other main product line, which some said was the main reason Monsanto bought Swann, comprised biphenyl and its chlorinated derivatives known as Arochlors, the now notorious PCBs. Biphenyl was produced by bubbling benzene vapor through large baths of molten lead at a temperature of around 800°C, as I remember. Liquid metal as a heating agent provided a rapid rise and close control of the temperature, thus minimizing formation of byproducts which mostly comprised ortho, meta and para diphenyl benzenes. The various Arochlors were characterized by their chlorine content which ranged from 10 per cent to 60 per cent by weight. One of their main uses was as transformer oils because they had very good heat transfer characteristics, would not burn and were very inert. Indeed, they were so inert and sticky that the chemists insisted that if Arochlors ever got into a laboratory some of it would remain forever! Because it was so inert we practically bathed in the stuff, never dreaming that it might be toxic. General Electric dumped PCB wastes into the Hudson River for many years and is now resisting Federal orders to dredge it all up. Recently, the press has reported growing anger in Anniston and threats of legal action as new surveys have shown the extent to which PCBs from the plant where I worked have contaminated the soil and waters around the town. My own substantial exposure seems to have had no ill effects, but I daresay that even today, sixty years later, the PCB content of my fatty tissue would horrify the EPA! There is no doubt that PCBs in the environment accumulate in the fatty tissues of mammals and fish. The evidence of resulting toxicity is less convincing. I remember a paper by an Australian chemist claiming that the only documented evidence of significant toxicity stemmed from two cases in which PCB's, used as the heat transfer medium in the heating coils of deep fat fryers, leaked into the fat. The resulting reaction produced highly toxic benzo-compounds whose identity I have forgotten. I do know that sixty years ago I practically bathed in the stuff while operating a pilot plant in which PCB was produced in the vapor phase by reaction of biphenyl with HCl and air over a copper catalyst in a variation of the old Deacon process for producing chlorine by the oxidation of HCl. Occasionally

I would get a slight skin rash, as did the other operators, but I have detected no lasting effects.

After I had been at Anniston for a year, James W. Mullen II., a young Ph.D. organic chemist from Princeton came aboard. He and I became good friends, sharing a growing dissatisfaction with the way things were being run at work. Jim insisted that someday he was going to start a research company and that when he did he wanted me to join him. Talking about what we might do was fun but it all seemed like pipe dreaming to me. Having become increasingly disenchanted with the work situation we both resigned and left Anniston on the same day in 1943. Jim went to Bell Labs in New Jersey. I took my family, which now included an 18 month old daughter, Marianne, to Wyandotte, Michigan where I had obtained a job in the research department of Sharples Chemicals, a relatively small producer of alcohols, amines, esters and other derivatives of amyl chloride produced by the chlorination of pentane. One day in June of 1945 a letter came from Jim saying that he had finally started his company and wanted me to join him. I took a train down to Richmond to find out more. At Bell Labs he had become involved in Project Bumblebee, a large scale effort by the Navy to develop a ramjet-powered anti-aircraft missile for the fleet. The ramjet depends upon ram pressure from high speed flight to compress air enough so that its expansion after heating by combustion could provide useful thrust. It was clear that supersonic flight velocities would be required to achieve substantial thrust but because there was little or no information on drag at such velocities, nobody knew whether that thrust could equal or exceed the drag. The Project name, Bumblebee, derived from a bit of aeronautical badinage to the effect that by all the laws of modern aerodynamics it is impossible for a bumblebee to fly, but the bumblebee doesn't know any aerodynamics so it goes ahead and flies anyway!

FROM FLAMES TO FLYING ELEPHANTS

Experiment, Inc., the name Jim Mullen gave his company, prospered over the next seven years. To watch and participate in its growth, from a total of three employees when I arrived to perhaps 45 or so when I left in 1952, was an exciting experience. Most of its business was based on R&D contracts with government agencies for which the central theme was combustion and propulsion but we did pursue some commercial ventures. There were some important personal dividends from my investment of time at Experiment. One was that I learned a little bit about the dynamics and thermodynamics of compressible flow, a newly popular area of research spawned by the advent of jet propulsion and sometimes referred to as "Aerothermodynamics", or "Aerothermochemistry" if changes in chemical composition were involved. Another personal dividend was the opportunity to do some research in combustion, both applied and fundamental. Jim was a strong believer in publishing our results. Thus the first research publication of my life was a paper on ignition in high speed flow, co-authored by Jim and a junior colleague, M. R.

Irby. That paper appeared in 1949, almost a decade after I left graduate school [1]. Consequent to several more papers, visits to other labs, and attendance at various meetings, I began to get some recognition in the Combustion Community. Moreover, in an about-face from my feelings at Yale, I found I enjoyed doing research and communicating with others about my findings and theirs.

My life in Berea had made me think that I would someday like to become a college professor. But I had also thought that I should get some of the real-world industrial experience that had seemed to help make Julian Capps, my chemistry professor at Berea, such an effective teacher. That desire to join a college faculty had faded somewhat at Yale but was revived a bit during my experience at Experiment, Inc. We obtained a number of provocative results on some properties of flames, in particular on stabilizing them in high velocity flows. Both Jim and the sponsors of our studies were happy to have us attend technical meetings and to publish results of our studies, (so long as they didn't involve classified information.) Consequently, I got to know, and become known by, a number of people in the combustion community. One result of this new and limited "notoriety" was an offer from Princeton University to serve as Director of Project SQUID, a program of pure and applied research in "those fields of science relating to jet propulsion" including combustion, fluid flow and heat transfer. Financed by the Office of Naval Research (ONR) and administered by Princeton, Project SQUID was named after the jet-propelled decapod of the sea. (That name had additional but unpremeditated relevance in that both the animal and its namesake secreted ink, the former in clouds that confused predators, the latter in print in reports that confused readers!) SQUID's organizational structure comprised a prime contract between ONR and Princeton, with subcontracts at any one time between Princeton and from 12 to 20 or so university and industrial laboratories. The overall purpose was to cultivate interest and activity by students, faculty and other investigators in combustion, fluid flow and heat transfer.

One reason for my interest in Princeton was that by then I had learned that research could be a lot more fun than it had been when I was a student at Yale. Another reason was that my Yale mentor, Gus Akerlof, had set up a laboratory in its newly established James Forrestal Center to develop, with Navy (but not SQUID) support, an electrical discharge process for the production of hydrazine, an attractive rocket propellant. Gus had once visited me in Richmond and in fact had put my name in the pot at Princeton when he learned that the previous director of SQUID was leaving. One of the "other reasons" for my interest was that Jim Mullen and many of his friends, were Princeton graduates. After seven years among those fiercely loyal alumni I had been brainwashed to the point of believing that maybe Old Nassau really was the "best old place of all." Moreover, as I have already noted, my life in Berea had long made me think that I would enjoy becoming a teacher and living in a college community.

At Princeton I would report directly to Sir Hugh Taylor, long-time

Chairman of Princeton's Chemistry Department, then Dean of the Graduate School and Chairman of the University Research Board. The position was "with rank of Professor" and the Dean assured me that I would be welcomed as an active member of the academic community. And so, after much soul searching, Magee and I decided to leave the "Cradle of the Confederacy" for what was and is often called "the northernmost southern town in the country". That epithet stemmed from the substantial numbers of Princeton students that over the years have come from below the Mason Dixon line. Before the Civil War it had become customary for students from the south to celebrate graduation by freeing their personal servants. An interesting result is that Princeton has a much larger black population than do other towns of comparable size in that part of the country.

My new job turned out to be very interesting and highly educational. I had to travel a lot but was able to visit many laboratories and meet well known scientists in many fields all over the country. I got to know the people in ONR pretty well and my circle of acquaintances included people in ARO (Army Research Office) and AFOSR (Air Force Office of Scientific Research) both of which were providing some support for Project SQUID and were represented on its Steering Committee. ONR had a branch in London (ONRL) that maintained a rotating corps of 12 to 15 Scientific Liaison Officers whose job was to promote interaction between scientists and engineers in Europe and America by visiting laboratories and attending scientific meetings. They would get to know the investigators, learn about what they were doing, help them establish contacts with their counterparts in America, arrange visits, find items of equipment, assist with travel arrangements, and the like. Research laboratories in Europe were beginning to recover from the trauma of World War II and they welcomed the communication links and the help with their needs that ONRL could provide. The SQUID office had occasionally been involved in making travel and visiting arrangements for some of the European scientists who came to America under ONRL's auspices. After I had been at Princeton for about three years I was invited to serve a year as an ONRL Liaison Officer in the areas of combustion and propulsion. My bosses in ONR Washington approved and I persuaded John Scott, who had just finished his Ph.D. in Aeronautical Engineering at Princeton, to mind the store at SQUID while I was gone. Thus, Magee and I and the children, then 9, 11 and 13, went to London for the calendar year of 1955. And what a wonderful year it was! The children still insist that living London was the high point in the Fenn family fortunes which have been going down hill ever since! I share some of those sentiments because that year was also a marvelous experience for the parents. The kids were in school so Magee was free to roam London during the day. I visited many laboratories in many countries, made many friends, and learned "more things than had been dreamt of in my philosophy."

Among these "things" was my stumbling on to something that was to set the stage for the rest of my scientific life. Up to that time flames and their behaviour were the focal point of almost all efforts to elucidate the kinetics and mechanisms of the important reactions in high temperature combustion. The general approach was mostly based on attempts to measure and interpret flame characteristics such as propagation velocity, flammability limits, ignition temperatures and the like. I had become convinced that flames, with their extremely high gradients of temperature and composition, were much too complex to be useful as stages for the study of those reactions. I often recalled a comment by Philip Rudnick, professor of Physics at Vanderbilt, who was active in the propulsion Panel of Project Bumblebee. He said that the more he learned about flames the more he was convinced that they were organisms whose study properly belonged in the province of biology! I'm sure that view strikes a responsive note in many investigators who have been exasperated by the vagaries of flame behaviour that seem to give them a life of their own. Indeed, the analogy can be extended. Flames, like organisms, are born, need fuel and oxidant for nourishment, grow and multiply to the extent that those necess-ities are accessible, and are poisoned unto death when immersed in their own waste products! Not only is their structure and behavior complex but the conversion of reactants to products involves sequences of so many reactions over such short distances that to determine how the composition and temperature changed with distance (i.e. time) in real flames, i.e. the relation between their "fine structure" and their behaviour, was not experimentally feasible with the tools and techniques then available.

Thus, for some time I had been somewhat naively musing about whether and how it might be possible to study chemical reactions by the same approach that physicists so successfully use in the study of nuclear reactions. They build accelerators to produce beams of atomic nuclei with energies of megavolts to gigavolts with which they bombard stationary targets comprising collision partners of interest. They then characterize the products from reactive collisions by analyzing the post-collision trajectories of the product particles as measured by appropriate detectors, e.g. bubble or cloud chambers. I had been introduced to the possible uses of molecular beam scattering experiments in chemistry by Professor I.(Izzy) Amdur, a colleague of the Professor Frederick Keyes at MIT whose elegant measurements on the transport properties of gases were being supported by Project SQUID. Amdur was using beam scattering experiments to characterize the intermolecular potentials of molecules. His approach was based on electrostatic acceleration of ions to relatively high energies and then neutralizing them by charge-exchange collisions with neutral molecules for which the cross-sections are much larger than those for momentum exchange. In this way he could produce beams of neutral molecules with translational energies equal to those of the incident ions. Space-charge effects give rise to the Childs-Langmuir law by which the maximum intensity (current density) of an ion beam is proportional to the square root of its energy. Consequently, neutral beams with intensities high enough to provide good signal/noise in a scattering experiment, must perforce have energies substantially higher than the intermolecular potentials that Amdur was trying to probe. His solution to this problem was to locate his detector so that it would "see" only the post-collision beam molecules whose trajectories differed by only a very small amount from their pre-collision values. This "small-angle scattering" approach allowed him to exploit the high intensities obtainable with beams of high energy molecules (in laboratory coordinates) in the study of molecule-molecule interactions at low energies in center-of-mass coordinates of the colliding partners. Unfortunately, that approach did not seem to lend itself to the study of reactive collisions.

When I went to ONR London in late 1954, nobody had successfully carried out reactive scattering experiments with molecular beams on any chemical reaction, even one with a negligibly small activation energy. Moreover, activation energies for many combustion reactions were known to be as high as 30 kcal per moltwo or more. The energies of beam molecules from conventional effusive sources are always limited by the temperature at which the source can be operated. The average translational energy of an oxygen molecule from a source at 3000 K would be only about 10 kcal/mol or 1.2 eV, (but that temperature is well above the melting point of most materials from which the source might be made!). Any polyatomic molecules of interest in combustion would completely decompose even at much lower temperatures. In sum, the collision energies that can be reached simply by raising beam source temperatures are severely limited. Even so, the possibility of bringing about reactive collisions by colliding a beam of reactant molecules with other gaseous molecules had been contemplated since the late 1920's. In spite of numerous attempts, no really convincing results had been obtained when we left for London at the end of 1954. However, shortly after we returned to Princeton in early 1956, the age of reactive scattering experiments with molecular beams dawned when Datz and Taylor reported that intersecting beams of K atoms and HBr molecules produced detectable amounts of KBr.[2] These results gave some substance to my musings but there was a fundamental problem with the idea of applying the same method to combustion reactions. Activation energies for K + HBr are negligibly small. For many of the most interesting reactions in combustion the activation energies were thought to range from 0.5 to 2.0 or more eV, much much higher than for the reaction studied by Datz and Taylor. Amdur's charge exchange technique could easily produce beams with much higher energies but because of space charge effects could not provide useful intensities at energies as low as a few eV. Nor did Amdur's ploy of looking only at small angle scattering seem applicable in the study reactive scattering. To produce beam molecules from classical effusive sources with energies of even 0.5 eV would require source temperatures of 3000 K. Neither materials of construction for the source, nor reactant molecules other than atoms, could endure such temperatures. The situation thus seemed hopeless.

One day in London I was calculating rocket exhaust velocities achievable with various propellants and suddenly realized that many reactant molecules at the velocities attainable with some such propellants would have translational energies as high as two or more eV. I began to fantasize about using micro-rockets in vacuum as molecular beam sources, anticipating by 30 years an experiment we actually carried out for another purpose! I then remembered that rocket visionaries had often touted hydrogen as an ideal propellant fluid for nuclear powered rockets because its of its low molecular weight. The limiting maximum convective velocity that can be reached by a gas expanding into vacuum is equal to $(Cp/MT_0)^{1/2}$ where Cp is its constant pressure heat capacity, M its molecular weight and T_0 its source temperature. Thus the maximum velocity reachable by helium is about 3.33 times that for argon. If the helium is "seeded" with one or two per cent argon, and if those argon atoms are swept along like dust particles in a wind storm, the translational kinetic energy of the argon atoms after free jet expansion of the mixture would be almost 10 times that of the helium atoms, if the two reached the same terminal flow velocity.

A few days later I was browsing in ONRL's library and ran across the now famous paper by E.W. Becker and K. Bier reporting the production of intense beams of hydrogen molecules by expansion of the gas from high pressure through a small converging-diverging nozzle into vacuum [3]. That paper referenced the also now famous paper by Kantrowitz and Grey in the Review of Scientific Instruments which proposed the use of a converging-diverging nozzle to produce a jet of high velocity molecules in a region of low pressure.[4] A small core portion of that jet would be admitted through a conical "skimmer" into a high vacuum region to form a jet of molecules. The "static" temperature of the molecules (as measured by a thermometer traveling with the gas and at the same velocity) would be very low so that the divergence of the molecular trajectories downstream of the skimmer would be very small and the velocity distribution in the direction of flow would also be very narrow. In sum, the jet would comprise a very intense beam of molecules, all traveling at nearly the same velocity! Another paper in the same issue of that journal reported a not very successful attempt by Kistiakowsky and Schlichter to reduce the Kantrowitz-Grey idea to practice.[5] Both of those papers also happened to be in the ONRL library and when I read them I began to take my musings seriously. (It is said that Kistiakowsky and Schlicter were so frustrated by troubles in their experiment that when Schlicter had suffered enough to deserve a degree, they vented their frustration by demolishing the apparatus with an axe!)

I later learned that twentyfive years earlier, T. H. Johnson, a Sterling Fellow at Yale, had produced intense beams of mercury atoms by expanding the vapor from a "boiler" at several hundred torr into vacuum [6]. Unfortunately this truly remarkable result had been ignored because it was at odds with the dogma from Stern's lab in Hamburg that raising the source gas pressure (density) would produce a sort of stagnant cloud of molecules at the exit of the source orifice. That mythical cloud would scatter a large fraction of the emerging molecules that might otherwise have contributed to beam intensity. Later in his career at the Franklin Institute Johnson he also found diffraction effects in the scattering of hydrogen from an LiF crystal, confirming the wave nature of particles much heavier than the electrons with which Davisson and Germer had earlier found diffraction effects. Unfortunately, some eight weeks earlier Estermann, Frisch and Stern had found the same result in Göttingen [7] so Johnson's work again went unnoticed while Stern got the Nobel Prize! An able investigator who was 25 years too early with one landmark experiment and 8 weeks too late with another, Johnson had every right to feel frustrated by the fickleness of fate.

The results predicted by Kantrowitz and Grey in 1951, having been already achieved by Johnson in 1929, were confirmed in the above-mentioned paper by Becker and Bier in 1954. They showed that convective flow of gas through a small orifice from high pressure into vacuum could indeed produce molecular beams with much higher intensities than could the effusive sources introduced and exploited to great advantage by Otto Stern and his colleagues. Moreover, these "convective" beams had much narrower velocity distributions than did their effusive cousins. Consequently, the achievable intensity within a narrow velocity interval, a highly desirable feature, could be very much higher with convective beams. What had not yet been shown, nor apparently considered, was that heavy molecules could be accelerated to suprathermal translational energies by expansion of a light carrier gas such as hydrogen or helium in which those heavy molecules were present a low levels.

When I returned to Princeton after my year in London, I found that John Scott, who had run the SQUID Office during my absence, was going to the University of Virginia as an assistant professor of Aeronautical Engineering. He was wondering what research he ought to work on and I told him about my musings on molecular beams from high pressure sources. I also told him that if he would submit proposal on such work to Project SQUID I would try my best to get support for it. When he arrived at Virginia he found a group that was already using beam scattering experiments to characterize the exchange of energy and momentum between molecules and surfaces. Their goal was to understand and predict surface heat transfer and drag for aerodynamic bodies at very high altitudes where flow was molecular rather than continuum. The high intensities achievable with beams from supersonic free jets (or "nozzle beams", as they came to be known) would be most useful in such surface scattering experiments. Moreover, the high translational energies that seemed achievable with the seeded beams I was dreaming about, would make possible the study of drag and heat transfer in molecular flow under conditions much closer to those expected in ultra high altitude flight than those obtainable with the effusive beams they had been using. As a result of their enthusiasm and mine, the first nozzle beam system in America was built in Charlottesville, VA. [8] I should really say, the first in America since the one T.H. Johnson built at Yale in 1929 that produced high intensity beams of mercury atoms but had been ignored. It now seems clear that the reason for Johnson's success was that the walls of his system were all water cooled enough to condense most of the incident mercury atoms. The net result was an effective pumping speed orders of magnitude higher than the nominal speed of his vacuum pumps or those of any of the other molecular beam systems then in existence. (Fraser in his 1932 book on "Molecular Rays" refers to a system then being built with the "exorbitant" pumping speed of

100 liters per second [9]. The nominal speed of each of two 32 inch pumps that we bought "off the shelf" 45 years later was 30,000 liters per second!) The real reason for the generally observed decrease in beam intensity with increasing source gas density was scattering due not to the molecules in a "cloud" at the source orifice exit but to those in the relatively high density of background gas throughout the system due to inadequate pumping speed.

Meanwhile, I was getting more and more intrigued with the problems and possibilities of beams from free jet sources and eager to become more directly involved. I had by this time learned that in spite of Dean Taylor's rosy assurances I could not become an active participant in academic research unless I had some direct association with a department. SQUID had been supporting research on two-phase flow by Shao Lee (Charlie) Soo, a young assistant professor of Mechanical Engineering. During a visit to Soo's lab one day I met Robert Drake, the new chairman hired to refurbish that department's image by expanding its research activities. I described my wild ideas on free jet beams and said: "If I write a proposal and get some support will your department provide a home for the project?" Bob had done his doctoral research in rarefied gas dynamics at UC Berkeley. Familiar with the molecular perspective on gas dynamics he immediately said yes. I prepared a proposal on which I was listed as a "Consultant" because I was not on the faculty and under university rules could not be a Principal Investigator. That role was assumed jointly by Bob Drake and Michele Boudart, my next door neighbor and close friend who was interested in chemical kinetics and catalysis and was on the faculty of Chemical Engineering. I had been discussing my beam ideas with him and he too found them intriguing. The proposal was submitted to NSF who sent it for review to Immanuel Estermann, a co-author with Stern and Frisch of the landmark paper on diffractive scattering of hydrogen from LiF, one of the two papers cited when Stern won the Nobel prize in 1943. Estermann had become Director of ONR's Material Sciences Division (whose Power Branch was responsible for Project SQUID) and was thus my ultimate boss. He took his reviewing seriously and spent a day in Princeton discussing the proposal with me. (The fact that his daughter was then living in Princeton was no doubt something of an inducement for him to make the trip from Washington.) He had visited Becker's laboratory in Marburg, Germany and was aware of the difficulties Becker had encountered in fabricating the tiny converging-diverging nozzles prescribed by Kantrowitz and Grey. As he took his leave he said: "John, I'm afraid this research may be too difficult to pursue with graduate students but I think it is important so I'm going to recommend support." He apparently was as good as his word because NSF gave us the money we had sought and we began work in the fall of 1960. As it turned out, Estermann's fears about the difficulties of making very small Laval (converging-diverging) nozzles were groundless. The Marburg group discovered empirically, that beam intensities became higher when the diverging section of the nozzle was cut off! [10] The lower intensities obtained with a diverging nozzle are due to viscosity effects which Kantrowitz and Grey had ignored in their analysis. Dave Miller's group in San Diego, later showed experimentally that for most practical purposes the jets from a simple orificea (hole in a flat plate), a simple converging nozzle, or a long tube, had almost identical properties as long as the flow was "choked", i.e. reached sonic velocity, at the exit plane of the nozzle, tube or orifice. [11]

ELECTROSPRAY WINGS FOR MOLECULAR ELEPHANTS

Nobel Lecture, December 8, 2002

by

JOHN B. FENN

Department of Chemistry, Virginia Commonwealth University, Richmond, Virginia 2384-2006, USA.

The grant from NSF in 1960 started us on the road that led to Stockholm in 2002. Co-Principal Investigator, Michel Boudart was then hosting Jacques Deckers, a young Ph.D. who had done his doctoral research on ions in flames with Van Tiggelen at the University of Louvain. Jacques had had experience with vacuum systems and was very excited about our beam project so Michel turned him over to my tender love and care, (or me to his, I'm not sure which!). With Bob Drake and Michel looking over our shoulders we began designing an apparatus. One of the lessons from the experiments of the Kistiakowsky and Becker groups was the need for lots of pumping speed. When the salesman from NRC showed us the performance characteristics of its 32 inch (in diameter) oil diffusion pumps, Jacques' enthusiasm overcame my reticence (or timidity) so the main chamber of our system became a long horizontal tank resting on and evacuated by two of those big pumps. In retrospect I think that in choosing such big pumps as the heart of our system we were either much wiser or more fortunate than we knew. Those who work with vacuum systems learn early that there is no such thing as too much pumping speed. Sooner or later they always wish they had more. It's nice not to have to worry about the vapor pressure of finger prints, even nicer to be able to sustain disasters. We arrived one Monday morning to find the apparatus completely filled with some 700 gallons of water from a failed cooling coil! We drained the system and easily decanted the water from the silicone pump oil. After cleaning, the interior surfaces of the pumps, pipes and chamber, all of which were made of mild steel, sported a beautifully uniform beige coat of rust! We then closed up the chamber and ran the forepumps for a couple of days, passing a small flow of air through the system to help dry things out. With fingers crossed we then started the diffusion pumps and got the best vacuum we had ever obtained! Figure 1 shows a picture of that apparatus after it was moved to Yale.

Meanwhile Dame Fortune had smiled at me in another way. Not long after the beam project started, Professor Soo, who had been teaching thermodynamics to the mechanical engineers, suddenly announced he would leave at the end of the first semester. Bob Drake, not knowing I had barely passed the

Figure 1. The "big" molecular beam system, with two "32" diffusions pumps and one "16" Jet Booster Pump, after it was moved from Princeton to Yale. The people, from left to right: S.P. Venkateshan and S. B. Ryali, both postdocs from India.

only course I had taken in the subject, asked me if I would accept a temporary appointment as a lecturer and teach the sophomore course on thermodynamics in the second semester. After a deep breath and a small prayer I said yes. Things must have gone pretty well because at the end of the term he asked if I would consider a permanent position on the faculty. All of a sudden I had the chance to realize what I had long hoped for, the opportunity to become a first class citizen in an academic community. Moreover, because my appointment as SQUID's director was "with rank of professor" I automatically became an instant full professor of Mechanical Engineering, a transmutation that would be much more difficult in today's world. I also continued to run SQUID until 1962 when, by mutual agreement of all parties, its headquarters moved to the University of Virginia where John Scott became director.

After an extensive effort in design and construction we finally assembled the molecular beam apparatus and got it working well. The first meaningful experiment was by Bob Koros, one of Michel Boudart's graduate students. He measured the sticking coefficient of water molecules on ice at liquid nitrogen temperatures. The simplicity and success of the experimental design were strong evidence of both the achievability and utility of high beam intensities from free jet sources. Bob simply suspended a small target from a calibrated helical spring of glass and measured the time-dependent extension of the spring with a cathetometer viewing the target through a window in the vacuum system. Nearby baffles, filled with liquid nitrogen, maintained the target at low temperature. The beam flux of water molecules, from a source containing at water vapor at pressure of a few torr, was sufficiently intense to de-

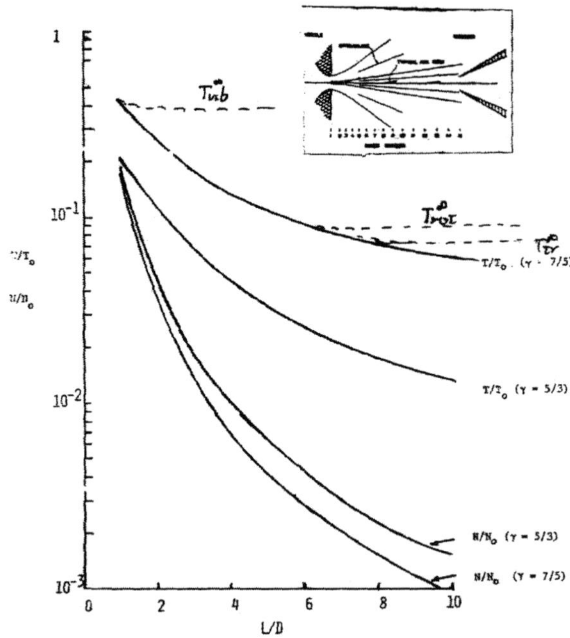

Figure 2. The axial distribution of gas properties in a supersonic free jet, normalized by their source values. The abscissa distance is in terms of nozzle diameters. The top two solid curves show the temperature distribution for a monatomic and a diatomic gas diatomic gas with rotational and vibrational degrees of freedom in equilibrim. Note that because the exchange of energy between translation and vibration is very slow, i.e. requires many collisions per quantum, the "vibrational temperature" (for a diatomic gas like nitrogen). "freezes" much earlier in the expansion than does the rotatational temperature. The bottom pair of curves show the axial distribution of gas density, also for a monatomic and a diatomic gas.

posit in a few hours enough mass of ice to extend the calibrated spring by measurable amounts. In a subsequent analogous experiment Jim Anderson, another Boudart student, found that oxygen molecules incident on a heated germanium wafer reacted to form a volatile GeO molecule. Above a threshold value that I've forgotten, the reaction probability of 0.04, was independent of the target temperature and the intensity of the incident beam. In that case we used the spring balance to measured the loss of mass in the germanium wafer. These preliminary experiments clearly demonstrated the utility of the apparatus whose design and performance were described and characterized in a 1963 paper [12].

Over the next four years at Princeton and the following twenty at Yale this equipment gave rise to some 65 papers. We confirmed the high intensities of beam molecules predicted by Kantrowitz and Grey and first realized by Becker and Bier. Moreover, our hope that the seeding technique could produce beams of molecules with supra thermal translational energies was abundantly realized by ourselves and others [13a]. Figure 2 shows some examples of the acceleration and the energies we obtained. Compargue and his group in France have since achieved translational energies as high as 45 eV.[13b]

The net result has been that studies of molecule-molecule and molecule-surface interactions, by ourselves and others, reached new levels of energy, utility and sophistication.

An originally unanticipated virtue of free jet expansions was their ability to produce and maintain populations of molecules under non-equilibrium but steady-state conditions which some investigators have characterized as "new states of matter!"[14] For present purposes it will be useful to show the distribution of conditions encountered by molecules of a gas undergoing a free jet expansion from a simple orifice into vacuum. Such expansions are "self-similar" in that they scale in terms of the orifice diameter. That is to say, at a distance of any particular number of diameters along the jet axis, the *equilibrium* state of a particular gas will be the same, no matter what the diameter of the jet may be. Thus, at 5 mm downstream from an orifice one mm in diameter the equilibrium state of the jet gas is the same as in a jet of the same gas at 5 m downstream from an orifice 1 m in diameter. As it happens, however, the expansion rate from small diameter orifices is so fast that the collision-induced processes and reactions involved in maintaining equilibrium in the jet gas may not be rapid enough for the gas to reach equilibrium before the gas density gets so low that the collision induced processes are arrested. The net result is that the terminal state of the jet gas may be far from equilibrium. This ability of free jet expansions from small nozzles to produce steady state conditions in gases that are far removed from equilibrium is one of the reasons that these jets have proved to be such interesting research tools.

Figure 2 shows the axial distribution of density (N/N_0) and temperature (T/T_0) in supersonic free jets from a monatomic and a diatomic gas where subscript "o" refers to the value in the source when the gas is at rest. The curves show the values first obtained by Owen and Thornhill from solutions of Euler's equation for monatomic species ($Cp/Cv = 5/3$) and for diatomic gases with rotational (but no vibrational) degrees of freedom ($Cp/Cv = 7/5$) [15]. We had no means of measuring the gas density "N" but we could and did obtain values of the centerline temperature at any point by locating the skimmer at that point, and "chopping" the extracted beam into pulses with a rotating shutter. The signal-vs-time output from a fast response detector located at a carefully measured distance downstream from the shutter was readily transformed into a velocity distribution and thus the translational temperature of the jet molecules at the skimmer entrance. From such velocity analysis at varying values of nozzle-skimmer distance we determined the distribution of translational temperatures of jet molecules along the jet axis. For monatomic and diatomic gases, e.g. argon and nitrogen, the experimental values fell right on the theoretical curves out to distances at which the jet gas densities became so low that the collision frequency was not high enough to maintain equilibrium. Because the jet expansion is adiabatic, and because vibrational relaxation times for many species are much longer than the sojourn time of the molecules in the jet, the vibrational energies for such molecules remain "frozen" at near the source values. Thus, as shown by a dashed line in Figure 2, the vibrational temperature for nitrogen "freezes" at an L/D value

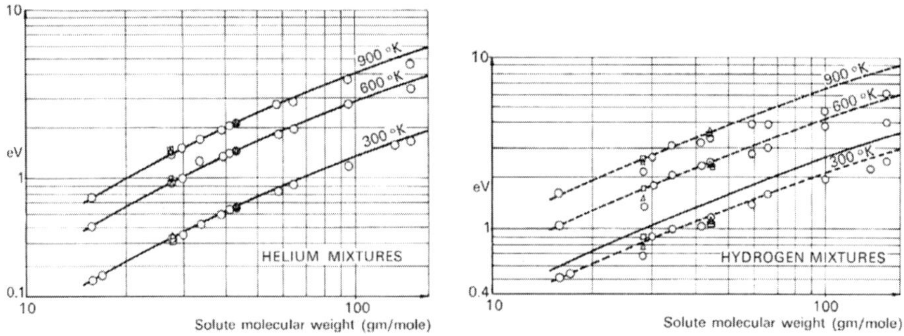

Figure 3. Some results on the acceleration of heavy molecules by a carrier gas comprising helium in 3a and hydrogen in 3b. Note that, as is clear in 3b for hydrogen, with increasing difference in molecular weight between seed and carrier species there can be appreciable slip resulting in lower velocities for the seed species than would be the case if the velocities of the two species remained in velocity equilibrium. The dotted lines in 3b assume no relaxation of hydrogen's rotational energy. The solid line at 300 K shows what would have happened if there were no slip and the rotational and translational "temperature" had remained in equililbrium. In Dole's experiments the differences in molecular weight between seed and carrier species were much greater than for any mixture in Figure 3. We estimate that the actual velocities of some of his large polymer molecules were as much as 40 per cent less than he thought they were.

of only 1.7 because in a single collision between two nitrogen molecules the probability of transferring a single quantum of vibrational energy into translational energy is only 1/5000. For rotational energy that probability doesn't "freeze" until an L/D value of about 6 is reached. Finally, even translational temperature becomes frozen at an L/D value of about 7.5, after with the flow is essentially collisionless. By means of an energy balance on the jet molecules it is thus possible to determine the relaxation rates for rotational degrees of freedom in polyatomic molecules.[16] It turns out that this approach is extremely effective for measurement of such rates at much higher gas temperatures than can be accommodated in the widely used instruments based measurements of the velocity of sound. In the present context the important result of our extensive studies of free jet expansion by velocity analysis is that the translational temperature distributions obtained analytically by Owen and Thornhill and shown in Figure 2 have been experimentally confirmed by ourselves and others. It will emerge in what follows that these findings played a crucial role in the development of Electrospray Ionization Mass Spectrometry.

Having developed the ability to measure velocity distributions at any point in the jet we, and subsequently others, were able to verify our hope that heavy molecules seeded at low levels into a light carrier gas could be accelerated to high velocities and thus high translational energy. [13a]. Figure 3 shows some early examples of the acceleration and the energies we obtained. Compargue and his group in France have since achieved translational energies as high as 8 eV [13b]. An important result of this seeding technique has been that studies of molecule-molecule and molecule-surface interactions, by ourselves and others, reached new levels of energy, utility and sophistication.

Free jet expansions have also revolutionized the optical spectroscopy of polyatomic molecules. Our studies showed that such expansions cool rotational degrees of freedom without affecting the vibrational modes. Thus, by performing absorption or emission spectroscopy on polyatomic molecules in a free jet, one can eliminate the rotational broadening of vibrational lines. Unfortunately, we were not knowledgeable enough to realize the spectroscopic implications of rotational cooling so it was Wharton, Smalley and Levy who provided the earliest demonstration of this effect, obtaining the first ever spectrum of NO_2 in which the vibrational lines could be clearly distinguished and identified without rotational broadening.[17] Now the free jets used in optical spectroscopy far outnumber those used as molecular beam sources in scattering experiments!

Because they can produce extremely low temperatures in gases at fairly high densities, free jet expansions can bring about supersaturation of almost any gaseous mixture for very short periods of time. Thus they have become sources of a wide variety of uncommon species such as HeBr and ArI and all kinds of aggregates, pure and mixed, even including helium. Consequently, by 1984 free jet expansions had began to play what has turned out to be a major role in the rapid development of so-called "cluster science and technology".[18] Indeed, the Scoles group at Princeton {19] and later the Toennies group at Goettingen [20] have found remarkable effects in the emission spectra of species contained in clusters of superfluid liquid helium where the absence of viscosity means that atoms can rotate freely!

In 1967, for a variety of reasons, I decided to accept an invitation to join the Chemical Engineering group in a newly organized "Department of Engineering and Applied Science" at Yale. The "beam machine" was dismantled in Princeton and reassembled at Yale in a somewhat different configuration. The picture in Figure 1 was in fact taken at Yale. The very large size of the apparatus had turned out to provide an unexpected dividend. Because our published results had many interesting implications in chemistry, many chemists visited our laboratory in both Princeton and New Haven. Frequently intimidated by the very size of our apparatus, they would listen to our story and say "Very nice, but not for us." The net result was that for a long time we enjoyed a relative absence of competition! In time, of course, we and others learned how to achieve similar results with much smaller and more compact gear so that free jet expansions have now became very commonplace tools in departments of chemistry and physics as well as engineering. It is perhaps appropriate to note that these supersonic free jets have played an important role in research that has helped at least five other investigators win Nobel prizes in Chemistry! [21]

In 1968, the year after I arrived at Yale, Malcolm Dole published his now famous paper describing his attempt to measure the molecular weights of oligomers of synthetic polymers based on electrospray ionization.[22] He chose to look at polystyrene molecules because relatively monodisperse samples were readily available over a wide range of molecular weights. That paper caught the eye of Professor Seymour (Sandy) Lipsky in Yale's Medical School.

With characteristic enthusiasm Sandy showed it to Csaba Horvath, one of my new Yale colleagues in Chemical Engineering who was working with Sandy on HPLC, then known as "High Pressure Liquid Chromatography." Sandy, who was also a mass spectrometrist, was quite excited because he thought that Dole's approach might make possible the mass spectrometric analysis of proteins and other complex molecules of biochemical importance. This cherished but elusive goal had long seemed a nearly impossible dream because all the then common methods for ionizing any atom or molecule depended upon an appropriately energetic gas phase encounter between a neutral molecule and and electron, photon or ion of some other atom or molecule. But large polyatomic molecules, especially the complex and fragile species of biological interest, could not be vaporized without extensive fragmentation and decomposition. With characteristic enthusiasm Sandy showed Dole's paper to Csaba. Noting the references to our Princeton papers Csaba told him that the John Fenn of those references had just moved to Yale so Sandy showed the paper to me. Always looking for new ways to apply our free jet technology I was readily seduced by Dole's results and Sandy's enthusiasm. Thus began a friendship with Sandy that lasted until his death and a romance with mass spectrometry that will doubtless last until mine. That romance led to the now popular *Electrospray Ionization (ESI)* which burst into prominence at about the same time as the other now most widely used method for ionizing large nonvolatile molecules, *Matrix Assisted Laser Desorption Ionization (MALDI)*, began to show its potential. A brief history of each of these techniques will be summarized in the next two sections of this report. MALDI's story will be told first even though ESI's was the first to begin.

IONIZATION BASED ON "ENERGY-SUDDEN" METHODS

One class of techniques for producing intact ions from complex and nonvolatile molecules was embodied in what might be called the "energy sudden" approach based on an idea originally proposed in 1974 by Beuhler *et al.* [23]. On the grounds of rate theory for unimolecular decomposition they argued that sufficiently rapid heating could vaporize complex molecules before they had time to decompose. In some proof-of-principle experiments on the vaporization of small peptides they showed that decomposition indeed decreased with increasing heating rates but they never were able to heat rapidly enough to eliminate substantial fragmentation. The several techniques based on this idea include so-called "Pyrolysis Mass Spectrometry" (PMS) [24], pre-MALDI versions of Laser Desorption (LD) [25], Fast Atom Bombardment (FAB) [26], Fast Ion Bombardment, (usually called SIMS for "Secondary Ionization Mass Spectrometry) [27], in which rapid energy addition is achieved by incidence of high energy atoms, ions, or photons and Plasma Desorption (PD) [28] in which the sudden energy comes from the decay of a radioactive isotope, usually Californium (^{252}Cf). Introduced by MacFarlane and Torgerson PD was the first of these techniques to gain appreciable use.[28] Its name stems from the idea that the energy released by the nuclear disintegration formed a small blob of plasma on the sample-bearing surface. As many as 30 or more analyte ions are ejected from that plasma when the analyte is dispersed in an appropriate matrix, typically nitrocellulose. The ions are then accelerated through a drift region so that their masses can be determined by time of flight (TOF) analysis for which a convenient zero-time marker is provided by back-scattered decay products of the Cf disintegration.

All these "Energy Sudden" methods depend on the nearly instantaneous achievement of high-energy density in a sample dispersed in or on a solid or liquid surface. Thus they can be regarded as extensions of the rapid heating idea of Beuhler *et al.* They all could produce intact ions from many complex molecules including peptides and small proteins. However, their procedures were complex and awkward, and because their processes were highly irreversible the yields of intact ions were very low. Even so, during the 70's and 80's they provided substantial accumulations of mass spectral data on a variety of complex, non-volatile molecules, giving rise to a literature much too large for review or comment here.

By far the most important legacy of these energy-sudden methods is so-called Matrix Assisted Laser Desorption Ionization or MALDI introduced in 1987 by Tanaka (29) and expanded in 1988 by Karas and Hillenkamp [30]. In MALDI the analyte molecules are dispersed on a surface in a thin layer of matrix, usually an organic acid. The energy of an incident pulse of laser photons is absorbed mostly by the matrix to form a jet of matrix vapor that lifts analyte molecules from the surface and by mechanisms still not well understood, transforms some of them into ions that are mostly singly charged. Because those ions are all produced at a well defined location in an exceedingly short time their mass analysis is most effectively achieved by Time-Of-Flight methods, as in the case of PD. MALDI is one of the two ionization methods for biomolecules that promise to dominate the MS scene for the foreseeable future. Recently MALDI-TOF in vacuo has been supplemented by MALDI at atmospheric pressure with subsequent introduction of the resulting ions into any of several types of mass analyzers.

IONIZATION BASED ON INTENSE ELECTRIC FIELDS

In 1966, when Dole first had the idea that led to his 1968 paper, there was little evidence of any useful solution to the problem of vaporizing large polar molecules. However, in 1951 the first step had been taken along one of two paths that ultimately led to successful "soft" ionization methods for complex molecules, though it was not fully appreciated at the time. In that year Mueller discovered that when a sharp metal point was at a sufficiently high voltage in vacuo, the electric field at the tip of that point could be intense enough to extract an electron from a nearby gaseous atom or molecule to form a positive ion [31], a discovery that led to his development of Field Ion Microscopy (32). The first use of such points as sources of ions for mass analysis was reported three years later by Inghram and Gomer [33], but only with gaseous molecules. The ion currents were too small to stir up much interest but they were larger than could be accounted for by the flux of molecules from the ambient gas into the region occupied by the point. Beckey then suggested that this apparent excess in ion formation rate could be accounted for by molecules that struck the surface of the wire at some distance from its tip and then arrived at the tip region by surface diffusion. In 1963 he showed that arrays of "whiskers" on thin wires could produce much higher currents than single needles and that non-volatile species present on the surfaces of such whiskers could migrate to the tip by surface diffusion to be desorbed as intact ions. Beckey went on to develop this discovery into "Field Desorption Ionization Mass Spectrometry" (FDIMS) and in 1977 published a book on the subject [34]. Emitter electrodes comprising wires with dense arrays of whiskers are now available off the shelf at relatively low cost but dosing them with sample and positioning them in the vacuum system are time-consuming and tedious tasks. Moreover, ion currents depend so strongly upon both the temperature of the emitter and the applied voltage that substantial manual skill is required to find and maintain the right combination for a particular sample. That control function might now be made much less tedious by computers but solving the control problem does not deal with an even more important difficulty. The ions from field desorption are released from high voltage into high vacuum. Consequently, their kinetic energies are so high that their mass analysis required relatively large and expensive magnetic sector instruments. For these and other reasons FDIMS a la Beckey has never been widely practiced. However, from the perspective of history's hindsight one can argue that the ability of intense fields at the surface of a tiny charged droplet, which play a vital role in Dole's ESI, is simply an obvious extension of the discoveries by Mueller and Beckey on the nature of Field Ionization at a sharp tip, i.e. a surface with a very small radius of curvature. Maybe so, but it is also quite clear that nobody recognized these possibilities at the time, and that Dole arrived at the ESI approach by a logic that depended in no way on the field ionization ideas of Mueller and Beckey. In other words Dole was a true pioneer in that his ESI ideas were breaking brand new ground.

Fields intense enough to desorb solute ions at useful rates from a condensed phase can also be achieved without the array of very sharp needles or "whiskers" required by Beckey's approach. In 1972 Evans and Hendricks found that a high voltage applied to the surface of a non-volatile conducting liquid in vacuo would produce a sharply pointed cone [35]. The field at the tip of such a cone was intense enough to desorb charge-bearing molecules and clusters of the liquid. If the liquid was a solution, the desorbed ions included solute species. These observations led to so-called "Electrohydrodynamic Ionization" (EHI} that was introduced to the mass spectrometry community in 1974 by Simons *et al.* [36]. In EHI a sample solution is electrosprayed from a small bore tube maintained at high potential relative to the surroundings, just as had been taught by Zeleny and adopted by Dole. The difference is that in Dole's ESI the droplets are dispersed into gas at near atmospheric pressure. In EHI the sample solution is dispersed into vacuum. In that case the solvent must have a very low vapor pressure in order to avoid "freeze drying". But as Dole had recognized, evaporation of solvent is the *sine qua non* for the formation of solute ions from charged droplets and such evaporation requires a source of enthalpy. This dilemma has never been completely resolved in EHI because no satisfactory method has yet been found by which the enthalpy needed to vaporize solvent from a droplet can be supplied to such a droplet in vacuo. Moreover, as in the case of FDI, any ions produced by EHI in vacuo have such high kinetic energies that their mass analysis generally requires magnetic sector instruments which are large, heavy and expensive. For these reasons EHIMS has never become widely used despite fairly extensive research that has been reviewed in detail by Cook [37].

In sum, Malcolm's Dole's paper in 1968 was the seed which, after an extended period of germination, ultimately blossomed into Electrospray Ionization (ESI), one of the two "soft" ionization methods that have made the precision, sensitivity and elegance of mass spectrometry readily available for the study of biomolecules and their reactions. The germination and development of Dole's ESI will be briefly described in what follows.

ELECTROSPRAY IONIZATION

The other of the two dominant methods for ionizing large molecules is Electrospray Ionization (ESI), the subject of the 1968 paper by Malcolm Dole and his colleagues [22] that Sandy Lipsky showed me in 1969. In fact Dole had first proposed his electrospray idea two years earlier (1966) at the International Symposium on Macromolecular Chemistry in Tokyo. Those two communications comprise a testament to Dole's remarkable acumen in realizing that the production and behaviour of highly charged droplets from solutions of non-volatile solute species in volatile solvents might produce intact gaseous ions of those species. Perhaps even more remarkable was the extent of his insight on the roles of ESI's several component processes. For example, he clearly recognized the need to disperse the charged droplets in a gas at relatively high (atmospheric) pressure. Such gas would provide the enthalpy required to vaporize solvent from those highly charged droplets. Moreover, collisions of the ions with molecules of that gas would reduce their initially high kinetic energies to those of the ambient neutral gas molecules, thereby avoiding the need for mass analyzers capable of accommodating high energy ions, e.g. magnetic sector instruments.

As most mass spectrometrists now know, the essence of Dole's idea was to take advantage of what happens during the evaporation of solvent from a droplet that has a net electric charge. In 1882 Rayleigh had analyzed the behaviour to be expected of such a droplet and found that as the solvent evaporated the density of charges on the droplet surface would increase to a critical value, now known as the "Rayleigh Limit", at which Coulomb repulsion would overcome surface tension. The resulting instability would cause the droplet to break up into a plurality of offspring droplets [38]. Thirty years later John Zeleny made the first reported observation of this "Rayleigh Instability" in pioneering experimental studies on the production and behaviour of

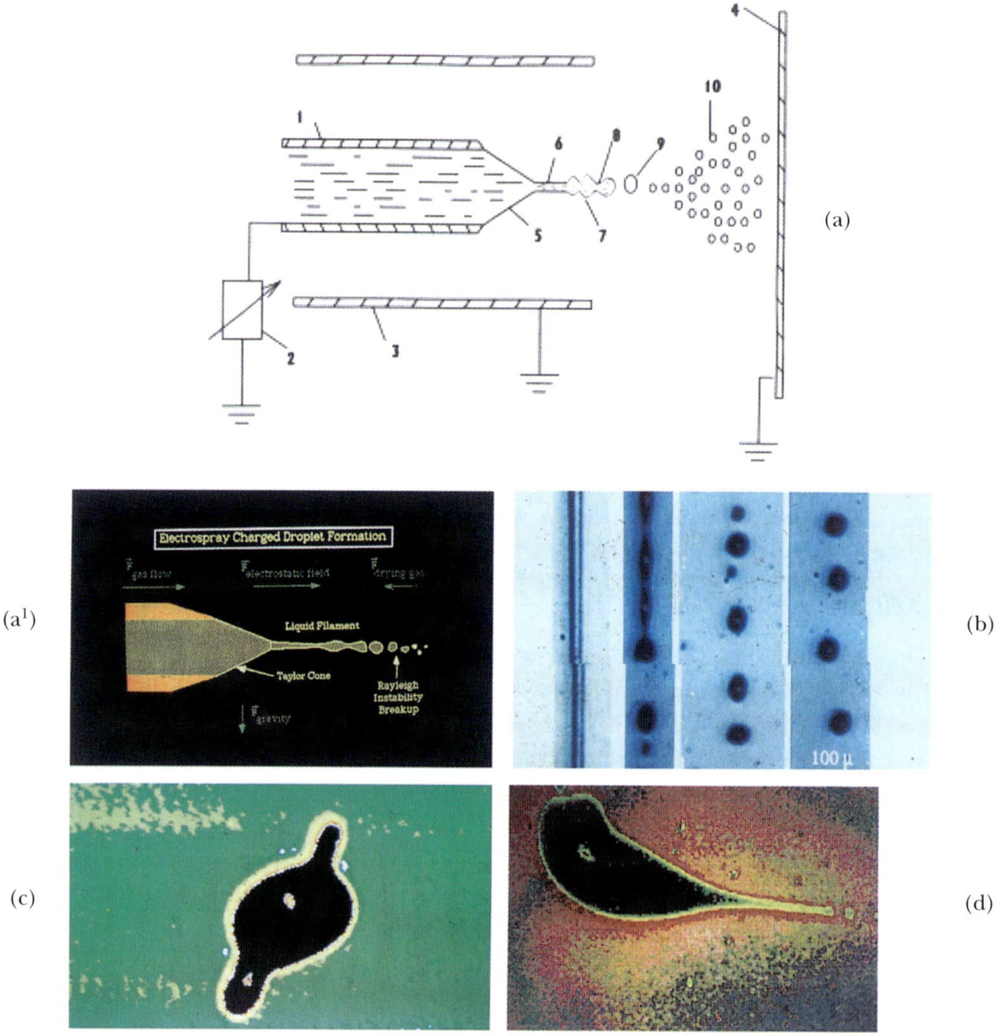

Figure 4. How a flow of conducting polar liquid is dispersed into highly charged droplets by an intense electric field at the exit tip of a small tube at high potential relative to an opposing electrode. Upper left hand panel *a* show schematically the conical meniscus formed by the emerging liquid as the result of competition between the surface tension of the liquid and forces due to the interaction of dipoles in the liquid with the applied field. This meniscus is known as a "Taylor Cone" because it was first characterized theoretically by G.I.Taylor. [41] When the liquid is electrically conducting, a small flow of the liquid emerges from the tip of the cone as a very narrow column or jet. Interaction between surface tension and viscosity produces so-called varicose waves which grow in amplitude until they truncate the jet into a sequence of segments of equal length that are molded by surface tension into droplets nearly uniform in size.[42] Because the droplets contain excess cations or anions on their surface, depending on the polarity of the field, they repel each other with the result that their trajectories diverge to form a conical "electrospray" of charged droplets. This sequence was captured in the photographs by Gomez and Tang shown in adjacent upper right hand panel *b*. If the liquid is volatile, evaporation shrinks the droplet thereby bringing the surface charges closer together until the resulting Coulomb repulsion overcomes surface tension and the droplets disrupt into a plurality of smaller droplets, a phenomenon characterized and predicted theoretically by Lord Raleigh in 1882. Two years more than a century later, Gomez and Tang captured on film, for the first time, the images in the lower two panels *c* and *d* showing a droplet in the throes of what is now known as a Rayleigh Instability, and sometimes referred to as a Coulomb explosion.

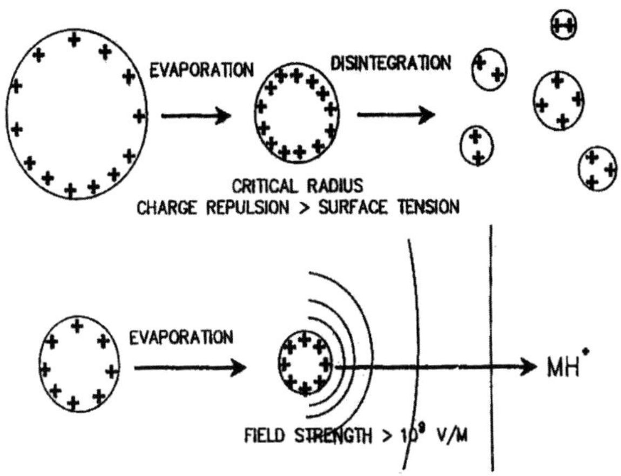

Figure 5. The upper sequence of sketches attempts to portray Dole's idea of how solute ions are formed. He assumed that the offspring droplets resulting from a first Rayleigh Instability would continue to evaporate solvent so that they too would undergo a Rayleigh Instability and disrupt. If the original solution were sufficiently dilute, a sequence of such evaporation-disruption episodes would ultimately produce ultlimate droplets so small that each would contain only one solute molecule. As the last of the solvent evaporated from the ultimate droplet the remaining solute molecules would retain some of the droplet charge and thus become a free gas phase ion. This scenario embodies what is sometimes referred to as the "Charged Residue Model" (CRM) for ES ion formation. The lower sequence of sketches portrays the so-called "Ion Evaporation Model" scenario later proposed by Iribarne and Thomson (46). It argues that before ultimate droplets containing only one solute molecules are formed, the field at a droplet's surface becomes sufficiently intense to lift a solute ion from the droplet surface into the ambient gas.

charged droplets carried out, first at the University of Minnesota and then at Yale [39]. He passed a stream of conducting volatile liquid through a small bore thin walled tube or "needle" (because it often comprised a short length of hypodermic needle tubing) maintained at a high potential relative to an opposing counter electrode. The resulting intense electric field at the tube tip dispersed the emerging liquid into a fine spray of charged droplets. Zeleny was able to see the break-up of charged droplets as solvent evaporated, just as Rayleigh had predicted. Figure 4a is a schematic representation of how the droplets are formed. The emerging liquid at the tube exit forms a conical meniscus now known as a "Taylor Cone" because its formation was predicted theoretically in 1964 by G. I. Taylor [41]. In truth, Taylor's treatment applied to a non-conducting liquid of neutral molecules with dipole moments for which there is no electric current and no liquid flow. Even so, the Taylor name is still applied even when the liquid is conductor because it contains anions and cations. In that case a thin jet emerges from the tip of the cone. As the result of another kind of instability, also first characterized by Rayleigh for uncharged liquids [42], the interaction between viscosity and surface tension produces so-called "varicose waves" on the surface of such a liquid jet. Those waves grow in amplitude until they truncate the jet into a series of uniform droplets as illustrated in Figure 4b. (That same kind of instability causes the break-up into droplets of a solid jet of water issuing from a garden hose or, on a smaller scale, the tiny stream of water from a tap that is incompletely turned off.) In the electrospray case the droplets all have excess charges of the same sign, (anions or cations, depending on the polarity of the applied field). Coulomb repulsion thus results in a divergence of their trajectories to form a conical "electrospray" of charged droplets. Figure 4b is a series of four photographs made by Tang and Gomez [40] illustrating, from left to right, the undisturbed jet, the appearance and growth of the varicose waves, the resulting droplets, and the beginning of the divergence of the droplet trajectories (due to Coulomb repulsion} to form a conical spray. We believe Figures 4c and 4d by Tang and Gomez are of historic significance as the first ever photographs showing droplets undergoing a Rayleigh instability when the Coulomb repulsion of the surface charges overcomes the surface tension that tries to contain the droplet liquid in a spherical shape.

Dole's scenario further assumed that the "offspring" droplets resulting from a Rayleigh instability would continue to evaporate until they too would reach the Rayleigh limit and break up into still smaller droplets. If the original solution were sufficiently dilute, a succession of such Rayleigh instabilities would ultimately produce droplets so small that each one would contain only one solute molecule. That lone molecule would retain some of its droplet's charge to become a free gas phase ion as the last of the solvent evaporated. The upper sequence in Figure 5 attempts to illustrate this process. (The lower sequence illustrates an alternative proposal by Iribarne and Thomson which will be described later.)

In the 1968 paper that described this scenario Dole also presented preliminary experimental results obtained in an attempt to reduce his idea to practice. As taught by Zeleny, a solution of polystyrene molecules was infused through a small bore tube maintained at high potential relative to a counter electrode. Dole was always generous and conscientious in acknowledging the work of others so it is curious that though he refers to Rayleigh's analysis of droplet instability he never mentioned Zeleny in any of his papers, even though his method of producing charged droplets seems to derive directly from Zeleny's experiments. His account of the electrospray experiments in his autobiography simply says: "I got this idea from learning about the electrospraying of paint on to automobile bodies while working as a consultant for a paint company in Chicago" [43]. Such "electrospraying" of paint consisted in maintaining a potential difference between the sprayer and the object being painted. The result was enough charge on the paint droplets to attract them toward the object being painted, thereby decreasing substantially the loss of paint to the surroundings by convection currents. Similar charging of droplets was practiced by "crop dusting" airplanes to decrease aggregation and increase adhesion of insecticides to the plants.

But Dole had another problem to solve. The mass spectrometers then available, even the most expensive magnetic sector instruments, were unable to "weigh" singly charged ions with masses larger than about ten thousand Daltons. The oligomer ions of interest to Dole could have molecular weights up to a million or more and he had apparently not taken into account the

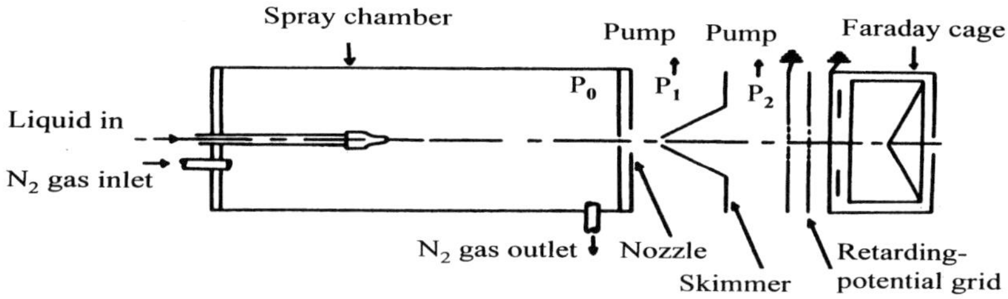

Figure 6. Schematic representation of Dole's original apparatus. The electrospray forms at the tip of the tube through which liquid enters the spray chamber concurrent with the nitrogen bath gas. Most of that gas leaves the chamber near the end plate containing the nozzle through which a small portion of the mixture of bath gas and ions enters the vacuum system in the form of a supersonic free jet. A small fraction of the ion-bearing jet gas passes though a skimmer into a second vacuum stage, thence through a retarding-potential grid into a Faraday cage connected to a galvanometer.

possibility of extensive multiple charging which then had not yet been discovered. The problem he thus faced was how to "weigh" such large ions after they had been produced. Moreover, as he also mentioned in his autobiography, he had found that the Magnetic Electron Multiplier (MEM) on his Bendix mass spectrometer did not respond to his ES ions because "Although these ions had the same kinetic energy as light ions their velocity decreased in the ratio of the square root of their mass. We proved the mass limitation of the MEM in a separate experiment." [44] That comment strongly suggests that the ions Dole had produced from polystyrene oligomers did not have extensive multiple charging. Therefore, their mass/charge ratios were too high for available analyzers so he had to find another method for "weighing" them. That is how our free jet studies entered the picture.

As noted earlier, we had shown that during such free jet expansions of mixtures comprising a very dilute solution of heavy "seed" molecules in a light carrier gas. e.g. argon in helium, the heavy seed species were accelerated like dust particles in a windstorm to almost the velocity that would be reached by the light carrier gas alone [13]. Consequently, the translational kinetic energy of the argon molecules after expansion was higher than would be achieved in a jet of pure argon by a factor roughly equal to the ratio of the molecular weights of the two species, i.e. 10 in the case of argon accelerated by helium. This acceleration effect was important to Dole because it seemed to provide a way to weigh the large polymer ions he hoped to produce. As shown in the sketch of his apparatus in Figure 6 (from his 1968 paper) a variable potential grid was placed between the skimmer (that passed a collimated, ion-containing beam from the free jet source into a vacuum chamber) and a sensitive galvanometer in a Faraday cage contained in that chamber. The only ions that could pass through the grid and reach the galvanometer were those with a translational kinetic energy ($mv^2/2$) exceeding the product of the charge on those ions and the potential on the grid. The translational velocity v of the stopped ions was equal to the readily calculated velocity reached by the carrier gas in the free jet. Thus, scanning the voltage applied to the grid should

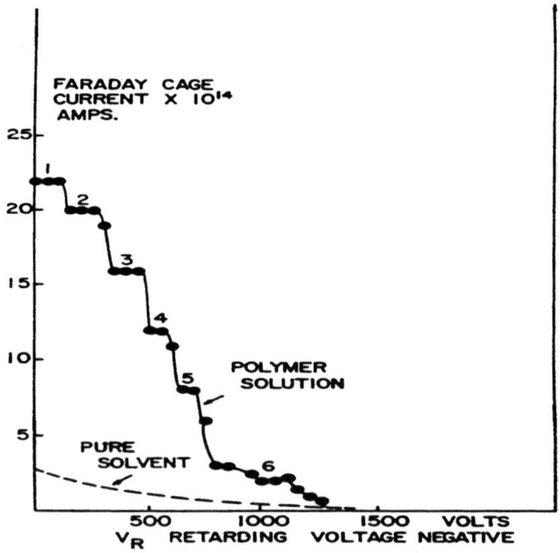

Figure 7. A current-voltage curve obtained by Dole when the electrosprayed solution comprised 0.005 wt per cent each of polystyrene samples having a nominal molecular weight of 51,000. The solvent comprised 3 parts of benzene to 2 parts of acetone by volume. The distances from the needle to the first aperture, the first aperture to the second aperture, and the second aperture to the nozzle were respectively 2, 3, and 3 inches.

produce a current-voltage curve with a decrease in current to the galvanometer at every value of the voltage corresponding to the energy of an ion species in the beam. Figure 7 shows the curve Dole obtained with a fractionated sample of polystyrene having a nominal Mr of 51,000 at a concentration of 0.01 % by weight in 60:40 (v:v) benzene:acetone. Also reported were results with polystyrene having a nominal Mr of 411,000 and with a mixture of the two. These results showed a certain consistency with expectations but, possibly for reasons to be described, were not convincing enough to persuade other investigators to confirm or extend them.

ELECTROSPRAY IONIZATION AT YALE

Early Efforts

A year or so after Sandy showed me Dole's paper I persuaded a new graduate student, Mike Labowsky, to repeat Dole's experiments in our much bigger vacuum system with much faster pumps than Dole had. We also had a better understanding of free jet expansions and realized that Dole had overlooked two of their features which were crucial in his experiments: (1) For the very large differences in molecular weight between the carrier gas and the seed species that he was using, there would have been a substantial amount of "slip" as suggested by the results in Figure 2. Thus, the actual velocities of his heaviest ions could have been as much as 50 per cent lower than he had reckoned. (2) During the adiabatic expansion that occurs in a free jet the temperature of the gas nose-dives. In the case of nitrogen, for example, at an axial distance downstream from the orifice of only ten nozzle diameters (ca 1 mm in Dole's experiments) the absolute temperature of the gas drops to about five per cent or less of its source value! Thus, the expanding gas rapidly becomes supersaturated with solvent vapor resulting from the evaporation of the ES droplets before the expansion. Ions are famously effective as condensation nuclei so that any ES ions in Dole's experiments must have been substantially resolvated to an unknown extent. Consequently, the actual masses of the unsolvated ions must have been much lower than Dole's measurements had indicated. In order to avoid such resolvation problems Mike modified the Dole arrangement so that (with reference to Figure 2) the bath gas of dry nitrogen was introduced at the exit-nozzle end of the spray chamber. Consequently, drying gas flowed "upstream", counter-current to the flux of charged droplets and ions from the spray needle, leaving the spray chamber through an exit port at its "entrance end", i.e. opposite to the end containing the nozzle leading into the vacuum system. Thus the solvent vapor from the evaporating droplets was carried out of the system so that the gas expanding into vacuum through the nozzle was free of solvent and any other "junk" that was not ionized. This counter-current flow of bath gas solved the problem of ion resolvation during the free jet expansion, greatly reduced system fouling, and has become a feature of the most widely used ESIMS systems. It will be described later in more detail. One can also avoid resolvation of ions during expansion by raising the temperature of the ion gas mixture to a value high enough so that after free jet expansion the gas temperature remains above "the dew point" for the ions. This "fix" was demonstrated by Chowdhury *et al.* [45] and has been incorporated in some commercial instruments. It has had limited use because, unlike a countercurrent gas flow, it does not prevent uncharged "junk" from entering the vacuum system. Much more frequent clean ing is therefore required.

As already mentioned, Dole's paper apparently failed to persuade other investigators to confirm his experiments which though simple in principle were really very difficult and demanding in practice. For example, large macroions incident on the first dynode of a multiplier do not produce secondary electrons unless their incident velocities are very high, either because of multiple charging or high accelerating voltages. As implied by reference 21b it seems highly likely that Dole's ions did not have many charges. Consequently, in his later experiments he used mobility measurements to characterize his macroions.

In 1969 at the age of 66, just two years before mandatory retirement at Northwestern, Dole had accepted an offer to become Welch Professor of Chemistry at Baylor University in Waco, Texas where he was assured there would be no mandatory retirement as there was at Northwestern. At Baylor he continued his electrospray experiments using mobility measurements rather than mass analysis to characterize the ions. He presented a short paper on that work at the 33rd Annual Meeting of American Society of Mass Spectrometery (ASMS) in San Diego in 1985 where I also gave a paper that included a discussion of the cooling and resultant resolvation that occurs in free jets. Dole came up to me afterward to introduce himself and to express his gratitude for finally understanding why there didn't seem to be any reasonable relation in his experiments between the size of the molecules and the mobilities of their ES ions. He said "Now I realize that we have been measuring the mobilities of highly solvated ions!" That was our only face-to-face encounter but we subsequently exchanged several letters and had a cordial relationship. I was delighted when he was later able to attend the first ASMS Workshop on ESIMS which was in Chicago. I didn't see him there because I had not been invited but by that time the "electrospray revolution" was well on its way so Dole had the satisfaction of realizing how fruitful his ideas had become before he passed on to his ultimate reward some months later. He would have been amazed if he had known that by 2001 the number of papers *per year* on ESI in the archival journals would reach 1500 and still be growing!

There was a hiatus in ESI studies at Yale after 1975 when, for understandable reasons, Mike Labowsky got discouraged. The retarding potential method of determining ion mass was most exasperating. Our only available means for measuring the very small ion currents was an old vibrating-reed electrometer that on its best days was crankier than a model T Ford! Mike became interested in the evaporation behaviour of the cloud of ES droplets and thus in the evaporation and combustion of a cloud of interacting fuel droplets, an important and challenging practical problem of interest to several of my colleagues in Chemical Engineering. He developed some unique computational techniques which together several important papers on this classic problem comprised his PhD thesis.

In 1982, Masamichi Yamashita, a young scientist I had met in Japan, came to my lab as a postdoc. When he arrived, "Gado" as he preferred to be called, asked for some suggestions about what he might work on. I mentioned that it might be interesting to take another look at ESI, but instead of trying to make ions out of big molecules, to explore whether Dole's idea would work with molecules of low enough mass for their ions to be analyzed with a small quadrupole mass spectrometer that had been a detector in some of our molecular beam experiments. It had an upper limit for mass/charge ratio of only 500 or so. Gado liked the idea and off he went. Not only was he an extraordinary experimentalist, with sound instincts and magic hands, he knew everything from the biology of sea urchins to the practice of digital electronics (already in 1982!). In a relatively short time he assembled what was really the world's first ESI Mass Spectrometer from the bits and pieces of molecular beam apparatus in our precious "junk pile". (My Scottish heritage of "Waste not, want not" makes it very difficult for me to throw anything away that might have some conceivable use!)

That first apparatus is shown schematically in Figure 8 and illustrates the arrangement for providing a counter-current flow of drying gas. We first tried electrospraying methanol-water mixtures and obtained spectra showing peaks corresponding, for example, to protons solvated by various combinations and numbers of methanol and water molecules that could be varied widely simply

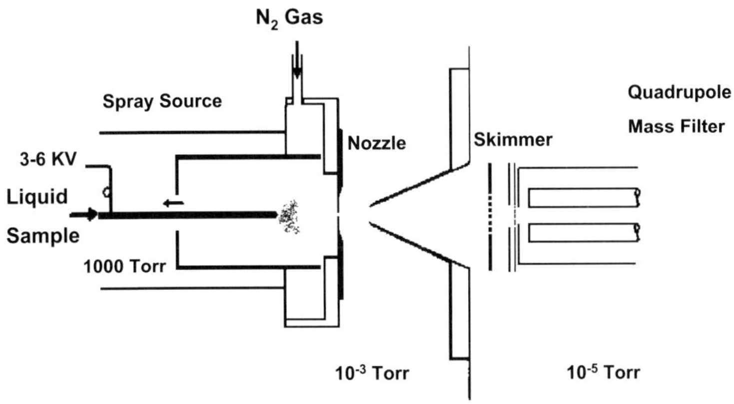

Figure 8. The first electrospray mass spectrometer built at Yale. Sample solution was sprayed from the hypodermic needle into a counter-current flow of dry nitrogen. The needle was at high potential relative to the cylindrical electrode and the end plate containing the orifice into the vacuum system. A center portion of the resulting free jet passed through the skimmer into a second vacuum chamber containing a quadrupole analyzer.

by adjusting the flow rate and/or temperature of the counter current drying gas. We then added some solutes to the solvent and were rewarded with the spectrum shown Figure 9a. It was obtained with a solution of several tetra-alkyl ammonium (and one phosphonium) halides at concentrations of a few parts per million. These results were especially gratifying because those solutes cannot be vaporized without extensive, even catastrophic decomposition. Beginning to get excited, Gado dissolved a vitamin B tablet in methanol-water and obtained the spectrum in Figure 9b showing a clearly defined peak for intact ions of every species in that tablet (i.e. with an Mr value under 500.)

The spectra in Figures 9a and 9b were exciting to us because they clearly demonstrated that ESI could produce intact ions of each of several species in a plurality of fragile molecules in a dilute solution while avoiding both fragmentation and interference between them. The spectra also raised second thoughts about the mechanism of ionization. We had been tacitly assuming that Dole's "Charged Residue Model" (CRM) accounted for the formation of the large ions he had found as well as the small ions that we had found in the results just described. In our early experiments, as well as in Dole's, there were always many more analyte (solute) molecules than charges in the ES flux. Consequently, as our results accumulated we had become increasingly uneasy about the adequacy of the CRM. The spectra in Figure 9 heightened that concern because we could not understand how the sequence of Rayleigh instabilities could produce, for each of several species in the same solution, singly charged ultimate droplets containing only one molecule of that species. Moreover, if the CRM model is to explain the results, the relative number of those droplets for each species would have had to be in direct proportion to the relative concentration of that species! How could the droplet know how to program its sequence of Rayleigh instabilities so as to produce a distribution of ultimate droplets such that the number of those ultimate

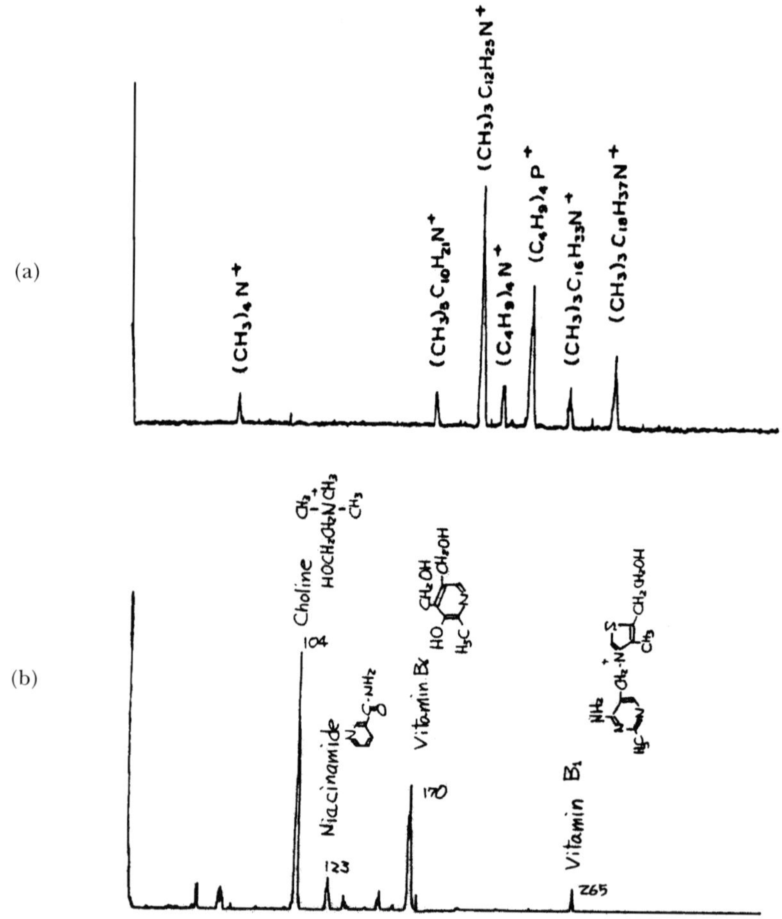

Figure 9. (a) The mass spectrum obtained with the apparatus of Figure 8 with a dilute solution (ppm) of quarternary ammonium halides and one quarternary phosphonium halide in 50-50 methanol water. (b) The mass spectrum obtained with a solution obtained by dissolving a vitamin B tablet in 50-50 methanol water.

droplets containing one molecule of any species x is always directly proportional to the concentration of species x in the original solution? Of course, that would happen if the droplet subdivision continued until all the droplets contained only one solute molecule. But that scenario somehow seems unlikely, especially when rapid evaporation from all droplets is occurring from the start. Moreover, if the subdivision process were to produce ultimate droplets for each ionizable species in a solution, one should then always find a spectral peak for every species in such a solution and that just doesn't happen.

These conceptual problems seemed to be resolved when we learned about some papers by Iribarne and Thomson, meteorologists at the University of Toronto [46]. Interested in the possibility that charged droplets of sea water might be a source of some of the ions found in the atmosphere, they had carried out experiments showing that evaporation of charged droplets could indeed produce gas phase ions of solute species in those droplets. The charged droplets in those experiments were produced by pneumatically nebulizing a conducting liquid. In such nebulization the droplets become charged as a result of statistical fluctuations in the distribution of anions and cations among those droplets so that roughly equal numbers of positively and negatively charged droplets are formed. In their first experiments only aerodynamic forces were used to nebulize the liquid, as in a perfume atomizer. Then they found that 3500 volts applied to an "induction electrode" (a cm away from the nebulizing zone) would result in all the droplets, and the ions from their evaporation, having the same sign, positive or negative, depending on polarity of that electrode. Iribarne and Thomson called their technique "Atmospheric Pressure Ion Evaporation" (APIE) a name which literally refers more to the mechanism of ion formation from charged droplets rather than to the method of producing the droplets. Moreover, both the earlier Electrospray Ionization (ESI) of Dole and the later Thermospray Ionization (TSI) of Marvin Vestal [47], also depend upon vaporization of charged droplets but their names do not identify any ion-forming mechanism. The essential differences between these three "spray" techniques consist mostly in the way the charged droplets are produced. Therefore, it seems more appropriate to give them names that relate to those characteristic differences, i.e. "ESI" for Electrospray Ionization, "ASI" for Aerospray Ionization, and "TSI" for Thermospray Ionization. These names relate specifically to the way droplets are produced in terms of the kind of energy used to produce them, (i.e. electrical, aerodynamic, and thermal) and will be used hereinafter.

The ASI technique of Iribarne and Thomson has never become has as widely used as ESI. However, their Ion Evaporation Mechanism (IEM) for ion formation from charged droplets, set forth in their papers, was a milestone contribution and has become a formidable competitor for Dole's Charged Residue Model (CRM) with each one having its champions. Both models assume that a sequence of evaporation steps followed by Rayleigh instabilities leads to smaller and smaller droplets. In the CRM this sequence continues until the droplets become so small that each one contains a single solute molecule that becomes an ion by retaining some of its droplet's charge as the last of the solvent evaporates. The upper sequence of steps in Figure 5 attempts to illustrate this process. The IEM holds that before the droplets become small enough to contain only one solute molecule, the field strength at the droplet surface becomes intense enough to "lift" a surface ion from the droplet into the ambient gas. In other words the surface field can evaporate an ion from the droplet surface. The lower sequence of steps in Figure 5 attempts to represent this mechanism. There is no unanimity in the ESI community as to which model is correct. The feeling in this corner is that under most circumstances the IEM is more consistent with more experimental observations than is the CRM. However, sometimes the CRM seems more likely to apply, especially in the case of very large analyte molecules.

A bit more needs to be said about TSI, the Thermospray Ionization

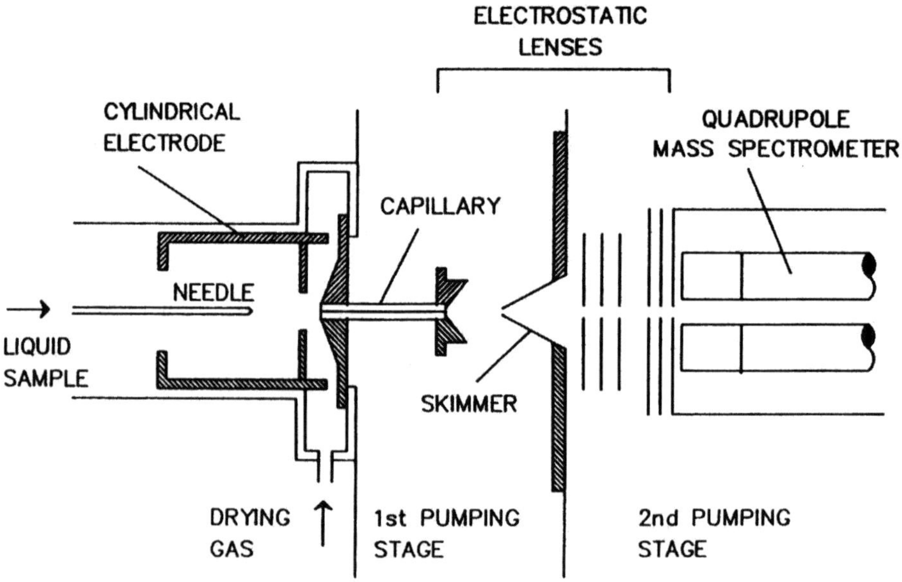

Figure 10a. Schematic diagram of the second ESIMS apparatus. It features the same counter current flow of drying gas shown in Figure 8 which excludes solvent vapor from the ion-gas mixture that enters the vacuum system by way of a free jet expansion. A novel feature of this apparatus is that the ion-gas mixture enters the free jet expansion at the exit of glass capillary 4 rather than through the simple flat-plate orifice of Figure 4. Each end of that capillary is metallized so that it can be maintained at any desired potential. Thus the field required to "electrospray" the sample solution can be achieved with the spray needle at while the metallized entrance of the capillary is at the required potential "below" ground. The ions entering the capillary are then in a potential well. The rapid flow of gas through the capillary then drags the ions out of that potential well to any desired potential at which the metallized exit end of the capillary is maintained. Thus, all external parts of the apparatus can be at ground, posing no hazard to an operator.

Technique introduced by Marvin Vestal and his colleagues beginning in 1978. The last of the charged droplet sources to appear, TSI became the first to be widely used, in part because of its effectiveness and in part because it was the first to become available commercially as an add-on to existing mass spectrometers. This sequence of events would seem to fulfill that passage in the scripture which says "The last shall be first." (That passage continues by saying "And the first shall be last", which apparently is being fulfilled by ESI because it was the first to be introduced and now would seem likely to be the last of the spray techniques to be abandoned, if it ever is!)

TSI consists in passing analyte solution through a capillary tube whose walls are hot enough to vaporize 90 per cent or more of the solvent. The resulting expansion produces the same kind of shear and acceleration forces on the liquid that occur in aerodynamic nebulization, e.g. as in ASI or a perfume atomizer. The result is a dispersion of droplets in vapor that emerges from the exit of the capillary as a supersonic jet into ambient solvent vapor at a pressure of 10 to 15 torr in a chamber whose walls are hot enough to maintain the vapor in a superheated state. The already mentioned statistical fluctuations in

Figure 10b. Photograph of the apparatus of Figure 10a showing (left to right) Chin Kai Meng (graduate student), Robert Dreyer (chromatographer from the Medical School) and Craig Whitehouse, (graduate student who did much of the design and assembly of the instrument).

the distribution of cations and anions give rise to equal numbers of positively and negatively charged droplets as in ASI. Evaporation of the droplets in the superheated vapor brings about the sequence of Rayleigh instabilities common to all the techniques based on charged droplets. That sequence leads to the formation of gaseous ions by either the CRM of Dole or the IEM of Iribarne and Thomson. The mixture of ions in vapor flows past an aperture through which ions of the desired polarity are driven by an applied field into the vacuum chamber housing a mass analyzer. For several years in the 1980's TSI sources became the preferred hyphen in LC-MS, but they have now been almost completely replaced by ESI sources.

Meanwhile we were achieving ever better results with ESI as we climbed the learning curve and Sandy Lipsky was keeping in close touch with our progress. He was a good friend of Brian Green of VG Analytics, a British manufacturer of mass spectrometers that has since evolved into what is now Micromass, Ltd. Brian would stop by New Haven and visit Sandy every few months and on most of those visits Sandy brought him by our laboratory. As our results with small molecules began to get better and better, Brian became more and more intrigued and asked us if the technique would work with larger molecules. We said we thought it might but couldn't be sure because our little quadrupole analyzer had an upper limit of about 500 for mass/charge ratio. Brian then arranged for VG to lend us a used quadrupole analyzer that could weigh ions with mass/charge ratios up to 1500 or so. That instrument was incorporated in a new system for which much of the design and

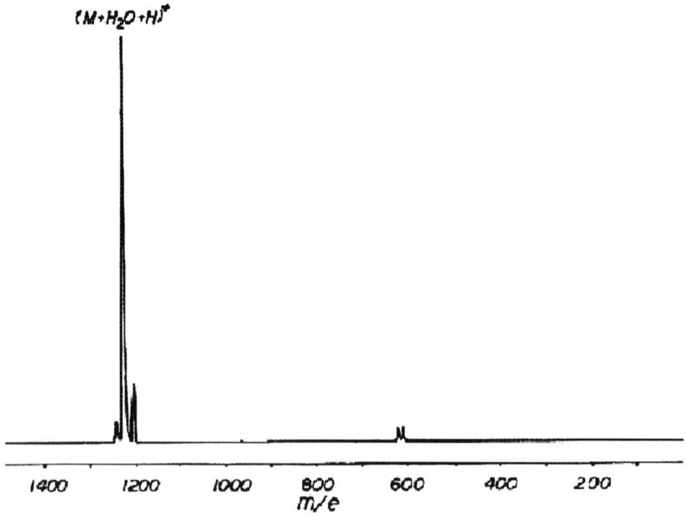

Figure 11a shows a mass spectrum obtained with a solution of the cyclic peptide, cyclosporin A, in 50–50 methanol-water. The dominant peak is for a singly charged molecule with one adduct water molecule. The small peak on the right is due to the bare ion. The tiny one on the left is due to the parent molecule with two adduct water molecules. (Because of the "polarity" of the chart recorder in these experiments, m/z values increase from right to left, the reverse of normal practice.

assembly was done by Craig Whitehouse. Craig had worked in Sandy's lab after finishing college and Sandy urged me to take him on for graduate study in Chemical Engineering. The design of this new system is shown schematically in Figure 10a and is pictured in Figure 10b. It has some novel features that deserve comment.

In light of the results obtained with Gado's machine, and with Csaba and Sandy hovering in the background, we increasingly viewed ESI as a possibly more potent hyphen in LC-MS than the already widely used TSI. An electric field sufficiently intense to disperse analyte solution emerging from the spray needle into a fine spray required a potential difference of up to several kilovolts between that tip and the orifice leading into the vacuum system. Thus, one was faced with a trilemma in LC-MS operation. Either (1) the LC system would have to be maintained at several kV above ground, (2) the MS at several kV "below" ground, or (3) the power supply would require enough capacity to maintain the required potential difference in spite of the leak to ground of current through the conducting path comprising the flow of liquid from the LC. Options (1) and (2) were not feasible but option (3) was being used in some systems. We then had a wild idea that turned out to be a very effective solution to this problem. We replaced the simple orifice into the vacuum system by a glass capillary tube as shown in Figure 10a. That glass tube was metallized at each end to provide an electrical contact. The inlet end of the capillary would be maintained at the required potential below ground so the ions entering the tube would be in a potential well. We guessed, correctly as it turned out, that the fairly high velocity flow of bath gas through the glass (di-

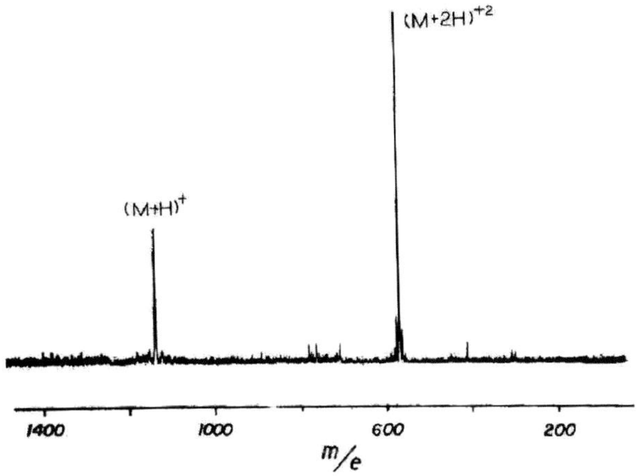

Figure 11b shows the spectrum obtained with a solution of gramicidin S, another cyclic peptide with a molecular weight close to that of cyclosporin A.

electric) tube into the vacuum system could drag the ions out of that potential well up to whatever potential might be desired at the metallized exit end of that tube. It worked like a charm. Indeed, the gas flow could raise the ions to an exit potential of several tens of kV, sufficient to inject the ions into a magnetic sector analyzer. At the same time all exposed external parts and surfaces of the instrument could be maintained at ground potential, thereby posing no hazard to an operator.

Eager to flex our muscles with the new apparatus we tried it out with a couple of cyclic peptide samples from friends in the Medical School, Cyclosporin A and Gramicidin S with molecular weights of 1184 and 1141 respectively. Figure 11a shows the spectrum obtained with a solution of 1.0 g/L of Cyclosporin A in 85:15 (v:v) acetonitrile:water. The dominant peak at 1203 corresponds to singly protonated parent molecules with one water of hydration. The small satellite peaks on either side are due to ions with one more and one less water molecule. The small triplet peak near mass 600 is from doubly protonated ions of the same species. (The right to left increase in mass/charge ratio is the opposite of custom due a quirk of the available chart recorder.) Figure 11b shows the spectrum for 1.0 g/L of Gramicidin S in 50:50 (v:v) methanol water. Here the peak is much larger for the ions that are doubly rather than singly protonated. These differences can be understood in terms of the IEM mechanism and the greater hydrophobicity of cyclosporin. A molecule of the latter can therefore be more easily removed from an aqueous environment can one of the former. Consequently, one charge can provide enough "lift" to remove the ion from the droplet surface. This observation seems to support the Ion Evaporation Model of Iribarne and Thomson. As the droplet shrinks, the charges get close enough together so that two of them can attach to a single gramicidin molecule thus providing more lift, in part because of the two charges and in part because the surface field is stronger, owing to the higher surface charge density.

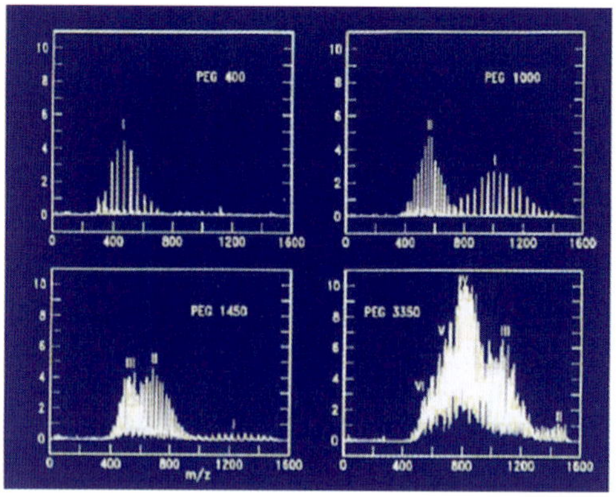

Figure 12a. Spectra obtained with the apparatus of Figure 10a for Poly(ethylene glycol)-oligomers with nominal molecular weights of 400, 1000, 1450 and 3350.

Peaks for doubly charged ions also dominated in similar spectra for the peptides bleomycin and "substance P". In the spectra for renin substrate and insulin chain B we found evidence of triple protonation. This propensity for multiple charging was quite provocative because the effective mass range of any analyzer increases by a factor equal to the number of charges per ion. In order to explore the extent to which such multiple charging can occur, and to identify and evaluate the controlling factors, we carried out an extensive study of ES ion formation with poly (ethylene glycol) (PEG) oligomers. Samples with a range of nominal molecular weights (Mr values) were kindly provided by Union Carbide and Chemical Corp. Each one comprised a mixture of oligomers with a Gaussian-like distribution of Mr's having a FWHM roughly 15 per cent of the nominal Mr of the most abundant oligomer. These species retain their chemical and structural similarity over a wide range of Mr values. It is known that Na^+ ions bind to the oxygen atoms in the oligomer chain with an energy of 2.05 eV and there is always enough sodium around so that spectra of samples from the manufacturer (Union Carbide and Chemical Corp) comprised oligomers with varying numbers of Na^+ adducts that increased with increasing molecular weight. (However, one can change these adducts by appropriate additions to the sample, e.g. to K^+ by adding KOH or KCl). Figure 12a shows mass spectra for PEG samples having nominal Mr values from 400 to 3350 for which the individual peaks are clearly resolvable in the original spectra and correspond to oligomers with varying numbers of Na+ adducts. Figure 12b shows spectra for samples with nominal Mr values of 8000 and 17,500 wherein the peaks are much too close together to be resolved (with our analyzer) because of the superposition of a wide distribution of oligomer sizes on a wide distribution of charge states for each oligomer.[48] A reviewer of that first paper on these results dismissed them outright, saying that the spectra in Figure 12b were "not mass spectra but were due to dirt in the system!" However, we have found over the years that

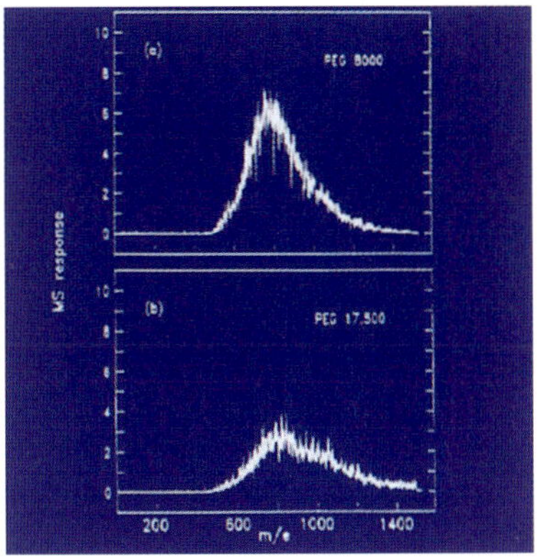

Figure 12b. Spectra obtained with the apparatus of Figure 10a for Poly(ethylene glycol)-oligomers with nominal molecular weights of 8000 and 17,500.

for polymers like these one can correctly and usefully assume that the m/z value at the maximum or peak value of the envelope of the overlapping peaks is a reasonably reliable estimate of the most probable m/z value for the most abundant oligomer in the sample. We used this assumption to great advantage in a later study which showed that intact ions could be obtained from PEG oligomers with Mr values of at least 5,000,000! [49] Others have since shown that ESI can produce intact ions of species with molecular weights of over 100 million! [50].

The message of these PEG experiments was clearly that to study the ESIMS behaviour of large molecules we needed samples of pure compounds in which all the molecules had the same molecular weight. Obvious examples of such compounds are natural peptides and proteins which for many other reasons, of course, are far more interesting and rewarding subjects of study than synthetic polymers. We began to get promising results with such compounds in late 1987 and presented the results shown in Figure 13 in June 1988 at the 36th ASMS Conference on Mass Spectrometry and Allied Topics in San Francisco. Inspection of those spectra shows that each peak in the spectrum for any one species differs from an adjacent peak only by one adduct charge, usually a proton in the case of these peptides and proteins. The first reviewer of a preliminary paper on these results asserted that they showed the method to be worthless for two reasons: (1) The total ion current was divided among so many peaks that the signal/noise ratio for any one peak would inevitably be low and (2) the multiplicity of peaks for each species in a mixture would inevitably make the spectra too congested to be interpreted. As I have often done, and still do, that reviewer underestimated the power of a computer. One day I commented to my then student, Matthias Mann, that each peak in one of these multiple peak spectra was really an independent measure of the parent ion mass. Therefore, there should be some way of averaging those independent values to get a more reliable and accurate measure of the parent

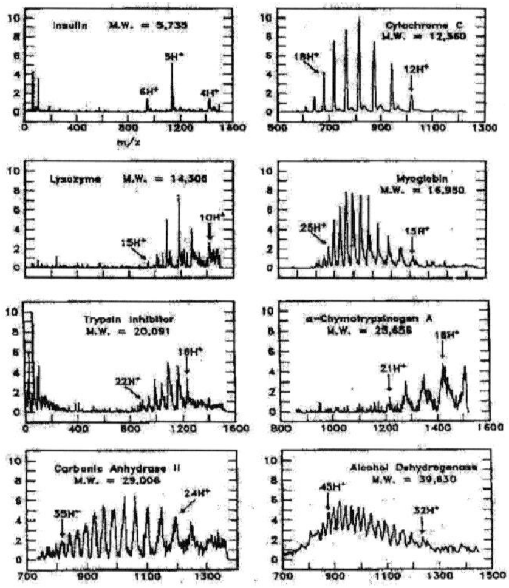

Figure 13. Spectra obtained with the apparatus of Figure 10a for a variety of protein molecules.

molecule mass than any single peak spectrum could provide. Two days later he had worked out a computer algorithm that transformed the multiple peaks into a single peak that would be obtained if all the ions had a single massless charge. The m/z value for that peak is thus the Mr value for that species. Figure 14 shows an enlargement of the spectrum for cytochrome C in the upper right hand corner of Figure 13. The insert shows the result of the deconvolution by the algorithm Matthias had worked out, i.e. the peak that would be obtained if all the ions had a single massless charge. The m/z value on the abscissa for the apex of that peak is thus the true molecular weight of the parent neutral molecule. It is noteworthy that the scale for the ordinate value, which represents the relative abundance of the ions in any peak, is the same on the insert as in the multi-peak spectrum. Clearly, values for both total signal and signal/noise are much higher in the deconvoluted spectrum than in the parent spectrum as measured. It is now taken for granted that computers can deconvolute what seem to be hopelessly complex mass spectra to obtain precise and accurate Mr values for very large molecules in an extremely complex mixture of species.

However, not everyone was impressed. A reviewer of a reputable journal to which we submitted a manuscript on the results with poly(ethylene glycols), refused to approve publication of the protein results on the grounds that the protein spectra clearly showed that the method was worthless! In the first place the ion currents would be distributed over so many peaks that signal/noise ratios would inevitably be too small. Moreover, the multiplicity of peaks for each species would make the spectrum of any mixture hopelessly complex and uninterpretable! But as I myself have done all too often, that critic underestimated the power of the computer. (In fairness I should admit that the manuscript he saw didn't include any discussion of the algorithm

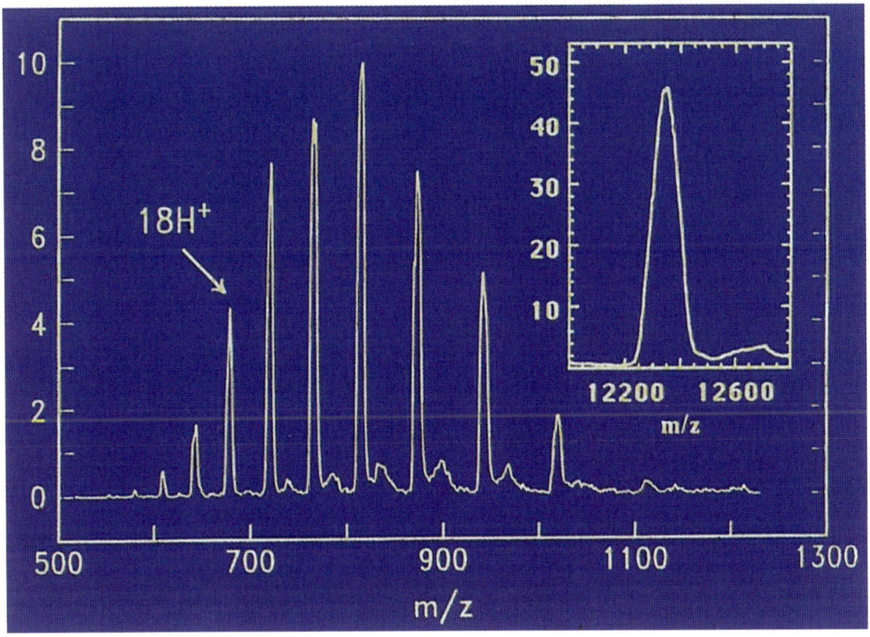

Figure 14. Spectrum for Cytochrome C , i.e. the right hand panel of Figure13. The insert shows the result of applying Matthias Mann's deconvolution algorithm which transforms the multiple peaks of the measured spectrum to what would be obtained if all the ions comprised the parent molecule with a single massless chargwe. Thus the abscissa value of the single peak is the molecular weight of the parent molecule. Note that the ordinate scale of relative abundance is the same for both the measured and deconvoluted spectra. Clearly both the effective signal and the signal/noise ratio are both much higher in the deconvoluted spectrum.

that had not yet then been worked out but the fact that such spectra could be obtained at all was surely worth publication at that time.

Funding agencies were also generally somewhat less than enthusiastic in response to our proposals. For example, when we submitted one for research to extend the results shown in Figures 9a and 9b, the reviewer simply said: "It is impossible for this man to obtain these results with his equipment!" I was reminded of the slogan once promoted by a company which said: "Difficult tasks we can finish quickly. The impossible ones take a little longer!"

There were only a handful of people at the session of the ASMS meeting at which we presented these results. Nor were there many questions or comments from the floor. However, when those results were published in 1989 [51], they began what has become a "flood" of papers and ignited what has been referred to as the "Electrospray Revolution!" Some idea of the dimensions of that flood can be gained from Figure 15 which was prepared by my good friend and former colleague, Prof. Juan de la Mora at Yale. It shows how the publications on ESIMS have been increasing with time. The sudden rise began in 1990, a year or so after we first presented the protein results and reaching over 1500 per year in 2001 and still climbing! (The first complete

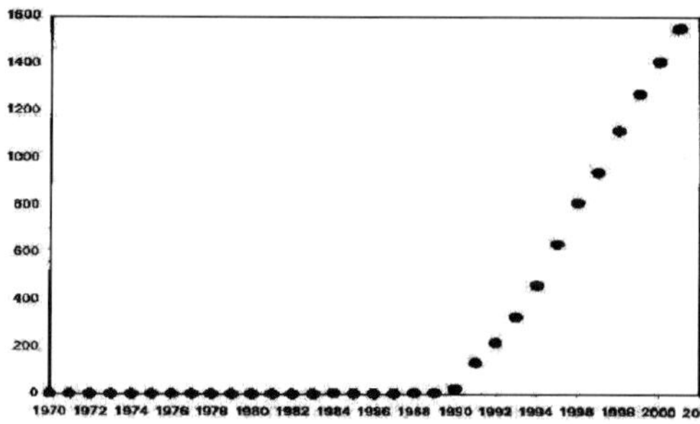

Figure 15. Number of publications on ESIMS from 1970 to 2002.

paper on the method and results appeared in the 10/6/89 issue of Science and is by far the most cited publication that ever came out of our lab (51). Moreover those numbers reflect only a small fraction of the total ESIMS activity, most of which is being carried out by the pharmaceutical companies and is not published.

In view of the flood of papers on ESIMS revealed in Figure 15, even a rudimentary summaray of what they have taught would clearly take much more than a little longer. Consequently, I will stop here with a final word of appreciation to the many who have made possible the preparation of this paper. I would especially emphasize the extent of my gratitude and thanks to my eager and able colleagues without whose efforts I would not be here today. Nor could I be writing this had I not had the support of numerous sponsors who over the years have provided the funds, (albeit sometimes unwittingly!) without which we could have done nothing. They include the Department of Energy, the National Science Foundation, the National Institutes of Health, the Army Research Office, the American Cancer Society, the du Pont Company, and the Jeffress Foundation. Finally, of course, I want to express my eternal gratitude to the Nobel Foundation and its marvelously helpful and gracious staff. They, along with what must have been generous and abundant approval from my peers and colleagues in the fellowship of scientists, have transformed my wildest dreams into a reality even more unbelievable than "flying elephants!"

REFERENCES

1. J. W. Mullen, M. R. Irby, J. B. Fenn, Third Internat'l Symp. on Combustion, Flames, & Explosive Phenomena, Williams and Wilkins, Baltimore (1949).
2. S. Datz and E. H. Taylor, J. Chem. Phys. 25, 395 (1956).
3. E. W. Becker and K. Bier, Z. Naturforsch, A9, 975 (1954).
4. A. Kantrowitz and J. Grey, Rev. Sci. Instrum. 22, 328 (1951).
5. G. B. Kistiakowsky and P. Schlichter, Rev. Sci. Inst. 22, 333 (1951).
6. T. H. Johnson, Nature 113, 745 (1927); Phys. Rev. 31, 103 (19280).

7. I. Estermann, O. Frisch and O. Stern, Zeit. f. Physik 73, 348,(1931).
8. J. E. Scott, Jr. and J.E. Drewry, Proc. 3rd. Intern. Symp. Rarefied Gas Dynamics, Paris 1963 Vol.I. p. 516, Academic Press, New York.
9. Fraser, R.G.J, "Molecular Rays", Macmillan, New York (1931).
10. E. W. Becker, K. Bier, and H. Berghoff, Z. Naturforsch, A10, 565 (1955).
11. H. R. Murphy and D.R. Miller, J. Phys. Chem. 88, 4474 (1984).
12. J. Deckers and J.B. Fenn, Rev. Sci. Inst. 34, 96 (1963.)
13a. N. Abuaf, J. B. Anderson, R.P. Andres, J. B. Fenn, Science 35, 997 (1967); cf. "Molecular Beam Engineering at Intermediate Energies", Entropie 18, 314 (1967).
13b. R. Campargue, J.B. Anderson, J.B. Fenn, B.B. Hamel, E.F. Muntz, J.R. White, in "Nuclear Energy Maturity" (P Zaleski, ed.p. 5, Pergamon Oxford (1975).
14 E. Kolb and D.R. Herschbach, J. Phys. Chem. 88, 4488 (1984).
15. P. L. Owen and C. K. Thornhill, Aero Res. Council (U.K.) R & M No. 2616.
16. R. J. Gallagher, J. Chem. Phys. 60, 3487 (1974).
17. R.E. Smalley, L. Wharton, and D.H. Levy, J. Chem. Phys. 63, 4977 (1975).
18. S. B. Ryali, J.B. Fenn, Ber. Bunsenges. Phys. Chem. 88, 4451 (1984).
19. S. Goyal, D.L. Schutt, G. Scoles, Phys. Rev. Lett. 69, 933 (1992).
20. M. Hartmann, R.E. Miller, J.P. Toennies, A. Vilesov, Phys. Rev. Lett. 75, 1566 (1995).
21. Dudley Herschbach, Yuan Lee and John Polanyi (1986); R.F. Curl, Jr., H.W. Kroto and Richard Smalley (1996); Ahmed Zewail (1999).
22. M. Dole, L.L. Mach, R. L. Hines, R. C. Mobley, R.C. Ferguson, M.B. Alice; J. Chem. Phys. 49, 2240 (1968).
23. R.J. Beuhler, E. Flanigan, L.J. Green, L. Friedman, J. Am. Chem. Soc. 96, 3990 (1974).
24. H.L.C. Meuzelaar, M.A. Posthumus, P.G. Kistemaker, J. Kistemaker, Anal. Chem. 45, 1546 (1973).
25. E. Unsoeld, F. Hillenkamp, R. Nitsche, Analysis, 4, 115 (1976); M.A. Posthumus, . P. G. Kistemaker, Ten Noever de Brauw Meuse, Anal. Chem. 45, 1546 (1973).
26. M. Barber, R.S. Baordolli, G.J. Elliot, R. D. Sedgwick, A.N. Tyler, J. Chem. Soc. Chem. Commun. 325 (1981).
27. A. Benninghoven, D. Jaspers, W. Sichtermann, Appl. Phys. 11, 35 (1976).
28. D.F. Torgerson, AR. P. Skowronski, R. D. Macfarlane, Biochem. Biophys. Res. Commun. 60, 616; B.U.R. Sundquist, R.D. Macfarlane, Mass Specttrom Reviews, 42, 421 (1988).
29. K. H. Tanaka, H. Wake, Y. Ido, S. Akita, Y. Yoshida and I. Yoshida, Rapid Commun. Mass Spectrom. 8, 2, (1988).
30. M. Karas, F. Hillenkamp, Anal. Chem. 60, 2299 (1988).
31. E.W. Mueller, Z. Physik 131, 136 (1951).
32. E.W. Mueller, Electronics and Electron Physics Vol XIII, 83ff Academic Press, N.Y. (1960).
33. M. Ingram and R. Gomer, J. Chem. Phys. 22, 1279 (1954).
34. H.D. Beckey, "Principles of Field Ionization and Field Desorption Mass Spectrometry," Pergamon Press, Oxford, 1977.
35. C.A. Evans, Jr., C.D. Hendricks, Rev. Sci. Instr. 43, 1527 (1972).
36. D.W. Simons, B.N. Colby, C.A. Evans, Jr., Int'l J. Mass Spectrom. Ion Phys. 15, 291 (1974).
37. K.D. Cook, Mass Spectrom. Revs. 5, 467 (1986).
38. Rayleigh, Lord, Phil. Mag. 14, 184 (1882).
39. J. Zeleny, Phys. Rev. 10, 1 (1917).
40. A. Gomez, K. Tang, Phys. Fluids 6, 404 (1994).
41. G.I. Taylor, Proc. Roy. Soc., A 280, 383 (1964).
42. Rayleigh, Lord, "The Theory of Sound", Chap. XX, Dover, N.Y. (1945).
43. M. Dole, "My Life in the Golden Age of America," P. 169 Vantage Press, NY (1989) 44. ibid. p. 167.
45. S. K. Chowdhury, V. Katta, B. T. Chait, Rapid. Commun. Mass. Spectrom. 1, 53 (1990).
46. J.V. Iribarne, B.A. Thomson, J. Chem. Phys. 64, 2287 (1976).

47. C. R. Blakley, M. J. McAdams, M.L. Vestal, J. Chromatog. 158, 264 (1978); Anal. Chem. 52, 1636 (1980). J. Am. Chem. Soc. 102, 5931 (1980).
48. S.F. Wong, C.K. Meng, J.B. Fenn, J. Phys. Chem. 92, 546 (1988).
49. T. Nohmi, J. B. Fenn, J. Am. Chem, Soc.114, 3241 (1992).
50. S. F. Fuerstenau, J. Mass Spectrom.Soc. Japan 51, 50 (2003).
51. J. B. Fenn, M. Mann, C. K. Meng, S.F. Wong, C.M. Whitehouse, Science 246, 6, 64 (1989).

Koichi Tanaka 田中耕一

KOICHI TANAKA

I was born in Japan on August 3, 1959. My natural mother died one month after I was born, apparently due to giving birth at an advanced age. Because my father was also physically frail, I was brought up by my uncle and aunt. When I use the words "father" and "mother" now, I am referring to the father and mother who raised me. Perhaps it was because of my easygoing character that I remained ignorant of my early circumstances until I was 18 years old, when my parents informed me of these circumstances. More likely, however, my blissful ignorance was due to the completely fair-minded upbringing I received by my parents, older sister and brothers, relatives and neighbours.

My parents operated a business selling and repairing carpentry tools in Toyama Prefecture. I grew up with images of my father sharpening saw-blade teeth and planes, images in which he was seemingly always busy working quietly with his hands. My mother was not only busy performing housework, but also making end-of-month collections and keeping the books of the family business until late at night. I have no memory of my parents encouraging me with words like 'study' or 'succeed.' There is no doubt, however, that the values they instilled in me were much more important than those suggested by these words. From my father, I learned the importance of working sincerely at things to which I had committed myself, and to persevere untiringly even in the face of little progress. My mother also stressed the importance of working quietly towards achieving my missions in life, without neglecting attention to details. My father passed away three years ago (1999).

Apart from my parents, I was also influenced by my grandmother. She often admonished me, using the expression "What a waste!" ("Mottai-nai!" in Japanese). While these words are approaching extinction these days, they used to be an integral part of the cultural values in Japan. My grandmother valued even the smallest of things. Once, when she noticed me crumpling up a sheet of paper to throw it away, she angrily reproached me, saying "What a waste. You could straighten that paper out and use it to blow your nose." It is no wonder the concept "What a waste!" is so ingrained in my psyche.

The prefecture name "Toyama" means "rich in mountains." Toyama is surrounded by the Tateyama Mountains to the east, other mountains to the south and west, and the Sea of Japan to the north. Just the sight of the Tateyama Mountains brings me a sense of calm. This area is blessed with bountiful nature, eliciting in me a feeling of awe towards all living things and a compelling interest in the mysteries of nature, so difficult to find in urban settings.

I was enrolled in the Hachininmachi Elementary School in Toyama City in

K.T. and my parents (taken about 1965)

Expo '70 held in Osaka, Japan

1966. I cannot say that I was a particularly diligent student, especially during the lower grades. One event, however, did make a lasting impression on me, and that was Expo '70, Japan's first world fair, held in Osaka in 1970. The Exposition site displayed the future of technology, which would actually be realized 20 to 30 years later. I truly felt the power of science and technology. Our teacher for the last three years of elementary school was Mr. Kyosei Sawagaki. He taught us not by having us memorize textbooks, but through the joy of performing scientific experimentation and discovering phenomena with our own eyes. One day our teacher showed us an experiment in which boric acid was first dissolved in water, and then re-crystallized. As I watched and experienced this incredible phenomenon, I blurted out, "It's starting to snow!" While this would have been considered incorrect as an answer in a test, my teacher cherished that response. That was when I discovered that learning could be enjoyable, and not just a painful experience.

I enrolled in Toyama Municipal Shibazono Middle School in 1972. This was not an especially elite middle school, but I did apply myself single-mindedly to my studies. My efforts enabled me to attain a ranking in the top ten percent of the class.

In 1975, I enrolled in Toyama Chubu High School in Toyama Prefecture. This high school is known for its competitive first or second ranking in Toyama Prefecture as a school for sending graduates on to "first-class" universities. Together with just about every other student in the school, I devoted a great deal of effort to studying for the university entrance examination. The tenacious character I've possessed since I was a small child propelled me to successfully meet this challenge, and I was able to safely gain acceptance to the university of my choice.

Upon receiving my notification of acceptance to the university, my parents

View of Tateyama Mountains (as high as 3,000 meters) from Toyama City

noticed that they were obliged to submit to the university, among other things, a copy of my official family register. After much mental anguish, they decided to inform me of the secret of my birth. The truth came as a considerable shock to me, and the trauma surged over me in waves for a long time afterwards. At the same time, however, it was a chance for me to assert my independence. The thought grew strong in me that since I had gone to the trouble of being born, I might as well be useful in helping people live long and healthy lives. And this thought has always resided in the back of my mind.

In 1978, I entered Tohoku University, into the Department of Electrical Engineering, Faculty of Technology. I suppose the reason I chose electrical engineering was because I had always been interested in electricity, involving myself in such projects as building radios from the time I was a child. Moreover, I thought that electrical-related skills would be useful upon graduation, and that it would be easy to find a job among the electronics-related businesses so active in Japan at that time. Perhaps as a reaction to the tremendous efforts I had made to pass the university entrance examination, I let up somewhat in my first- and second-year studies at the university. As a result, my grades suffered in German class, and I was forced to repeat the year. Aware that I had placed a burden on my parents, from that point on, I diligently applied myself to my studies. In my senior year at the university, we were obliged to complete a graduation research project, and mine was entitled "Absorption of a Plane Wave by an Impedance-Loaded Dipole Array Buried in a Lossy Medium." The objective of the research was to reduce the ghost effect of television broadcast waves by placing an array of line-shaped conduc-

K.T., students and teachers of Adachi Laboratory at Tohoku University (1982) (taken by Prof. Saburo Adachi)

tors in front of a building to prevent the reflection of electric waves from the building. I was guided in my research by Professor Saburo Adachi. Needless to say, my present research endeavors have practically no relevance to that subject.

The Faculty of Technology of Tohoku University is renowned for its tradition of practical studies. For instance, historical individuals associated with the university include Professor Kotaro Honda for his contributions in metallurgical engineering, Professor Hidetsugu Yagi, the inventor of the well-known Yagi antenna (patented in 1940), and Professor Jun-ichi Nishizawa for his opto-electronics and semiconductor research. In such an environment, I was able to study things that could be of immediate usefulness to the world. That learning experience undoubtedly served me well when I eventually entered the work force.

When it came time to find employment, I set my sights on becoming an engineer at a home electronics manufacturer, a field that was closely related to my major at university. I took an entrance examination for employment at one such company, but failed the examination. Despite my desire to be an electrical engineer, I am sure my answers to the electricity-related questions were somewhat less than acceptable. Even though I failed, however, I still hold that company in high regard.

At that point, I decided that there was no need to focus solely on electrical engineering, especially because I had only two years of electrical-related knowledge. I turned to my mentor, Professor Adachi, and he was kind enough to introduce me to Shimadzu Corporation. I learned from

Shimadzu's employment literature that the company was manufacturing X-ray devices and other types of medical equipment. This struck a chord with me, rekindling the possibility that I might yet satisfy my desire to help suffering people, albeit indirectly. I decided to take the employment examination, and that time, I passed without problem. However, instead of being assigned to the area of medical equipment, I learned I was to be involved in research and development of analytical instrumentation. Fortunately, that field also held interest for me, and it became the central area of my work.

Kyoto city is home to Shimadzu's main factories and research divisions, and has fostered six out of Japan's nine Nobel Prize laureates in scientific fields, including Prof. Hideki Yukawa (selected in 1949), Professor Shin-ichiro Tomonaga (1965), Prof. Ken-ichi Fukui (1981), Prof. Susumu Tonegawa (1987), Prof. Ryoji Noyori (2001), and myself. Probably no one can explain exactly why so many laureates hail from Kyoto, but the knowledge of this fact did have somewhat of an effect on me. I might even have mused that I could reach to such a level if I were to put my heart and soul into the effort.

I joined Shimadzu Corporation in 1983, and was immediately assigned to the Central Research Laboratory, a new department which had been established in 1980. At that time, there were three laboratories, for mechanical-, chemical- and electrical-related research, respectively. I was assigned to the laboratory for electrical systems-related research, and joined a team charged with developing component technology for analytical instruments. This team comprised Dr. Tamio Yoshida, Mr. Yoshikazu Yoshida, Mr. Satoshi Akita, Mr. Yutaka Ido, the one member who joined the company with me, and myself. We were a very young team, with an average age in the 20's, with me being the youngest of all.

Even before I joined the company, Shimadzu had been engaged in research involving the mapping the elements on the surface of a semiconductors and metals after ionizing the surfaces using an Nd-YAG laser beam concentrated to a diameter of about one micrometer. An instrument that was to be competitive with a German product was developed, but since its performance did not significantly exceed that of the other product, it was not placed on the market. However, rather than give up at that point, our team consulted with a product manager in the Scientific Equipment Division to find a more constructive alternative. We decided to utilize the instrumentation we had been using in our research together with the accumulated laser-related technology we had developed, and we shifted the focus of our work to the field of mass analysis of biological-related substances. That research and development began in 1984. Since Mr. Akita was already a talented designer of electrical circuits and systems, I assumed the responsibility of acquiring data from the chemical experiments. This would be a first-time ever experience for me to irradiate substances with a laser, and the first time to perform chemical experiments away from the educational context of school. However, I undertook this project thinking that it would be very interesting, drawing on the inquisitiveness that had guided me since my elementary school days.

My specific task was to search for a matrix that would enable non-destruc-

K.T. and my colleagues when LDI-TOFMS was developed in 1983–1986 (from left, Tamio Yoshida, Yoshikazu Yoshida, from right, Yutaka Ido, Satoshi Akita in Dec/2002 in Stockholm) (taken by Kyodo News)

tive ionization of macromolecules by efficiently absorbing the laser energy. The laboratory stocked hundreds of substances that were candidates for the matrix. Without having studied the theory of ionization coupled with my paucity of knowledge in chemistry, I single-mindedly, day-after-day, repeated trial and error experiments with the candidate substances. Continuing this regimen tirelessly, I felt like I had become one with these substances and the instrument. However, a solution was not easily found.

One day in February of 1985, instead of using Cobalt Ultra Fine Metal Powder (UFMP) as a matrix, I mistakenly used a glycerin-UFMP mixed matrix. I noticed this mistake immediately, but I thought, "Mottai-nai!" at the idea of throwing the mixture away. If I allowed the glycerin to evaporate, I thought that I could still make use of the UFMP, so I placed the mixture in the vacuum chamber to dry it out. Thinking that I could even speed up the drying process by irradiating the mixture with the laser, I switched on the laser beam. On top of that, anxious to confirm the elimination of the glycerin as soon as possible, I kept the spectrometer running and monitored the results. And then, I noticed a signal peak I had never before seen mixed in with the noise peaks. I think because up to that point I had always felt annoyance at the sight of that noise wave data in the experiments, I noticed a slight difference. The signal peak that I had never seen before now appeared at the same position no matter how many times I ran the experiment.

K.T. and LDI-MS rig (1985). Matrices are in glass bottles (bottom)

K.T. and poster presented at the Second Japan–China Joint Symposium on Mass Spectrometry (1987) (taken by Prof. Robert Cotter)

Thinking about it now, this "monumental blunder" was the start of it all. From then on, noticeable progress was evident every time I ran an experiment. This contrasted entirely with the situation that continued day-after-day until a few days prior, when I would say to myself each time I failed, "Well now I know that method won't work." During that productive period of time I was truly happy as an engineer. Most of the work performed by a development engineer results in failure. However, the occasional visit of success provides just the excitement an engineer needs to face work the following day.

To say that I discovered the method of ionization does not take into account the complete story. It is necessary to point out that because the signal was so minute, the discovery could not have been made if it were not for the sensitivity and high performance of the instrument. Technologies for the mass separation mechanism, the detector and the signal processor were developed by Dr. Tamio Yoshida, Mr. Yoshikazu Yoshida, Mr. Yutaka Ido and Mr. Satoshi Akita of the research team. Many advancements that exceeded the technological levels at that time were incorporated into these components. Rather than to mention only my excellence, I believe it is more appropriate to say that the overall support of the research team was excellent. We were also probably influenced by the climate within Shimadzu Corporation, which provided a large degree of freedom for this type of research to a team composed of such young members.

The product was developed based on our research results. These were announced for the first time outside the company at the Annual Conference of the Mass Spectrometry Society of Japan held in Kyoto in May 1987. At that time, the mass number that we had been able to measure had already exceeded 48,000. However, this announcement did not cause much of a stir in

K.T., my wife and Taro (Golden Retriever) in 1996 (taken by Mr. Mitsuo Ikegami)

the world of mass spectrometry. We, on the other hand, were just satisfied that we had achieved our goal and were able to announce it publicly. We were not at all troubled by the lack of reverberating excitement in the world of mass spectrometry. In September 1987, the Second Japan-China Joint Symposium on Mass Spectrometry was held in Takarazuka, Japan, and it was there that we announced our results in English for the first time (at that time we could measure mass numbers in the range of 72,000). There is a double significance here, in that not only were the research results written in English, I actually presented the results in English for the first time. Although my English was far from good, my meaning was well enough understood by Professor Cotter for him to make the results known around the world.

Our first product, the "LAMS-50K," was put on the market in 1988. I visited many research laboratories at universities and companies to introduce and explain the product. I also performed analysis on many samples provided by potential customers. However, because the performance fell short of the expectations of customers performing analysis of biological samples, and because the instrument was so expensive, we were not able to sell any of these instruments in Japan. Ultimately, we were able to sell only one instrument, and that was to the City of Hope's Beckman Research Institute in the US, where the instrument's performance received a satisfactory evaluation (1990). After that, manufacture of this instrument was discontinued, and I was obliged to join the development team for Gas Chromatograph Mass Spectrometer (GCMS) instruments, another kind of mass spectrometer.

In May of 1989, Mr. Yoshikazu Yoshida and I won a Research Award from the Mass Spectrometry Society of Japan for "Research on Macromolecular Ion Detection using a Laser Ionization Time of Flight Mass Spectrometric

My colleagues in Kratos Analytical Ltd. in 2001

Method." That was the only acknowledgement I had ever received publicly before being selected for the Nobel Prize in Chemistry 2002. I was so pleased to receive that first award for the research in which I had participated.

In January of 1992, I was transferred temporarily to KRATOS Analytical Ltd. (a subsidiary company of Shimadzu Corporation) in Manchester, England for a period of one year. This was my first experience to reside overseas, and it was a bewildering experience from various viewpoints, the most prominent being language and culture. To begin with, I was to act as a liaison between the two companies. However, I also joined the development team charged with developing compact MALDI mass spectrometers (KOMPACT MALDI series). Although I returned to Japan midway through the development project, I subsequently continued my cooperation through involvement in domestic sales of the product, as well as performing MALDI technology-related research. There are many engineers who dislike being involved in sales efforts through education of potential customers. But as long as I am an engineer, I think it is important to assure that customers obtain satisfaction with a product that I have had a hand in developing. Moreover, because such contact with customers widens one's own knowledge, this knowledge cannot but help in focusing on desirable features in development of the next product.

In May of 1995, at the age of 35, I married Yuko Ikegami who also grew up in my native Toyama Prefecture. Thirty-five years of age is generally considered to be a bit late for a first marriage in Japan. Because I have always had a tendency to tense up whenever speaking to a woman, I never had good karma when it came to women. What's more, being so involved with work all the time, I had little time to think of marriage. I suppose the correct interpretation is that by the time I noticed, I was 35 years old. However, even for someone like me, my joy was undeniable when my wife said to me "There is nothing about you I don't like." In front of her, I lost my awkwardness and I am now able to behave naturally. Furthermore, after getting married, I became much more stable. I am normally too embarrassed to express my thanks verbally, but I want to take this opportunity to tell my wife "Thank you."

From April of 1997 to May of 2002, I was posted once again to the UK for the purpose of developing a new type MALDI-MS called a "Matrix Assisted Laser Desorption Ionization Quadrupole Ion Trap Time of Flight Mass Spectrometer (MALDI-QIT-TOFMS)" (Model Name: AXIMA-QIT). The theory and results were presented to the public for the first time at the Annual Conference of the American Society for Mass Spectrometry in 1999.

One fond remembrance I have occurred in May of 2002. Just before leaving the KRATOS premises, I was summoned to the boardroom. When I entered the room, there was a surprise farewell party all prepared for me. I received as a present a bright red uniform with the number 7 printed on it, along with my name. Of course, number 7 is the number of the renowned football player David Beckham, a member of the world famous Manchester United football team, whose stadium is located very near the KRATOS home office.

In May of 2002, I returned to Japan to begin preparing for sales of the new AXIMA-QIT product. I performed marketing to widen the sales opportunities of the instrument, and also continued feasibility studies on analytical methods for biological macromolecules beyond those for proteins. However, my destiny changed suddenly and dramatically with one telephone call from Sweden received on October 9, 2002.

Until writing these passages, I have never looked back on my life as I am doing so now. I have always thought that I was doing what I desired to do, doing what was interesting to me. However, upon reflecting on my life in this way, it seems that my life is a product of my relationship with such factors as my birth, my family, teachers, friends, companions at work, and even the business world, geographical regions, the natural environment as well as my cultural environment. I am sure that receiving this Nobel Prize will also have an effect on my life. However, just as when there is an unexpected result in an experiment which brings a pleasant surprise, I hope that I will be able to continue enjoying my life in a natural manner after receiving this prize. It is my desire to keep walking through life as an engineer, and to continue producing results that are useful to society.

Finally, I'd like to express my thanks to everyone who gave me the opportunity to walk freely through this life. "ARIGATO GOZAIMASU!"

THE ORIGIN OF MACROMOLECULE IONIZATION BY LASER IRRADIATION

Nobel Lecture, December 8, 2002

by

KOICHI TANAKA

Shimadzu Corporation, 1 Nishinokyo-Kuwabaracho, Nakagyo-ku, Kyoto 604-8511, Japan.

INTRODUCTION

When I was a student at university, my major field of study was Electrical Engineering. I did not specialize in chemistry in my studies at university, and my knowledge of chemistry overall was somewhat less than rich. Under those circumstances, upon entering the company, I applied myself wholeheartedly to the study of chemistry to challenge and overcome a barrier in analytical chemistry considered to be impenetrable. I am, of course, very pleased to receive the Nobel Prize and further, especially pleased that the Prize serves to acknowledge my efforts and successes in the field of analytical chemistry, a field that while rather inconspicuous among the general population is indeed very important and useful to the world.

I am employed by the analytical instrument manufacturer, Shimadzu Corporation. I was assigned to a team charged with the task of designing and building a completely new mass spectrometer. The achievement of developing "a soft desorption ionization method for mass spectrometric analysis of biological macromolecules", for which I was chosen to receive this prize, was one aspect of that project. At that period in the 1980's, four of my colleagues, whom I will introduce shortly, in addition to myself, were each in charge of developing a part of the instrument. I was responsible for developing the sample preparation and ionization technologies. Success of the instrument would not have been assured unless the other technologies were suitably provided, even given the excellence of the ionization technology. First, I will briefly describe the progress of development of the technologies at that time, other than that of ionization.

MASS SEPARATION: TOF TECHNOLOGY

In mass spectrometry, it is necessary to use some method to separate the generated ions according to mass, and then perform detection of the ions. One of these methods is called *Time-of-Flight* Mass Spectrometry (TOF-MS), which involves separating ions according to mass by measuring their respective flight times.

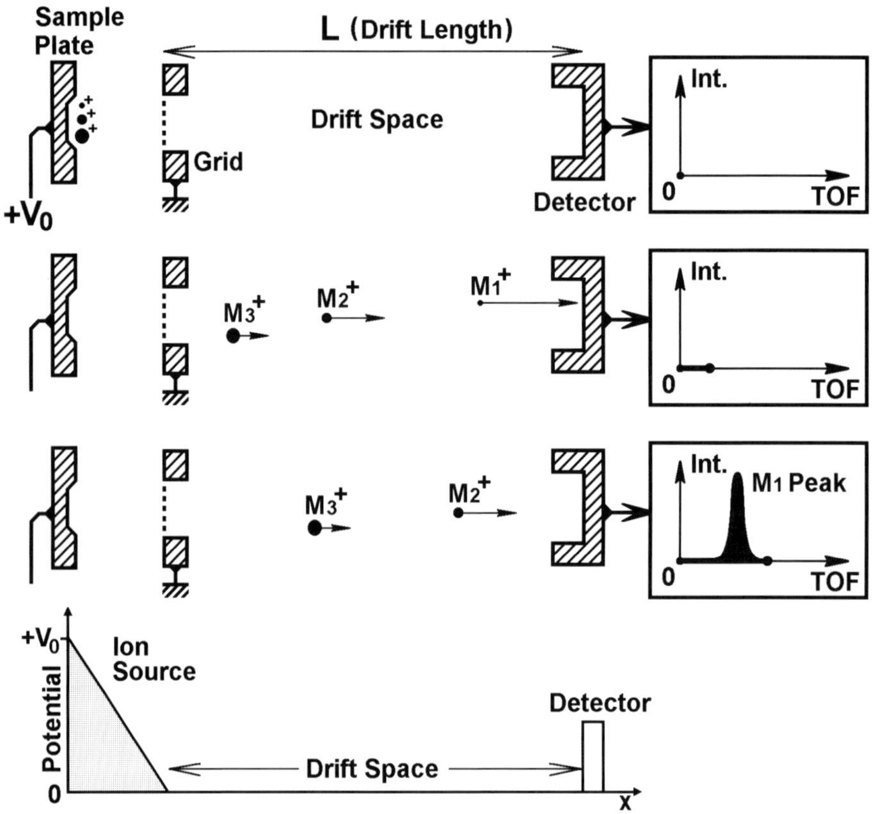

Figure 1. Schematic of Time of Flight Mass Spectrometry.

$$zV_0 = \frac{1}{2}mv^2$$

$$v = \sqrt{\frac{2zV_0}{m}} \quad \text{-----------} \; (1)$$

$$t = \frac{L}{v} = L \cdot \sqrt{\frac{m}{2zV_0}} \quad \text{-----------} \; (2)$$

$$t' = L \cdot \sqrt{\frac{m}{2(zV_0 + \varepsilon)}} \quad \text{-----------} \; (2)'$$

z: charge of ion, m: mass of ion, v: velocity of ion,
L: length of drift space, t: flight time of ion in drift space,
t′: flight time of ion in drift space with initial kinetic energy ε

Figure 1 provides the simplest representation of the principle behind TOF-MS. This assumes that equally charged ions of differing mass, M_1, M_2 and M_3 are generated instantly at the same time and starting position. Assume that M_1 is the lightest of the ions. The ions are generated at an electrical potential

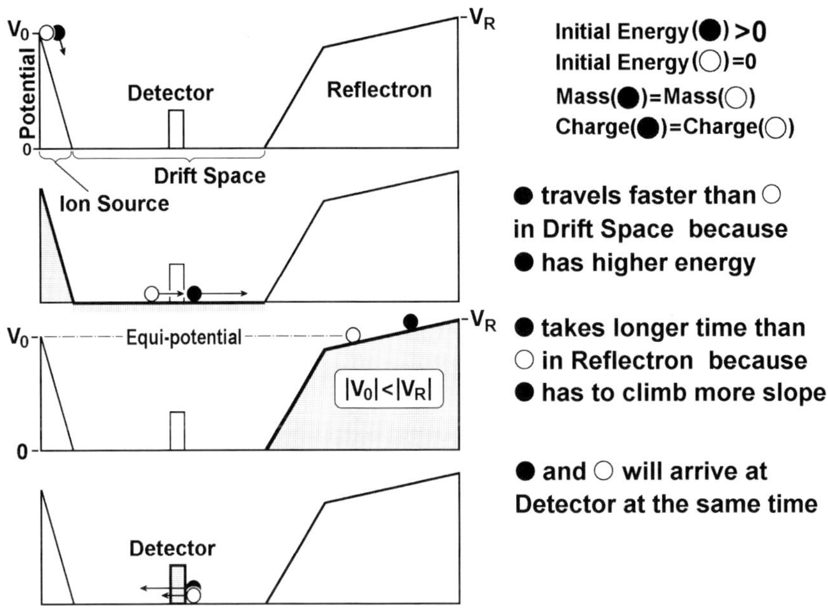

Figure 2. Mass Resolution Improvement by Reflectron.

V_0, and have a potential energy of zV_0, where z is the charge of the ion. If all of the potential energy is converted to kinetic energy, the velocity v is expressed by Equation (1), based on the law of conservation of energy. In other words, the lighter ion travels at a higher speed, reaching the detector first.

This explanation is based on ideal conditions (Equation (2)). Since the ions actually possess a distribution of initial velocities, that is, initial kinetic energy ε, their arrival time t' at the detector will show some distribution, even among ions of equal mass (Equation (2)'). This results in diminished mass resolution.

REFLECTRON TECHNOLOGY

To improve the resolution, it would be best if ions of the same mass and charge arrive at the detector at the same time, even if their initial kinetic energies are different. Specifically, a device referred to as a *Reflectron*, first described by Boris Mamyrin[1] in 1973, is used to reflect the ions midway in their flight back toward the detector. This turning back of the ions has the effect of adjusting the flight times of the ions in conjunction with their initial kinetic energies, to ensure that ions of the same mass and charge reach the detector at the same time (Figure 2).

The isochronism of the pendulum is very well known. Stated simply, even if the swing amplitude of the pendulum varies, the cycle times remain equal as long as the cord length of the pendulum and the force of gravity stay the same. Applying this principle, an ideal Reflectron was developed. In this case, the cord length corresponds to the ion mass, the swing amplitude to the initial kinetic and potential energy, and the cycle time to the time of flight, re-

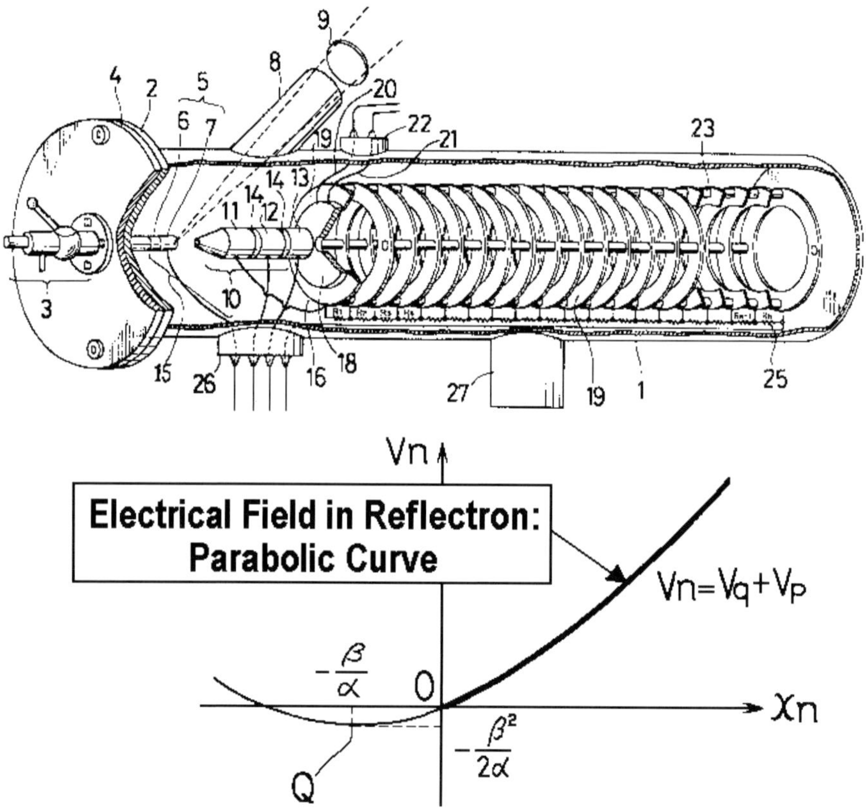

Figure 3. Mass Resolution Improvement by Ideal Reflectron (Patent: US4625112).

spectively. One method of this Reflectron technology was developed by one of my colleagues at that time, Mr. Yoshikazu Yoshida (Patent number: US4625112 issued in 1986) (Figure 3).

TIME DELAYED ION EXTRACTION

The principle behind TOF-MS assumes that ions are generated instantaneously, that is, in an infinitely short generation time. The time width of a laser pulse is extremely small, in the order of a few nanoseconds in the case of the N_2 laser. However, when the laser intensity is high enough to greatly exceed the ion generation threshold, as shown in Figure 4, ions may continue to be generated even after the completion of laser irradiation. In the TOF model, the mass resolution decreases due to the wider time width of ion generation.

One of my colleagues at that time, Dr. Tamio Yoshida, succeeded in improving the mass resolution, as shown in Figure 5, by drawing out the time from ion generation to ion extraction. This is accomplished by maintaining equivalent voltages at the sample plate and the grid to prevent the ions from being extracted until they are all generated, and then by creating a potential

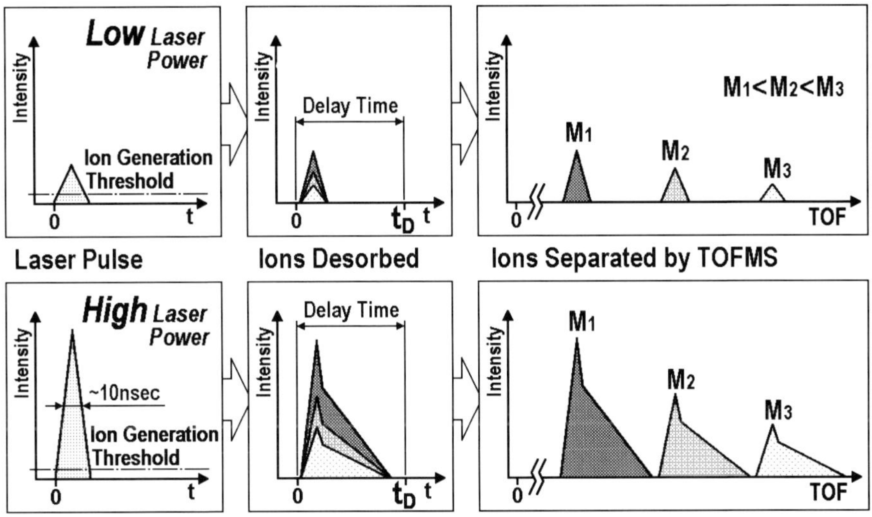

Figure 4. TOF Mass Spectrum Measurement with Low and High Laser Power.

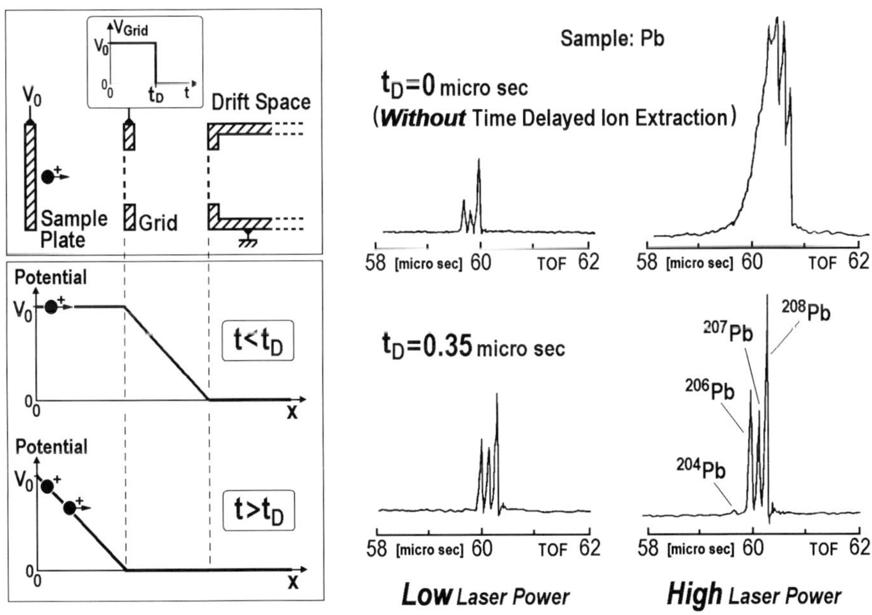

Figure 5. Mass Resolution Improvement by Time Delayed Ion Extraction.

difference after t_D to cause extraction of all the ions at the same time (filed in Japan Patent Office [S60035097] in 1985). This method is the same in principle as the *time-lag focusing* method which was originally designed by William Wiley and Ian McLaren[2] to reduce the effect of initial spatial and kinetic energy distributions, but was now applied specifically to minimize the effect of the prolonged laser-induced ion generation time.

In addition, there is a heightened possibility that particles given a charge,

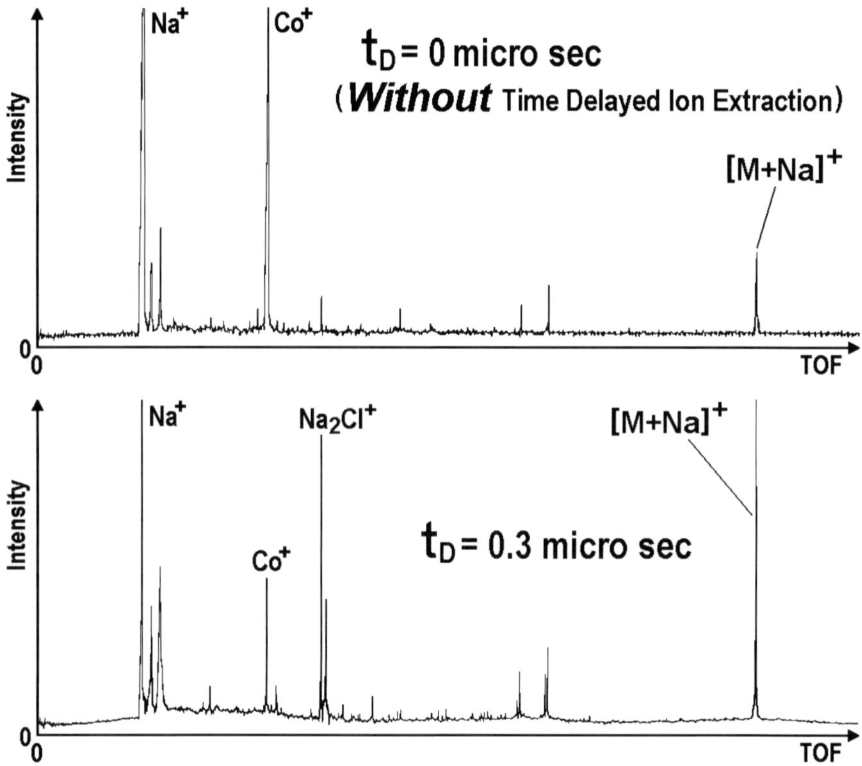

Figure 6. Molecular Ion Increase by Time Delayed Ion Extraction.

for example, protons and cations, will collide and react with normally neutral, that is, uncharged sample molecules. Such processes are generally referred to as *chemical ionization,* as originally described by Burnaby Munson and Frank Field.[3] The quantity of molecular ions ($[M+Na]^+$ in Figure 6) can be expected to increase.

ION DETECTION TECHNOLOGY

The term "detector" here refers to the conversion of ions separated by the mass separation mechanism into a flow of electrons, in other words, an electrical signal. The maximum mass number of ions that had been detected in the early 1980's was in the range of several thousand. In principle, the greater the molecular weight (M.W.) of the ion, the slower the velocity, so the ion-to-electron conversion efficiency could even drop to zero.

The simplest method of raising sensitivity is to accelerate the ions even further after mass separation occurs (this is called *post acceleration*). As shown in Figure 7, the higher the post acceleration voltage, the greater the intensity of the ions, and further, the higher the mass number, the greater will be the effect, especially when a high voltage is applied. Another method is the ion-to-

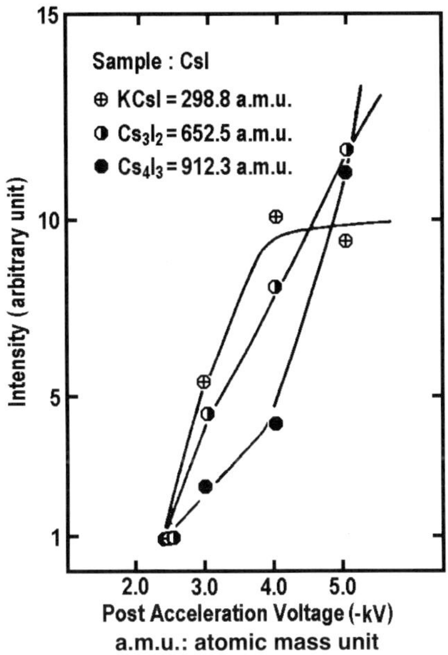

Figure 7. High Sensitivity Ion Detection by Post Acceleration.

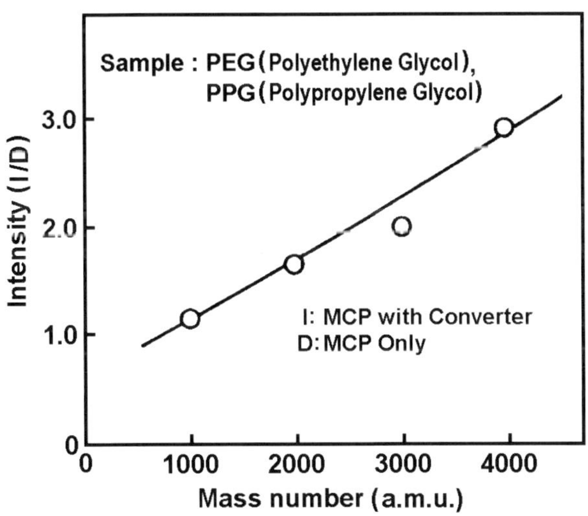

Figure 8. High Sensitivity Ion Detection by Ion-to-Electron Conversion.

electron conversion method. As shown in Figure 8, the intensity increases using this method as well.

One of my colleagues at that time, Mr. Yutaka Ido, developed a detector (Figure 9) that obtained the effects of both of these methods, enabling sufficiently sensitive detection even when ions of especially high molecular weight were generated.

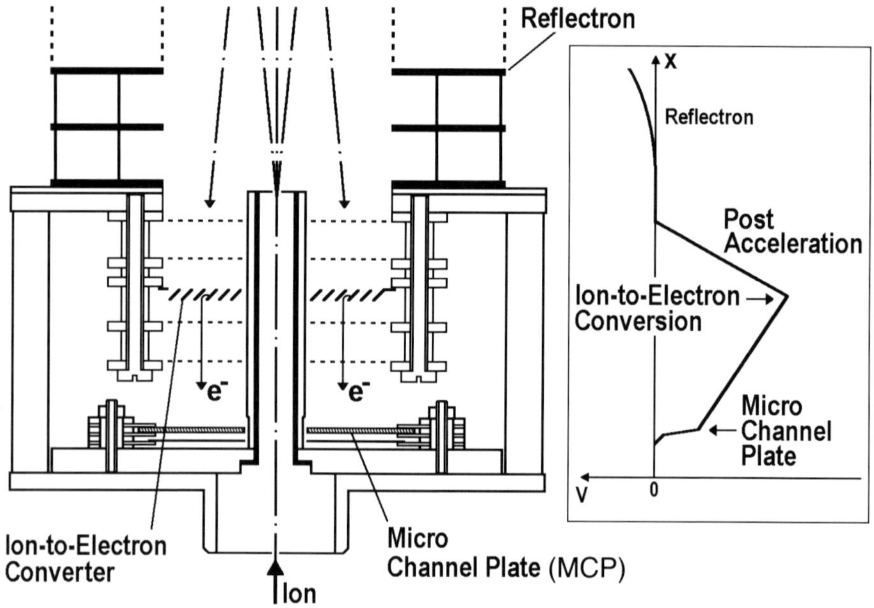

Figure 9. Macromolecule Detection by Post Acceleration and Ion-to-Electron Conversion.

MEASUREMENT TECHNOLOGY

ADC method

Even with improved detector sensitivity, it can be assumed that the intensity of macromolecular ions would be weak. In particular, when performing a single, one-shot measurement using a minute quantity of sample, successful quantitative analysis may not be possible without a high enough signal-to-noise (S/N) ratio. For this reason, a spectrum accumulation circuit is considered to be indispensable. To satisfy the demand for an instrument capable of high-speed measurement, a TOF spectrum measurement accumulation circuit was developed within our group.

One of my colleagues at that time, Mr. Satoshi Akita, developed an analog-to-digital converter (ADC) circuit adopting a *pipeline technique* to achieve the required speedup in measurement. The newly developed ADC circuit provided a maximum 1 kHz processing speed with real-time accumulation measurement at a maximum time resolution of 10 nsec and 24 bit intensity resolution (Figure 10).

Moreover, to enable measurement of both low mass high intensity ions as well as high mass low intensity ions, an amplifier that could perform gain switching within one microsecond was developed and adopted to provide a wider dynamic range.

TDC Method

At that time, ADC circuits had a time resolution of 10 nsec. It was our opinion that even this resolution would not be enough. Mr. Yoshikazu Yoshida devel-

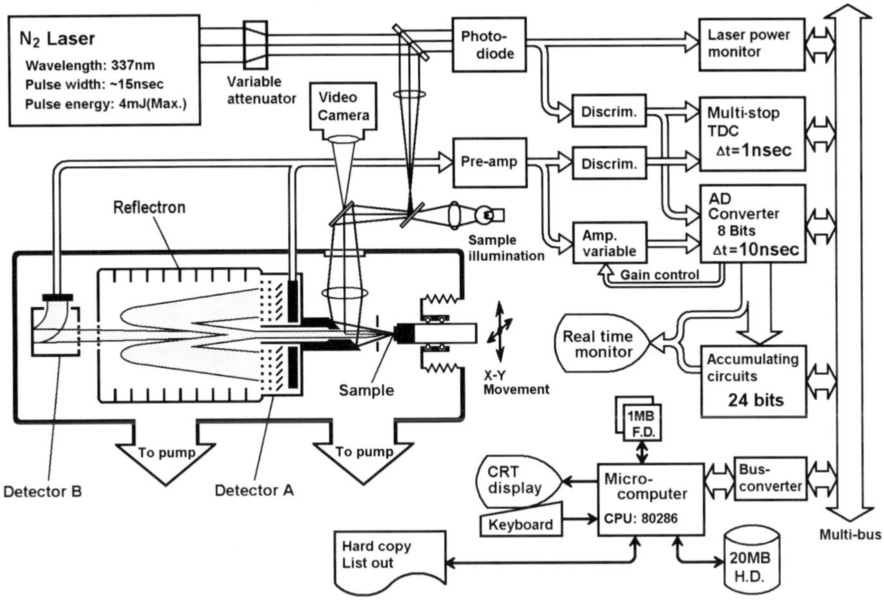

Figure 10. Schematic of Shimadzu LDI-TOFMS.

oped a time-to-digital converter (TDC) circuit that could measure with great accuracy the flight time of the ions while measuring a maximum of 255 events after one start signal (Multi-Stop-TDC), even though peak intensity information was sacrificed. As a result, we were able to measure the time of flight at a maximum time resolution of 1 nsec (Figure 10).

All of these crucial developments further emphasize the certainty that if even one of our group of five were not there, we would not have been able to achieve measurement of macromolecular ions. I am truly happy that the technologies we developed in the 1980's have been improved through their use in various instruments, and that they have played a role in bringing TOF technology into the mainstream of MS from the 1990's.

IONIZATION TECHNOLOGY

Long before the 1980's, there had been investigation into the use of optical energy, not limited to laser light, to achieve ionization of organic compounds.[4] However, for the most part, this ionization consisted of stripping electrons from compounds in the gas phase.[5] Because this was limited to compounds that could be vaporized without being decomposed, very few compounds could be ionized. When a solid phase organic compound was irradiated with a laser, the organic compound absorbed the laser light, providing enough energy for desorption. Further, when the positive and negative charges of the particle are not in balance, it could be measured as an ion. However, in this case, since the laser light is absorbed directly into the analyte,

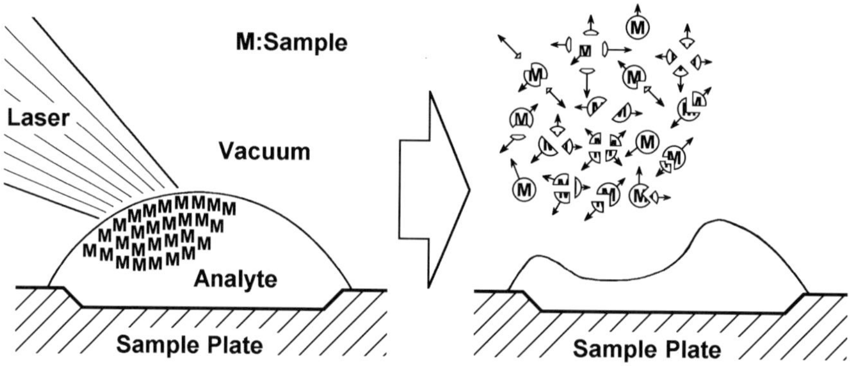

Figure 11. Laser Desorption Ionization (LDI) from Solid or Liquid Phase.

the molecular bonds may be broken due to the increased internal energy. In other words, there is an increased risk that measurement would occur in the decomposed state (Figure 11). Because we wanted to measure the un-decomposed, intact molecular ion, this was unsatisfactory.

At that time, the molecular weight of compounds that could be ionized by laser was about 1,000 Da, and ionization of compounds having a molecular weight exceeding 10,000 Da was generally considered by chemists at the time to be impossible. However, not being a specialist in chemistry, I was unaware of the existence of this widely-held belief.

RAPID HEATING TECHNOLOGY

In the 1980's, "desorption to gas phase by rapid heating" gained attention through the efforts of several groups that had obtained some degree of encouraging results. These include Franz Röllgen[6, 7] of Germany, Mamoru Ohashi[8] of Japan, and Louis Friedman[9], Don Hunt[10], Burnaby Munson[11] and Robert Cotter[12] of the USA. A simple description of the principle is depicted in Figure 12. In short, if a given compound "AB" is heated, it is assumed that "AB" will be released to the gas phase as intact "AB" as well as fragmented "A" and "B". Respectively applying the well-known Arrhenius Equation (Equation (3)) will give the logarithmic expressions shown in Equation (4).

Here, for compounds that are thermally labile and decompose easily, the rate constant for decomposition "K_D" is larger than that for vaporization "K_V" at low temperatures. Because the activation energy for vaporization "E_V" is higher than that for decomposition "E_D", the slope of the vaporization reaction "K_V" in Figure 12 is steeper than that of the decomposition reaction.

If this is always correct, at high temperature, where 1/T is small, vaporization should be further promoted. In other words, reaching the highest possible temperature as quickly as possible should provide a high degree of vaporization.

USE OF UFMP

To perform this rapid heating, various heating methods were devised.[9] However, a high enough temperature could not be achieved quickly enough to obtain intact vaporization of macromolecules, such as proteins. Since the time width of a pulse laser itself is between a nanosecond to several microseconds, rapid heating would seem possible using a focused beam to generate energy at high density and high speed. However, this would require a medium to enable its conversion into thermal energy.

In Japan at that time in the 1980's, *Ultra Fine Metal Powder* (UFMP) received much attention in the field of metallurgy due to its usefulness in efficient production of alloys, and was referred to as *Japanese Powder*. UFMP, as its name indicates, is an extremely fine metallic powder with particle diameters measuring a few tens of nanometers. When compared to bulk metal, the particle diameter of UFMP is about the same as a wavelength, increasing the possibility that scattered light could be taken up internally (Figure 13). In other words, light energy could be absorbed more efficiently. Even visually, it is black in color. Moreover, the possibility that thermal energy would be dispersed and lost decreases due to the inter-particle spaces.

Vaporization: "AB" $\xrightarrow{K_V}$ "AB", Decomposition: "AB" $\xrightarrow{K_D}$ "A"+"B"

Arrhenius Equation: $K = F \exp(-E/RT)$ ------------------- (3)

K: Rate constant F: Frequency factor E: Activation energy
R: Gas constant T: Absolute temperature

$\mathbf{\ln K_V = \ln F_V - E_V/RT}$, $\ln K_D = \ln F_D - E_D/RT$ ----------- (4)

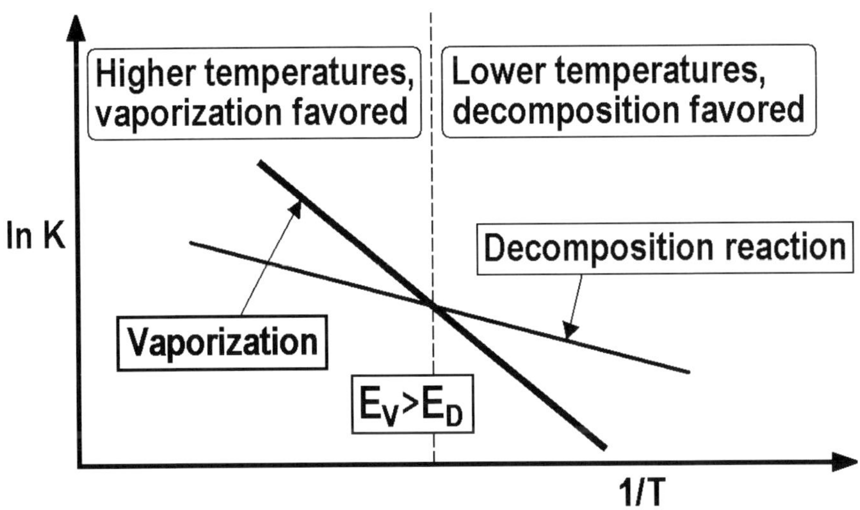

Figure 12. Vaporization vs. Decomposition.[9]

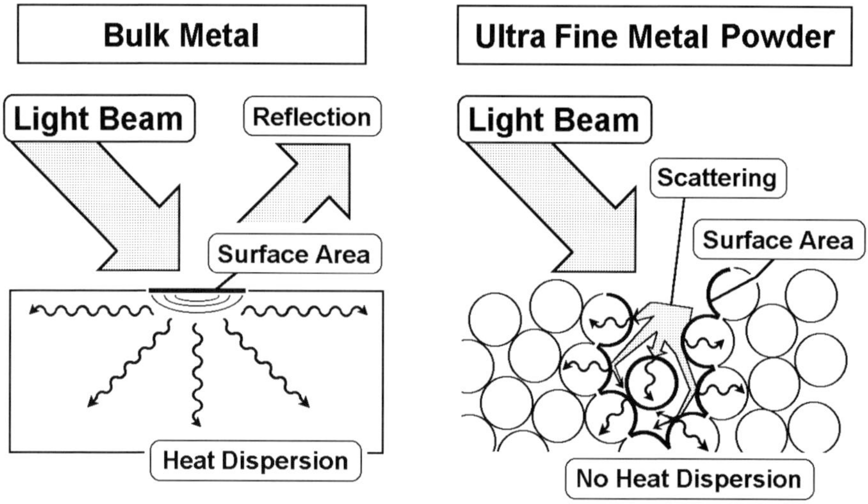

Figure 13. Differences between Bulk Metal and Ultra Fine Metal Powder.

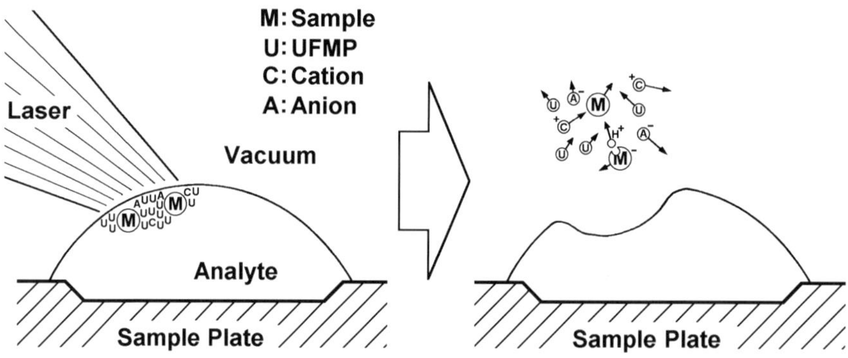

Figure 14. Molecular Ion Desorption by using UFMP Matrix.

Mixing this with an organic compound should allow the laser light to be efficiently absorbed into the UFMP, and to achieve a high UFMP temperature very quickly without the dispersion and loss of heat. This would efficiently achieve heating of the sample in the mixture. It was Mr. Yoshikazu Yoshida who proposed using UFMP as a *matrix** (Figure 14) (filed in 1985, Patent number: JP01731501 issued in 1993).

There was at that time an interest in using laser desorption for the analysis of polymers, such as in the reports by Charles Wilkins[13] of the USA. The MS spectrum shown in the upper part of Figure 15 is obtained when the sample, consisting of a polymer mixture of polyethylene glycols (PEG), is irradiated with a laser. The lower spectrum of Figure 15 shows that of the sample mixed with UFMP. With PEG alone, fragment ion peaks of the PEG molecule are dominant, and no PEG2000 peaks are observed. On the other hand, not on-

* In the ionization method, a substance used to assist ionization is called a "matrix."

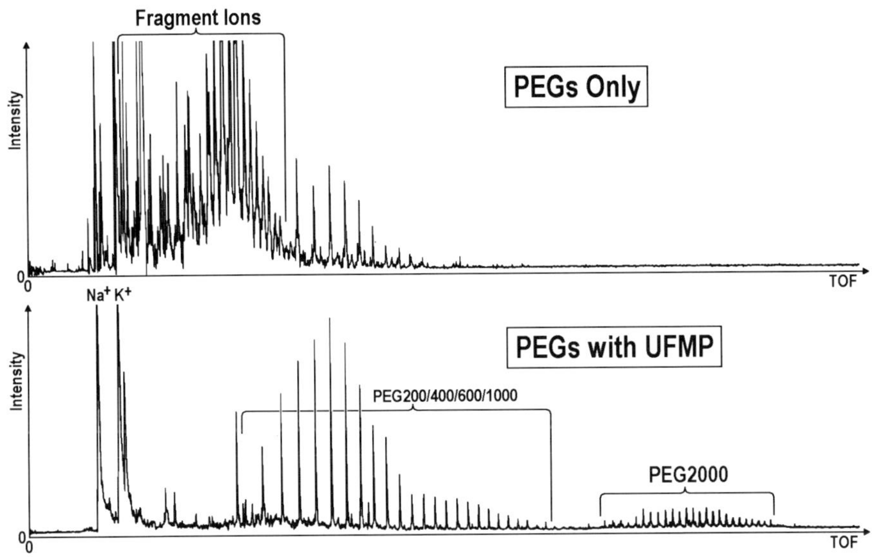

Figure 15. Polyethylene Glycols Spectra with and without UFMP Matrix.

ly are there almost no fragment peaks observed when PEG is mixed with UFMP, PEG2000 peaks have become clearly evident.

With the use of this matrix, the molecular ions from many organic compounds that could not be measured previously could then be measured. However, this technique could not be applied to compounds with molecular weights in the tens of thousands according to our experimental results.

GLYCERIN MATRIX TECHNOLOGY

At that time in the 1980's, the *Fast Atom Bombardment* (FAB) MS method was widely used to achieve ionization in thermally labile compounds. The FAB method was developed by Michael Barber and colleagues[14], and David

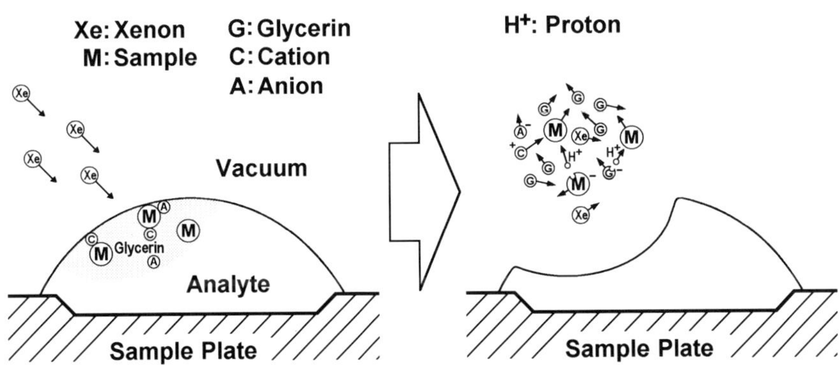

Figure 16. Molecule Desorption/Ionization using Glycerin Matrix in FABMS.

Surman and John Vickerman[15] of the UK, and involved accelerating neutral particles such as Xenon at high speed so as to collide with the sample and generate ions. In this method, the solid sample was maintained in a liquid state by using glycerin (glycerol) as a matrix (Figure 16).

It is thought that since glycerin is a liquid at room temperature, it releases the solid sample from its crystalline state to dissolve in the liquid, thereby assisting ionization. We used glycerin as a matrix just as it was used in FAB, but because the 337 nm laser wavelength that we used was only slightly absorbed into the glycerin, no significant effect was seen on the enhancement of molecular ion generation in organic compounds. However, this matrix was useful in improving reproducibility due to its effectiveness in providing a uniform mixture.

UFMP GLYCERIN MIXTURE TECHNOLOGY

Here it can be said that I had come up against a brick wall. Soon after that, while preparing many trial and error UFMP suspensions, changing the organic solvent and concentration repeatedly to obtain even the slightest improvement in data, I committed a monumental blunder, which was followed by a series of fortuitous, arbitrary decisions. I was using ordinary acetone as a UFMP suspension, and one day, by mistake, (1) I used the glycerin instead of the acetone. Reasoning that it would be wasteful to discard the expensive UFMP, (2) I decided to use the "ruined" preparation as the matrix solution. Because glycerin gradually evaporates in a vacuum, I knew it would eventually disappear. However, instead of waiting, (3) I decided to speed up vaporization of the glycerine by continuing to irradiate the laser. On top of that, because I wanted to see the result as soon as possible, (4) I was monitoring the acquisition of the TOF spectrum. In this way, only after at least four factors fell into place could I observe a phenomenon never before seen.

I think that the sample first measured was Vitamin B_{12}. I have tried recently to locate the original spectrum, but unfortunately I couldn't find it. Because this sample itself can absorb laser light with an extremely high degree of efficiency, up to then, only the fragments could be measured with high sensitivity. However, by using the mixed UFMP glycerin matrix, at the position where unfragmented ions (molecular ions) should have appeared, peaks which seemed to be noise were observed. No matter how many times I performed the measurement, these peaks were present. Molecular ions were observed even after changing the sample.

After that, as a result of progressing to higher mass ions while optimizing the concentration, laser power and other parameters, I was able to measure an ion having a molecular weight of 35,000 Da (Figure 17) (1985) and another ion having a mass number exceeding 100,000 (Figure 18) (1987). The patent application for this ionization method was submitted in Japan in 1985, and was subsequently issued in 1993 (Patent number: JP01769145).

WHY DID THE MACROMOLECULE BECOME IONIZED WITHOUT DECOMPOSING?

Use of UFMP conveniently allows high temperatures to be reached at high speed. However, even if this is mixed with a typical organic compound, it is difficult to achieve sufficient mixing. Even when viewing the mixture in an electron microscope-generated photograph, the lack of uniformity is obvious (Figure 19, top). When glycerin is added to the mixture, however, the uniformity becomes evident (Figure 19, bottom). Moreover, glycerin may also serve to release the sample from its crystalline state. The ability to measure mole-

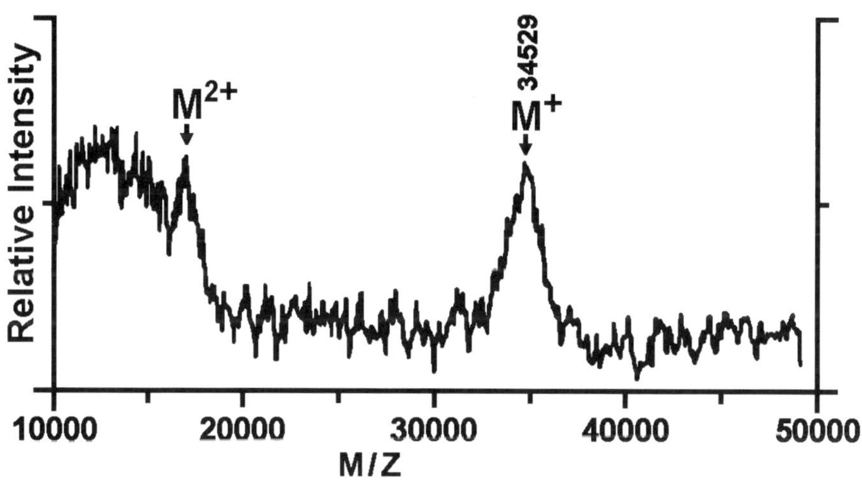

Figure 17. Molecular Ion Measurement of Carboxypeptidase-A (M.W.: ~35k Da).

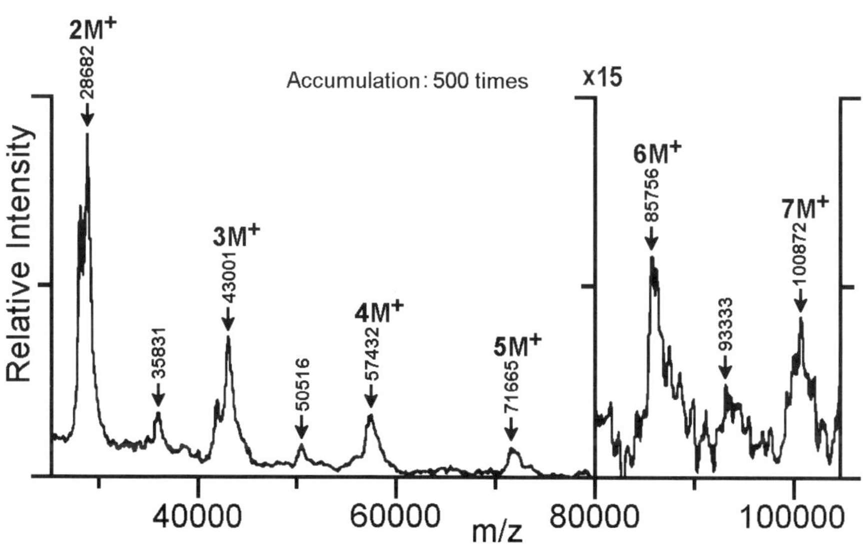

Figure 18. Cluster Ion Measurement of Lysozyme (M.W.: 14,306 Da).

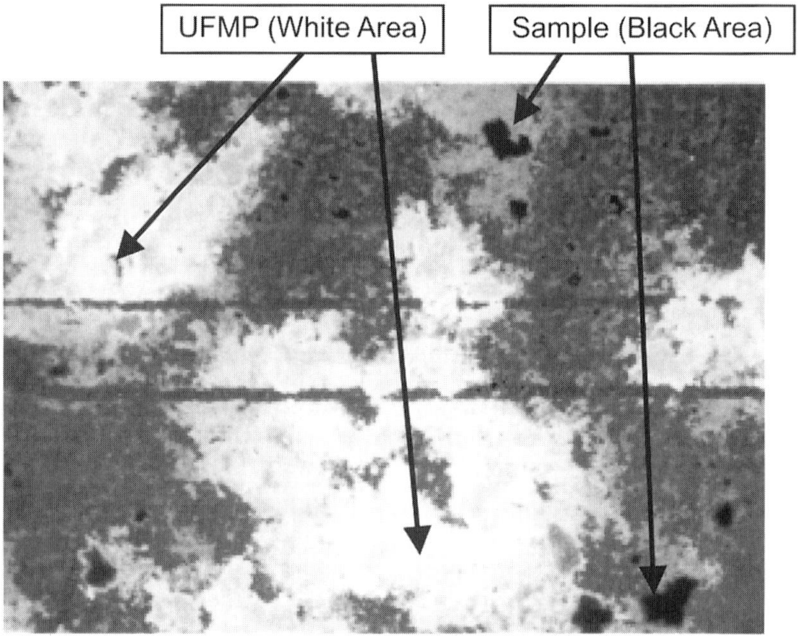

Sample UFMP Mixture without using Glycerin

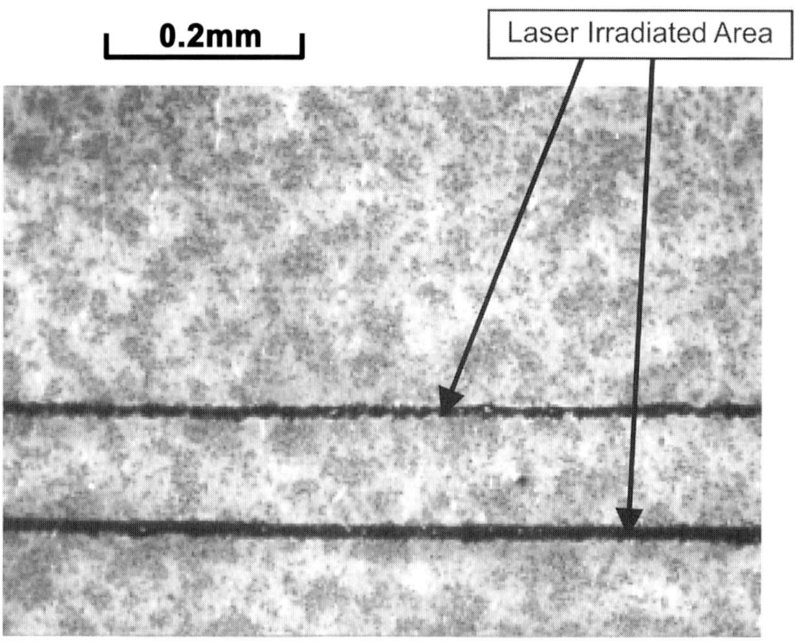

Sample UFMP Mixture with Glycerin

Figure 19. Electron Microscope Images of Sample and Matrix Mixture.

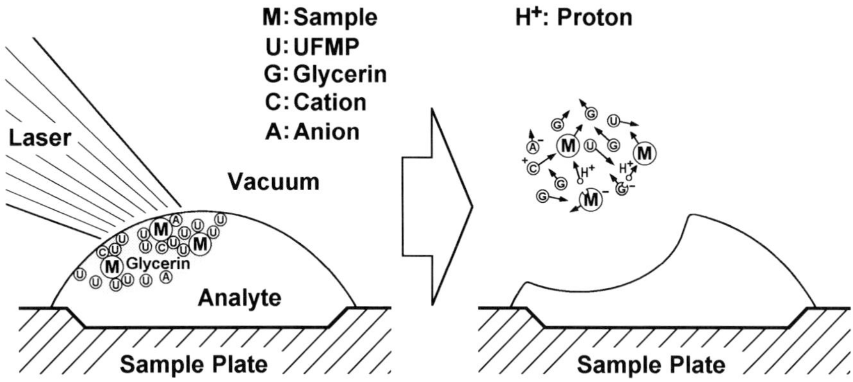

Figure 20. Macromolecule Desorption/Ionization using UFMP Glycerin Mixed Matrix.

cular ions from organic compounds even exceeding a molecular weight of 10,000 Da may indeed be due to these effects (Figure 20).

Here I have suggested the principles by which laser light irradiation is able to generate huge molecular ions. However, these principles are not necessarily correct because they have not yet been fully proven scientifically. Even with the knowledge of all the developments in *MALDI* (Matrix Assisted Laser Desorption / Ionization) technology that followed my discovery, I cannot yet claim to have correctly grasped the theoretical principles behind all of these phenomena.

However, I am an engineer, an engineer employed in a business enterprise. So even if the principles are unverified, their application takes priority if they are useful and practical.

COMMERCIALIZATION AND PUBLIC ANNOUNCEMENT

If a business organization does not consider that a technology is basically saleable, development would not be advanced and there would be no announcement regarding the technology outside of the company. If this technology had not been made into a product, it would have been buried inhouse and forgotten. For example, similar technology might have been developed somewhere else in the world in the 1980's, but if it had not been introduced to the international scientific community, it would not have been further developed to become a useful technique worldwide.

Because this basic technology that we developed at the Central Research Laboratory of Shimadzu Corporation was to be incorporated into a product, I was transferred to the business division. However, this technology was not made known outside the company until it was decided to release the product. The five members of our group made the first announcement outside the company at the Annual Conference of the Japan Mass Spectrometry Society held in Kyoto in May of 1987 (Oral and poster presentations [16–20] were conducted in Japanese). The announcement caused some reverberations, however, almost no one believed that this technology would become so useful. Of course, we were just pleased and satisfied to confirm that we had made scientific progress, and further, that we had presented it to the public.

TECHNOLOGY TRANSFER TO EUROPE AND THE USA

In September of 1987, I participated in the Second Japan-China Joint Symposium on Mass Spectrometry that was held in Takarazuka, Japan, and presented the result for the first time in English.[21] At that time in the 1980's, *Plasma Desorption* MS (PDMS) was widely used as a method of measuring macromolecular compounds such as proteins.[22] During his presentation there, Professor Robert Cotter stated that "PDMS can measure higher masses than *Laser Desorption/Ionization* (LDI) MS." Professor Cotter was an authority with respect to TOF-MS even at that time, but given my own recent results, which I would present the next day, I could not agree with his statement. I apparently indicated to Professor Cotter that I thought he must be mistaken, and that he should look at my data. Professor Cotter tells me that he was indeed impressed with the high mass results. He later saw my poster presentation and he and Dr. Catherine Fenselau forwarded details of our ionization method and copies of the spectra to other researchers in Europe and the USA.

At that time, Dr. Takekiyo Matsuo, an assistant professor at Osaka University who was attending the symposium, strongly recommended that I write a formal paper in English. I reluctantly agreed, and in 1988, our contribution was published in Rapid Communications in Mass Spectrometry (RCM)[23], and is referred to as one of the sources of the MALDI technique.

MANUFACTURE AND THEN TERMINATION OF SALES

The analytical technology and the instrument became a commercial product. In order to sell the instrument, I traveled around Japan visiting many customers, and actually analyzed more than two hundred samples. However, the reaction was almost always, "There is little practical use." I then looked further afield and there was only one potential customer on whom we could pin our hopes. That was Dr. Terry D. Lee of the City of Hope's Beckman Research Institute in the USA. This became the first and only instrument of our commercialized product that was sold, and we eventually halted development and sales of the instrument. Although the instrument could do what no other instrument was capable of, it seemed to have little practical use. Perhaps, it is fair to say that the sensitivity, accuracy and resolution required improvement for use in pharmaceutical and medical diagnostic applications.

DEVELOPMENT OF MALDI TECHNOLOGY AND ITS USE IN VARIOUS FIELDS

My research and development efforts on MALDI technology remained inactive until 1992 when I was transferred temporarily to KRATOS Analytical (a subsidiary company of Shimadzu Corporation from 1989) in Manchester, England. During that period, our technology that had been conveyed to Europe and the USA was steadily being improved upon and reformed through the efforts of Professor Cotter, Professor Fenselau and many other researchers throughout the world.

Special acknowledgement is also warranted for the meritorious efforts of both Professor Michael Karas and Professor Franz Hillenkamp [24–26] of Germany. Were it not for their tremendous and continuous efforts and the achievements of thousands of other researchers, it is certain that subsequent development of this technology that has benefited so many people around the world would not have advanced.

CONCLUSION

Development of this Macromolecule Ionization technology was not accomplished by one genius or gifted person. It is a victory of teamwork, and I believe it is the result of a lot of honest activity by many individuals seeking to advance a good technology.

The history of chemistry is long and it is natural that the field is wide and deep. Because of this, there is a tendency to believe that chemistry-related technology is already well developed. It should be mentioned here that I strongly believe that building new theories and bringing them to useful roles in the world requires the total mobilization of knowledge in specialized fields. However, I think it can be said that the technique that I discovered and the subsequent technologies that originated throughout the world broke through the common knowledge of chemistry. If I had been caught within the grasp of this knowledge, I would not have been able to make this type of unique discovery. Even though I cannot claim to possess extraordinary intellectual abilities, and further, my academic specialization was in a field other than that of chemistry, there is enough proof that I was able to contribute to the development of chemistry-related technology. With this in mind, it would give me great pleasure if many people around the world, and especially engineers working in private business enterprises, would mobilize their courage and hold onto their dreams in the pursuit of scientific innovation for the benefit of all living beings.

ACKNOWLEDGEMENT

I would like to express my thanks to Mr. Ralph Nesson for his tremendous efforts in helping me prepare for my Nobel Lecture presentation, and for his invaluable insight and patience in helping me express my thoughts in

English. I would also like to thank Prof. Robert Cotter, Prof. Catherine Fenselau, Dr. Chris Sutton and Dr. Andrew Bowdler for many stimulating discussions.

REFERENCES

[1] Mamyrin, B.A., Karataev, V.I., Shmikk, D.V., Zagulin, V.A. "The Mass-Reflectron, a New Nonmagnetic Time-of-Flight Mass Spectrometer with High Resolution" *Sov. Phys. -JETP* Vol.**37** No.1, pp.45–48 (1973).

[2] Wiley, W.C., McLaren, I.H. "Time-of-Flight Mass Spectrometer with Improved Resolution" *Rev. Sci. Inst.* Vol.**26** No.12, pp.1150–1157 (1955).

[3] Munson, M.S.B., Field, F.H. "Chemical Ionization Mass Spectrometry. I. General Introduction" *J. Am. Chem. Soc.* Vol.**88** No.12, pp.2621–2630 (1966).

[4] Terrenin, A., Popov, B. "Photodissociation of Salt Molecules into Ions" *Physik Z. Sovjetunion* Vol.**2**, pp.299–318 (1932).

[5] Lubman, D.M., Kronick, M.N. "Mass Spectrometry of Aromatic Molecules with Resonance-Enhanced Multiphoton Ionization" *Anal. Chem.*Vol.**54** No.4, pp.660–665 (1982).

[6] Giessmann, U., Röllgen, F.W. "Ion Formation by $[Li]^+$ Ion Attachment in High Electric Fields" *Org. Mass Spectrom.* Vol.**11**, pp.1094–1100 (1976).

[7] Röllgen, F.W., Giessmann, U., Heinen, H.J., Reddy, S.J. "Field Ion Emitters for Field Desorption of Salts" *Int. J. Mass Spectrom. Ion Phys.* Vol.**24**, pp.235–238 (1977).

[8] Ohashi, M., Yamada, S., Kudo, H., Nakayama, N. "In-beam Electron Impact Mass Spectrometry of Amino Sugars" *Biomed. Mass Spectrom.* Vol.**5** No.10, pp.578–581 (1978).

[9] Beuhler, R.J., Flanigan, E., Greene, L.J., Friedman, L. "Proton Transfer Mass Spectrometry of Peptides. A Rapid Heating Technique for Underivatized Peptides Containing Arginine" *J. Am. Chem. Soc.* Vol.**96**, pp.3990–3999 (1974).

[10] Hunt, D.F., Shabanowitz, J., Botz, F.K., Brent, D.A. "Chemical Ionization Mass Spectrometry of Salts and Thermally Labile Organics with Field Desorption Emitters as Solids Probes" *Anal. Chem.* Vol.**49** No.8, pp.1160–1163 (1977).

[11] Hansen, G., Munson, B. "Surface Chemical Ionization Mass Spectrometry" *Anal. Chem.* Vol.**50** No.8, pp.1130–1134 (1978).

[12] Cotter, R.J., Fenselau, C. "The Effects of Heating Rate and Sample Size on the Direct Exposure/Chemical Ionization Mass Spectra of Some Biological Conjugates" *Biomed. Mass Spectrom.* Vol.**6** No.7, pp.287–293 (1979).

[13] Wilkins, C.L., Weil, D.A., Yang, C.L.C., Ijames, C.F. "High Mass Analysis by Laser Desorption Fourier Transform Mass Spectrometry" *Anal. Chem.* Vol.**57** No.2, pp.520–524 (1985).

[14] Barber, M., Bordori, R.S., Sedgwick, R.D., Tyler, A.N. "Fast Atom Bombardment of Solids (F.A.B.) : A New Ion Source for Mass Spectrometry" *J. Chem. Soc., Chem. Commun.*, pp.325–327 (1981).

[15] Surman, D.J., Vickerman, J.C. "Fast Atom Bombardment Quadrupole Mass Spectrometry" *J. Chem. Soc., Chem. Commun.*, pp.324–325 (1981).

[16] Yoshida, Y., Tanaka, K., Ido, Y., Akita, S., Yoshida, T. "Development of Laser Ionization Time of Flight Mass Spectrometer I — Curved Electric Field Ion Reflector"(in Japanese) *35-kai Shitsuryo Bunseki Rengo Toronkai, Yoshishu*, pp.12–13 (1987).

[17] Akita, S., Tanaka, K., Ido, Y., Yoshida, Y., Yoshida, T. "Development of Laser Ionization Time of Flight Mass Spectrometer II — High-speed, High-accuracy Measuring Circuit for TOF Spectrum"(in Japanese) *35-kai Shitsuryo Bunseki Rengo Toronkai, Yoshishu*, pp.14–15 (1987).

[18] Ido, Y., Tanaka, K., Akita, S., Yoshida, Y., Yoshida, T. "Development of Laser Ionization Time of Flight Mass Spectrometer III — Sensitive Ion Detection in High Mass Range"(in Japanese) *35-kai Shitsuryo Bunseki Rengo Toronkai, Yoshishu*, pp.20–21 (1987).

[19] Tanaka, K., Ido Y., Akita, S., Yoshida, Y., Yoshida T. "Development of Laser Ionization Time of Flight Mass Spectrometer IV — Generation of Quasi-Molecular Ions from High Mass Organic Compound"(in Japanese) *35-kai Shitsuryo Bunseki Rengo Toronkai, Yoshishu*, pp.22–23 (1987).

[20] Yoshida, T., Tanaka, K., Ido, Y., Akita, S., Yoshida, Y. "High Molecular Ion Detection using Laser Ionization TOF Mass Spectrometer"(in Japanese) *35-kai Shitsuryo Bunseki Rengo Toronkai, Yoshishu*, pp.30–31 (1987).

[21] Tanaka, K., Ido, Y., Akita, S., Yoshida, Y., Yoshida, T. "Detection of High Mass Molecules by Laser Desorption Time-of-Flight Mass Spectrometry" *Proceedings of the Second Japan-China Joint Symposium on Mass Spectrometry*, pp.185–188 (1987)

[22] Cotter, R.J. "Plasma Desorption Mass Spectrometry: Coming of Age" *Anal. Chem.* Vol.**60** No.13, pp.781A–793A (1988).

[23] Tanaka, K., Waki, H., Ido, Y., Akita, S., Yoshida, Y., Yoshida, T. "Protein and Polymer Analyses up to m/z 100 000 by Laser Ionization Time-of-Flight Mass Spectrometry" *Rapid Commun. Mass Spectrom.* Vol.**2** No.8, pp.151–153 (1988).

[24] Karas, M., Bachmann, D., Hillenkamp, F. "Influence of the Wavelength in High-Irradiance Ultraviolet Laser Desorption Mass Spectrometry of Organic Molecules" *Anal. Chem.* Vol.**57** No.14, pp.2935–2939 (1985).

[25] Karas, M., Hillenkamp, F. "Laser Desorption Ionization of Proteins with Molecular Masses Exceeding 10 000 Daltons" *Anal. Chem.* Vol.**60** No.20, pp.2299–2301 (1988).

[26] Karas, M., Ingendoh, A, Bahr, U., Hillenkamp, F. "Ultraviolet-Laser Desorption/Ionization Mass Spectrometry of Femtomolar Amounts of Large Proteins" *Biomed. Environ. Mass Spectrom.* Vol.**18**, pp.841–843 (1989).

K. Wüthrich

KURT WÜTHRICH

I was born in Aarberg, Switzerland, on October 4, 1938, and during my childhood I lived in the small town of Lyss in the Berner Seeland. At the time this was a rural area of farmland, forests and rivers. The roots of the Wüthrich family are in an even more rural, mountainous area, the farming village of Trub in the Emmental near Bern. My mother's family owned the Restaurant "Bären" and a bakery in Lyss. My grandfather, Otto Kuchen, enjoyed fishing and hunting, and his jugged hare dish was a widely known fall season delicacy at the "Bären". My interests during childhood were largely influenced by our living in an old farmhouse, where my second grandfather, Jakob Wüthrich, had been a farmer. Although my father, Herrmann Wüthrich, took up an occupation as an accountant, he remained very much attached to his upbringings and our family produced a wide range of farming goods. My mother, Gertrud Wüthrich-Kuchen was the true center of our family life. In addition to raising me and my two younger sisters, Elisabeth and Ruth, she did marvelous things in the kitchen, tended our big garden, raised fowl, and was involved in various activities in the community.

My intense contacts with the rural environment of plants and animals awakened my interest in natural science at an early age. In particular, I acquired a thorough knowledge of the behavior of all sorts of water animals, mostly through observations made while enjoying all aspects of work and fun with a private trout river. On rare occasions I still enjoy fishing trips, and I am

With my parents, Herrmann and Gertrud Wüthrich-Kuchen, and my two sisters Elisabeth and Ruth in the garden of our home in Lyss, Switzerland, 1944.

At the Mercury Bay Game Fishing Club in Whitianga, New Zealand, 1987.

a member of the Mercury Bay Game Fishing Club in Whitianga, New Zealand, which lists Ernest Hemingway and Zane Grey among its all-time membership. With regard to my professional life, I had set my mind on becoming a forest engineer. Although I subsequently changed my mind in this regard, I still enjoy tending the family forest, which now contains trees that were planted by three generations of our family starting with my grandfather.

My formal training toward an academic profession started in 1952, when I transferred from the village schools in Lyss to the Gymnasium in the nearby "bilingue" city of Biel/Bienne. During the Gymnasium years my interests widened beyond forestry and fishing. We had the good fortune that our science and language teachers were either former University professors, who had left their academic positions elsewhere in Europe during the Second World War and found a haven in Biel, or followed the then common practice of using a teaching assignment at Gymnasium level as a stepping-stone for an academic career. At age 14 to 18 we were a group of seven students specializing in "natural sciences" who were thus trained in mathematics and physics at university level, and I happily accepted the challenge. According to my mother, it was during those years that I got used to working through the nights.

As a postdoctoral student in Berkeley, 1966.

Another focus was the French language, French literature, and French theatre and movies, which was largely motivated by the fact that the composition of our class as well as our teachers represented the bilingual character of Biel/Bienne. The Gymnasium Biel was informally attached to the Swiss Federal School of Sports and Gymnastics in nearby Magglingen, and thus my interest in competitive sports was awakened. These three areas all play an important role in my life up to the present days. Physics and mathematics are key activities in my professional life, professional visits in Paris and "les provinces" are combined with the sampling of French food, wine and culture, and I not only obtained the "Eidgenössisches Turn- und Sportlehrerdiplom" as one of my University degrees, but also played in a competitive soccer league well beyond the age of 50.

By now I can look back on 40 years of intense involvement with techniques referred to as "magnetic resonance spectroscopy". At the outset in 1962 and throughout my graduate studies there was electron paramagnetic resonance (EPR spectroscopy). EPR was complemented during my postdoctoral training from 1965–1967 by nuclear magnetic resonance (NMR) spectroscopy applied to chemical physics projects, and since the fall of 1967 I have used NMR for studies of biological macromolecules. From there it was a sinuous avenue that led by 1984 to the NMR method for protein structure determination in solution. Our results were occasionally met with doubts and disbelief, so that considerable moral strength and perseverance was at times called for.

During my student years from 1957–1962, NMR spectroscopy was just being introduced as an analytical tool in chemistry, molecular biology was not yet established as an independent discipline, and the initial three-dimensio-

With my wife Marianne, son Bernhard and daughter Karin in front of our apartment building in Greifensee, Switzerland, 1972.

nal protein crystal structures were just emerging. My education at the University of Bern could thus not possibly cover the areas of our current research. The faculty and the student classes in Bern were small in numbers, with three physics students and seven chemistry students starting in 1957. From my curriculum in chemistry, physics and mathematics, I best remember intense work in linear algebra, classical mechanics, chemical thermodynamics, physical chemistry of synthetic polymers, and preparative biochemistry of proteins and nucleic acids. This combination turned out to be an excellent foundation for my later scientific activities. The last two years of formal education, from 1962 to 1964, were spent at the University of Basel, majoring in sports and getting a Ph.D. in chemistry. Studying sports included about 25 weekly hours of intense physical exercise as well as premedical courses in human anatomy and physiology. Combined with experience gained from observations made on myself in the pursuit of competitive sports, this provided an additional dimension to my education. The subject of my Ph.D. thesis in inorganic chemistry with Professor Silvio Fallab was the catalytic activity of copper compounds in autoxidation reactions, and for this project the availability of a state-of-the-art EPR spectrometer in the Physics Institute was a great opportunity.

Studying natural sciences has always been a lot of fun for me, but nonetheless my mind was quite solidly set on a career as a high school teacher with a heavy involvement in sports. In parallel to my studies in natural sciences, I extensively yielded to what I thought to be my vocation. Thus, during the years

With, from left to right, Karin, Marianne and Bernhard in the desert near Tucson, AZ, USA, 1984.

1957–1962, I spent part of each winter as a ski instructor in Swiss mountain resorts. From 1959 to 1965, I had part-time jobs in high schools, first teaching physics at the Kantonsschule Solothurn, then chemistry at the Gymnasium Biel, and finally gymnastics at the Mädchengymnasium in Basel. These teaching assignments also had an important impact on my personal life. In 1961, while on my job as a ski instructor in the resort town of Saanenmöser in the Berner Oberland, I met my wife, Marianne Briner, who at the time was an elementary school teacher. We were married in 1963, and Marianne then joined me in studying sports at the University of Basel, graduating with the "Eidgenössisches Turn- und Sportlehrerdiplom" and specializing in modern dance. After the graduate student and postdoctoral years we started a family, with our son Bernhard Andrew being born in 1968 in Berkeley Heights, NJ, USA, and our daughter Karin Lynn joining us in 1970 in Greifensee near Zürich, Switzerland.

After finishing my graduate studies I spent another year in Basel concentrating on EPR studies of metal complexes in solution. In the spring of 1965 we moved to the USA, where I joined Professor Robert E. Connick at the University of California, Berkeley, for postdoctoral training. We used NMR spin relaxation measurements of ^{17}O, ^{2}H and ^{1}H in addition to EPR for studies of the hydration of metal ions and metal complexes. The Berkeley period was devoted to intensive work on the theory of nuclear spin relaxation, group theory and quantum mechanics, which was motivated by Bob Connick's weekly group seminar, a graduate course on "Group Theory and Quantum

Bob Connick in my office at the ETH Zürich, 1981.

Mechanics" by Professor Michael Tinkham, and an intense collaboration with another Swiss postdoc, Alex von Zelewsky, who soon thereafter accepted the chair of inorganic chemistry at the University of Fribourg in Switzerland. Over the years, Marianne and I returned at regular intervals to Berkeley, to renew the friendships of the 1960s and revive fond memories.

In October 1967 I joined the Biophysics Department of Dr. Robert G. Shulman at Bell Telephone Laboratories in Murray Hill, New Jersey. I was given responsibility for the maintenance of what was one of the first superconducting high resolution NMR spectrometers, which operated at a proton resonance frequency of 220 MHz, and I was otherwise free to use this instrument for "research on protein structure and function". Due to my background, my interest was focused on metal centers rather than on polypeptide chains, and all my initial projects in high resolution NMR had to do with hemoproteins. Using blood sampled from my arm in the first aid station, a Japanese colleague at Bell Labs, Dr. Tetsuo Yamane, prepared "hemoglobin (KW)", and within a few months we found entirely new avenues of deriving information on structure–function correlations from the NMR spectra of hemoglobin and other hemoproteins. These projects were a lucky choice: with the limited sensitivity and spectral resolution of the instrumentation available

Visiting the University of California, Berkeley, 1997. Dudley Herschbach happened to visit on the same day. Lunch at 'Chez Panisse' was running late, so our host, Alex Pines, carried the desserts along to the seminar room.

in 1968, the special spectral properties of hemoproteins were a great asset for successful NMR applications. Many years later, the unique NMR spectral features that enabled the early work with these metalloproteins had an important role in various aspects of the development of the NMR method for three-dimensional protein structure determination.

In October 1969 I returned to Switzerland to join the ETH Zürich. From the start I was equally well equipped with NMR and EPR instrumentation as previously at Bell Telephone Laboratories, and during the following 32 years the ETH provided us in regular intervals with the most advanced NMR equipment. Until 1975 I was working with a small group of students, a chemical engineer, Rudolf Baumann, who has stayed with me throughout all these years, and a postdoctoral associate with a physics Ph.D. in solid state EPR, Dr. Regula Keller, who largely took responsibility for the research with hemoproteins from 1970 to 1982. In 1973, Gerhard Wagner decided to do his graduate work with me. Gerhard then stayed with the group until 1987, pursuing a classical European academic career with Habilitation before settling as a Professor at Harvard Medical School. Being able to keep outstanding junior scientists as research associates over extended periods of time was a special privilege enjoyed by senior faculty in the traditional "European system",

A dinner party during a visit with Professor Setsuro Ebashi (second from the right) in Okasaki, Japan, 1996. Mrs. Ebashi is next to Marianne, and Kuniaki Nagayama, who started 2D NMR with proteins as a postdoctoral fellow with Richard Ernst and myself in the late 1970s, is on the left wing.

and the continued presence of Rudolf, Regula and Gerhard during most of my initial 15 years in Zürich was a key factor for success with our research program.

In Zürich, we continued research on hemoproteins with the use of NMR and EPR spectroscopy, where the biochemical work was mostly done by groups outside of the ETH who joined us for collaborative projects, and the spectroscopic work was done by Regula Keller, myself and a succession of graduate students. In addition, we started a program of systematic studies on the application of NMR techniques with polypeptides and small proteins. Spirits were kept high by successful studies of cyclic peptides in collaboration with the Head of the Institute of Molecular Biology and Biophysics, Professor Robert Schwyzer, the observation of unexpectedly well-resolved and long-lived NMR lines of amide protons in the small protein basic pancreatic trypsin inhibitor (BPTI), and the discovery of "ring flips" in BPTI. On the main line of research, which should lead to a method for protein structure determination in solution, there was only little progress. In 1975, in an attempt to survey the state of the field of NMR spectroscopy with biological macromolecules, I wrote the monograph *NMR in Biological Research: Peptides and Proteins.* There were two principal conclusions from this venture that should greatly affect the continuation of our work plan. First, I fully realized that we really had been extremely fortunate in choosing hemoproteins as a focus for our early NMR efforts. Second, it became clear that attempts of the early 1970s to derive *de novo* three-dimensional protein structures from conformation-dependent proton chemical shifts was not a promising approach, independent of whether these shifts were caused by intrinsic or extrinsic diamagnetic or paramagnetic probes. We thus had to look for novel avenues for NMR structure determination, where hemoproteins with their unique NMR-spectral properties could be an ideal testing ground for new ideas.

Shortly after I had learned my lessons from writing the 1976 monograph, the conditions under which I could pursue my work evolved in quite important ways. After working for more than 5 years with a small group of students

The President and the Secretary General of IUPAB with their wives during a diplomatic mission in China, 1983. President Professor Richard Keynes and Ann Keynes are on the left and right, respectively.

and research associates from the environs of Zürich, and being able to spend long hours of my own time at the bench and on the NMR spectrometers, I found myself suddenly surrounded by more than 20 postdoctoral fellows and students from all over the world. At around the same time, I also started to travel quite extensively in all parts of the world, with a first visit to India at the end of 1974, and a first "round-the-world" trip including stops in the USA and in Japan in the fall of 1975. The visits to India and Japan resulted in new, lasting friendships with local colleagues, and also in attracting a number of most talented postdoctoral fellows to Zürich. Ever since, professional travel has become an important part of my activities. Over the years this also included visiting faculty appointments at the University of California, Berkeley, Cornell University in Ithaca, NY, Johns Hopkins University in Baltimore, MD, the California Institute of Technology in Pasadena, CA, the Scripps Research Institute in La Jolla, CA, RIKEN in Tokyo, Japan, and the University of Edinburgh, UK. Spending part of my time in these places of highest standards added greatly to my quality of life as well as to the progress of our research in Zürich. The international aspect of my activities got a special boost in 1975, when – out of the blue – I was elected to membership in the Council of the International Union of Pure and Applied Biophysics (IUPAB). There was little work involved in this assignment, but in 1978 my IUPAB affiliation changed to being its Secretary General, and with this I also became a member of the "General Committee" of the International Council of Scientific Unions (ICSU) and of the ICSU Standing Committee on the Free Circulation of Scientists. During the six-year term as Secretary General the demands on my time were thus quite heavy. Fortunately, Marianne agreed to run the IUPAB

At dinner during an ICSU General Committee meeting in Munich, Germany, 1985. To my left is Professor Jorge Allende from Chile, to my right is Professor Raymondo Villegas from Venezuela.

office. This made things easier, since she would travel with me and we dealt with the IUPAB business in makeshift offices temporarily installed in hotels all around the world. The sunny side was that I got to know many prominent scientists, whose names I had previously mostly known from the textbooks. For example, structural biology was represented in the IUPAB Council from 1978–81 by Britton Chance, Henryk Eisenberg, David Phillips, Frederic Richards and Akiyoshi Wada, a true center of excellence! In the business meetings as well as in the social gatherings, we spent much of our time discussing the latest research advances long before they appeared in print. There was a particularly close collaboration with the IUPAB Presidents during my tenure as Secretary General, Professor Setsuro Ebashi and Professor Richard Keynes. Richard Keynes is a great-grandson of Charles Darwin. During IUPAB-related joint travel in Europe and the Far East in 1982/83, I listened to a more and more enjoyable but seemingly endless series of presentations of his "Darwin Lecture" commemorating the 100th anniversary of Darwin's death; in return, Richard lived through a heavy dose of biomolecular NMR spectroscopy.

Through my association with ICSU and IUPAB, I also got involved in entirely novel business. Most notable in hindsight were negotiations during the period 1980–1983 about joint adherence of China and Taiwan in international science organizations. We eventually defined terms and conditions for adherence to IUPAB of both "The Biophysical Society of China located in Beijing, China" and "The Chinese Biophysical Society located in Taipei, China". This involved extensive, highly formal correspondence, as well as

Addressing the audience during the opening session of an IUPAB/ICSU/UNESCO-sponsored 'Winter School on Magnetic Resonance in Biology and Medicine' in Cairo, Egypt, 1986. To my left is Professor I. El Gohari, who has over the years been the Egyptian delegate to numerous IUPAB functions.

visits and personal negotiations with Government and Academy officials in both countries. I also participated in IUPAB and ICSU programs of support for scientists in developing countries, and I organized summer schools and symposia in Africa, the Far East and Latin America. This all greatly influenced my outlook to the world. Although each year the IUPAB-related activities and my research-related travel kept me out of my laboratory for several months, the effect on our research was overall highly beneficial. As a bonus, I gained experience in directing a research group at a distance, and my junior associates could test their own initiatives during my absences.

With all the new talent assembled in my group by 1976, we started to develop new NMR experiments and novel algorithms for the structural interpretation of NMR data, which eventually resulted in the NMR method for protein structure determination. This included the identification of the nuclear Overhauser effect (NOE) as a NMR parameter that can be related in an unambiguous way to three-dimensional macromolecular structures. We made used of the outstanding resolution of parts of the hemoprotein ^{1}H-NMR spectra for calibrating NOE distance measurements with the then-available one-dimensional (1D) NMR techniques. In addition to Regula Keller, Sidney Gordon, a sabbatical visitor, made a key contribution with the introduction of the 1D "transient NOE" experiment. Subsequently, the NOE had a key role in

With Professor Chen-Lu Tsou, enjoying the beautiful scenery of Wuxi, China, during an IUPAB/ICSU-sponsored Summer School on Biophysics, 1992.

the approach used for obtaining sequence-specific assignments of the many hundred to several thousand NMR lines in a protein. The "sequential assignment strategy" was initially implemented by Gerhard Wagner and a diploma student, Andreas Dubs, using 1D NOE and spin decoupling experiments. In parallel with the 1D NMR investigations on NOEs and NMR assignment, the development of two-dimensional (2D) NMR techniques for macromolecular studies had been started in 1976 as a joint project with Professor Richard R. Ernst, (Nobel prize in Chemistry, 1991). In 1977 the first 2D NMR spectrum of a protein was recorded, and by 1980 we had assembled four 2D NMR experiments that were then used for the initial protein structure determinations: COSY (2D correlated spectroscopy), SECSY (2D spin-echo correlated spectroscopy), FOCSY (2D foldover-corrected correlated spectroscopy) and NOESY (2D nuclear Overhauser enhancement spectroscopy). It was a lot of fun at the time to decide on these acronyms! Soon my group started to use 2D NMR experiments in daily practice, and the experience from more than a decade of one-dimensional NMR spectroscopy with proteins was happily and profitably married with the new potentialities of 2D NMR.

By 1982, complete sequence-specific assignments had been obtained for a small protein, BPTI, and for the polypeptide hormone glucagon bound to lipid micelles. This was published in a series of four 1982 papers. Although the first one of these papers already outlines the presently used protocol for protein structure determination by NMR, it took two more years of intense work on metric matrix distance geometry algorithms and their implementation in efficient software packages before the first NMR structure determination of a globular protein, bull seminal protease inhibitor (BUSI), could be completed. A large number of brilliant junior scientists working in my group from 1976 to 1985 contributed directly or indirectly to this result: Gerhard Wagner was involved in each step of the project; Kuniaki Nagayama and Peter Bachmann devised the first generation of 2D NMR experiments for studies of biological macromolecules and wrote the software needed to

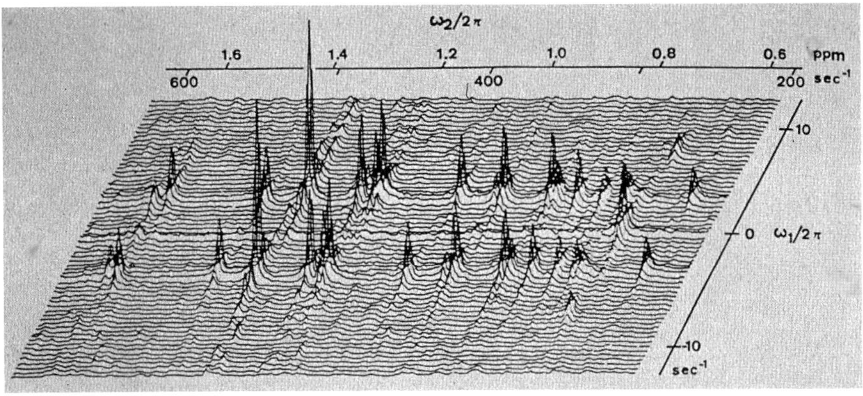

The first two-dimensional proton NMR-spectrum of a protein, recorded by Kuniaki Nagayama, 1997.

handle such data with the then-available limited computational facilities; Anil Kumar recorded the first 2D NOESY experiment with a protein; Gerhard Wider made key contributions to 2D NMR spectroscopy and to the sequential assignment method; Werner Braun, Martin Billeter and Timothy Havel started a tradition in my laboratory of theoretical work on the structural interpretation of NMR data; Peter Strop prepared BUSI and worked on its resonance assignments; finally, Michael Williamson and Timothy Havel actually solved the structure of BUSI. They all, and many additional students and postdoctoral associates from the "heroic period" 1976–85 have in the meantime started highly successful independent academic careers.

The completion of the first NMR structure of a protein brought new, unexpected challenges. When I presented the structure of BUSI in some lectures in the spring of 1984, the reaction was one of disbelief and suggestions that our structure must have been modeled after the crystal structure of a homologous protein. Apparently the structural biology community had thoroughly adjusted to the role of NMR as a method that could provide some exotic supplementary data, but which would not be suitable for *de novo* structure determination at atomic resolution. The criticism raised had two major consequences. The first one resulted from a discussion with Professor Robert Huber (Nobel laureate in Chemistry, 1988) after a seminar in Munich on May 14, 1984. Robert proposed to settle the matter by independently solving a new protein structure in his laboratory by X-ray crystallography and in my laboratory by NMR. For this purpose, each one of us received an ample supply of the α-amylase inhibitor Tendamistat from Hoechst AG. Virtually identical three-dimensional structures of Tendamistat were obtained by NMR in solution and by X-ray diffraction in single crystals, which settled matters once and for all. This was particularly comforting in the context of the fact that the subsequently solved NMR structure of metallothionein was completely different from an independently solved metallothionein crystal structure (it took six years before the crystal structure was redetermined and found to coincide

Tim Havel and Mike Williamson in front of a computer displaying the three-dimensional structure of the protein BUSI that we had just solved, 1984.

with the NMR structure!). The second consequence was that I asked for a sabbatical leave and ended up in Wengen, a beautiful mountain resort in the Berner Oberland. This was possible because I was also finishing my 6-year term as the Secretary General of IUPAB in the summer of 1984. Considering the critical reaction to the initial NMR structure determinations, I felt that it was important to document our work in a complete and detailed fashion. I thus had good reasons to honor my commitment of writing a monograph on the Baker Lectures, which I had delivered in 1983 at Cornell University. As I spent much of the time alone in Wengen, with my family joining me for weekends and vacation periods, work progressed well. "NMR of Proteins and Nucleic Acids" covers primarily work in my research group during the period 1977–84. It also turned out that directing my research group at a distance was surprisingly successful, and since the manuscripts were typed in Zürich from my handwritten notes, it helped that even ordinary mail was still reliably delivered within one day within Switzerland. It was therefore an easy decision for me to extend the stay in Wengen from the originally planned 6 months to 18 months. Besides the deskwork, important occupations in Wengen were skiing in the winter, and jogging and mountain climbing in the summer. According to my diaries I did not miss a single day of skiing from December 1, 1984 to April 10, 1985. This made up for having stayed away from the ski slopes during the 14 years from 1971 to 1984 because of my other professional activities. I also returned to the skiing outstation of the Federal Sports School in Mürren for a much-needed overhaul of my skiing technique, and participated in the organization of the famous Lauberhorn ski race in Wengen.

View from my office in our home in Wengen, Switzerland, during my first-ever sabbatical, 1984–1986.

In the spring of 1986, after a second winter of skiing in Wengen, I had thoroughly cleaned up the backlog of unpublished material, in addition to having finished work on the Baker Lectures monograph. Protein structure determination by NMR had by then found its believers, as documented by the fact that the first printing of my new book was sold out within a few weeks. For us, a new chapter had to be opened, and we established contacts with biochemists and molecular biologists for the real test of the NMR technique in applications to biologically interesting systems. By 1990, a collaboration with Professor Walter Gehring of the Biocenter at the University of Basel yielded structure determinations of the *Antennapedia* homeodomain and its complex with the operator DNA. Using this structure as a platform, additional NMR experiments provided entirely novel insights into the role of hydration water for specific DNA recognition. A NMR structure determination of the cyclophilin A–cyclosporin A complex was pursued as a joint project with two former graduate students, Hans Senn, and Hans Widmer, and their research team at Sandoz AG. It had immediate practical impact, since the structure of the bound immunosuppressant turned out to be very different from the only other structural information available at the time, *i.e.*, crystal and NMR structures of free cyclosporin A. It was, for all involved, a completely unexpected and for many reasons surprising result!

In yet another exciting collaboration, with Professor Rudi Glockshuber at the ETH Zürich, we completed a structure determination for the C-terminal half of the mouse prion protein in April 1996, barely 10 days after the BSE-crisis in Great Britain broke into the open. With this timing, the prion protein structure had high visibility also in the popular media. In 1997 we succeeded to characterize the structure of the intact prion protein, and found that the N-terminal half of the molecule forms a highly flexible, extended "tail". The prion protein thus presented a striking illustration of the unique power of NMR to characterize partially structured polypeptide chains. Others among the more than 70 protein structure determinations completed in my laboratory functionally relate to enzymology, toxicology, chaperone-mediated protein folding, and intercellular signaling.

The biological and biomedical projects pursued during the past 16 years with the use of the NMR technique have added and still add greatly to the quality of my professional life. In these endeavors, the quite extreme specialization needed to maintain a high standard of structure determination breaks open in that I learn about an ever-increasing range of biological systems and biomedical problems. I feel very fortunate that my field of specialization thus leads me to an education in biology from people who have high standards, and who sometimes even tend to consider me as one of their own.

NMR STUDIES OF STRUCTURE AND FUNCTION OF BIOLOGICAL MACROMOLECULES

Nobel Lecture, December 8, 2002

by

KURT WÜTHRICH

Eidgenössische Technische Hochschule Zürich, CH-8093 Zürich, Switzerland, and The Scripps Research Institute, 10550 N. Torrey Pines Rd., La Jolla, CA 92037, USA.

INTRODUCTION

Nuclear magnetic resonance (NMR) spectroscopy is unique among the methods available for three-dimensional structure determination of proteins and nucleic acids at atomic resolution, since the NMR data can be recorded in solution. Considering that body fluids such as blood, stomach liquid and saliva are protein solutions where these molecules perform their physiological functions, knowledge of the molecular structures in solution is highly relevant. In the NMR experiments, solution conditions such as the temperature, pH and salt concentration can be adjusted so as to closely mimic a given physiological fluid. Conversely, the solutions may also be changed to quite extreme non-physiological conditions, for example, for studies of protein denaturation. Furthermore, in addition to protein structure determination, NMR applications include investigations of dynamic features of the molecular structures, as well as studies of structural, thermodynamic and kinetic aspects of interactions between proteins and other solution components, which may either be

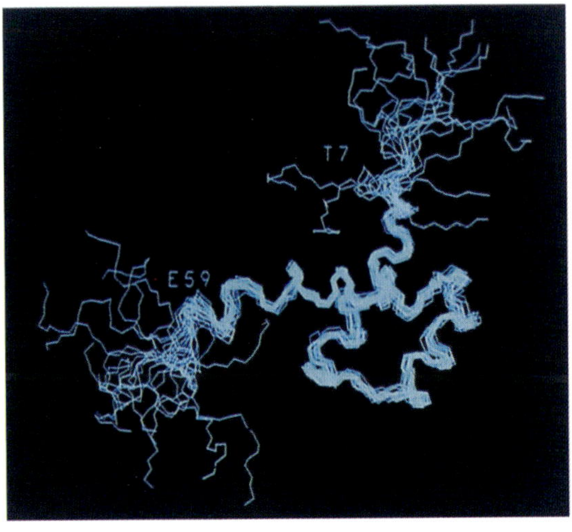

Figure 1. NMR structure of the *Antennapedia* homeodomain (1). A bundle of 20 superimposed conformers represents the polypeptide backbone. For the polypeptide segment 7–59 the tight fit of the bundle indicates that the structure is defined with high precision, whereas the two chain ends are disordered.

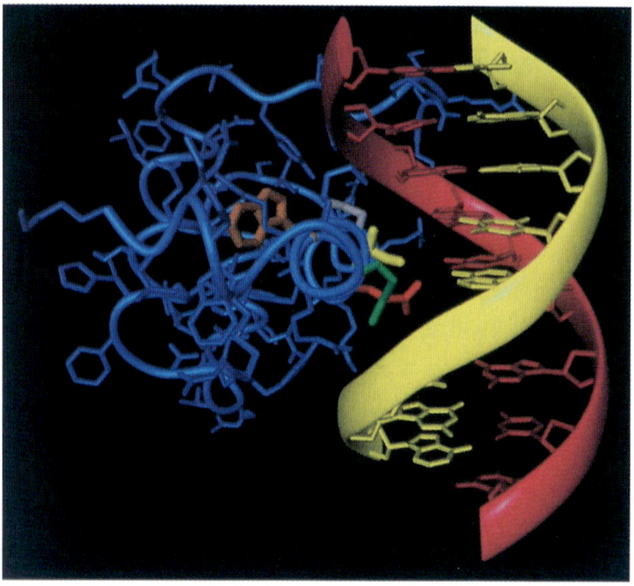

Figure 2. NMR structure of the complex formed between the *Antennapedia* homeodomain (blue, with functionally important residues in red, yellow, green and brown) and its operator DNA (yellow and red) (2).

other macromolecules or low molecular weight ligands. Again, for these supplementary data it is of keen interest that they can be measured directly in solution.

The NMR structure of the *Antennapedia* homeodomain (Figure 1) illustrates one of the exciting features of being able to perform structural studies in solution. The polypeptide chain in this protein is only partially folded, with both chain ends showing pronounced disorder (1). In the complex with its operator DNA (Figure 2) the N-terminal chain end is located in the minor groove of the DNA, where it adopts a well-defined structure (2). This mode of intermolecular recognition illustrates the functional importance of partially structured polypeptide chains. Mammalian prion proteins are an even more striking example of partial polypeptide folding, since the three-dimensional structure of the benign "cellular" form (PrP^C) includes a flexibly disordered 100-residue tail linked to the N-terminal end of a globular domain (Figures 3 and 4) (3). Considering that the mechanism of transformation of PrP^C into the aggregated, disease-related form of mammalian prion proteins is still subject to speculation, the observation of this flexible tail has been highly intriguing.

Partially folded polypeptide chains are usually difficult to crystallize. Furthermore, if crystals can be obtained, the chain segments that are disordered in solution will either be ordered by intermolecular contacts in the crystal lattice, or they will not be visible by diffraction methods. As a consequence, NMR has in many cases been the only method capable of providing structural information on partially folded polypeptides. Although a standard protocol for NMR structure determination provides a static picture of the unstructured chain segments (Figures 1 and 4) (4), additional NMR experi-

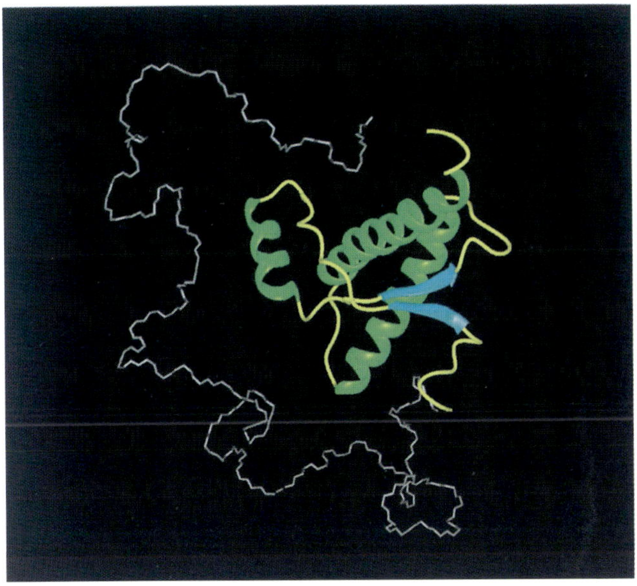

Figure 3. NMR structure of the bovine prion protein (3). In the C-terminal globular domain of residues 126–230, α-helices are green, an antiparallel β-sheet is blue, and non-regular secondary structure is yellow; the "unstructured" N-terminal tail of residues 23–125 is white.

ments can provide information on the frequencies of the rate processes that mediate transitions between discrete states of the molecule within the conformation space spanned by the static bundle of NMR-conformers (5, 6).

The ability of NMR techniques to characterize macromolecular structures and their intermolecular interactions with high spatial and temporal resolution has long attracted keen interest. This article reports on experience gained with NMR studies of proteins in my laboratory during the past 35 years (7).

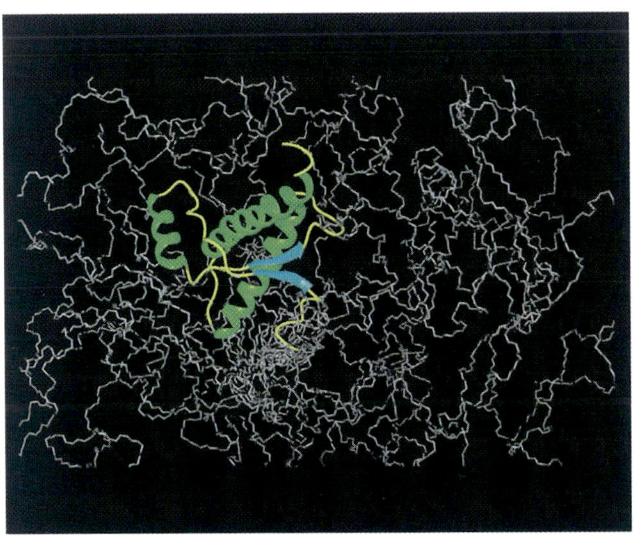

Figure 4. Visual impression of the variation of the bovine prion protein structure (Figure 3) during a time period of about 1 nanosecond. The superposition of 20 snapshots illustrates that the globular domain maintains its mean geometry, whereas the tail undergoes large-scale changes with time.

NMR SPECTRA OF PROTEINS IN SOLUTION

When I joined the field of biomacromolecular NMR spectroscopy in 1967, important qualitative NMR features of amino acids and proteins had already been noted and tentatively rationalized (8–12). Most important, it had been well documented that the spectrum of a globular protein is more complex than the sum of the NMR lines from the constituent amino acid residues in the polypetide chain (Figure 5), and the differences between the ^{1}H NMR spectra of folded and unfolded ("random coil") forms of a polypeptide chain had been tentatively explained by different interactions with the solvent (Figure 6). The spectral analysis was primarily focused on the positions of the individual NMR lines in the ^{1}H NMR spectrum, as given by the "chemical shift", δ, in parts per million (ppm) relative to a reference compound (Figure 5). Although the chemical shift is primarily determined by the covalent structure of the amino acid residue, it can also be significantly affected by the interactions with the solvent. Therefore, the exclusion of the solvent water from the interior of a globular protein (Figure 6) causes the chemical shifts of the core residues to be different from those of the water-exposed amino acid residues, so that even NMR lines originating from multiple residues of the same amino acid type can be distinguished. This "conformation-dependent chemical shift dispersion" was found to be sufficiently large to enable ^{1}H NMR studies of protein denaturation (Figure 7) (12). This then indicated the exciting prospect of using NMR for detailed studies of protein folding, and in particular for distinguishing between two-state and multiple-state folding and unfolding transitions. For me, the observations and ideas illustrated by Figures 5 to 7 also showed that one would need the ability of resolving and analyzing highly complex NMR spectra (upper trace in Figure 5) in order to obtain information about physiologically active, folded forms of proteins. This was a clearly defined problem that appeared sufficiently challenging to be attractive for my own research.

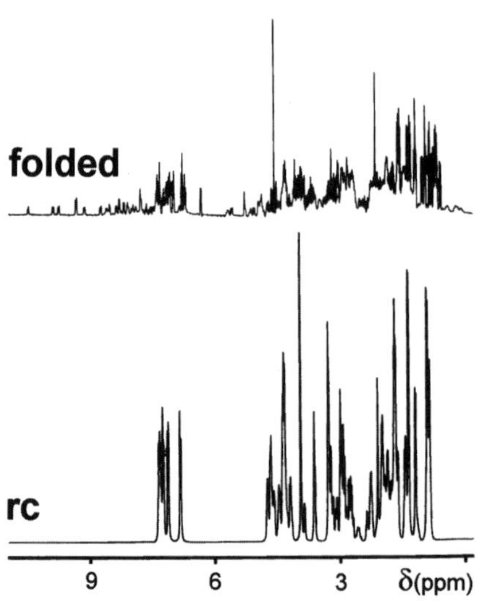

Figure 5. One-dimensional (1D) ^{1}H NMR spectra of the small protein bovine pancreatic trypsin inhibitor (BPTI, M ≈ 6000). Top: Experimental spectrum of folded, active BPTI in a freshly prepared ^2H_2O-solution. Bottom: Simulated spectrum for the unfolded, random coil form of the BPTI polypeptide chain.

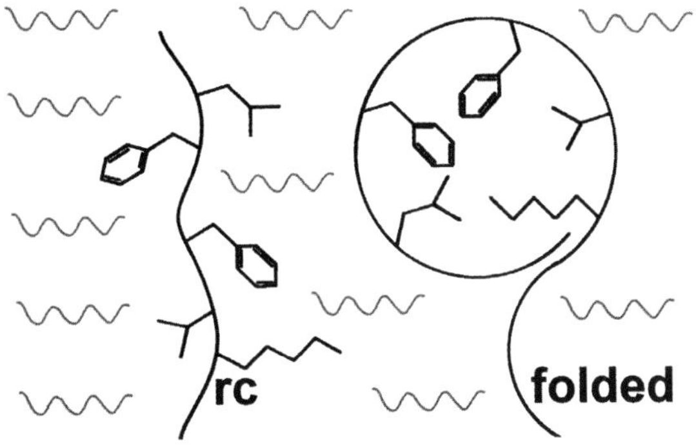

Figure 6. Scheme illustrating that the solvent (blue) has free access to all parts of a random coil polypeptide chain (rc), whereas it is excluded from the core of a folded globular protein.

In attempts at rationalizing in more detail the observations of Figures 5 and 7, it had been suggested that local magnetic fields generated by "ring currents" in the aromatic amino acid side chains (Figure 8) should cause outstandingly large conformation-dependent chemical shifts of hydrogen atoms located near the rings in the three-dimensional protein structure. This was qualitatively confirmed by comparison of the largest observed deviations from random coil chemical shifts (4) in hen egg white lysozyme (Figure 7) with calculations of the "ring current shifts" based on the then-available low-resolution crystal structure of this protein (12). It had also been pointed out that hemoproteins could be expected to have particularly large ring current shifts for hydrogen atoms located near the heme groups (Figure 9) (10,12).

For my start in the field of NMR with proteins, the prediction of unique, large heme ring current shifts seemed to be an attractive feature of hemo-

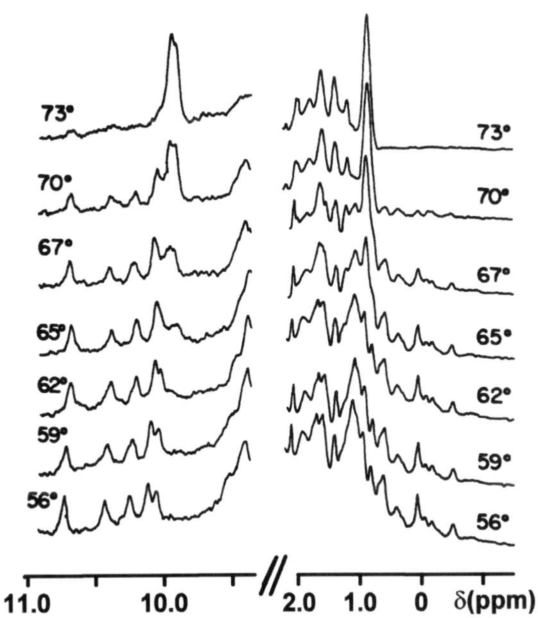

Figure 7. Low-field and high-field regions of 1D ^{1}H NMR spectra of the protein lysozyme at different temperatures, reflecting the transition from the folded form at 56 °C to the random coil form of the polypeptide chain at 73 °C (220 MHz, H_2O-solution) (adapted from ref. 12).

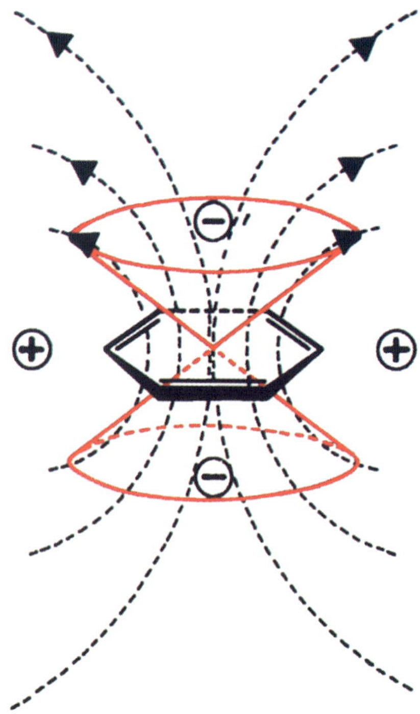

Figure 8. Local "ring current field" around aromatic rings in solution induced by an external, static magnetic field. The shape of the ring current field is indicated by the red double-cone and by broken magnetic-field-lines. The minus-sign indicates that the NMR lines of hydrogen atoms located inside the cone in the three-dimensional protein structure are shifted "upfield" (to the right in the spectra of Figure 5), whereas for atoms outside of the cone the shifts are "downfield".

proteins, and the presence of the heme iron (Figure 9) was appealing in view of my background in inorganic chemistry. As an additional advantage, low-resolution crystal structures for several hemoproteins were already available in 1967. It turned out that for several years nearly all my research projects were focused on the heme iron and its coordinatively linked ligands, and on the non-bonding interactions of the heme groups with their immediate environment in the hemoproteins.

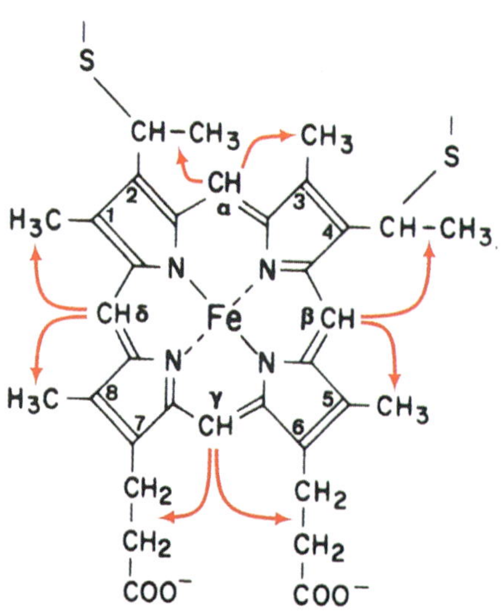

Figure 9. Chemical structure of heme c, which is the prosthetic group in cytochromes c. The red arrows connect groups of hydrogen atoms in the covalent structure of heme c that are separated by a sufficiently short distance to be connected by nuclear Overhauser effects (NOE).

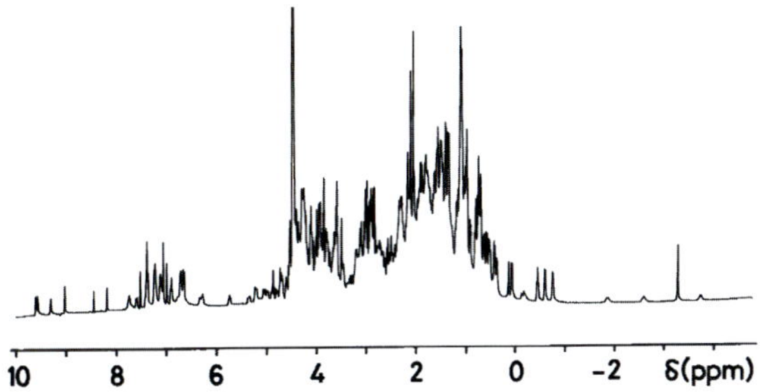

Figure 10. 1D ^{1}H NMR spectrum (360 MHz, 52°C, solvent ^2H_2O) of horse ferrocytochrome c (M ≈ 12000).

As anticipated, the large heme ring current field generates a small number of well-separated resonance lines in the spectra of folded, globular hemoproteins (Figure 10). For the paramagnetic states of hemoproteins, additional well-resolved lines result from interactions with the unpaired electrons of the heme iron. Although they represented less than 3% of all the hydrogen atoms in any given hemoprotein, the well-resolved resonances were of crucial importance, because these lines could be selectively irradiated in 1D ^{1}H NMR experiments. For example, performing some simple ligand exchange and redox reactions with cytochrome c (Figure 11) resulted in the identification of one of the axial ligands bound to the heme iron, which had not been seen in the low resolution X-ray crystal structure available in 1969.

The experiments in Figure 11 (13) started with the oxidised form of cytochrome c. As was then learned through this study (see below), ferricyto-

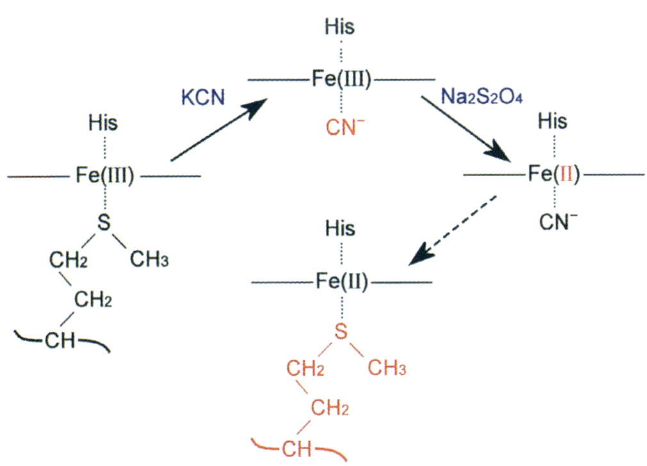

Figure 11. Three chemical reaction steps with cytochrome c that were used to identify methionine as one of the axial ligands of the heme iron (13). The roman numerals indicate the oxidation state of the heme iron, and His stands for the amino acid residue histidine. The solid and broken arrows indicate rapid and slow reaction steps, respectively. Colour code: black, heme iron with ligands, where the horizontal line through Fe represents a side view of the heme c ring (Figure 9); blue, chemicals added to the protein solution; red, new structural features resulting from the preceding reaction.

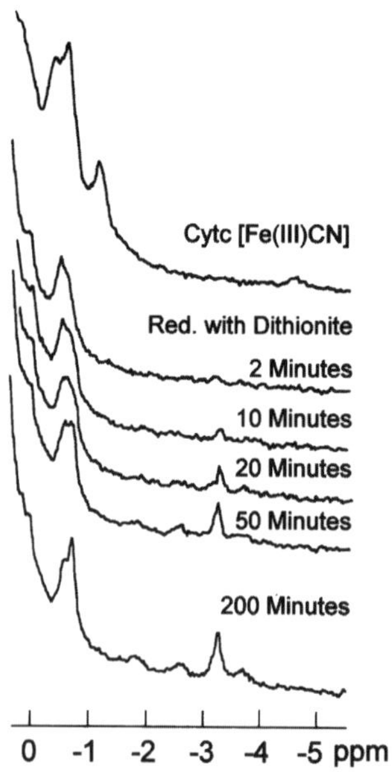

Figure 12. Changes with time in the 1D ^{1}H NMR spectrum of cytochrome c after reducing the ferric cyanide complex with dithionite (220 MHz, 9°C, solvent ^2H_2O).

chrome c contains a methionine side chain of its polypeptide chains in one of the axial heme iron coordination sites. Upon addition of KCN to the protein solution, this methionine is replaced in the binding site on the heme iron by a cyanide ion, and therefore moves out of the heme ring current field. The completion of this reaction can readily be monitored, since different patterns of well-resolved resonance lines are present in the ^{1}H NMR spectra of ferricytochrome c and its cyanide complex, respectively. After reduction of the heme iron to the ferrous state, the cyanide complex is thermodynamically less stable than the native form of cytochrome c, and in a slow reaction the cyanide in the axial coordination site of the heme iron is again replaced with the natural methionine ligand. The return of the axial methionine into the heme ring current field was monitored in real time (Figure 12) by the appearance of its typical high-field lines in the ^{1}H NMR spectrum (Figure 10). This experiment resulted in the identification of the nature of this axial ligand in the native, active form of cytochrome c, and further yielded information on the relative thermodynamic stabilities of the four different structures in Figure 11.

As a sideline, comparison of the figures 10 and 12 illustrates the influence of various experimental factors on the appearance of the NMR spectra. The higher resonance frequency and the high temperature used in the experiment of Figure 10, which was recorded around 1980 with Fourier transform spectroscopy, resulted in narrower and more clearly separated peaks in the region 0 to -5 ppm than the experiment in Figure 12, which was recorded in 1968 with continuous-wave NMR spectroscopy at low temperature.

Other early experiments based on the observation of well-resolved resonances in the ^{1}H NMR spectra of hemoproteins resulted in the characterization of conformation changes during the oxygenation of myoglobin and hemoglobin (14, 15), and to detailed characterization of the electronic structure of the heme groups in different classes of hemoproteins (16–18).

THE WAY TO NMR STRUCTURES OF PROTEINS

Around 1970 there was a lot of enthusiasm in my laboratory and elsewhere about the success of experiments of the types illustrated by Figures 5 to 12. During the following years, however, there was little further progress toward an NMR method for *de novo* structure determination of proteins. In hindsight this can readily be rationalized: Early successful structural interpretations of NMR data invariably supplemented a previously known low resolution X-ray crystal structure of the same protein. In spite of the high symmetry of the ring current fields (Figure 8) and other sources of local conformation-dependent chemical shifts, including natural paramagnetic centers and extrinsic "paramagnetic shift reagents" (17, 19), the observed chemical shifts could thus in some instances yield exciting new information. In a *de novo* protein structure determination, however, the high symmetry of the local magnetic fields would lead to ambiguities in the structural interpretation of the resulting chemical shifts. Different approaches were therefore called for, and eventually a NMR method for protein structure determination could be based on the following four principal elements:

1. Measurement of NOE upper distance limits as conformational constraints

Nuclear Overhauser effects (NOE) are due to dipolar interactions between different nuclei. The intensity of the NOE is related to the product of the inverse sixth power of the internuclear distance and a correlation function, $f(\tau_c)$, which describes the modulation of the dipole–dipole coupling by stochastic rate processes, with an effective correlation time τ_c:

$$\text{NOE} \; \alpha \; \frac{1}{\langle r \rangle^6} \cdot f(\tau_c) \qquad [1]$$

Although the NOE is a common phenomenon for all combinations of closely spaced nuclear spins, NOEs between pairs of hydrogen atoms are of prime interest for structural studies. A ^1H–^1H NOE is related to the "through-space" distance between a pair of atoms that are either not at all linked by covalent bonds (intermolecular NOE), or that may be far apart in the amino acid sequence of a polypeptide chain. ^1H–^1H NOEs can be observed in double-irradiation 1D NMR experiments as the fractional change in intensity of a NMR line that results from preirradiation of another resonance (inset in Figure 13). Although NOE distance measurements had successfully been used for studies of small organic molecules (20), including oligopeptides (21), and observation of NOEs in proteins had also been reported (22, 23), it was not clear whether the desired distance information could actually be obtained in proteins (20, 24, 25). This uncertainty arose because the Brownian motions of large structures in solution are slow, with long effective correla-

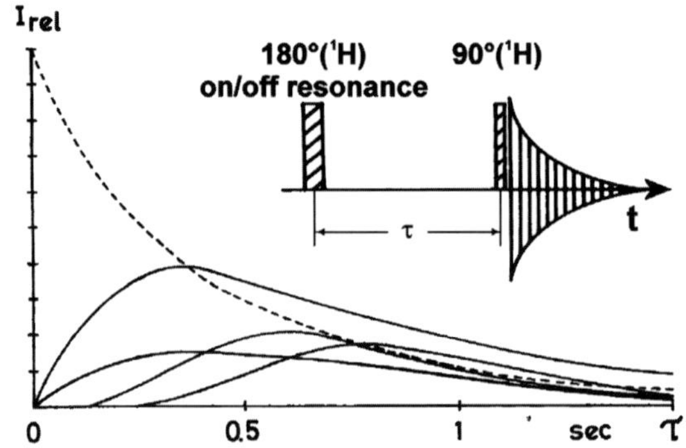

Figure 13. NOE build-up curves observed with 1D transient [¹H,¹H]-NOE measurements (26). The relative intensities, I_{rel}, of the pre-irradiated resonance line (broken curve) and of the lines experiencing NOEs (solid curves) are plotted *versus* the duration of the mixing period, τ. The experimental scheme used to record transient NOE spectra (inset in the upper right) includes a selective 180° radio-frequency pulse applied to the "pre-irradiated" ¹H NMR line, and a non-selective 90°-pulse applied to the entire ¹H NMR spectrum. The two pulses are separated by the mixing period, τ, and followed by signal acquisition. When working with the crowded spectra of proteins, one records the difference between two spectra obtained with the pre-irradiation 180°(¹H)-pulse applied on- and off-resonance, respectively.

tion times, τ_c, for this stochastic process, and proteins contain a dense network of hydrogen atoms. With the combination of these two features, "spin diffusion" could partially or fully deteriorate distance measurements based on ¹H–¹H NOE experiments (24 – 28).

In the solution NMR experiments discussed here, spin diffusion arises as a consequence of the dependence of the NOE on the inverse sixth power of the internuclear distance, since magnetization transfer between two spins through multiple short steps may be more efficient than a one-step transfer over the longer, direct distance (Figure 14). We used 1D transient NOE difference experiments (inset in Figure 13) (26) and 1D truncated-driven NOE (TOE) difference experiments (27) to record "NOE build-up curves" (Figure 13), which then provided a basis for ¹H–¹H distance measurements in macromolecules. These studies showed that even more favourable conditions for NOE distance measurements can be found in macromolecules, which have long effective correlation times for the modulation of dipole–dipole couplings, than in low-molecular-weight compounds, for which the condition of "extreme motional narrowing" applies (24, 28).

Analysis of ¹H–¹H NOE build-up curves recorded with 1D transient NOE experiments (Figure 13) is based on the consideration that during the early phase of the mixing period, direct magnetization transfer from spin 1 to spin 3 (Figure 14) will result in a linear increase of the magnetization on spin 3 with time. In contrast, transfer to atom 3 *via* an intermediate step through atom 2 will have a lag period, even though for long mixing periods, τ, the extent of the magnetization transfer by such "spin diffusion" may for individual pairs of hydrogen atoms exceed that of the direct transfer (Figure 13).

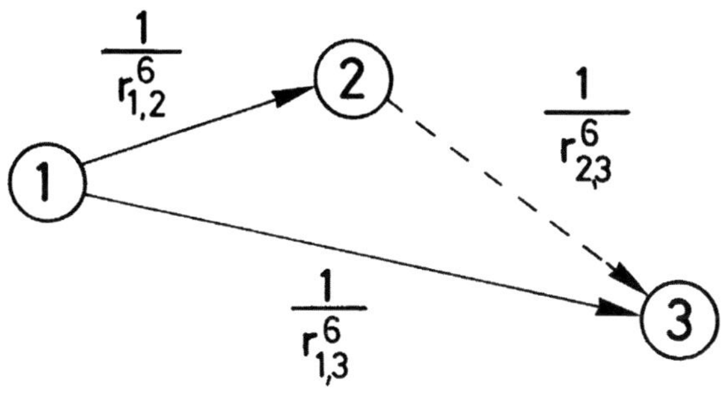

Figure 14. Spin diffusion. Transfer of magnetization between two hydrogen atoms 1 and 3 in the presence of additional hydrogen atoms goes through two competing paythways, *i.e.*, the direct NOE across the distance $r_{1,3}$, and two- or multiple-step spin diffusion *via* intervening hydrogen atoms.

Corresponding considerations apply for the analysis of NOE build-up curves recorded with truncated-driven NOE experiments (27). With proper selection of the duration of the mixing period, one can thus measure highly selective ^{1}H–^{1}H NOEs between distinct pairs of hydrogen atoms, or groups of chemical shift-equivalent hydrogen atoms in proteins in solution.

There is a second criterion that needs to be satisfied for obtaining selective NOEs with 1D NMR experiments, *i.e.*, at least one of the two lines that are connected by the NOE must be sufficiently well resolved (*i.e*, separated from all other lines in the spectrum) to enable selective radio-frequency pre-irradiation (inset in Figure 13). The previously discussed well-resolved lines in the ^{1}H NMR spectrum of ferrocytochrome c (Figure 10) thus had again an important role in enabling us to study the spin physics in the interior of this globular protein with 1D ^{1}H NMR experiments, as well as to obtain novel structural information. The figure 15 shows a series of highly selective 1D truncated-driven NOE measurements, which were used to determine ^{1}H NMR assignments for the heme group in a c-type cytochrome (Figure 9) (29).

2. Sequence-specific resonance assignments

Similar to the situation in heme groups (Figure 9), there are closely spaced pairs of hydrogen atoms in neighbouring residues of a polypeptide chain (Figure 16). These can be connected by the observation of "sequential NOEs". The figure 16 illustrates that NOE-based ^{1}H NMR assignments for a polypeptide chain can conceptually be considered as a 2-step process. Each amino acid residue represents a "spin system" (4), *i.e.*, it consists of an array of hydrogen atoms including an amide proton (H^N), an α-proton (H^α), and the side chain protons, which can be connected by steps over three or less covalent bonds through the observation of scalar spin–spin ("through-bond") couplings (see (4) for the exceptions represented by proline and the aromatic side chains). In contrast, hydrogen atoms located in sequentially neighbouring amino acid residues are separated by at least four covalent bonds. Pairs of neighbouring residues in the sequence can therefore only be connected *via* NOEs manifesting short through-space distances, such as $d_{\alpha N}$ and d_{NN}. Suitable combinations of intraresidual ^{1}H–^{1}H connectivities established by scalar spin–spin couplings, and inter-residue connectivities established by sequential NOEs enable progressive resonance assignments while "walking

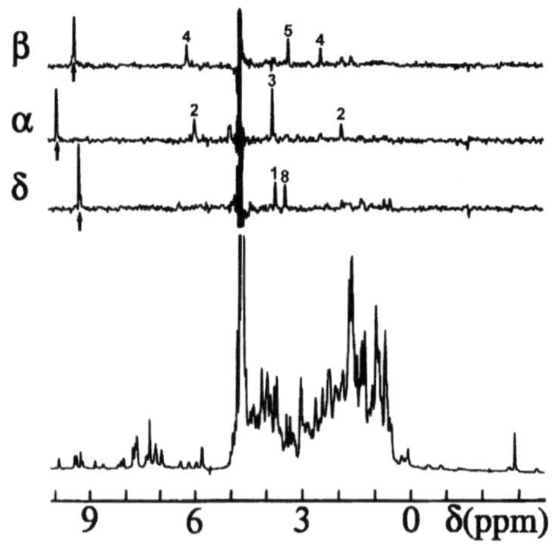

Figure 15. ^{1}H NMR assignments for the heme group of a c-type cytochrome using ^{1}H–^{1}N NOEs (see Figure 7). Bottom trace: 1D ^{1}H NMR spectrum of ferrocytochrome c-551 from *Pseudomonas aeruginosa* (M ≈ 12000). Upper traces: three 1D truncated-driven NOE difference spectra obtained with selective pre-irradiation (indicated by arrows) on the hydrogen atoms β, α and δ (see Figure 9), respectively (360 MHz, 35°C, ^2H_2O-solution). The NOE peaks are identified with numbers indicating the substituents attached to the corresponding porphyrin ring carbons (Figure 9).

along the polypeptide backbone". In other words, the spin systems of two neighbouring residues can be connected by the intervening H^α–H^N or H^N–H^N sequential NOE connectivities (Figure 16). 1D double-resonance experiments with selective irradiation of the well-resolved amide proton resonances between 8 and 10 ppm in folded BPTI (Figure 5) thus yielded assignments for most of the residues in the regular secondary structure elements (Figure 17) (30). Further assignments were not possible, because the other regions of the polypeptide chain were not represented by well-separated ^{1}H NMR lines that could have been selectively irradiated in 1D NMR experiments.

3. Two-dimensional (2D) NMR

With the introduction of 2D NMR experiments, and subsequently 3D and 4D NMR experiments, NMR studies of biological macromolecules evolved from intellectually stimulating science to a practical approach for protein structure determination. The foundations of multi-dimensional NMR have been presented in the Nobel lecture by Richard R. Ernst (31). Here, I only want to

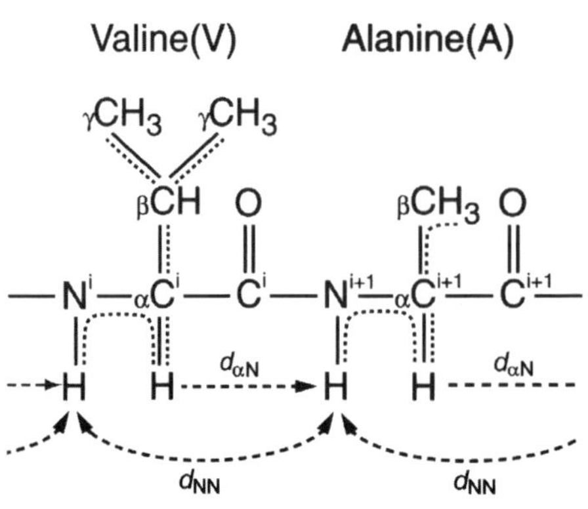

Figure 16. Sequential ^{1}H NMR assignment of proteins. The drawing shows the chemical structure of a –valine–alanine– dipeptide segment in a polypeptide chain. The dotted lines connect groups of hydrogen atoms that are separated by at most three chemical bonds and can therefore be connected using scalar spin–spin couplings. The broken arrows link pairs of hydrogen atoms in neighbouring amino acid residues that are separated by short through-space distances, $d_{\alpha N}$ and d_{NN}, and can therefore be connected by "sequential NOEs".

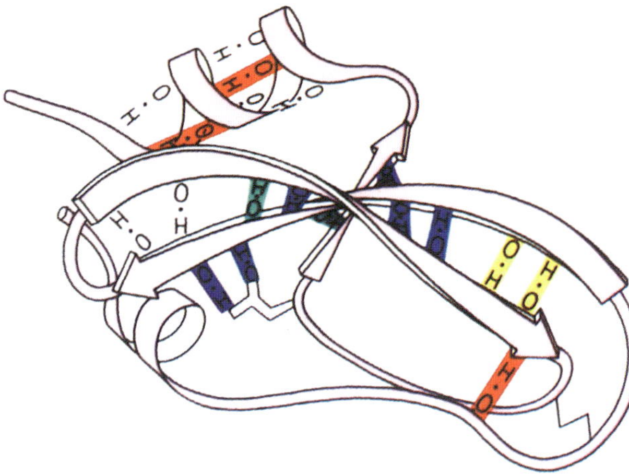

Figure 17. Sequence-specific resonance assignments for BPTI that were obtained using 1D ^{1}H NMR experiments (30). The polypeptide backbone is shown, and the assigned residues are identified by indication of hydrogen bonds with their amide protons (the colour code indicates variable exchange rates of the amide protons with the solvent; drawing by Jane Richardson, 1980).

comment on two crucial consequences of multi-dimensional NMR for studies of proteins. First, 2D ^{1}H NMR enables the recording of selective interactions between pairs of hydrogen atoms, or groups of chemical shift-equivalent hydrogen atoms, without selective irradiation of individual resonance lines. It thus enables a detailed analysis of the entire ^{1}H NMR spectrum of a protein, which contrasts with the 1D NMR situation of being limited to using only a small number of resolved lines at the periphery of the spectrum. Second, the dispersion of the resonances in a two-dimensional frequency plane affords greatly improved separation of the individual peaks. For example, in Figure 18 the intense lines on the diagonal from the lower left to the upper right correspond to the 1D NMR spectrum, and the weak "cross peaks" in the plane outside of the diagonal manifest selective NOEs between pairs of

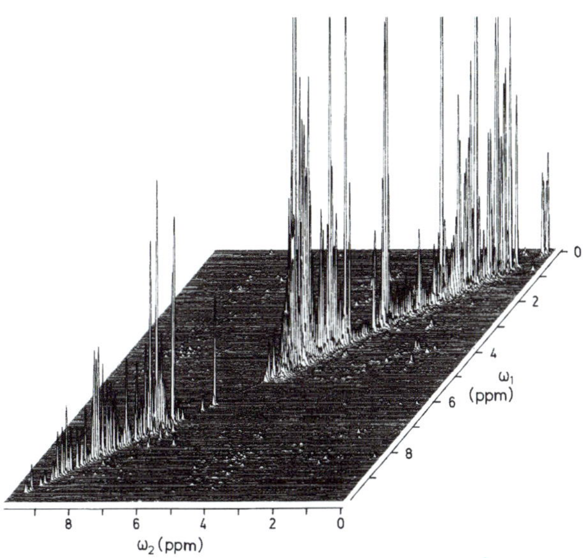

Figure 18. Two-dimensional (2D) [^{1}H,^{1}H]-NOE spectroscopy ([^{1}H,^{1}H]-NOESY). A stacked plot representation of a spectrum of the small protein bull seminal proteinase inhibitor IIA (BUSI IIA, M ≈ 6000) is shown (500 MHz, 45°C, H_2O-solution).

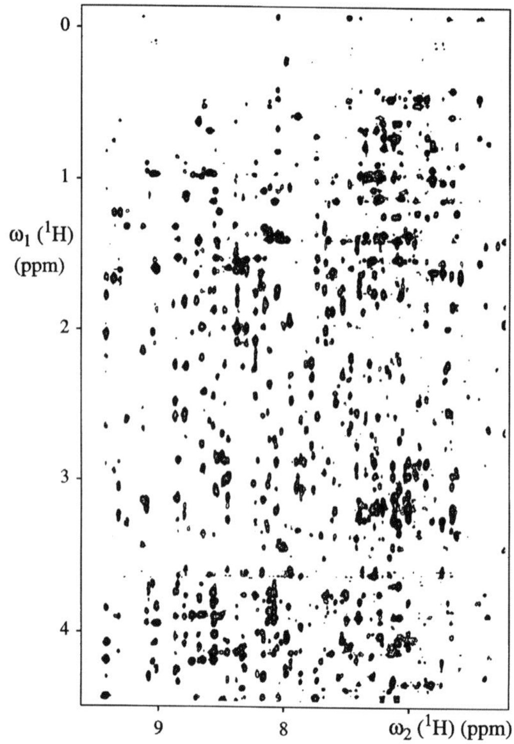

Figure 19. 2D [^{1}H,^{1}H]-NOESY spectrum of the plant pathogenesis-related protein P14A (M ≈ 15000). A contour plot of the spectral region [ω_1(^{1}H) = 0 – 4.3 ppm, ω_2(^{1}H) = 6.3 – 9.5 ppm] is shown (750 MHz, 30°C, H_2O-solution).

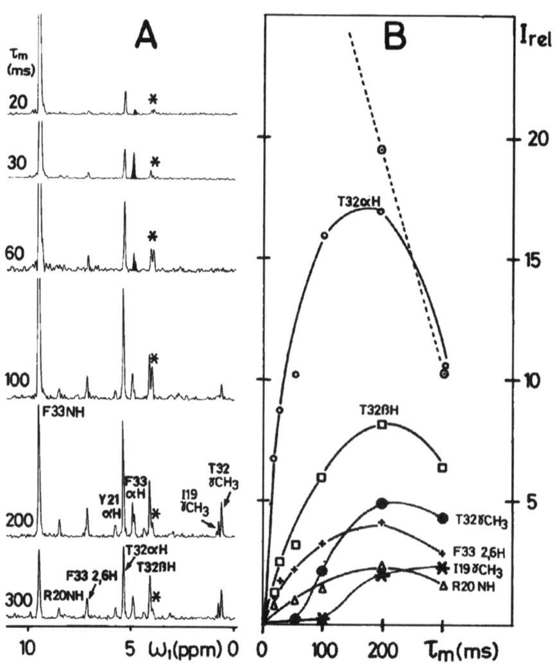

Figure 20. Measurement of NOE build-up curves using 2D [^{1}H,^{1}H]-NOESY experiments. (A) Six NOESY spectra of BPTI recorded with different mixing times, as indicated on the left. The same 1D cross-section along the ω_1 frequency axis through the diagonal peak of the Phe 33 amide proton is plotted for each spectrum. (B) NOE build-up curves obtained from the analysis of the data in (A); same presentation as in Figure 13, with the relative intensities of the NOESY cross peaks plotted versus the length of the mixing time, τ_m, and the broken line representing the decay of the magnetization on the diagonal peak of the Phe 33 amide proton.

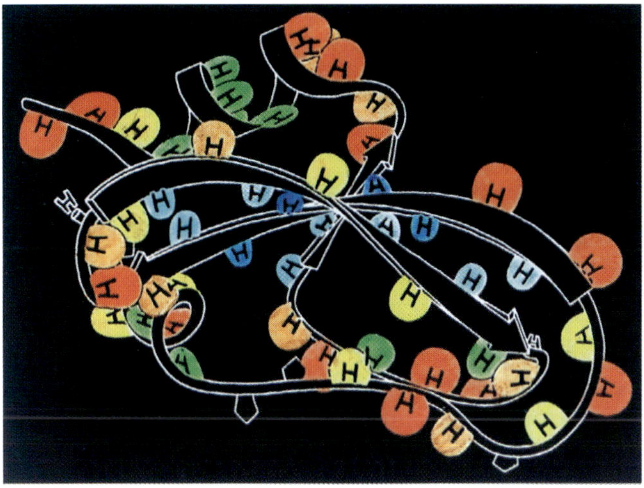

Figure 21. Complete sequence-specific resonance assignments for BPTI obtained using 2D NMR experiments (34). Assigned residues are identified by coloured patches covering their amide protons. (The colour code indicates variable amide proton exchange rates; drawing by Jane Richardson, 1982).

hydrogen atoms, which are represented in the NMR spectrum by two distinct (but not necessarily "resolved") chemical shift positions along the diagonal. The greatly improved separation of the individual cross peaks is best seen in contour plots of 2D NMR spectra (Figure 19), which is the presentation used for detailed analysis. The 1D decoupling experiments for the identification of scalar spin–spin couplings were thus replaced by 2D correlation experiments, such as COSY, SECSY and FOCSY, and the 1D NOE difference experiments were replaced by 2D [^{1}H,^{1}H]-NOESY, which yielded similar NOE build-up curves for each NOESY cross peak (Figure 20) as the 1D transient NOE technique (Figure 13) (4, 32, 33). Using these 2D ^{1}H NMR experiments, complete sequence-specific resonance assignments could be obtained for the protein BPTI (Figure 21) (34).

The "NOESY-COSY connectivity diagram" in Figure 22 (4, 35) makes use of the fact that standard 2D [^{1}H,^{1}H]-COSY and 2D [^{1}H,^{1}H]-NOESY spectra are symmetrical with respect to the diagonal peaks. Combining the upper left half of a NOESY spectrum with the lower right half of a COSY spectrum therefore enables a straightforward visualization of assignments by a succession of $d_{\alpha N}$ sequential NOEs and intraresidual H^N–H^α scalar coupling connectivities. The data were recorded in a freshly prepared solution of BPTI in ^2H_2O, where only the resonances of the slowly exchanging amide protons are seen (35). The assignments start with the COSY cross peak identified by a black square and, as indicated by the arrows, go clockwise to the sequentially preceding isoleucine residue, and counter-clockwise to a sequence of four residues. It is customary that the data leading to the ^{1}H NMR assignments of a protein are collected in a plot *versus* the amino acid sequence (Figure 23). The figure 23 shows that most of the sequential connectivities are independently documented by two or three different sequential NOEs (4, 36), and that possible remaining gaps in the assignment pathway are readily recognized in this presentation.

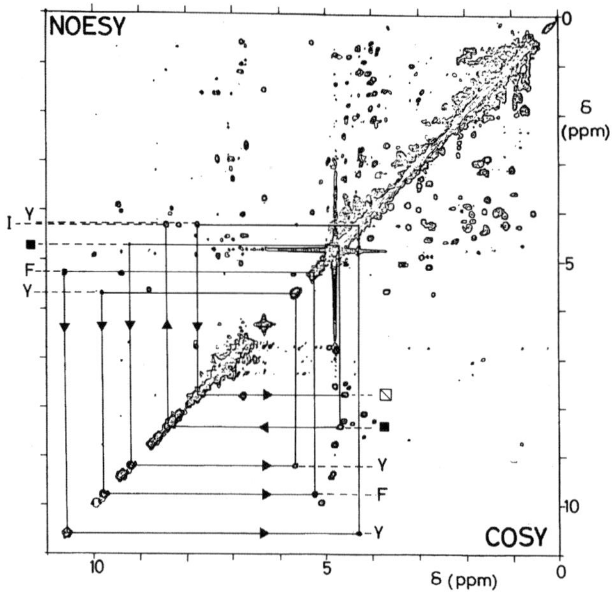

Figure 22. NOESY-COSY connectivity diagram for sequential ^1H NMR assignments using the $d_{\alpha N}$ sequential NOEs. The data was recorded with BPTI in ^2H_2O solution. (see text).

4. Structural interpretation of NOE distance constraints

A polypeptide chain with 100 amino acid residues has a length of about 400 Å, whereas NOE-observable distances are shorter than about 5 Å. Observation of a NOE between a pair of hydrogen atoms with assigned chemical shift positions therefore enforces the formation of a ring-like structure (Figure 24). A successful structure determination generates three-dimensional arrangements of the polypeptide chain that simultaneously contain all

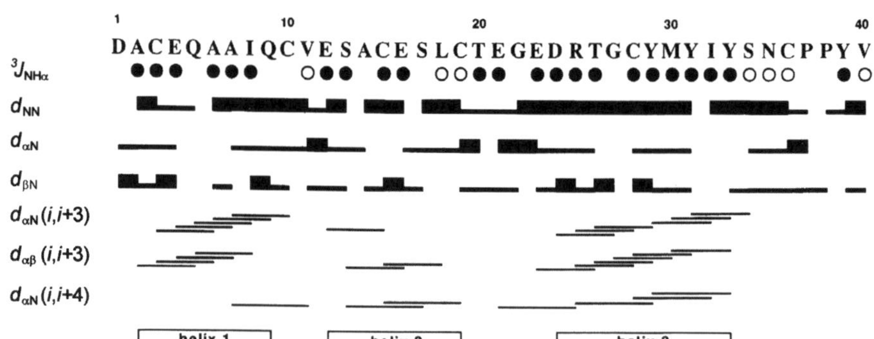

Figure 23. Standard presentation of ^1H NMR data leading to sequence-specific resonance assignments and the identification of regular secondary structures in proteins. Experimental data plotted *versus* the amino acid sequence are shown for the small pheromone protein E*r*-1 from *Euplotes raikovii* (M ≈ 4500). $J_{HN\alpha}$ are scalar spin–spin coupling constants, with small and big values indicated by filled and empty circles, respectively. d_{NN}, $d_{\alpha N}$ and $d_{\beta N}$ are distances manifested in sequential NOEs, where strong and weak sequential NOEs are indicated by a thick or a thin line, respectively. Small values of the distances $d_{\alpha N}(i,i+3)$, $d_{\alpha\beta}(i,i+3)$ and $d_{\alpha N}(i,i+4)$ are observed by medium-range NOEs linking the given atom types between residues spaced as indicated in the parentheses, and by the short horizontal lines. The locations of three α-helices are indicated at the bottom.

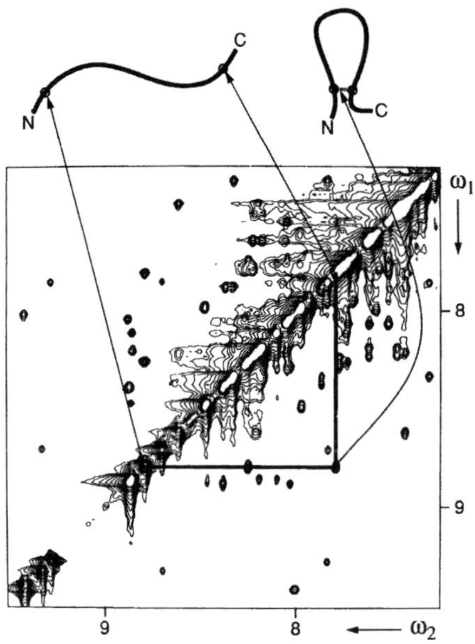

Figure 24. Scheme indicating the relations between an experimental 2D [^{1}H,^{1}H]-NOESY spectrum, a polypeptide with the chain ends indicated by N and C, sequence-specific assignments for two hydrogen atoms in the polypeptide chain indicated by circles, and the NOE upper distance constraint derived from the NOESY cross-peak connecting the chemical shift positions of the two assigned hydrogen atoms (see text).

the small and large circular structures imposed by the ensemble of all NOESY cross peaks.

Partial structure determination has been be obtained by an empirical approach for the identification of regular secondary structures in polypeptide chains, which relies on recognizing distinct patterns of NOEs (4, 37). For example, in Figure 23 a succession of strong sequential d_{NN} NOEs in combination with the observation of medium-range NOEs in the same polypeptide segment identifies three α-helical structures, which are independently also indicated by successions of small values for the scalar spin–spin couplings $^3J_{HN\alpha}$ (4, 38).

For the calculation of complete three-dimensional protein structures from NMR data (39), it was quite clear from the outset that an input of quantitative NOE distance measurements would be difficult to obtain. The observed NOEs depend on the proton–proton distance, *r*, as well as on the effective rotational correlation times, τ_c (Equation [1]). Since for each pair of hydrogen atoms the effective τ_c-value is governed not only by the overall rotational molecular tumbling (Brownian motions), which depends on the size and shape of the protein as well as on the viscosity of the solvent, but can also be affected by intramolecular motions, $f(\tau_c)$ may vary for different pairs of hydrogen atoms in a protein molecule. Additional ambiguities could arise from partial quenching of individual ^{1}H–^{1}H NOEs by competitive spin relaxation processes, for example, spin diffusion (Figure 14), chemical or conformational exchange, and interactions with other nuclear or electronic spins. Furthermore, as a result of rapid intramolecular mobility, a given NOE may be the result of sampling over a range of distances between the two hydrogen atoms of interest (40). In view of these intrinsic limitations for efficient quantitative NOE distance measurements, we decided to use a constant value of the correlation

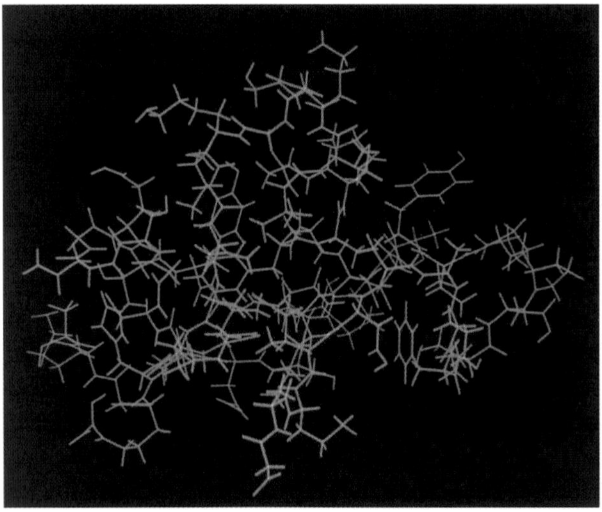

Figure 25. NMR structure of BUSI IIA (43).

function, $f(\tau_c)$ (Equation [1]), for all ^1H–^1H combinations in a protein, and to derive only upper limits on the ^1H–^1H distances from the NOE measurements. In practice, the input for a structure calculation then consists of allowed distance ranges, which are bounded by a NOE upper limit of 3.0 to 5.0 Å, depending on the intensity of the NOE, and a lower limit of 2.0 Å, which represents the sum of the van der Waals radii of the two NOE-connected hydrogen atoms. Although each individual entry in the input data thus has only limited precision, this procedure is robust and can conceptually account for the influence of intramolecular mobility in most of the situations that are commonly expected for the structured parts of globular proteins.

For the initial globular protein structure calculations from NMR data (Figure 25), we used a metric matrix distance geometry algorithm to search for molecular geometries that are consistent with the ensemble of all experimentally determined NOE distance constraints (41–43). Each such calculation ends with the minimization of an error function, and the residual error function value represents a straightforward measure for the success of having found a molecular geometry that satisfies the experimental input data. In view of the aforementioned distance-range format of the input, it is further of keen interest to evaluate the uniqueness of the calculated structure. To this end, the structure calculation is repeated with identical input data but different boundary conditions, and the uniqueness of the resulting NMR structure is judged from the tightness of the fit among the resulting ensemble of conformers. Typically, about 100 conformers are generated, and a sub-group of the 20 conformers with the smallest residual error function values is selected to represent the NMR structure of the protein. The average of the pairwise root-mean-square distances (RMSD) calculated for this bundle of conformers (Figure 1) is then taken as a measure for the precision of the structure determination. Visually, a tight fit of the bundle of conformers indicates regions where the structure is defined with high precision by the NMR data, whereas structurally disordered polypeptide segments show a large dispersion among the members of the bundle, as exemplified by the two chain ends of the *Antennapedia* homeodomain in Figure 1. In the absence of long-range NOE distance constraints, a properly functioning algorithm for the structure calculation will sample the entire conformation space that is accessible with the given length of the polypeptide chain, as exemplified by the "unstructured" tail of the bovine prion protein in Figure 4.

NMR IN STRUCTURAL BIOLOGY

Standard protocol for NMR structure determination of biological macromolecules

The protocol for NMR structure determination includes the preparation of a homogeneous protein solution, the recording and handling of the NMR data sets, and the structural interpretation of the NMR data (Table 1). The techniques used in 1984 for the structure determination of bull seminal proteinase inhibitor IIA are listed in Table 1; the four steps of the structural interpretation (III in Table 1) were performed separately, although the result of the first round of constraint collection and structure calculation was subsequently used for additional checks on the sequence-specific resonance assignments as well as on the collection of conformational constraints. Since 1984, the protocol outlined in Table 1 has been used in over 3'000 NMR structure determinations of proteins and nucleic acids, and greatly improved experimental techniques have been incorporated into this general scheme.

Major advances in the experimental techniques for NMR structure determination were spurned by the introduction of methods for the production of recombinant proteins labelled with stable isotopes, in particular

Table 1. Standard protocol for NMR structure determination of proteins

Step [a]		BUSI IIA [b]
I	Sample preparation	Protein isolated from natural source; natural isotope distribution; 16 mM solutions in H_2O and in ^2H_2O, respectively
II	NMR spectroscopy	2D ^1H NMR
IIIa	Resonance assignments	Sequential NOEs
IIIb	Conformational constraints	[^1H,^1H]-NOEs, $^3J_{HN\alpha}$, $^3J_{\alpha\beta}$
IIIc	Structure calculation	Metric matrix distance geometry
IIId	Structure refinement	Restrained energy minimization

[a] The structural interpretation of the NMR data, III, is somewhat arbitrarily divided up into four steps; in practice, one goes though multiple cycles of collection of conformational constraints (IIIb) and structure calculation (IIIc), and the completion of the sequence-specific assignments (IIIa) as well as the structure refinement (IIId) may also be part of this iterative approach.

[b] This column lists the techniques used in the first structure determination of a globular protein in 1984 (43).

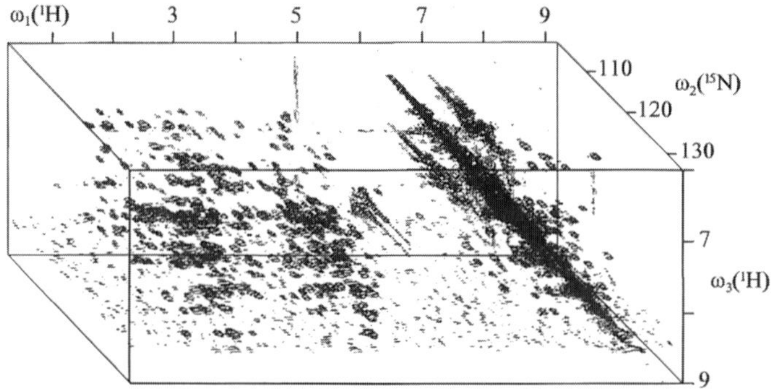

Figure 26. Three-dimensional (3D) ^{15}N-resolved [^1H,^1H]-NOESY spectrum (600 MHz, 28°C, H_2O-solution) of the DNA-binding domain of the P22 c2 repressor (M ≈ 10000, uniformly ^{15}N-labeled).

carbon-13, nitrogen-15 and deuterium (44, 45). For example, this opened the way for efficient use of heteronuclear NMR techniques with proteins, such as 3D [^1H,^{13}C,^{15}N]-triple resonance experiments, 3D ^{13}C- or ^{15}N-resolved [^1H,^1H]-NOESY (Figure 26) (46, 47), and the use of heteronuclear filters (48). Important advances have also been made with the methods of structure calculation, where the c.p.u. time needed for the calculation of a small protein structure has been reduced from about one day in 1984 (41, 42) to a few seconds (49). Currently, intense work is focused on the automation and combined execution of the individual steps in the structural interpretation of the NMR data (50, 51). It is beyond the scope of this article to present in detail the wide range of beautiful novel experimental approaches developed by the community of macromolecular NMR spectroscopists during the past 15 years, which now enable studies with ever more complex systems (Figures 1–4).

Globular protein structures in solution

The static picture of a protein molecule obtained from the standard protocol for structure determination (Table 1) typically shows variable precision of the structure determination along the polypeptide chain, as manifested by the variations in the closeness of the fit among the bundle of conformers used to represent the NMR structure (Figure 1). Even in proteins where the entire polypeptide chain is part of the global fold, increased disorder is observed toward the periphery of the surface side chains (Figure 27). This pronounced surface disorder, which typically also involves the ends of the polypeptide chain, is in most instances the only significant difference between corresponding globular protein structures in single crystals and in solution (5). In part the increased disorder in solution arises because of a scarcity of packing constraints, when compared with the protein core, and a concomitant scarcity of NOE distance constraints. With the additional use of NMR spin relaxation measurements (24, 28, 33, 52), one can distinguish between static disorder, which would then presumably arise from the scarcity of constraints,

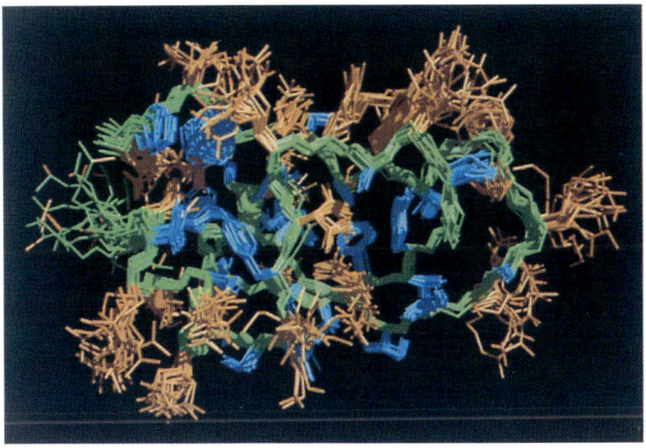

Figure 27. NMR structure of BPTI represented by a bundle of 20 conformers superimposed for best fit of the polypeptide backbone. The polypeptide backbone is green, core side-chains are blue, and solvent-accessible surface side-chains are red.

and dynamic disorder, with intramolecular motions on the nanosecond and sub-nanosecond time scale. Overall, quite independent of the dynamics issue, the observation of partially folded polypeptide chains in solution (Figures 1, 4 and 27) is important complementary information to the data that can be obtained by studies in crystals. It is also the main reason why the quality of a NMR structure determination is not usually characterized by a single, global parameter (4–6).

An important extension of the characterization of proteins in solution resulted from high resolution NMR studies of protein hydration. Thereby the location of hydration waters is determined by the observation of NOEs between water protons and hydrogen atoms of the polypeptide chain (53). Because of the dependence of the NOE on the inverse sixth power of the ^{1}H–^{1}H distance, only one layer of hydration water molecules is observed (Figure 28). For the hydration studies, the dependence of the NOE intensity

Figure 28. Molecular model of hydrated BPTI in H_2O solution. The drawing shows an all-heavy-atom-presentation of one of the conformers in Figure 27 (yellow) covered with a layer of hydration water molecules (dotted blue spheres).

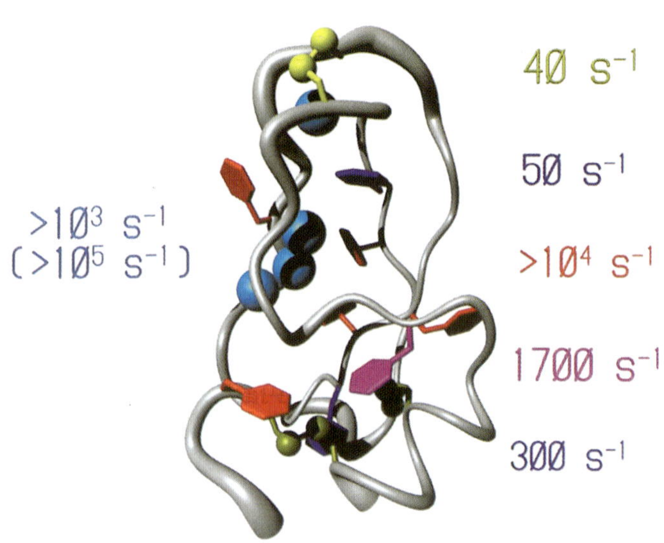

Figure 29. Intramolecular rate processes in BPTI. The polypeptide backbone is represented by a grey spline function through the α-carbon positions, with the thickness of the line representing the spread of the bundle of 20 conformers in Figure 27. Individual rate processes and their frequencies are indicated with the following colour code: dark blue, magenta and red: ring flips of phenylalanine and tyrosine; yellow, exchange between the R- and S-chiral forms of the disulfide bond at the top; cyan, interior hydration water molecules, with indication of the exchange rates with the bulk water.

on the correlation function describing the stochastic modulation of the dipole–dipole coupling between the interacting protons (Equation [1]) has a key role. The value of $f(\tau_c)$ may be governed either by the Brownian rotational tumbling of the hydrated protein molecule, or by interruption of the dipolar interaction through translational diffusion of the water molecules relative to the protein surface, whichever is faster. On this basis it could be established that surface hydration of peptides and proteins is characterized by very short residence times of the water molecules in the hydration sites, in the range from about 20 to 300 picoseconds at 10°C. This result presents an intuitive rationale for the generally observed dynamic disorder of the protein surface structure (Figure 27), and indicates that the dehydration of the polypeptide surface will hardly ever be a rate-limiting step either in protein folding or in intermolecular interactions with proteins.

The BPTI crystal structure contains four interior hydration water molecules. These are an integral part of the molecular architecture and are inaccessible to the solvent in a rigid model of the three-dimensional structure (Figure 29). The chemical shift dispersion anticipated for these four water molecules on the basis of the considerations in Figures 5–8 was not observed. This degeneracy of the chemical shifts of bound water and bulk water was found to be due to rapid exchange of water molecules in and out of the protein molecule, with an upper limit on the life-time of about 1 millisecond (53; as indicated in Figure 29 by the number in parentheses, the actual life times for individual waters may be significantly shorter). Rapid exchange of interior hydration water molecules appears to be a general property of globular proteins, and was also observed for water molecules located in protein–DNA interfaces, for example, in the DNA complex with the *Antennapedia* homeodomain (Figure 2).

Another intriguing NMR observation of internal protein mobility are the 180°-ring flipping motions of phenylalanine and tyrosine (54). Observation of these "ring flips" on the millisecond to microsecond time scale (Figure 29) was a genuine surprise for the following reasons: in the refined X-ray crystal structure of BPTI reported in 1975, the aromatic rings of phenylalanine and tyrosine are among the best-defined side chains, with the smallest temperature factors. In each ring the relative values of the temperature factors for the individual atoms increase toward the periphery, so that the largest positional uncertainty is indicated for the peripheral ζ-carbon atom on the symmetry axis through the C^{β}–C^{γ} bond, rather than for the four δ- and ε-ring carbon atoms, which undergo extensive movements during the ring flips. Theoretical studies then resolved this apparent contradiction: the crystallographic temperature factors sample multiple rotation states about the C^{α}–C^{β} bond, but they do not manifest the ring flips because the populations of all non-equilibrium rotational states about the C^{β}–C^{γ} bond are vanishingly small. Although the flipping motions about the C^{β}–C^{γ} bond have low frequencies (Figure 29), they are very rapid 180°-rotations connecting two indistinguishable equilibrium orientations of the ring. Similar to the exchange of internal hydration waters, the ring flip phenomenon is a general feature of globular proteins, manifesting ubiquitous low-frequency internal motions that have activation energies of 60 to 100 kJM^{-1}, amplitudes larger than 1 Å, activation volumes of about 50 $Å^3$, and involve concerted displacement of numerous groups of atoms. Combined with sequence-specific NMR assignments, these experiments provide high spatio-temporal resolution for the description of rate processes in proteins. In Figure 29 this is illustrated by a mapping of the frequencies for ring flips and water exchange onto the NMR structure of BPTI.

In addition to the ring flips and the exchange of internal hydration water molecules, the figure 29 includes data on the exchange of a disulfide bond between the R and S chiral states. In contrast to the other two phenomena, this rate process connects two different molecular structures.

Although all the data collected in Figure 29 have been known for more than a decade, and some of them for nearly three decades, no widely accepted functional interpretation of these low-frequency motional processes has been advanced. The same holds for the conformational equilibria manifested by the "protection factors" governing the amide proton exchange rates in folded proteins (Figures 17 and 21). Quite possibly these NMR measurements are ahead of their times, and represent a source for future novel insights into structure–function correlations in proteins.

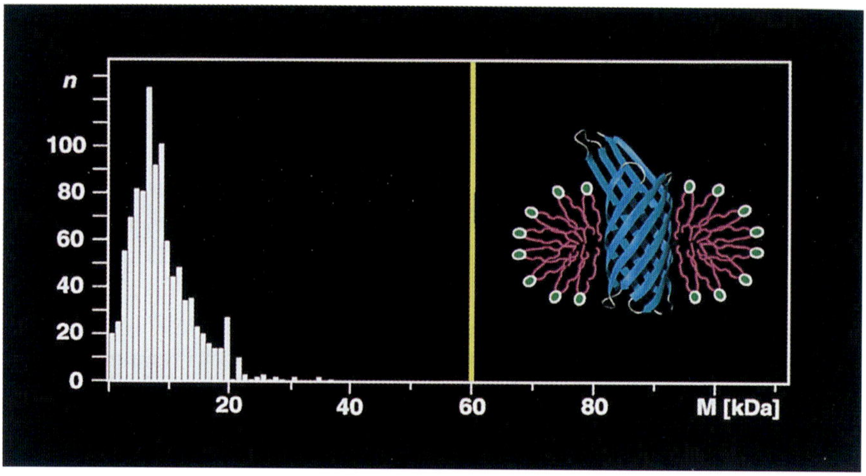

Figure 30. NMR structure determination and molecular weight. The horizontal axis covers the molecular weight-range 0 – 110'000. The left side shows the molecular weight distribution of the NMR structures in the protein databank (December 2000). On the right, the structure of the protein OmpX from *E. coli* (M ≈ 18000) in DHPC micelles (M ≈ 70 000 for the mixed micelles) indicates along the horizontal axis the approximate molecular weight range for NMR structure determination of mixed micelles with membrane proteins that is presently accessible with the NMR techniques of Figures 31 – 33.

OUTLOOK TO NMR APPLICATIONS IN STRUCTURAL AND FUNCTIONAL PROTEOMICS

With the availability of a rapidly increasing number of completely sequenced genomes, new challenges arise for the methods used for three-dimensional structure determination. On the one hand, "structural genomics" initiatives in several leading research centers focus on the development of technology for high-throughput structure determination to generate a comprehensive atlas of protein folds, so that remaining gaps could be filled by structure prediction methods. There is clearly a lot of room to further enhance the efficiency of each step of the NMR structure determination procedure (Table 1). On the other hand, we face the situation that newly determined protein structures should enable us to predict novel functions, whereas in classical structural biology one encounters more typically the challenge of rationalizing known functions on the basis of the three-dimensional structure. This section describes some recent work in our laboratory that may eventually contribute to future strategies for the discovery of new physiological functions from molecular structure data.

It has been widely recognized that supplementing the determination of new protein folds with data on intermolecular interactions may provide a key for the identification of unknown gene functions. Since efficient use of conventional NMR spectroscopy in solution had been limited to particle sizes with molecular weights up to about 30'000 (Figure 30, left side), a new chal-

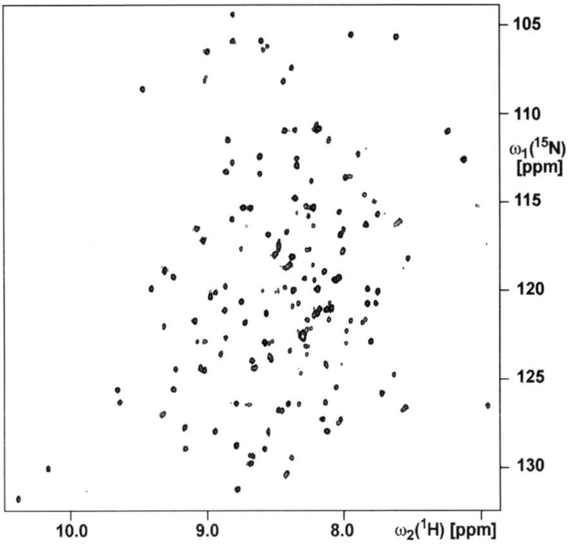

Figure 31. Transverse relaxation-optimized [^{15}N,^{1}H]-correlation spectroscopy (TROSY). A spectrum of the uniformly ^{2}H- and ^{15}N-labeled membrane protein OmpX reconstituted in DHPC micelles is shown (750 MHz, 30°C, H_2O-solution).

lenge for solution NMR techniques then arose from the fact that the supramolecular structures resulting from interactions of two or several proteins, or of other macromolecular components tend to have high molecular weights. Although a 30'000-Dalton size limit allowed to work with a large pool of physiologically interesting proteins, it was hardly compatible with extensive use of NMR for studies of such supramolecular structures. For example, this size limit would severely narrow down the range of potential receptor systems accessible to NMR in drug discovery projects (55, 56), restrict studies of protein–nucleic acid complexes (Figure 2) to a small number of systems with modest size, and prevent the use of solution NMR for studies of membrane proteins, since these have to be reconstituted and solubilized in mixed micelles with detergents or lipids (Figure 30, right side). A few years ago this limitation was successfully challenged, since the size-range for applications of solution NMR techniques could be significantly extended through the introduction of transverse relaxation-optimized spectroscopy (TROSY) (57). As an illustration, the figure 31 shows a [^{15}N,^{1}H]-TROSY correlation spectrum of a membrane protein reconstituted in detergent micelles. Sharp, well-separated peaks are obtained in spite of the large size of the mixed micelles (Figure 30, right side). With the use of the TROSY-principle, the more complex NMR experiments needed for a structure determination with [^{2}H,^{13}C,^{15}N]-labelled proteins can also be obtained with high quality (58), so that solution NMR can now be used for *de novo* membrane protein structure determination (59).

The improved quality of the NMR spectra of large structures with the use of TROSY can be qualitatively rationalized by the following considerations. As has previously been indicated in the discussion on NOEs (Equation [1]), the appearance of solution NMR spectra is intimately related to the effective cor-

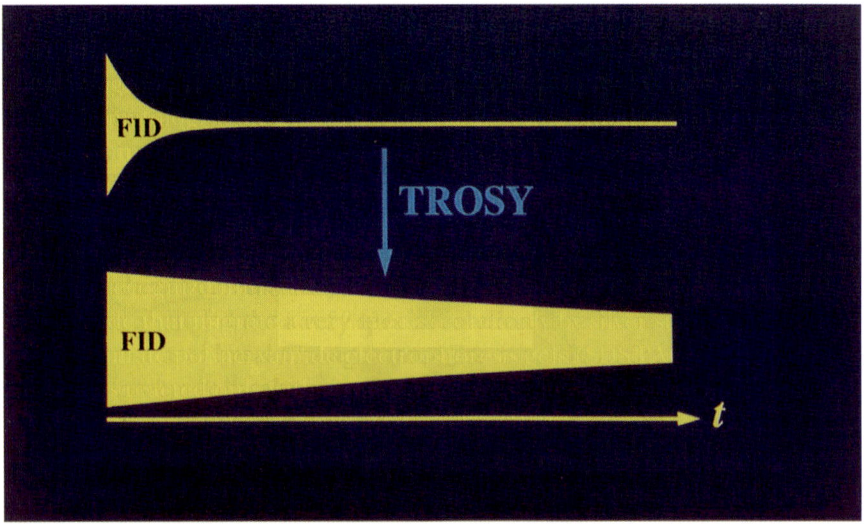

Figure 32. Schematic drawing of the free induction decay (FID) for a large protein with and without the use of TROSY. The period covered along the time axis is of the order of 100 milliseconds. The FID represents the primary recording of the NMR data in the "time domain", from which the frequency domain spectrum (for example, Figures 18, 19 and 31) is obtained by Fourier transformation (31, 33).

relation time, τ_c, which characterizes the thermal motions of the molecule considered. The increase of the τ_c-values in larger structures results in line broadening due to rapid transverse spin-relaxation. For example, if the spectrum of Figure 31 had been recorded with conventional NMR techniques, most of the resonance lines would not be individually resolved due to severe line broadening, and one would have experienced a severe loss of sensitivity for detection of the NMR signals. The reduced sensitivity can be readily appreciated by examination of the "free induction decays" (FID) in the time domain data. The figure 32 shows that rapid loss of magnetization in a conventional NMR experiment with a large structure can be slowed down by the use of TROSY, which corresponds to the reduction of the line width in the frequency domain spectrum. The impact on the sensitivity is visualized in Figure 33 with a simplified scheme for a 2D correlation experiment, which includes two magnetization transfer periods, and the evolution and acquisition periods. During evolution and acquisition, the system is not subjected to external perturbations, and loss of magnetization occurs at a rate determined by the "transverse relaxation time". It is readily apparent that rapid loss of magnetization in a conventional NMR experiment (upper trace in Figure 32) leads to weak or even vanishing signal intensity at the end of the evolution period, and accordingly the sensitivity for detection of the signal during the acquisition period is very low. In contrast, one can obtain much improved sensitivity with the use of TROSY (lower trace in Figure 32), since plenty of signal intensity will be preserved at the end of the evolution period, and the signal can be recorded with high signal-to-noise ratio during a large portion of the ac-

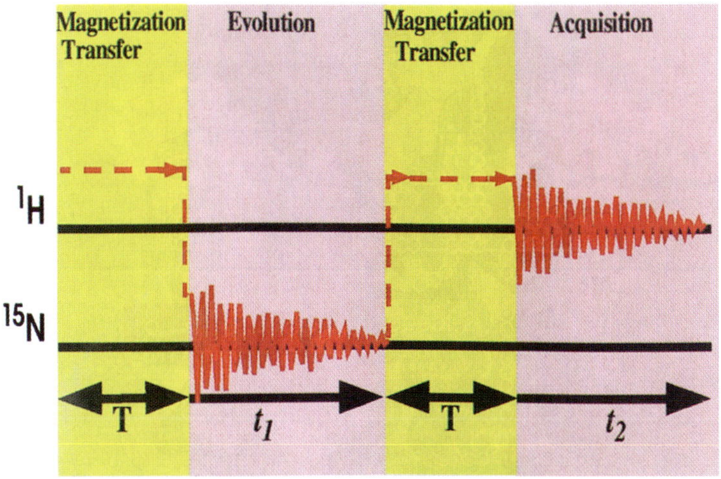

Figure 33. Basic features of a 2D [^{15}N,^{1}H]-correlation experiment. On the left, ^{1}H and ^{15}N indicate radio-frequency channels for irradiation of these isotopes. Yellow shading identifies the $^1H \rightarrow {}^{15}N$ and $^{15}N \rightarrow {}^1H$ magnetization transfer periods, T, and pink shading the evolution and acquisition periods, t_1 and t_2. During t_1 and t_2 the decay of the magnetization (FID) is schematically indicated. The overall duration of the experiment, $2T+t_1+t_2$, can be of the order of 10 to 1000 milliseconds, depending on the size of the structure studied and the intended purpose of the measurement.

quisition period. Following such considerations for minimizing the loss of magnetization during all four time periods indicated in Figure 33, and with additional optimization of the magnetization transfer techniques (60), solution NMR spectra have by now been recorded for structures with molecular weights up to 870'000. In the spectrum of the co-chaperonin GroES bound to the chaperonin GroEL (Figure 34), resonance lines that provided novel in-

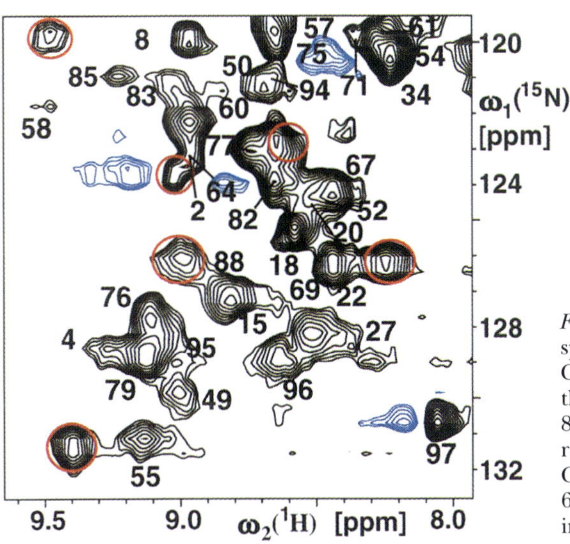

Figure 34. [^{15}N,^{1}H]-correlation spectrum of the co-chaperonin GroES (M ≈ 70000) bound to the chaperonin GroEL (M ≈ 800 000). The spectrum was recorded with a [^{15}N,^{1}H]-CRIPT-TROSY experiment (60, 61). The red and blue colouring is explained in the text.

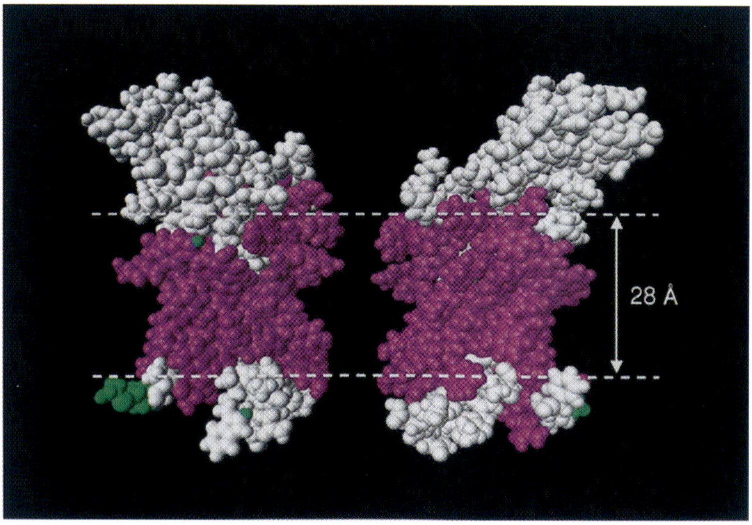

Figure 35. NMR structure of the membrane protein OmpX reconstituted in DHPC micelles. Front and back views of a space-filling model of OmpX are shown, with pink and green coloring identifying the surface area of the protein that is in contact with DHPC molecules in the mixed micelles. The two broken horizontal lines indicate the thickness of the lipid phase in the *E. coli* cell membrane.

formation on structural and dynamic features of the GroES–GroEL interface are colored in red and blue (61).

The principal contributions to the rate of transverse spin-relaxation can be traced to two different types of interactions, *i.e.*, dipole–dipole coupling of the observed spin with other, nearby spins, and chemical shift anisotropy (CSA) (28). These interactions are modulated by the stochastic rotational motions in solution, and as a consequence the rate of transverse relaxation increases for larger structures with slower Brownian motions. TROSY exploits constructive interference between the two relaxation mechanisms, and actually uses CSA-relaxation at high fields to cancel the dipolar relaxation (57). In this way the appearance of the NMR spectrum is effectively uncoupled from the Brownian motions, which then enables the recording of solution NMR spectra with large structures.

It remains to be seen how these new NMR techniques will be employed most profitably in the future. Intriguing possibilities include that NMR can now be employed in drug discovery projects with very large receptors (55, 56). Combined with suitable isotope-labelling strategies, TROSY-based NMR techniques have also been shown to provide a powerful approach for investigations of intermolecular interactions in supramolecular structures with two or several macromolecular components (62, 63). In these applications, a detailed structural interpretation of the NMR data will in most instances be dependent on the availability of an independently determined atomic-resolution structure for one or multiple components, which may have been obtained either by NMR in solution or by diffraction methods in single crys-

dents, Research associates, Technical staff 1970–2002

llimann
eraldi
on
eller
nann
zel
nn
rathwohl
r
el
rz
gonski
rber
chumacher
gner
mmbach
boa
s
Marco
Gremlich
wn
z
. Frey
ösch
i
poulou
erwein
nann

Klaus Loth
Kuniaki Nagayama
Stephen J. Perkins
Sidney L. Gordon
Lotti Rutz
Andreas Dubs
Daniel Picot
Werner A. Braun
Heinrich Roder
Albert Eugster
Gerhard Wider
Christoph Briner
Chrit Moonen
Adela Chrzeszczyk
Markus Meier
Subbaraya Ramaprasad
Nobuhiro Gö
Mitiko Gö
Johannes Schaffner
Markus Schmidiger
Hans Senn
Wayne E. Steinmetz
Anil Kumar
Yvonne Hunziker
Sadato H. Yabuki
Edith Huber
Fredi Stucki
Olle Teleman
Martin Billeter
Slobodan Macura
Kong-Hung Lee
Ramakrishna Hosur
Alexandre Arseniev

Petr Štrop
Dominique Marion
Erik R. P. Zuiderweg
Françoise Guerlesquin
Arthur Pardi
Kenneth R. Metz
Timothy Havel
Michael P. Williamson
Paula Y. Watnick
Martin Hassler
David Neuhaus
Peter Schultze
Hans Widmer
Werner Leupin
Satoshi Ebina
Reto Marani
Walter J. Chazin
Michael H. Frey
Winfried Denk
Sven Hyberts
Gottfried Otting
Allen D. Kline
Alexander M. Labhardt
Gaetano Montelione
Chantal Wynants
Thomas Schaumann
Phuntsok Ernst
Qiwen Wang
Erich Wörgötter
Norbert Müller
Luzi Jenny
Bernardo Celda
Rolf Grütter

Daniel Brühwiler
Manuel Glauser
Oscar D. Redwine
Francesco Riva
Hydeyuki Haruyama
Daniel Baumann
Bennett Th. Farmer II
Christoph Weber
Hua Qingxin
Peter Güntert
Francesco Avilés
Tai-he Xia
Leonard P. M. Orbons
Barbara Messerle
Craig D. Eccles
Yan-qiu Qian
Dario Neri
Hervé Darbon
Hugo Amacker
Wolfgang Kaissl
Ivan Vranesic
Kandala V. R. Chary
Daniel Braun
Beat Werner
Patrick Sodano
Martin Bos
Maria Paz Sevilla
Agostino Fede
Josep Vendrell
Thomas A. Szyperski
Kurt Berndt
Claus Spitzfaden
Jonathan M. Moore

Berthold von Freyberg
Edvards Liepinsh
Johanna Blatter
Miguel S. Col
John F. O'Connell
Barbara Leiting
John H. Bushweller
Tadakazu Maeda
Martin Schick
Walfrido Antuch-Garcia
Peter Luginbühl
Christian Bartels
Volker Dötsch
Siggi Mronga
Jan Johansson
Gregg Siegal
Oliver Schott
Nikolaus Schaefer
Gabriel Hänggi
Reto Koradi
Helena Kovacs
Marcel Ottiger
Ralph Daumke
Matej Grilc
Serge Altmann
Maurizio Pellecchia
César Fernández-Estrabao
Christian Mumenthaler
Oliver Zerbe
Bénédicte Prêcheur
Hideo Iwai
Oliver Ohlenschläger
René Brunisholz

Gerhard Frank
Walter Sidler
Konstantin Pervushin
Rosmarie Hug
Cyrill Arnet
Roland Riek
Francisco López-Garcia
Martin Wahl
Jihui Wu
Ralf Glaser
Aizhuo Liu
Michael Salzmann
Alexander C. Sobol
Fred F. Damberger
Chih-yen King
Bogdan Banecki
Yves-Laurent Viollier
Ralph Zahn
Michel Hochuli
Shin-ichi Tate
Eliane Ulrich
Jörg Pieper
Tian Y. Tsong
Thorsten Lührs
Christine von Schroetter
Roberto Fattorusso
Helmut Grubmüller
Reinhard Wimmer
Martha Geier
Sandra Zangger
Torsten Herrmann
Kai Schnee

Jocelyne Fiaux
Luigi Calzolai
Regitze Vold
Reto Horst
Donghan Lee
Koba Adeishvili
Dominikus A. Lysek
Sophie Bonjour
Christian Hilty
Morten Hohwy
Rochus L. J. Keller
Giorgio Pioda
Lars Dreier
Beat Vögeli
Touraj Etezady-Esfarjani
Christian Schorn
Francesco Fiorito
Pascal Bettendorff
Simone Hornemann
Vicent Esteve Moya
Valmir Fadel
Kent A. Baker
Hakim Tafer
Alvar Gossert
Lucas Nivon
Sebastian Hiller
Ida Arm
Yolanda Auchli
Daniel Pérez
Marianne Wüthrich
Isabelle Allen
Gabriela Arm

Figure 36. My collaborators at the ETH Zürich from 1970 to 2002.

tals. Applications of the new NMR techniques for *de novo* determination of large structures appears to be particularly attractive for, but not limited to, nucleic acid–protein complexes and small membrane proteins reconstituted in soluble detergent or lipid micelles. The characterization of a membrane protein structure has also been extended to include the detergents in the mixed micelles (Figure 35) (64), since the areas of the protein surface in contact with the detergent molecules could be delineated by the observation of intermolecular NOEs between hydrogen atoms of the protein and the detergent. In these particular mixed micelles this surface area is virtually identical to the protein surface in contact with the lipid phase in the biological membrane. Since this leaves both ends of the β-barrel of OmpX freely accessible to the aqueous solvent, this reconstitution system should also be suitable for functional studies of the membrane protein with NMR.

This last section of my presentation certainly indicates only a narrow range of potential further developments of solution NMR techniques and their applications in structural biology and structural proteomics. I look forward with great expectations to the future evolution of this awesome and beautiful technique, which has given me so many years of joy and excitement in studies of the molecules of life.

ACKNOWLEDGMENTS

From 1970 through 2002, 229 students, postdoctoral research associates, and technical and administrative staff worked with me at the ETH Zürich. I am deeply indebted to all of them for their enthusiasm and dedication, which resulted in the work summarized in this article. Their names are collected in Figure 36, and individual contributions can be recognized in the reference list. Although human minds stand behind all progress in science, the success of our research projects also depended critically on financial resources. I would like to acknowledge the ETH Zürich, the Swiss National Science Foundation, the Kommission für Technologie und Innovation (KTI), Bruker-Biospin AG, and the Scripps Research Institute in La Jolla, CA, USA, for their support.

REFERENCES

1. Qian, Y.Q., Billeter, M., Otting, G., Müller, M., Gehring, W.J. and Wüthrich, K. (1989) *Cell* **59**, 573–580. The structure of the *Antennapedia* homeodomain determined by NMR spectroscopy in solution: comparison with prokaryotic repressors.
2. Otting, G., Qian, Y.Q., Billeter, M., Müller, M., Affolter, M., Gehring, W.J. and Wüthrich, K. (1990) *EMBO J.* **9**, 3085–3092. Protein–DNA contacts in the structure of a homeodomain–DNA complex determined by nuclear magnetic resonance spectroscopy in solution.
3. Lopez Garcia, F., Zahn, R., Riek, R. and Wüthrich, K. (2000) *Proc. Natl. Acad. Sci. USA* **97**, 8334–8399. NMR structure of the bovine prion protein.
4. Wüthrich, K. (1986) *NMR of Proteins and Nucleic Acids*, Wiley, New York.
5. Wüthrich, K. (1995) *Acta Cryst.* D **51**, 249–270. NMR–this other method for protein and nucleic acid structure determination.
6. Wüthrich, K. (1995) *NMR in Structural Biology: A Collection of Papers by Kurt Wüthrich*, World Scientific, Singapore.
7. Wüthrich, K. (2001) *Nature Struct. Biol.* **8**, 923–925. The way to NMR structures of proteins.
8. Saunders, M. and Wishnia, A. (1958) *Ann. N.Y. Acad. Sci.* **70**, 870–874. Nuclear magnetic resonance spectra of proteins.
9. Jardetzky, O. and Jardetzky, C.D. (1958) *J. Biol. Chem.* **233**, 383–387. Proton magnetic resonance spectra of amino acids.
10. Kowalsky, A. (1962) *J. Biol. Chem.* **237**, 1807–1819. Nuclear magnetic resonance studies of proteins.
11. Mandel, M. (1965) *J. Biol. Chem.* **240**, 1586–1592. Proton magnetic resonance spectra of some proteins.
12. McDonald, C.C. and Phillips, W.D. (1967) *J. Amer. Chem. Soc.* **89**, 6332–6341. Manifestations of the tertiary structures of proteins in high-frequency nuclear magnetic resonance.
13. Wüthrich, K. (1969) *Proc. Natl. Acad. Sci. USA* **63**, 1071–1078. High resolution proton nuclear magnetic resonance spectroscopy of cytochrome c.
14. Shulman, R.G., Ogawa, S., Wüthrich, K., Yamane, T., Peisach, J. and Blumberg, W.E. (1969) *Science* **165**, 251–257. The absence of "heme–heme" interactions in hemoglobin.
15. Davis, D.G., Lindstrom, T.R., Mock, N.H., Baldassare, J.J., Charache, S., Jones, R.T. and Ho, C. (1971) *J. Mol. Biol.* **60**, 101–111. NMR studies of hemoglobins VI: heme proton spectra of human deoxyhemoglobins and their relevance to the nature of co-operative oxygenation of haemoglobin.
16. Kurland, R.J., Davis, D.G. and Ho, C. (1968) *J. Amer. Chem. Soc.* **90**, 2700–2701. Paramagnetic proton NMR shifts of metmyoglobin, methemoglobin, and hemin derivatives.
17. Wüthrich, K. (1970) *Structure and Bonding* **8**, 53–121. Structural studies of hemes and hemoproteins by nuclear magnetic resonance spectroscopy.
18. Shulman, R.G., Glarum, S.H. and Karplus, M. (1971) *J. Mol. Biol.* **57**, 93–115. Electronic structure of cyanide complexes of hemes and heme proteins.
19. Wüthrich, K. (1976) *NMR in Biological Research: Peptides and Proteins*, North Holland, Amsterdam.

20. Noggle, J.H. and Schirmer, R.E. (1971) *The Nuclear Overhauser Effect*, Academic Press, New York.
21. Gibbons, W.A., Alms, H., Bockman, R.S. and Wyssbrod, H.R. (1972) *Biochemistry* **11**, 1721–1725. Homonuclear INDOR spectroscopy as a means of simplifying and analyzing proton magnetic resonance spectra of peptides and as a basis for determining secondary and tertiary conformations of complex peptides.
22. Gupta, R.K. and Redfield, A.G. (1970) *Science* **169**, 1204–1206. Double resonance NMR observation of electron exchange between ferri- and ferrocytochrome *c*.
23. Balaram, P., Bothner-By, A.A. and Dadok, J. (1972) *J. Amer. Chem. Soc.* **94**, 4015–4017. Negative nuclear Overhauser effects as probes of macromolecular structure.
24. Solomon, I. (1955) *Phys. Rev.* **99**, 559–565. Relaxation processes in a system of two spins.
25. Kalk, A. and Berendsen, H.J.C. (1976) *J. Magn. Reson.* **24**, 343–366. Proton magnetic relaxation and spin diffusion in proteins.
26. Gordon, S.L. and Wüthrich, K. (1978) *J. Amer. Chem. Soc.* **100**, 7094–7096. Transient proton–proton Overhauser effects in horse ferrocytochrome c.
27. Wagner, G. and Wüthrich, K. (1979) *J. Magn. Reson.* **33**, 675–680. Truncated driven nuclear Overhauser effect (TOE): a new technique for studies of selective ^{1}H–^{1}H Overhauser effects in the presence of spin diffusion.
28. Abragam, A. (1961) *Principles of Nuclear Magnetism*, Clarendon Press, Oxford.
29. Keller, R.M. and Wüthrich, K. (1978) *Biochem. Biophys. Res. Comm.* **83**, 1132–1139. Evolutionary change of the heme c electronic structure: ferricytochrome c-551 from *Pseudomonas aeruginosa* and horse heart ferricytochrome c.
30. Dubs, A., Wagner, G. and Wüthrich, K. (1979) *Biochim. Biophys. Acta* **577**, 177–194. Individual assignments of amide proton resonances in the proton NMR spectrum of the basic pancreatic trypsin inhibitor.
31. Ernst, R.R. (1992) *Angew. Chemie –Int. Ed.* **31**, 805-823. Nuclear magnetic resonance Fourier transform spectroscopy.
32. Wider, G., Macura, S., Anil Kumar, Ernst, R.R. and Wüthrich, K. (1984) *J. Magn. Reson.* **56**, 207–234. Homonuclear two-dimensional ^{1}H NMR of proteins: experimental procedures.
33. Ernst, R.R., Bodenhausen, G. and Wokaun, A. (1987) *Principles of Nuclear Magnetic Resonance in One and Two Dimensions*, Oxford University Press, Oxford.
34. Wagner, G. and Wüthrich, K. (1982) *J. Mol. Biol.* **155**, 347–366. Sequential resonance assignments in protein ^{1}H nuclear magnetic resonance spectra: basic pancreatic trypsin inhibitor.
35. Wagner, G., Anil-Kumar and Wüthrich, K. (1981) *Eur. J. Biochem.* **114**, 375–384. Systematic application of two-dimensional ^{1}H nuclear magnetic resonance techniques for studies of proteins 2: combined use of correlated spectroscopy and nuclear Overhauser spectroscopy for sequential assignments of backbone resonances and elucidation of polypeptide secondary structures.
36. Billeter, M., Braun, W. and Wüthrich, K. (1982) *J. Mol. Biol.* **155**, 321–346. Sequential resonance assignments in protein ^{1}H nuclear magnetic resonance spectra: computation of sterically allowed proton–proton distances and statistical analysis of proton–proton distances in single crystal protein conformations.
37. Wüthrich, K., Billeter, M. and Braun, W. (1984) *J. Mol. Biol.* **180**, 715–740. Polypeptide secondary structure determination by nuclear magnetic resonance observation of short proton–proton distances.
38. Pardi, A., Billeter, M. and Wüthrich, K. (1984) *J. Mol. Biol.* **180**, 741–751. Calibration of the angular dependence of the amide proton–C^a proton coupling constants, $^3J_{HNa}$, in a globular protein: use of $^3J_{HN\alpha}$ for identification of helical secondary structure.
39. Wüthrich, K., Wider, G., Wagner, G. and Braun, W. (1982) *J. Mol. Biol.* **155**, 311–319. Sequential resonance assignments as a basis for determination of spatial protein structures by high resolution proton nuclear magnetic resonance.

40. Braun, W., Bösch, C., Brown, L.R., Gō, N. and Wüthrich, K. (1981) *Biochim. Biophys. Acta* **667**, 377–396. Combined use of proton–proton Overhauser enhancements and a distance geometry algorithm for determination of polypeptide conformations: application to micelle-bound glucagon.
41. Havel, T.F. and Wüthrich, K. (1984) *Bull. Math. Biol.* **46**, 673–698. A distance geometry program for determining the structures of small proteins and other macromolecules from nuclear magnetic resonance measurements of intramolecular ^1H–^1H proximities in solution.
42. Havel, T.F. and Wüthrich, K. (1985) *J. Mol. Biol.* **182**, 281–294. An evaluation of the combined use of nuclear magnetic resonance and distance geometry for the determination of protein conformations in solution.
43. Williamson, M.P., Havel, T.F. and Wüthrich, K. (1985) *J. Mol. Biol.* **182**, 295–315. Solution conformation of proteinase inhibitor IIA from bull seminal plasma by ^1H nuclear magnetic resonance and distance geometry.
44. Wagner, G. (1993) *J. Biomol. NMR* **3,** 375–385. Prospects for NMR of large proteins.
45. Kay, L.E. and Gardner, K.H. (1997) *Curr. Opin. Struct. Biol.* **7**, 564–570. Solution NMR spectroscopy beyond 25 kDa.
46. Bax, A. and Grzesiek, S. (1993) *Accounts Chem. Res.* **26**, 131–138. Methodological advances in protein NMR.
47. Wider, G. (1998) *Prog. Nucl. Magn. Reson. Spectr.* **32**, 193–275. Technical aspects of NMR spectroscopy with biological macromolecules and studies of hydration in solution.
48. Otting, G. and Wüthrich, K. (1990) *Q. Rev. Biophys.* **23**, 39–96. Heteronuclear filters in two-dimensional [^1H,^1H]-NMR spectroscopy: combined use with isotope labelling for studies of macromolecular conformation and intermolecular interactions.
49. Güntert, P. (1998) *Q. Rev. Biophys.* **31**, 145–237. Structure calculation of biological macromolecules from NMR data.
50. Moseley, H.N.B. & Montelione, G.T. (1999) *Curr. Opin. Struct. Biol.* **9**, 635–642. Automated analysis of NMR assignments and structures for proteins.
51. Herrmann, T., Güntert, P. and Wüthrich, K. (2002) *J. Biomol. NMR* **24**, 171–189. Protein NMR structure determination with automated NOE-identification in the NOESY spectra using the new software ATNOS.
52. Luginbühl, P. and Wüthrich, K. (2002) *Progr. Nucl. Magn. Reson. Spectr.* **40**, 199–247. Semi-classical nuclear spin relaxation theory revisited for use with biological macromolecules.
53. Otting, G., Liepinsh, E. and Wüthrich, K. (1991) *Science* **254**, 974–980. Protein hydration in aqueous solution.
54. Wüthrich, K. and Wagner, G. (1975) *FEBS Lett.* **50**, 265–268. NMR investigations of the dynamics of the aromatic amino acid residues in the basic pancreatic trypsin inhibitor.
55. Shuker S.B., Hajduk P.J., Meadows R.P. and Fesik S.W. (1996) *Science* **274**, 1531–1534. Discovering high-affinity ligands for proteins: SAR by NMR.
56. Pellecchia, M., Sem, D.S. and Wüthrich, K. (2002) *Nature Rev. Drug Disc.* **1**, 211–219. NMR in drug discovery.
57. Pervushin, K., Riek, R., Wider, G. and Wüthrich, K. (1997) *Proc. Natl. Acad. Sci.* USA **94**, 12366–12371. Attenuated T_2 relaxation by mutual cancellation of dipole–dipole coupling and chemical shift anisotropy indicates an avenue to NMR structures of very large biological macromolecules in solution.
58. Salzmann, M., Pervushin, K., Wider, G., Senn, H. and Wüthrich, K. (1998) *Proc. Natl. Acad. Sci. USA* **95**, 13585–13590. TROSY in triple-resonance experiments: new perspectives for sequential NMR assignment of large proteins.
59. Fernández, C., Hilty, C., Bonjour, S., Adeishvili, K., Pervushin, K. and Wüthrich, K. (2001) *FEBS Lett.* **504**, 173–178. Solution NMR studies of the integral membrane proteins OmpX and OmpA from *Escherichia coli.*
60. Riek, R., Wider, G., Pervushin, K. and Wüthrich, K. (1999) *Proc. Natl. Acad. Sci. USA* **96**, 4918–4923. Polarization transfer by cross-correlated relaxation in solution NMR with very large molecules.

61. Fiaux, J., Bertelsen, E., Horwich, A. and Wüthrich, K. (2002) *Nature* **418**, 207–211. NMR analysis of a 900K GroEL–GroES complex.
62. Pellecchia, M., Sebbel, P., Hermanns, U., Wüthrich, K. and Glockshuber, R. (1999) *Nature Struct. Biol.* **6**, 336–339. Pilus chaperone FimC–adhesin FimH interactions mapped by TROSY-NMR.
63. Frickel, E.M., Riek, R., Jelesarov, I., Helenius, A., Wüthrich, K. and Ellgaard, L. (2002) *Proc. Natl. Acad. Sci. USA* **99**, 1954–1959. TROSY-NMR reveals interaction between ERp57 and the tip of the calreticulin P-domain.
64. Fernández, C., Hilty, C., Wider, G. and Wüthrich, K. (2002) *Proc. Natl. Acad. Sci. USA* **99**, 13533–13537. Lipid–protein interactions in DHPC micelles containing the integral membrane protein OmpX investigated by NMR spectroscopy.

Chemistry 2003

PETER AGRE

"for the discovery of water channels"

and

RODERICK MacKINNON

"for structural and mechanistic studies of ion channels"

THE NOBEL PRIZE IN CHEMISTRY

Speech by Professor Gunnar von Heijne of the Royal Swedish Academy of Sciences.
Translation of the Swedish text.

Your Majesties, Your Royal Highnesses, Ladies and Gentlemen,

In the days of Alfred Nobel, the learned academies used to entertain and educate the public by holding open demonstrations explaining the latest scientific advances. This tradition has been largely – and perhaps unfortunately – forgotten. So let us try to revive the public demonstration of science, if only for a brief moment.

The demonstration I have in mind is a simple one, and only requires that you do something that is in any case particularly fitting for a Nobel Prize ceremony: to think. But only for exactly 5 seconds!

So, please start thinking, for 5 seconds.... Thank you!

Let us now reflect briefly on what has just happened, in each and every one of us. First, a sudden increase in the activity of the brain when you started to wonder what this is all about – should I really think at this point in the ceremony? – then, cascades of nerve signals when you were actually thinking, and finally a return to the normal resting state. And all this thinking ultimately relied on one of the simplest chemical compounds you can imagine: ordinary salt – sodium, potassium and chloride ions – streaming back and forth across the walls of your nerve cells, thereby generating the signals that activated your mind. And not even very much salt – a rough estimate is that the total amount of salt spent during these five seconds in each one of us was no more than a few grains. Only a fistful of salt to set a whole Concert Hall thinking!

And while all this brain activity was occupying our minds, our kidneys worked on quietly, as they always do, reabsorbing water from the urine to the blood. But in this case, the volumes of water transported are too big, even during five seconds, to be suitable for a demonstration from the podium.

This year's Nobel Prize in Chemistry is all about salt water, and the biochemical mechanisms that control where, when, and how often ions and water are let into or out of the cells in our body. Mechanisms that the two Laureates – Peter Agre and Roderick MacKinnon – have elucidated down to the atomic level.

Agre's was a "serendipity discovery": while working on a completely different problem, he stumbled across a protein in red blood cells that he could soon show was the water channel researchers had been looking for in vain for well over a century. His unexpected discovery opened a whole new field of study.

MacKinnon, on the other hand, decided at an early stage that he should try to do what was then thought impossible: to determine the three-dimensional structure of ion channels at atomic resolution. He bet his career on this vision – and succeeded to an extent that probably surprised even himself.

There is a lesson here, I believe: There is no one way to do science, and our support system must be sufficiently well funded and versatile to prepare the ground for both unexpected serendipity and focused, often risky, attacks on central scientific problems.

Peter Agre and Roderick MacKinnon stand for decisive contributions to the biochemistry of cell membranes, but their discoveries also have an almost tangible aesthetic component. Their work has uncovered an amazing "economy of design" in the atomic structures of the water and ion channels that is breathtaking in its simplicity and perfection. Indeed, after seeing these molecular machines, you find yourself thinking, "Of course, this is how it must be, this is how it must work!" What more could we ask of science?

Professor Agre, Professor MacKinnon,

Your fundamental discoveries concerning water and ion channels are singular achievements that have made it possible for us to see these exquisitely designed molecular machines in action at the atomic level. The biochemical basis for the transport of water – the most abundant and primordial substance of life – and ions – these tiny, mundane and yet absolutely essential constituents of the living world – can now be understood in unparalleled detail. On behalf of the Royal Swedish Academy of Sciences, I wish to convey to you our warmest congratulations, and I now ask you to step forward to receive the Nobel Prize in Chemistry from the hands of His Majesty the King.

PETER AGRE

ANCESTRAL ORIGINS

In 1862, Abraham Lincoln signed the Homestead Act, a bill opening one half million square miles of territory in the western United States for settlement. The Homestead Act offered new arrivals from other countries the opportunity to stake and develop farms of 160 acres by simply working the land for five years. Although they were only in their teens or early twenties, my great grandparents individually left their villages in Norway and Sweden between 1875 and 1885 and migrated to western Minnesota and South Dakota. Similar to the protagonists in the epics of Ole Rolvaag and Vilhelm Moberg, they worked the rich farmland, married, raised families, and achieved prosperity unattainable at that time in Scandinavia.

Things changed for my parents' generation. My father, Courtland Agre, and his two sisters grew up in Wallace, a hamlet in eastern South Dakota where his parents ran the general store. Wallace was also the hometown of Hubert Humphrey, and while my aunt Pearl remembers baby-sitting young Hubert, my father, an ardent Republican, claimed that they never met. To accommodate their educational needs, my grandparents moved the family to a larger town and ultimately to Minneapolis where Dad earned his B.S. and Ph.D. in chemistry from the University of Minnesota. He contributed to the U.S. effort in World War II by working as an experimental polymer chemist for the 3M company. My mother, Ellen Swedberg, was the sixth of eight children. Her upbringing was more severe. She was only five years old when her mother died; later her father lost their farm in Twin Brooks, South Dakota during the Depression. At age 18, Mother moved to St. Paul in order to support herself. Despite eleven years difference in age, Dad and Mom met at a Lutheran church social, fell in love, and married. I was never certain how much their families approved, since even small differences in geographical origins are taken very seriously by Scandinavians. The Agres were Norwegian (Osterdalen and Trondelag), while the Swedbergs had mixed origins – Swedish (Skåne) and Norwegian (Telemark).

CHILDHOOD

Following WWII, my parents moved to Northfield, a town 40 miles south of Minneapolis where Dad was recruited to the chemistry department at St. Olaf College. Dad was energetic and, with the help of his St. Olaf students during

Courtland Agre in his lab at St. Olaf College.

the summers, he built our house across the street from the college athletic fields and meadows. We could look up at the college from our living room. As was the tradition, Mom had babies and took care of the family. Preceded two years by my sister Annetta, I was born on January 30, 1949 and received the anglicized name of my grandfather, Peder. My closest sibling, James (Jim), was born one year later, followed by Paul, Ruth, and Mark. We had an idyllic childhood. Grandmother Agre lived nearby and coaxed us to speak rudimentary Norwegian in return for cookies and other bribes ("Jeg liker Bestemor's mat!"). Northfield was in many ways a new-world enclave of pre-Ibsen Norway with 19th century religious and socially conservative values. My friends had family names like Lunder, Finholt, Berglund, and Fredriksen. We schoolchildren all sat on the hillside waving our Norwegian flags when King Haakon visited St. Olaf. He was chauffeured from the train station to the college in the only Cadillac in town – owned by the local plumber. We always had lutefisk for Christmas dinner, after which Dad read from the Norwegian Bible. During the summers, he welcomed us into his laboratory at St. Olaf where he rigged simple "experiments" for us such as changing the color of solutions containing indicator dye by adding acid or base. As a youngster, it was obvious to me that I would follow my father's career path, since he was my greatest hero.

Life changed for our family while I was in the third grade. Grandmother

Agre died, and Dad decided to take a sabbatical year at the University of California. Dad had high aspirations, and through the American Chemical Society he became acquainted with renowned scientists. Berkeley was an amazing change from Northfield, and Jim and I attempted, with limited success, to demonstrate our Norwegian athletic and scholastic superiority to the smart and culturally heterogeneous Berkeley youngsters. It was at this time that my brother Paul was recognized as mentally retarded, and my sister Ruth began to exhibit her lifelong personality disorder with lack of impulse control. While they never fretted openly, these problems must have caused my parents profound heartache.

Following the year in Berkeley, we returned to Minnesota. Always eager for a challenge, Dad accepted a professorship at Augsburg College, a small Norwegian Lutheran college in Minneapolis with a chemistry department in need of help. We lived in a beautiful, large brick house on the banks of Lake Nokomis and went to the public schools. My inconsistent academic performance was usually well tolerated, as I tried to amuse my school classes with my practical jokes and amateurish wit. As in Northfield, my teachers were the kindest and nicest people imaginable. It is impossible to overestimate the importance of a teacher in the life of a child, and my all-time favorite was my sixth grade teacher Richard Hughes whose kindly personality and gentle sense of humor inspired in us the idea that learning is wonderfully fun. This impression was shared by many. My classmate Julia Lofness remains a close personal friend and still cheers me by retelling events from our days in Mr. Hughes' class.

My brother Jim and I spent many wonderful summers working on dairy farms in Wisconsin owned by Mom's cousins, and as members of our local Boy Scout troop. Scouting was a particularly important activity for us, and through the generous instruction of our Scoutmasters, Harold Neuendorf and Francis McMahon, we learned the resourcefulness needed to camp out even in Minnesota winters. One of our happiest times was in 1964 when Jim and I received our Eagle Scout Awards together – Dad carried that snapshot in his wallet for the rest of his life. Dad's presence was always palpable. Needing a medical doctor to perform physical examinations before summer camp, Dad always arranged for one of his former St. Olaf students to serve. One summer, Dr. Charles Mayo, grandson of the Mayo Clinic founder, examined the boys of Troop 185. Also as a Scout, I developed a deep interest in the culture of the Ojibway Indians of Northern Minnesota and explored the Canadian wilderness by canoe – an activity that I still undertake with family members each summer. In retrospect, another remarkable experience was the several-days visit to our home by Dad's friend Linus Pauling who presented lectures on physical chemistry (subject of his 1954 Nobel Prize in Chemistry) and the dangers of nuclear arms (subject of his 1962 Nobel Peace Prize). In person, Pauling was jovial, confident, and more engaging than anyone we had ever met. Dad always raved that Pauling's accomplishments were the greatest, and he instilled in us his view that the Nobel Prize is the closest thing to the Holy Grail.

HIGH SCHOOL AND COLLEGE

My years at Theodore Roosevelt high school were notable for reasons other than academics. Like other adolescents, I developed a strong attraction to girls and discovered that my parents' ideas were not the least bit interesting. I earned money as a concession salesman at the Minnesota Twins and Vikings games. At this time I developed a lifelong love of cross country skiing and long distance bicycling with my close friend and classmate Tom Page, an aspiring artist and fellow adventurer. Following my junior year in high school, I went on a camping trip through Russia in a group led by Horst Momber, a young language teacher from Roosevelt. This permanently sparked great enthusiasm for international travel in me. After my return, I gravitated in a Bohemian direction. Fancying myself as some sort of Bolshevik, my senior year in high school went badly in terms of conforming with my family's expectations. Rebelling against the establishment, my friends and I self-published an underground newspaper, *The Substandard* – a parody of the *Roosevelt High School Standard.* Our foray into newspaper work caused much delight amongst everyone except the school administration. Facing dismissal, I withdrew from school in the winter of 1967 when my grade in chemistry had dropped to a "D." Nevertheless, I finished my high school degree in night school, and I studied Russian language at the University of Minnesota during the day. I worked the evening shift and drove a truck for a factory making dummy land mines and parts for military equipment destined for the war in Vietnam.

This experience in the real world was far less appealing than I had anticipated, but Dad's faculty position offered a logical solution. I swallowed my pride and enrolled at Augsburg where I majored in chemistry in preparation for a career in medicine. Living at home, my social life was restricted to bicycle rides along the Mississsippi River and around the lakes. My closest friends Tom and Julia remained in Minnesota for college and were generous sources of support. My second year at Augsburg was enlivened when brother Jim and I organized and played for the varsity soccer team. I also worked part time delivering flowers for a local florist. Although I was not sympathetic to the Lutheran affiliation of the school, I developed my first real academic self-discipline at Augsburg. I greatly benefited from excellent faculty, especially John Holum, the organic chemistry professor, and from the warm and friendly personalities of the other chemistry majors. Unlike many campuses, Augsburg lacked an overly competitive pre-med atmosphere, and all eight of the premedical chemistry majors in the class of 1970 were accepted into medical school. My brothers Jim and Mark also attended Augsburg and went on to medical school. Annetta graduated from Augsburg before marrying and raising her family; her daughter Christina has also become a medical doctor.

EARLY MEDICAL CAREER AND MARRIAGE

The long, cold Minnesota winters instilled in me a fascination for exotic far off places; I aspired toward a career in tropical diseases and world health problems. Johns Hopkins is highly regarded in these areas of research, and I was ecstatic when I was accepted into their medical school. Having enough credits to finish Augsburg early, I left Minnesota in the winter of 1970 to travel alone throughout Asia for several months. This was an eye-opening experience, and after hitchhiking throughout Japan and Taiwan, I journeyed to Laos where I traveled up the Mekong River and explored the hills of northern Thailand by motorcycle. Determined to visit Angkor Wat, I entered Cambodia and eventually arrived in Pnom Penh just as the U.S. Air Force launched its 'Parrots Beak' offensive against Vietcong east of the city. Advised by the U.S. Embassy to depart immediately, I boarded an evacuation flight to Vietnam where I viewed the U.S. forces in and around Saigon for several days before leaving for the calm and verdurous Malay highlands. Following a counterculture passage through Ceylon, India, Pakistan, Afghanistan, and Iran, I eventually arrived in Istanbul, weakened from multiple traveler's ailments including viral hepatitis.

I began my medical studies at Johns Hopkins in September 1970. After being footloose in Asia, I found it difficult to concentrate on medical school, and my best efforts seemed barely enough to earn passing grades. I concluded that my presence at Hopkins fulfilled some sort of affirmative action quota for a Scandinavian student from Minnesota or the Dakotas. My medical school roommate, Vann Bennett, was from Hawaii, and he was thoroughly committed to surfing and laboratory science. Unlike most Hopkins medical students, Vann and I led severely ascetic lives, inhabiting a series of dilapidated rooms. We enjoyed touring the countryside by bicycle, and I made some long rides through Mexico and Canada. I maintained my counterculture image by always making my return trips to Minnesota by hitchhiking.

Johns Hopkins introduced me to two defining events in my life: commitment to biomedical research and meeting my future wife, Mary. It was known in 1970 that most traveler's diarrhea was caused by a protein similar to the cholera toxin secreted by certain strains of *E. coli.* I began a summer project to purify the *E. coli* toxin in the lab of Brad Sack in the Infectious Disease group at Johns Hopkins. At Vann's suggestion, I performed some of the biochemical studies in the laboratory of Pedro Cuatrecasas in the Pharmacology Department. The project fascinated me so much that I stayed in Pedro's lab during my final year in medical school, 1974, and put in an additional year as a postdoctoral fellow. I eventually succeeded in purifying the toxin, but of larger importance was my discovery that I wanted to make biomedical research my life's work. This was in part due to the fascinating scientists in the group. Looking back, it seems unlikely that such a passionate but disparate group of scientists could form a team and become close personal friends. Led by Pedro, a cosmopolitan intellectual with Spanish and South American roots, our group included a Palestinian from Lebanon, a conservative Jew from Brooklyn, a French psychiatrist, a Swiss-Indian chemist, a Canadian Rhodes Scholar, a Spanish anti-Franco activist, and a strikingly handsome Italian film actor who attracted the notice of every woman at Hopkins as he elucidated the molecular basis of femininity by performing the first purification of the estrogen receptor.

During these lab years, I met Mary Macgill, an attractive and intelligent young woman who was working in a Hopkins neurovirology lab. Mary and I immediately fell in love and became inseparable. Her calm and wise approach to life balanced my frenetic nature. From our first days together, Mary always encouraged me in my scientific career, and she often brought food to the lab late at night so we could keep working. Despite my ragamuffin appearance, Mary's family received me warmly. Representing a fascinating mix of blueblood lineage and farm-raised practicality, both sides of her family were among the early settlers in Maryland. Her mother, who had died a year before I met Mary, was a direct descendant of Pocahontas, and her father was a distinguished trial judge in Howard County, Maryland. We were married on March 29, 1975.

POSTGRADUATE TRAINING

My goal was to develop into an independent research scientist studying clinical problems at the laboratory bench, but I felt that postgraduate residency training in internal medicine was necessary. Mary and I settled in Cleveland Heights, Ohio, from 1975 to 1978 while I completed clinical training at Case Western University Hospitals of Cleveland. The CWRU Hospitals were led by Charles C. J. Carpenter, Jr., a distinguished physician and former member of the cholera group at Hopkins. I found the clinical years to be emotionally and physically draining. Nevertheless, I enjoyed the patients and took pride when elderly African-American patients referred to me as their "jive doctor."

During this time, Pedro's group had moved to the Wellcome Laboratories in Research Triangle Park, North Carolina. I accepted a clinical fellowship in hematology and oncology at the nearby University of North Carolina (UNC), and Mary and I moved to Chapel Hill in late June 1978. I supplemented the meager fellowship stipend by working as a physician at the army hospital at Fort Bragg (home of the 82nd Airborne Division). Mary and I purchased a small cottage in the country and two Irish Setter puppies. Our daughter Sara was born in September; Claire was born a year and a half later. Our family life was Spartan but delightful. We made many bicycle trips to neighboring farms and camping trips in the Smokies with our adorable little daughters – Sara a towhead with beautiful blue eyes and an engaging smile, and Claire with a thick shock of red hair and many clever but often mischievous ideas.

My clinical duties taking care of cancer patients at UNC were frequently depressing, but I enjoyed attending the outpatient clinic with John Parker, a clear-sighted Professor of Medicine who always encouraged me to study red cell membranes. Vann had also joined the Wellcome Laboratories, and he developed techniques to study the principal structural proteins from red blood cell membranes – spectrin, which formed the girders of the membrane skeleton, and ankyrin, the attachment protein that he discovered. Campbell McMillan, a UNC pediatric hematologist, had identified two little girls in his clinic with an extreme form of hereditary spherocytosis, a disorder of red cell shape causing increased fragility. Spherocytosis is generally accompanied by

John C. Parker.

only mild anemia, but the sisters were nearly lethally affected. Together with Vann, I evaluated the membranes and found them to be grossly deficient in spectrin, and we reported the study in the *New England Journal of Medicine.* This strongly reinforced my desire to pursue biomedical research.

RETURN TO BALTIMORE AND HOPKINS FACULTY

When Vann returned to Johns Hopkins in 1981 as a faculty member, I applied for an NIH career development grant to join him for more research training. This did not make sense financially, as it meant a 40% reduction in salary and loss of my staff position in the Wellcome Labs and adjunct faculty position at UNC. I consulted with my friend John Parker who generously reassured me that the opportunity at work with Vann at Hopkins would be important for me. Encouraged by Mary, we sold our lovely cottage, packed up our station wagon and a rental truck, and returned to Baltimore in August 1981. We scraped together funds and purchased a brick Dutch colonial house next to the Stoneleigh Elementary School and a large park and forest. The location was fortunate; I could commute the 6 miles to Hopkins every day by bicycle. To supplement our resources, I moonlighted as a ringside physician at professional boxing matches.

The Department of Cell Biology at Johns Hopkins was founded and directed by Tom Pollard, an engaging young scientist with remarkable energy and enthusiasm. I continued working with Vann and reported a series of studies of red cell membranes. I would have been happy to have remained in Vann's lab, but he recognized my need for an independent faculty position. In early autumn 1983, I received a faculty offer from the Hematology Division at Brigham and Women's Hospital at Harvard Medical School – a program regarded as the best in the country. I accepted the offer verbally, but Mary was pregnant with our third child, and while she remained supportive, she strongly desired to remain in Baltimore near her family. The Hematology Division at Johns Hopkins was in a transitional mode and exhibited minimal interest in me. However, Victor McKusick, then-Chairman of the Department of Medicine at Hopkins, interceded and I was offered an Assistant Professorship but with a much smaller start-up package.

This offer from Hopkins coincided with the blackest period in our family life. In the 9th month of her pregnancy, Mary experienced a premature separation of the placenta. After induction of labor, our daughter Lydia was born in severe distress. Although she survived, our baby never recovered. Like her sisters, Lydia was a gorgeous child, but her spontaneous movements were few. She moaned softly whenever she was repositioned in her crib, but she always relaxed when we held her. Devastated and expecting a need for permanent home nursing assistance, I accepted the offer at Hopkins, since it would provide the supplemental health care insurance needed for Lydia's care. Although Mary always exhibited the utmost determination to provide the best possible life for Lydia, I wavered badly and only with great difficulty was I able to accept our situation. We were again crushed when Lydia died four days before Christmas 1983.

We had the advantage of broad support from family and friends but struggled to cope. Mary stayed home to care for Sara and Claire while I set up a tiny lab in the old Blalock Building in Johns Hopkins Hospital. I was able to hire a technician, Andy Asimos, a bright-eyed young man from a Greek family who applied to work for two years before going to medical school. Andy and I continued the spherocytosis studies, establishing that the level of spectrin deficiency correlated with the clinical severity of the disease in new reports published in *Nature* and the *New England Journal of Medicine.*

Although the importance of the blood group antigen Rh was universally recognized, I was surprised to learn from Wendell Rosse, our friend at Duke Medical Center, that the molecular identity of Rh was completely unknown. The existence of 32 kDa polypeptides in red cells from Rh(D) individuals was reported by two European groups. We initiated a new project on the Rh blood group antigen in the hope that isolation of the polypeptide would allow us to define the components of the Rh antigen. Andy and I injected a series of rabbits with the partially purified 32 kDa Rh polypeptide. When Andy left for medical school, Barbara Smith, a former blood bank technologist, replaced him and set about purifying the Rh polypeptide. We soon recognized that we actually had isolated two membrane proteins – the 32 kDa Rh and a second protein of 28 kDa. My research expanded and we focused our major attention on Rh but continued to dabble with the 28 kDa protein.

Our son Clarke was born in 1985 – an adorable little boy with a dimple in

Mary and Clarke.

his left cheek. Mary and I took Clarke with us on a motor trip through Ireland the next year, and our family life gradually returned to normal. Mary organized camping trips for us to the national parks in the U.S and Canada each summer and Appalachian Trail hikes and visits to farms on the weekends. The tent shook as we played monkey pile, a game where I wrestled with the children and tickled them until they giggled themselves speechless. The children loved camping, and I made up crazy bedtime stories about a pirate named Cinco Grumpy and his brother Buffalo Grumpy who rode a horse named Snapping Pony. Sara and Claire attended Stoneleigh Elementary, played on recreation league soccer teams, and greatly enjoyed the synchronized swim program at the community pool. Mary always played the major role of parent. She organized an aluminum-recycling program for the elementary school and volunteered to teach science to the children at Stoneleigh. She made friends with the McKnights, a neighboring family with three children close in age to our own. I met the father, Steve, at a neighborhood function when Claire pointed him out as "Gracie's Dad." Steve had done pioneering work on DNA transcription factors and had already established himself as a major figure in molecular biology. I was inspired to join his lab at the Carnegie Institution as a sabbatical worker for the 1988–1989 academic year. This proved to be enormously productive – I learned DNA technology and made a worthwhile contribution to the understanding of transcription factor-DNA interactions. By coincidence, another child was born to each family in 1989 – Johnny McKnight and Anne Carlyle Agre (Carly).

DISCOVERY OF THE AQUAPORINS AND GROWTH OF OUR CHILDREN

Upon returning to our lab, I felt that we should explore the 28 kDa protein further. Barb and postdoctoral Fellow Brad Denker worked to develop a simple method for purifying the 28 kDa protein from the membranes of human red blood cells based upon its relative insolubility in N-lauroylsarcosine, a detergent that extracted the other membrane proteins. The 28 kDa protein was found to be amazingly abundant in red cell membranes (approximately 200,000 copies per cell) and behaved physically like a membrane-spanning homotetramer. Other scientists had failed to observe the protein because it did not react with the usual protein stains. The 28 kDa protein was astonishingly abundant in renal proximal tubules and in thin descending limbs of the loop of Henle. Determination of the N-terminal sequence of the purified protein showed it to be related to a series of proteins from diverse sources – lens of cow eye, fruit fly brain, bacteria, and plants.

These clues intrigued me. I wondered if the 28 kDa protein might be some sort of channel. I discussed the protein with numerous scientists who were as puzzled as I and who could not provide any insight. In April 1991, we stopped in Chapel Hill while returning from a camping trip to Florida – a compromise vacation during which we took the children to their choice of park (Disney World) and our choice (the Everglades). I visited Vann, who had joined Duke as a Howard Hughes Investigator. I then dropped by UNC to visit John Parker. Although it was late in the afternoon, and he was visibly tired from a long day in the clinic, John sat quietly as I described the mysterious 28 kDa protein. After listening thoughtfully, John leaned forward in his chair and suggested that the protein might be a channel for water – something he stated that physiologists regarded as controversial since no one had ever been able to define the molecular identity of such a protein.

John's idea yielded the breakthrough that opened a new field. Greg Preston, a Postdoctoral Fellow with strong molecular biological experience, cloned the cDNA for the 28 kDa protein. We were also extremely fortunate to have Bill Guggino, Hopkins Department of Physiology, as a collaborator. Greg expressed the protein in frog eggs isolated from *Xenopus laevis* – a useful expression system known to have low water permeability. Greg tested our hypothesis on October 9, 1991. Although the oocytes expressing the 28 kDa protein and the control oocytes looked the same, when he transferred them from isotonic solution to distilled water, an amazing difference was immediately obvious – all six control oocytes were unaffected whereas all six 28 kDa oocytes immediately exploded like popcorn. Greg ran into my office almost speechless, and we both celebrated with joy. We knew from the first experiment that the 28 kDa protein conferred water permeability and must be the long-sought water channel.

PROFESSIONAL FULFILLMENT AND HOME FRONT AMUSEMENTS

Our lab had always refrained from keeping our studies secret. We openly shared the information of our studies with Joseph Handler and his group in the Hopkins Nephrology Division. Joe considered this a major breakthrough and congratulated us; however, he warned me to publish the information as soon as possible. He was aware of other scientists with large groups who were attempting to isolate or clone the putative water channel. Word got out quickly, and within a month investigators from high-powered laboratories in the U.S., Japan, and Europe requested our plasmid and information about our studies. Intensely nervous, Greg and I worked through the Christmas holidays in marathon sessions to perform every conceivable control.

We submitted the paper on February 4, 1992, but I lay awake at night worrying that my openness would cause us to be scooped. My departmental chairmen Jack Stobo, Department of Medicine, and Tom Pollard, Department of Cell Biology/Anatomy, reassured me that we had taken the ethical high road and would not be denied credit for the discovery. I subsequently learned that other groups had immediately focused all their attention on the 28 kDa protein after hearing about our unpublished observations. We succeeded in attaining priority when our paper was accepted for publication in the April 17 issue of *Science*. Greg was selected to present the study as a plenary lecture at the American Society of Clinical Investigation. Several colleagues joined us for lunch before Greg's talk, and the name "aquaporin" emerged. We all agreed that it was perfect, and our 28 kDa protein became officially designated as AQP1.

We initiated a series of collaborations with laboratories outside of Hopkins, and these scientists soon became close friends of my lab and family. The collaborations yielded a rich set of studies that catapulted us into prominence within the membrane transport field. Mark Knepper, at the NIH Laboratory for Kidney and Electrolyte Metabolism, introduced us to his junior colleague, Søren Nielsen, at the University of Aarhus, and his former mentor, Arvid Maunsbach. With Søren in the lead, we defined the sites of expression of AQP1 in kidney, brain, capillaries, and other tissues by high resolution light microscopy and immunogold electron microscopy. Using AQP1 protein that Barb Smith purified to homogeneity, our collaborators Mark Zeidel at Harvard and Suresh Amubudkar at Hopkins defined the biophysical functions of AQP1 reconstituted into synthetic liposomes. Ueli Aebi, a former Hopkins colleague, introduced us to Andreas Engel at the M. E. Müller-Institute at the Biozentrum of the University of Basel in Switzerland. Andreas and his grad student Tom Walz prepared membrane crystals containing our highly purified AQP1 protein and determined the structure by negative staining electron microscopy, atomic force microscopy, and electron crystallography.

The next two years (1992–94) proved critical to my academic development and our laboratory's research. My lab had temporarily resided within the Cell Biology Department, but when space constraints emerged, we were faced with the need to relocate. This was not initially popular with our team, and signs emblazoned "Hell No – We won't go!" appeared in the lab. Fortunately, Dan Lane, Chairman of the Department of Biological Chemistry, offered me the lab once occupied by the renowned biochemist, Albert Lehninger. This was a godsend, and I have held a faculty position in the Department of Biological Chemistry ever since. Promotion to full professor came in 1993. Also at this time, I assisted Tom Pollard in the launching of a new Ph.D. grad-

Agre children: Clarke standing, Claire left, Sara right, Carly front.

uate program. With funding from the Markey Charitable Trust, the Johns Hopkins Graduate Program in Cellular and Molecular Medicine (CMM) was initiated to permit talented students in the laboratories of clinical and basic science departments to attain fundamental research skills and in-depth understanding of clinical diseases. Our first students matriculated in September 1994.

These years brought the realities of adult life. Mary was always very close to her father, but his health had declined and he died at age 81 while we were on a camping trip in California. Rather than returning for his funeral, we joined with the children in a special Jamie-remembrance at our campsite at Yosemite. John Parker was tragically stricken with a malignancy and died soon after at age 58. My father had been diagnosed with prostate cancer and succumbed at age 82. Mary and I both became the senior generation of our families. Fortunately, Mary's sister Sally moved nearby and added immensely to our lives and the lives of our four children. Sara and Claire became artists and scholars in high school. Clarke became a Cub Scout. Carly enjoyed special status as the "baby of the family" and kept us all amused with her wacky sense of humor. Sometimes her siblings teased Carly too much. The children all enjoyed our family dogs – especially when they misbehaved, but Carly experienced an emotional meltdown when Sara once misinformed her that the rabbit that our dog had killed was the Easter Bunny!

GETTING THE WORD OUT – A SCIENTIFIC ITINERANT

The decade of 1994–2003 proved exhilarating but exhausting. We continued at a fast pace in our own laboratory and in key collaborations with outstanding laboratories in Europe; together we generated more than 100 publications. Of these, a few stand out in my mind as particularly notable achievements.

The structural studies begun with analysis of recombinants in our lab by Greg Preston and Jin Sup Jung led to proposition of the hourglass model – a unique structure with four subunits each containing an individual aqueous pore formed from two hemi-channels oriented at 180° to each other that overlap within the bilayer. The electron crystallographic studies with Andreas Engel were expanded to include Yoshinori Fujiyoshi and his group in Kyoto. These studies of biologically active membrane crystals revealed electron densities at 3.8Å, and merging with the recombinant studies yielded the first high resolution structure of a human membrane channel. X-ray crystallography and molecular dynamics simulations by other scientists soon advanced the structure further.

The identification of humans lacking AQP1 protein resulted from co-localization of the structural gene with the Co blood group antigens on human chromosome 7 by Chulso Moon and demonstration of the Co polymorphism by Barb Smith and Greg Preston. Co-null individuals are extremely rare, and knockout mutations were identified in the gene encoding AQP1 (hence, AQP1-null). Although AQP1-null individuals are extremely rare, they feel perfectly well. Nevertheless, definitive studies by Landon King elucidated a major defect in renal concentration after thirsting and a marked delay in vascular water permeability.

Cloning the AQP4 homolog in brain by Jin Sup Jung led to a major new chapter in the recognition of water transporters at the blood brain barrier by Søren Nielsen and Ole Petter Ottersen. The structural basis of AQP4 localization resulted from a pivotal observation by John Neely and led to the recognition by Mahmood Amiry-Moghaddam that AQP4 contributes detrimentally to the onset of brain edema.

AQP5, the aquaporin from secretory glands, was cloned by Surabhi Raina and was shown by Landon and Søren to have unique distribution in glands and alveolar pneumocytes. While multiple clinical roles are anticipated, functional demonstration of AQP5 in sweating may have profound relevance to body temperature regulation. AQP6 was shown by Masato Yasui to be unique amongst aquaporins in that it freely transports anions within acid secretory cells of renal collecting ducts. Experiments by Jen Carbrey and Dan Gorelick predict that AQP9 is essential for conversion of glycerol to glucose in liver during starvation. Barry Rosen and his group at Wayne State identified arsenite transport through AQP9, thereby explaining the hepatotoxicity of the compound. Multiple studies by Giuseppe Calamita, Melanie Bonhivers, Jen Carbrey, David Kozono, Mario Borgnia, and Eric Beitz have documented fascinating roles for aquaporins in micro-organisms.

The astonishing interest in our work demonstrated by groups throughout the U.S. and abroad provided a direct means of showcasing the efforts of our young scientists and collaborators. During the decade, I undertook an extensive series of lecture trips, making over 250 presentations at universities, institutes, and scientific meetings on five continents. While the opportunity to speak directly to scientists around the world was important to establish visibility for the aquaporin field, the chance to meet and encourage hundreds of students and postdoctoral fellows was a most wonderful experience, since these young people represent the future of science.

Our family life accommodated my extended absences. We undertook special family wilderness camping trips and each child accompanied me on some of the international travels. However, missing birthday celebrations, anniversaries, and graduations is something I am not proud of. Nevertheless, all members of the family have done very well. Sara graduated from Colgate University and is an administrator at the University of Virginia in Charlottesville. Engaged to Jason Watson, we look forward to her wedding this coming summer. Claire graduated from Duke University and taught science to 8-year-old children in Italy. She is now studying landscape architecture at Harvard Graduate School of Design in Cambridge. Clarke earned his Eagle Scout Award and received numerous honors for art and photography. He is a student at Hampshire College in Amherst MA. Carly sometimes expresses dismay for being the only child left at home, but her interests in wilderness adventures and studio arts follow the interests of her siblings. Mary took graduate school courses in education and became a pre-school teacher. When we share observations, it is clear that her 3-year-old pupils bite each other more often than my 25-year-old graduate students do, but the difference is quantitative and not absolute.

"IT'S UNBELIEVABLE"

Prior to the acclaim resulting from the aquaporin studies, my greatest academic recognition was the 1991 Young Investigator Award from the Eastern Section of the American Federation of Clinical Research. The presentation at a small meeting in New York City was delightful. My children, however, concluded that I, at age 42, must be "the world's oldest young investigator." Although I never considered myself a physiologist or a nephrologist, scientists in those fields generously embraced our studies on aquaporins and showered our lab with accolades including the 1999 Homer Smith Award. The international recognition of our work was made shockingly real when I received an invitation to present the aquaporin story as part of the Jubilee Symposium in Stockholm preceding the 100th anniversary of the Nobel Prizes in December 2001.

The possibility that I might someday actually win a Nobel was raised by my father during his terminal illness. Not being superstitious, I chose to disregard the notion, but pleasant Nobel daydreams must occasionally enter the minds of many scientists. It became a vivid reality when the phone rang at

Agre lab on Oct. 8, 2003.

5:30 a.m. on October 8, 2003. The Royal Swedish Academy of Sciences and the Nobel Chemistry Committee awakened me with extraordinary news. A deluge of journalists descended upon our house, and only with difficulty was Mary able to contact my mother to share the news. Consistent with her modest Scandinavian farming origins, she responded calmly, "That's very nice, but tell Peter not to let it go to his head." When I arrived at the laboratory, celebrations had already begun. The telephone message system had filled up, but the excited voice of my Norwegian colleague was special. "Peter, Peter, we just heard the news. It's unbelievable!" Then, after a short pause, the voice returned. "I mean it's very believable, but it's wonderful!" Within the next few days, I was contacted by hundreds of scientific colleagues, relatives from the U.S. and Scandinavia, friends from childhood, classmates from grade school through medical school – many whom I had not seen in years. For me, the chance to renew these bonds is perhaps the best part of winning a Nobel Prize.

AQUAPORIN WATER CHANNELS

Nobel Lecture, December 8, 2003

by

PETER AGRE

The Johns Hopkins University School of Medicine, Baltimore, MD 21205, USA.

Thank you very much. I am humbled, I am delighted; I am honored. This is every scientist's dream: to give the Nobel Lecture in Stockholm. But I would not be honest if I did not tell you that I am having a little anxiety being on this platform. I have lectured a number of times in Sweden, and I thought I would share with you some events preceding a special lecture that I gave here a few years ago. Arriving at Arlanda Airport, I waited in line at the Pass Control behind a group of businessmen in suits with briefcases. I heard the first in line asked by the control officer to state the purpose of his visit to Sweden. When the man replied "business," the officer approved and stamped his passport. One at a time, each stepped forward and was asked the same thing; each answered "business" and was approved. Eventually it was my turn, and I was dressed in rumpled clothes after spending the night in the Economy Minus section of an SAS jetliner. The officer asked me the purpose of my visit, and I said "I am here to give the von Euler Lecture at Karolinska Institute." The officer immediately looked up, stared at me, and asked, "Are you nervous?" At that point I became intensely nervous and said "Yes, I am a little nervous." The officer looked up again and stated "Well, you should be!" So if the lecturers look a little nervous, the problem is at Arlanda.

INTRODUCTION

I am going to talk about aquaporin water channels. We have studied these proteins for several years, and we now understand that they explain how water crosses biological membranes. Water is commonly regarded as the "solvent of life," since our bodies are 70% water. All other vertebrates, invertebrates, microbes, and plants are also primarily water. The organization of water within biological compartments is fundamental to life, and the aquaporins serve as the plumbing systems for cells. Aquaporins explain how our brains secrete and absorb spinal fluid, how we can generate aqueous humor within our eyes, how we can secrete tears, saliva, sweat, and bile, and how our kidneys can concentrate urine so effectively. These proteins are fundamental to mammalian physiology, but they are also very important in the lives of microorganisms and plants.

I wish to discuss the background in order to give credit to the individuals who were in this field long before we joined the field. With the recognition of the lipid bilayer as the plasma membranes of cells back in the 1920's, it was correctly proposed that water could move through the membrane simply by diffusing through the lipid bilayer. The current view is that the lipid bilayer has a finite permeability for water, but, in addition, a set of proteins exists that we now refer to as "aquaporins." Their existence was suggested by a group of pioneers in the water transport field who preceded us by decades – people including Arthur K. Solomon in Boston, Alan Finkelstein in New York, Robert Macey in Berkeley, Gheorghe Benga in Romania, Guillermo Whittembury in Venezuela, Mario Parisi in Argentina – who by biophysical methods predicted that water channels must exist in certain cell types with high water permeability such as renal tubules, salivary glands, and red cells (reviewed by Finkelstein, 1987).

The difference between diffusional and channel-mediated water permeability is fairly distinct. Diffusion is a low capacity, bidirectional movement of water that occurs in all cell membranes, whereas the membranes of a subset of cells with aquaporin proteins have very high capacity for permeation by water. This permeability is selective, since water (H_2O) crosses through the membranes with almost no resistance, while acid, the hydronium ion (H_3O^+) does not permeate the proteins. This distinction is essential to life. As I will discuss later, our kidneys reabsorb 99% of the water from primary filtrate to prevent us from becoming dehydrated. If our kidneys reabsorbed water plus acid, we would develop systemic acidosis. The movement of water is directed by osmotic gradients, so aquaporins are not pumps or exchangers. They form a simple pore that allows water to rapidly pass through membranes by osmosis, a phenomenon that we all learned about in middle school. There are also other differences between diffusion and channel-mediated water transport. No inhibitors are known for simple diffusion. In contrast, mercurials were discovered by Robert Macey to inhibit water transport in red cells but water permeability was restored by treatment with reducing agents (Macey and Farmer, 1970). These observations predicted that water channels must be proteins with sulfhydryls and characteristically low Arrhenius activation energy.

A number of investigators using very logical approaches attempted to identify the water channel molecule; identification proved a very difficult problem. Water is ubiquitous and cannot be modified to include photo-activated side chains. Attempts by expression cloning were unsuccessful. Isotopic mercurials labeled several membrane proteins – the anion exchanger (band 3) by Solomon and a group of several proteins (band 4.5) by Benga. Nevertheless, none of the proteins were isolated, reconstituted, and shown to transport water (reviewed by Agre *et al.*, 1993a).

DISCOVERY OF AQP1

The field was essentially stuck, but following the well known scientific approach known as "sheer blind luck," we stumbled upon the protein that is the answer to the question: do water channels exist? Looking through our notebooks for the earliest studies that showed there was such a protein, I found an autoradiograph prepared by my first lab assistant, Andy Asimos. I was at that time a hematologist, and we were studying the Rh blood group antigens. We were attempting to raise antibodies in rabbits to the denatured partially purified Rh polypeptide. The rabbits vigorously produced antibodies that reacted with a polypeptide of approximately 30 kDa, and a series of higher order oligomers and glycosylated proteins. We were very excited about this, but we failed to recognize initially that our antibody did not react with the core Rh polypeptide that migrated at 32 kDa, seen clearly by silver staining of sodium dodecyl sulfate polyacryamide electrophoresis gels (SDS-PAGE). Instead, our antibodies reacted only with a 28 kDa polypeptide that we had mistakenly believed to be a breakdown product of the larger Rh polypeptide. Thus, the 28 kDa was an unrelated protein that merely contaminated our Rh preparations.

A number of features about the 28 kDa protein created a large amount of interest. Postdoctoral fellow Brad Denker and lab assistant Barbara Smith isolated the protein by a simple approach based upon detergent solubility. Silver staining of SDS-PAGE revealed a discrete band of 28 kDa in detergent insoluble extracts. No one had seen this protein before, since it failed to stain with the conventional protein stains such as Coomassie blue. The protein was then purified in large amounts from human red cell membranes (Denker *et al.*, 1988; Smith and Agre, 1991).

The 28 kDa protein was strikingly abundant. With approximately 200,000 copies per red cell, it was one of the major proteins in the membrane. In a way, this may be compared to the surprise one would have if, while driving along a road in northern Sweden, one came upon a town with 200,000 people not on the map – it certainly got our attention. Moreover, the protein had features suggesting that it was a tetrameric membrane-spanning protein – suggesting that it was a channel, but a channel for what? The purified protein also provided us the N-terminal amino acid sequence that we used for cDNA cloning. Using our antibody, we looked at several other tissues and found the protein is also strikingly abundant in human kidney. We observed staining over the apical and basolateral membranes of proximal renal tubules and the descending thin limb of the loops of Henle, but we were still frustrated by our failure to recognize what the protein's function might be.

In science, one should use all available resources to solve difficult problems. One of our most powerful resources is the insight of our colleagues. I spoke to a dozen or more well-known biochemists and physiologists about this protein, but none could predict what its function might be. My clinical mentor, John C. Parker (1935–1993) at the University of North Carolina at Chapel Hill, was the first to suggest to me that red cells and renal tubules were exceedingly permeable to water. He recommended that we consider a role in membrane water transport. While John did not live to see our later studies, he did live to see our initial discovery and we celebrated together.

Postdoctoral fellow Gregory Preston cloned the cDNA from an erythroid library (Preston and Agre, 1991). The coding region corresponded to a 269

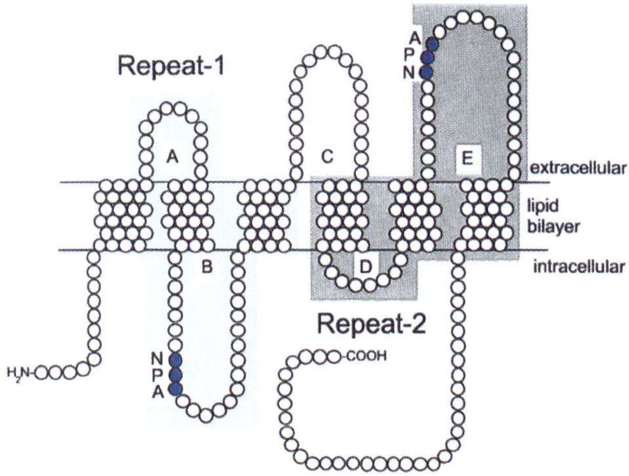

Figure 1. Membrane orientation of AQP1 predicted from primary amino acid sequence. Two tandem repeats each have three bilayer-spanning domains; the repeats are oriented 180° with respect to each other. The loops B and E each contain the conserved motif, Asn-Pro-Ala (NPA).

amino acid polypeptide, predicted by hydropathy analysis to have six bilayer-spanning domains. Interestingly, the amino terminal half (repeat-1) and the carboxy terminal half of the molecule (repeat-2) were genetically related – about 20% identical. Two loops B and E were more highly related to each other, and each contained the signature motif – asparagine, proline, alanine (NPA) [Fig. 1]. Examining the genetics database, we recognized several sequence-related DNAs from diverse sources: lens of cow eyes, brains of fruit flies, bacteria, and plants. Nevertheless, none was functionally defined.

These clues heightened our suspicion that the 28 kDa protein was a water transporter, so we tested for possible water transport function in collaboration with our colleague Bill Guggino at Johns Hopkins. We used oocytes from the frog *Xenopus laevis.* This convenient expression system was particularly useful, since frog oocytes have very low water permeability and survive even in fresh water ponds. Control oocytes were injected with water alone; test oocytes were injected with 2 ng of cRNA encoding our protein. After three days of protein synthesis, the oocytes appeared essentially identical. Then we stressed the oocytes by transferring them to distilled water, and an amazing difference was immediately apparent. Having exceedingly low water permeability, the control oocytes failed to swell. In contrast, the test oocytes were highly permeable to water and exploded like popcorn [Fig. 2] (Preston *et al.*, 1992). This result produced much celebration in the laboratory; the celebrations continue still. The protein was christened "aquaporin" and is now officially designated "AQP1," the first functionally defined water channel protein (Agre *et al.*, 1993b).

We confirmed the function of this protein by studying the purified AQP1

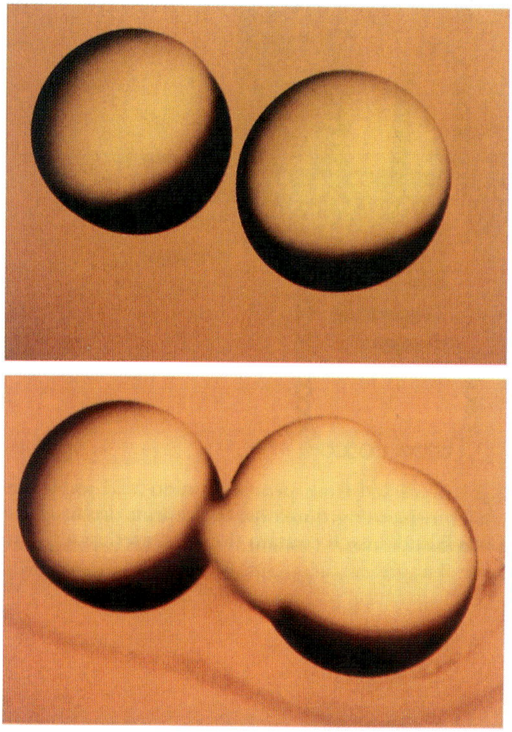

Figure 2. Functional expression of AQP1 water channels in *Xenopus laevis* oocytes. Control oocyte (left) was injected with water; AQP1 oocyte (right) was injected with cRNA. The oocytes were transferred to hypotonic buffer. After 30 seconds (top) the AQP1 oocyte has begun to swell; after 3 minutes (bottom), the AQP1 oocyte has exploded. Modified and reprinted from Science with permission (Preston *et al.*, 1992).

reconstituted into synthetic lipid vesicles of ~0.1 micron diameter prepared by our colleague Suresh Ambudkar at Johns Hopkins (Zeidel *et al.*, 1992). These simple membrane vesicles were examined by freeze fracture electron microscopy by our colleague Arvid Maunsbach, from the University of Aarhus. When lipid was reconstituted without protein, the membrane surfaces were smooth; however, membranes reconstituted with AQP1 contained many intramembraneous particles 0.01 micron diameter (Zeidel *et al.*, 1994). We tested the membranes for water permeability in collaboration with Mark Zeidel at Harvard Medical School. Using stopped flow transfer to hypertonic buffer, the simple liposomes shrank, reaching equilibrium in about one half second; this is believed to represent the baseline water permeability. When membranes reconstituted with AQP1 were examined, the shrinking occurred much more rapidly, reaching equilibrium in about 20 milliseconds. The channel-mediated flow of water was confirmed, since it was inhibited with mercurials. We calculated the Arrhenius activation energy (<5 kcal/mol), and we determined the unit permeability to be ~$3x10^9$ water molecules per subunit per second. Importantly, we attempted to measure proton permeation of AQP1, but despite massive water permeability, acid permeation was not detected. These studies verified that we had, in fact, isolated the long-sought water channel protein.

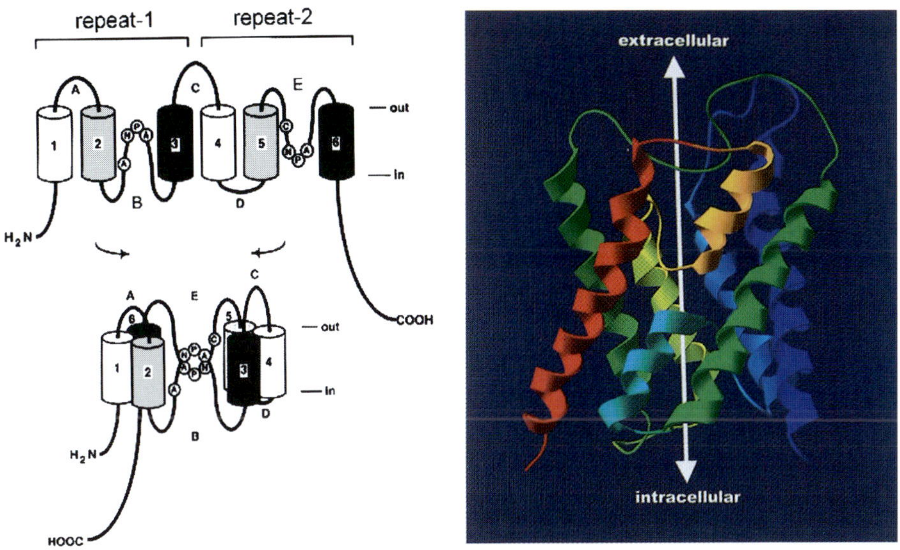

Figure 3. Hourglass model for membrane topology of AQP1 subunit.

Left panel – Schematic folding of loops B and E overlap within the lipid bilayer to form a single aqueous pathway.

Right panel – Ribbon model of three dimensional structure of AQP1 subunit confirms hourglass with single aqueous pathway. Modified and reprinted with permission from *Journal of Biological Chemistry* (Jung *et al.*, 1994b) and *Journal of Clinical Investigation* (Kozono *et al.*, 2002).

STRUCTURE OF AQP1

Subsequent efforts were devoted to identifying the mercurial inhibitory site predicted by the studies of Macey. Mercurials react with free sulfhydryls in the amino acid cysteine. Four cysteines are found in the AQP1 polypeptide, but only the residue in loop E (Cys-189 proximal to the second NPA motif) is inhibited by mercurials. We altered the AQP1 sequence by site-directed mutagenesis and expressed the recombinants in oocytes for water permeability studies. Mutation of this residue to serine (Cys-189-Ser) resulted in full water permeability without mercurial inhibition. When we then replaced the alanine in the corresponding position of loop B with a cysteine (Ala-73-Cys), the protein exhibited mercurial sensitive water permeability (Preston *et al.*, 1993). Substitutions elsewhere in the AQP1 failed to produce this behavior. This suggested to us that loops B and E in opposite parts of the molecule must somehow form the aqueous pore. The model that we concocted turned out to be schematically correct and was termed "the hourglass." The ancient timepiece allows sand to run from upper chamber to lower chamber; if inverted, the sand will flow in the opposite direction. Six bilayer spanning domains were predicted to surround a central domain containing loop B, dipping into the membrane from the cytoplasmic surface, and loop E, dipping into the membrane from the extracellular surface [Fig. 3 left and right]. The overlap of loops B and E was predicted to form a single aqueous pore down through the center of the molecule with the NPA motifs juxtaposed and the mercurial inhibitory site alongside (Jung *et al.*, 1994b). The AQP1 protein is tetrameric with a central pore in each subunit. Thus, AQP1 is structurally unlike ion channel proteins where four subunits surround a single central pore, as discussed by Rod MacKinnon in his lecture.

We then sought to establish the high resolution structure of AQP1 in collaboration with Andreas Engel and his group at the Biozentrum in Basel. We were later joined with Yoshinori Fujiyoshi and his group at Kyoto University. Human red cell AQP1 protein was purified by Barb Smith in our lab; Andreas's student Tom Walz reconstituted it into synthetic membranes at very high protein concentrations. Under these conditions, the AQP1 protein forms remarkably symmetrical arrays referred to as membrane crystals. By measuring the water permeability, we confirmed that the function was 100% retained, giving us confidence that the structure we deduced would be the biologically relevant structure (Walz *et al.*, 1994).

Using the techniques of cryoelectron microscopy and atomic force microscopy, our colleagues in Basel and Kyoto undertook studies that provided an electron density map of human AQP1 at 3.8 Å. Models were developed using constraints established by our site-directed mutagenesis studies (Murata *et al.*, 2000). When a single AQP1 subunit is viewed in cross-section in the middle of the membrane bilayer, a single aqueous pore is visible with hydrophobic residues lining most of the circumference of the wall, but the two highly conserved asparagines of the NPA motifs are aligned on the other wall. When viewed in longitudinal section, a narrow aqueous pore is visible with the two highly conserved asparagines. We faced the question of how a simple pore with no moving parts could allow permit rapid transit of water with no movement of protons. Exciting contributions by X-ray crystallographers in the groups of Robert Stroud, at UCSF, and Bing Jap, at Lawrence Berkeley Laboratory, solved the atomic structures of the *E. coli* glycerol facilitator GlpF (Fu *et al.*, 2000), and bovine red cell AQP1 (Sui *et al.*, 2001) at 2.2 Å resolution. These observations permitted molecular dynamics simulations by Bert de Groot and Helmut Grubmüller at the University of Göttingen (2001) and the group of Klaus Schulten at University of Illinois at Urbana-Champaign (Tajkhorshid *et al.*, 2002). We now understand how this protein facilitates the movement of water but not protons. If water were simply H_2O, it would be a gas, which it is in vapor, but water in bulk solution exists as $(H_2O)_n$ with hydrogen bonding between adjacent water molecules causing it to be a liquid. If water existed as a column of aligned water molecules in contact with each other, protons would rapidly jump along by a mechanism known as the Grotthus effect – also referred to as a "proton wire."

The hourglass structure of the AQP1 molecule has an extracellular vestibule and an internal vestibule where water is in bulk solution. These vestibules are separated by a span of approximately 20 Å that is so narrow that water moves through in single file while protons are reflected. Near the top of this span the channel reaches its narrowest constriction – 2.8 Å. Thus, the

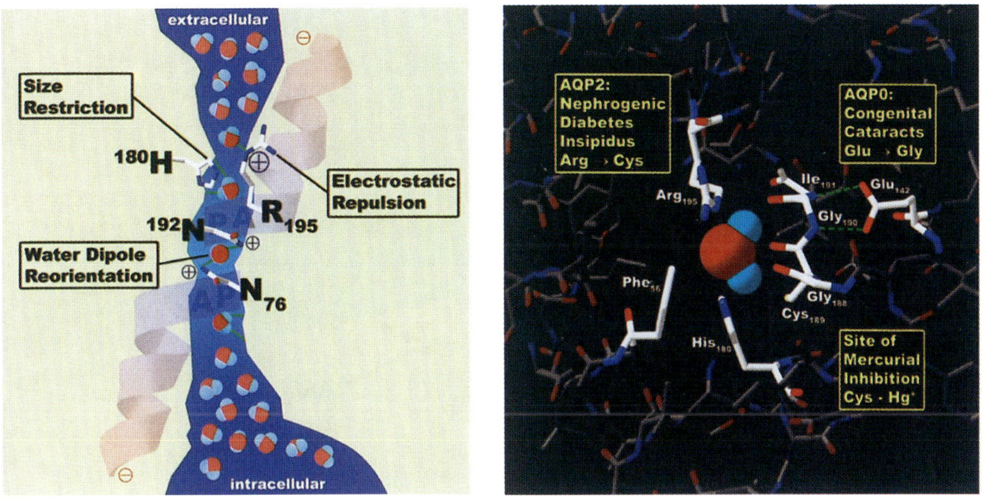

Figure 4. Functional representation for selective water flow through AQP1 subunit and residues involved in human disease.

Left panel – Schematic of sagittal cross-section of AQP1 reveals bulk water in extracellular and intracellular vestibules of hourglass. These are separated by a 20Å span where water passes in single file with transient interactions with pore-lining residues that prevent hydrogen bonding between water molecules (bold colors). Two structures are believed to prevent permeation by protons (H_3O^+): electrostatic repulsion is created by a fixed positive charge from pore-lining arginine (R195) at a 2.8Å narrowing in the channel; water dipole reorientation occurs from simultaneous hydrogen bonding of water molecule with side chains of two asparagines residues in NPA motifs (N192 and N76). Two partial positive charges at the center of the channel result from orientation of two non-membrane spanning alpha helices distal to the NPA motifs.

Right panel – Space filling representation of water molecule at cross-section of narrowest region of pore: AQP1 site of mercurial inhibition in pore-lining cysteine (Cys189), AQP2 mutation in nephrogenic diabetes insipidus (Arg to Cys); AQP0 mutation in congenital cataracts (Glu to Gly). Reprinted with permission from *Journal of Clinical Investigation* (Kozono *et al.*, 2002).

pore is just large enough to accommodate a single water molecule. At this site, the side chain of a perfectly conserved arginine following the NPA motif in loop E forms a fixed positive charge, and a conserved histidine on the other wall forms a partial positive charge – together they serve to repel protons. Further along, another barrier to protons exists where a single water molecule will transiently undergo a transient dipole reorientation as it simultaneously forms hydrogen bonds with the side chains of the two asparagines in the juxtaposed NPA motifs. In addition, the non-bilayer spanning alpha helices in the distal ends of loops B and E contribute partial positive charges that further serve to block proton conduction [Fig. 4 left] (reviewed by Kozono *et al.*, 2002). This also shows how mercurials inhibit water flow through AQP1. The side chain of Cys-189 lies along the pore, so if it is blocked by a mercurial, the channel will be occluded [Fig. 4 right].

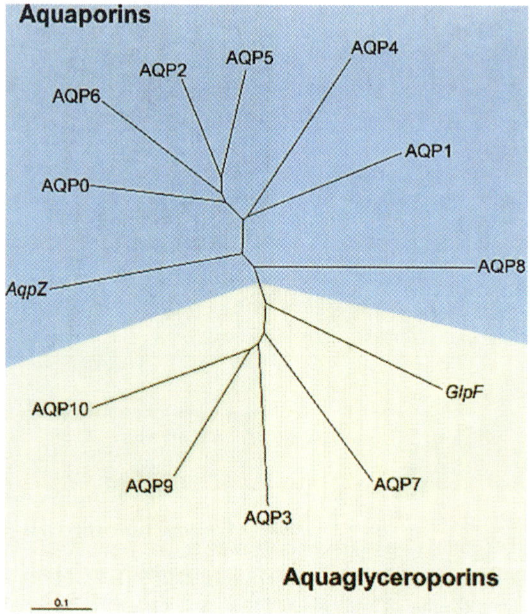

Figure 5. Human aquaporin gene family contains two subsets. Homologs freely permeated by water (classical aquaporins, blue) or water and glycerol (aquaglyceroporins, yellow) are represented. *E. coli* has one aquaporin (AqpZ) and one aquaglyceroporin (GlpF). Reprinted with permission from *Journal of Physiology* (Agre *et al.*, 2002).

THE AQUAPORIN AND AQUAGLYCEROPORIN PROTEIN FAMILY

While we were pursuing studies of AQP1, several other research groups from around the world became interested in what is now known to be a large family of related proteins. The combined efforts of these labs have led to the molecular identification of 12 mammalian aquaporin homologs, and several hundred related proteins have been recognized in other vertebrates as well as invertebrates, plants, and unicellular micro-organisms. The mammalian homologs may be loosely clustered into two subsets [Fig. 5]. The first is referred to as "classical aquaporins", since they were initially considered to be exclusive water pores. The second is referred to as "aquaglyceroporins", since they are permeated by water plus glycerol. Interestingly, *E. coli* has one member of each – AqpZ (Calamita *et al.*, 1995), and GlpF, isolated by other investigators much earlier. Together, the mammalian aquaporins and aquaglyceroporins are now known to contribute to multiple physiological processes that occur during our daily lives.

AQP1 PROTEINS IN KIDNEY

We were very fortunate to collaborate with Søren Nielsen from the University of Aarhus, a superb microscopist with particular expertise in kidney. Each of our two kidneys contains approximately one million nephron units working in concert to eliminate toxic wastes and provide the appropriate pH and ionic composition of our circulating blood. Every day, approximately 180 liters of plasma are filtered through the capillary tufts in the glomeruli. Water and certain solutes in this primary filtrate are then reabsorbed as the fluid makes its way through the tubules. The proximal convoluted tubules and descending thin limbs of the loop of Henle exhibit constitutively high water permeability and are responsible for the counter-current mechanism for renal concentration. This coincides exactly with the sites where AQP1 is expressed. The ascending segments are permeated by certain ions and small molecules but lack the AQP1 protein and exhibit low water permeability. Multiple nephron units empty into collecting ducts which exhibit low water permeability in the unstimulated state. Collecting ducts are known to become highly permeable to water when stimulated by the antidiuretic hormone, vasopressin, that is released from brain in response to dehydration.

Immunohistological analysis of rat kidney by Søren revealed extremely strong labeling of proximal tubules and descending thin limbs with our affinity purified anti-AQP1 (Nielsen *et al.*, 1993c). The apical brush border stains intensely; our analyses indicate that AQP1 comprises ~4% of the total brush border membranes. Immunogold electron microscopy showed that AQP1 is largely present at the apical brush border membrane surface, where water reabsorption is known to occur, but the protein is minimally present within the cell body [Fig. 6 left]. Thus it became clear that AQP1 is not actively trafficking from intracellular sites to the plasma membrane and back to intracellular sites. AQP1 is also abundant in the basolateral membranes of these epithelial cells. Thus water is reabsorbed in a transcellular route from the tubular lumen through the apical membrane of the tubular epithelium and out through the basolateral membrane into the interstitium where it is absorbed into capillaries and venules that are also rich in AQP1 protein [Fig. 6 right]. In normal circumstances, water does not flow between cells, since tight junctions form a barrier. The direction of the water flow through these epithelia, from primary filtrate back into the vascular space, is determined by small, standing osmotic gradients created by the solute transporters, and water follows passively.

AQP1-NULL HUMANS

Our student Chulso Moon in the Graduate Program in Human Genetics at Johns Hopkins was studying the structural gene encoding human AQP1 which we located on the short arm of human chromosome seven by fluorescence in situ hybridization (Moon *et al.*, 1993). We looked in Victor McKusick's atlas of human genetic linkages; our friend Colvin Redman at the New York Blood Center pointed out that this location coincides with the site of an interesting blood group system – the Colton antigens. Being a hematologist, I am very much interested in blood group antigens. While hundreds of blood group antigens are known, they represent multiple polymorphisms in approximately 25 different blood group systems, each system being a differ-

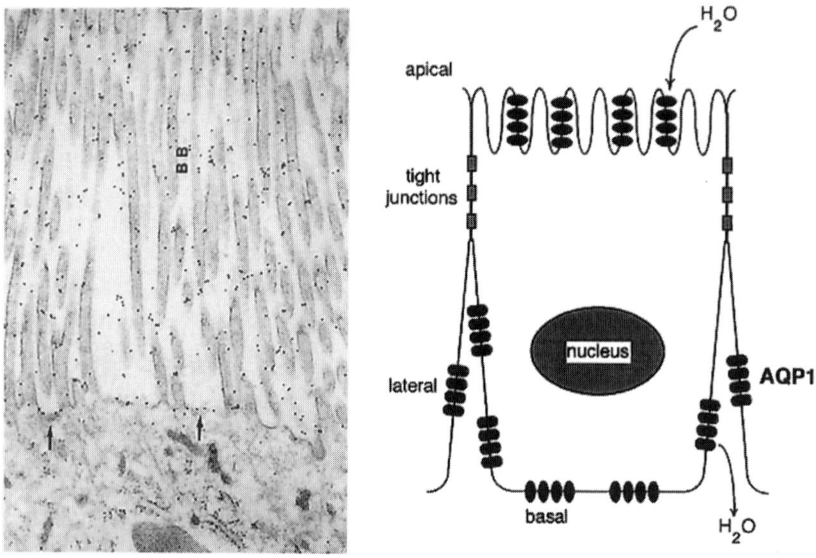

Figure 6. Subcellular localization of AQP1 in proximal tubules from rat kidney.
Left panel – Anti-AQP1 immunogold electron microscopy reveals labeling of apical brush border (BB) but not cell body. 25,000 x.
Right panel – Schematic representing transcellular passage of water: entry through AQP1 in apical brush border; exit through AQP1 in basolateral membranes. Reprinted with permission from *Journal of Cell Biology* (Nielsen *et al.*, 1993c).

ent genetic locus. So the coincidental localization of the *AQP1* gene and a human blood group antigen was a very strong clue that they correspond to the same molecule. The Co blood group antigens are not of large clinical significance, since most of us have the same Co blood type – approximately 90% of us are homozygous for Co^a, almost 10% are heterozygotes carrying both Co^a and Co^b, and only 0.3% are homozygous for Co^b. In collaboration with investigators from the International Blood Group Referencing Laboratory in Bristol UK, we found the Co antigens correspond to an alanine-valine polymorphism at residue 45 which resides in an extracellular location of the AQP1 protein (Smith *et al.*, 1994).

Surprisingly, in the entire world, the blood group referencing laboratories have identified only a handful of individuals who lack both Co antigens. We turned our attention to the Co-null individuals, and, with the help of John Moulds and Jean Pierre Cartron, we obtained DNA, blood, and urine from three different Co-null kindreds. The probands were all women who apparently during childbirth became sensitized to the Co antigens on the red cells of their babies. We hypothesized that these individuals carried mutations in their genes encoding, AQP1. We found that the probands from each kindred were homozygous for a unique disruption of the *AQP1* gene (Preston *et al.*, 1994). They do not express AQP1 in red cells, renal tubule cells, or presumably in choroid plexus, ciliary epithelium, or any other tissues where the protein is normally found – hence AQP1-null. The probands, however, did not describe any clinical illness. Their unique blood types make it impossible for them to receive transfusions from any other individual, so each has cryopreserved blood in local blood banks for emergencies or if a transfusion is needed during surgery.

Their lack of clinical symptoms suggested that AQP1 is either not important in humans or is important only in certain physiological states or clinical settings. Studies of mice bearing targeted disruption of the gene encoding AQP1 by Alan Verkman at UCSF and collaborating scientists at NIH indicated that AQP1 is important for renal concentration (Schnermann *et al.*, 1998; Chou *et al.*, 1999). Our colleague Landon King at Johns Hopkins designed a series of clinical evaluations to probe for the importance of the human AQP1 protein. Two AQP1-null individuals were studied during inpatient evaluations at the General Clinical Research Center at Johns Hopkins Hospital. Water permeability is clearly reduced in AQP1-null red cells, but all baseline studies were found to be normal. Then Landon performed a simple test – he thirsted the patients. Every night during our sleep we become a little dehydrated; upon arising in the morning that first taste of orange juice is so delicious. To protect against dehydration, our kidneys respond by maximally concentrating our urine, and all normal individuals concentrate to approximately 1,000 mosM. The AQP1-null individuals were found to have urine concentrations reaching no higher than 450 mosM after overnight thirsting, and prolonged thirsting to 24 hours followed by administration of the antidiuretic hormone, vasopressin, and infusion of hypertonic sodium chloride yielded no further response (King *et al.*, 2001). In modern life with free access to fluids and air conditioning, the need to concentrate above 450 mosM is not essential; however, the AQP1-null individuals would be at a profound disadvantage if they were to reside in an environment with restricted access to fluids. Certainly during human evolution, our progenitors did not always have free access to water. The patients are described as having a mild form of nephrogenic diabetes insipidus that is usually subclinical but physiologically distinct.

Søren Nielsen also established that AQP1 protein is extremely abundant in human capillaries in both the luminal and abluminal endothelial membranes (Nielsen *et al.*, 1993b; King *et al.*, 1996). Landon King in collaboration with Robert Brown measured the capillary fluid permeability in AQP1-null humans (King *et al.*, 2002). Using a technique developed by Wayne Mitzner and Elias Zerhouni at Johns Hopkins employing high resolution computerized tomography of lung, they performed scans at baseline. Focusing upon small bronchioles of 2–3 mm diameter, the cross sectional areas of the adjacent venules were measured and the wall thicknesses determined. The scans were repeated after rapid infusion of three liters of saline. All normal individuals and AQP1-null individuals exhibited comparable evidence of vascular engorgement following saline infusions. We then looked at the wall thickness of the small breathing tubules and found that all normal individuals developed an increase in the wall thickness – representing release of fluid from the venules and accumulation in the soft tissues surrounding the bronchioles, a situation referred to by clinicians as incipient pulmonary edema. The interstitial edema appears in the peribronchiolar soft tissues in all normal individuals, but the AQP1-null individuals are resistant to this, indicating to us that they have a very significant reduction in vascular water permeability. Whether this may be beneficial or deleterious will depend on the clinical setting. At the time of birth, lungs turn from a secretory organ to an absorptive organ. We hypothesize that the inability of newborn infants to move fluid from the pulmonary intersitium into the vascular space may be very harmful, possibly explaining why the AQP1-null state is such a rare phenotype.

OTHER AQUAPORINS – AQP2 IN COLLECTING DUCT

Within weeks of our original studies, other groups began to search for other aquaporins. Particular interest in renal collecting ducts led Sei Sasaki and his colleagues at Tokyo Medical and Dental University to isolate a cDNA encoding AQP2 (Fushimi *et al.*, 1993). Søren Nielsen and others established that AQP2 is primarily expressed in renal collecting ducts where regulated water permeability is known to occur (Nielsen *et al.*, 1993a). Although my laboratory was not involved in the studies of AQP2, I wish to share exciting information published by several other investigators because I believe this has very large potential clinical significance.

In a definitive study, Søren Nielsen and Mark Knepper from NIH looked at AQP2 in collecting ducts isolated from rats (Nielsen *et al.*, 1995). The ducts were perfused while water permeability was measured. Some of the specimens were then sectioned and stained with anti-AQP2, and by immunogold electron microscropy they found that very little of the AQP2 protein was found in the apical plasma membrane of collecting duct principal cells, but most of the protein appeared to be in membranes of intracellular vesicles [Fig. 7 upper left]. When the isolated perfused tubules were pulsed with physiological concentrations of vasopressin (the physiological antidiuretic hormone released from brain), the water permeability was increased approximately five-fold, coinciding with the expected physiological increase. When examined by immunogold electron microscopy, the AQP2 protein had relocated to the apical plasma membrane [Fig. 7 lower left]. When vasopressin was removed, the AQP2 protein was re-internalized and the collecting duct water permeability decreased.

These observations closely reflect the known variations in renal collecting duct water permeability occurring in our daily lives. On Friday afternoons, our students celebrate the end of the week at the departmental beer party (TGIF = Thank God it's Friday). A student drinking 2 liters of beer has excess fluid in his system; the alcohol inhibits release of vasopressin from brain. This causes AQP2 to remain in intracellular vesicles, so the apical membranes of renal collecting duct principal cells are not able to reabsorb water from the filtrate – the diuretic state. The first place the student visits following TGIF is the toilet where he empties his bladder of a large volume of dilute urine. The next morning, the student lies in bed in a moderately dehydrated state with high vasopressin levels in his circulation. This causes AQP2 to relocalize to

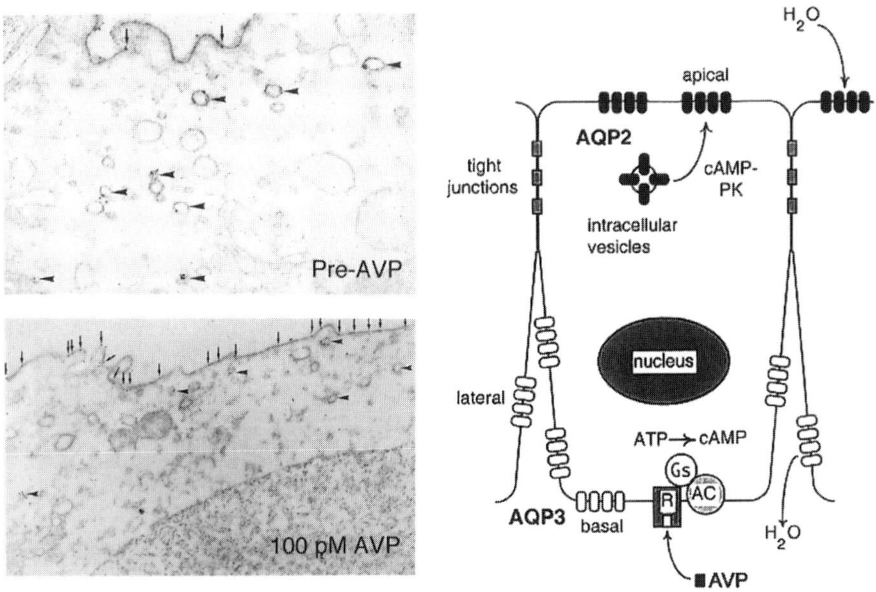

Figure 7. Subcellular localization of AQP2 in collecting duct principal cells from rat kidney. Left panel – Anti-AQP2 immunogold electron microscopy of isolated collecting duct from rat kidney. Before vasopressin (AVP) stimulation, most labeling occurs over intracellular vesicles (top). 25,000 x. After stimulation with 100 pM AVP, most labeling appears at apical membrane (bottom). 15,000 x.

Right panel – Schematic representing AVP-dependent exocytosis of AQP2 to apical membrane of principal cell providing transcellular passage of water: entry through AQP2 in apical membrane; exit through AQP3 (or AQP4) in basolateral membranes. Modified and reprinted with permission from *Proceedings of the National Academy of Sciences of the USA* (Nielsen *et al.*, 1995) and *Kidney International* (Nielsen and Agre, 1995).

the apical plasma membranes, facilitating maximal reabsorption of water from the urine. Thus, when he arises, his urine is maximally concentrated – the antidiuretic state. These cycles may be repeated multiple times even within a single day.

In summary, vasopressin binds to V2 receptors at the basolateral membrane of renal collecting duct causing activation of a G-coupled adenylyl cyclase cascade resulting in protein kinase A phosphorylation of residue 256 on the C-terminus of AQP2. The protein is then inserted into the apical plasma membrane, as established by Nielsen and Knepper, the groups of Deen in Nijmegen, Dennis Brown in Boston, Sasaki in Tokyo, Anita Aperia in Stockholm, and other groups. Water in the primary urine enters the apical membrane through AQP2 and exits from the basolateral membranes through other aquaporins, AQP3 or AQP4, following a transcellular pathway from the lumen into the vascular space (reviewed by Nielsen *et al.*, 2002).

Inherited defects in AQP2 cause very rare but severe clinical problems – such has been the finding of Carel van Os, Peter Deen, and their colleagues in Nijmegen the Netherlands (Deen *et al.*, 1994). These patients are profoundly affected, making up to 20 liters of urine per day in some cases.

Acquired defects in the level of AQP2 expression are commonly found in clinical medicine. When AQP2 levels are too high, fluid retention occurs, as shown by the Nielsen and Knepper group as well as by Robert Schrier and his colleagues at the University of Colorado. Presumably this is happening in patients sustaining myocardial infarctions; such patients often suffer fluid retention causing inadequate pulmonary function. Similarly, some women during pregnancy experience pathological retention of fluid – causing hypertension and neurological dysfunction. The opposite situation results when AQP2 levels are too low, resulting in inadequate concentration of urine. This result is found after reversal of urinary obstruction, as seen in the work of Jørgen Frøkiaer in Aarhus, and in patients with nocturnal enuresis, as seen in the work of Ivana Valenti in Bari. Thus, AQP2 is very important in clinical medicine (reviewed by Nielsen *et al.*, 2002).

AQP6 IN ACID SECRETORY CELLS

At this point we thought we knew a lot about aquaporins, but then we were surprised. AQP6, a protein genetically closest to AQP2, was found to be a very poor water channel and is never present in the plasma membrane. Double label immungold electron microscopy by Søren Nielsen and Tae-Hwan Kwon showed that AQP6 co-localizes with H^+ -ATPase in intracellular vesicles of alpha intercalated cells in renal collecting duct. Upon stimulation of acid secretion, H^+ -ATPase becomes exocytosed to the plasma membrane, but AQP6 is fully retained in within the cell. A second surprise was that AQP6 is a very good anion channel with particularly high permeation by nitrate when activated by low pH (Yasui *et al.*, 1999). These cells control acid secretion, so while AQP6 is clearly a genetic member of the aquaporin family, it apparently has a very different physiological function, as proposed by our colleague Masato Yasui at Johns Hopkins.

AQP0 IN LENS

Another member of the aquaporin family is AQP0, expressed only in fiber cells of lens. First recognized more than 30 years ago and labeled MIP (major intrinsic protein), the physiological function was unknown. AQP0 has a fairly low water permeability, but the protein may have a second function as a cell-to-cell adhesion molecule, as proposed by Andreas Engel and his group.

Shomi Bhattacharya, from Moorefield's Eye Hospital in London identified two large kindreds with congenital cataracts of dominant inheritance with complete penetrance. While cataracts are fairly common among elderly individuals, they are very rare in small children. Each family bore a missense mutation in a structurally important residue, and the physical nature of the opacities was quite different (Berry *et al.*, 2000). Members of the family with an arginine replacing an essential threonine (Thr-138-Arg) have numerous opacities throughout the lens. Members of the second family with a glycine replacing an essential glutamate (Glu-134-Gly) have a single lamellar opacity in the center of the lens. This distribution corresponds to the perimeter of the lens at the time of birth, indicating that the mutation is deleterious only at birth (Francis *et al.*, 2000). This example illustrates how human biology is so much more complicated than represented by simple gene disruption studies performed on mice. Lens fiber cells survive for decades, so it is plausible that subtle polymorphisms in the AQP0 protein may be a risk factor for development of cataracts as individuals grow older.

AQP4 IN BRAIN

This protein is expressed in multiple tissues, including brain (Jung *et al.*, 1994a). Unlike peripheral capillaries that are quite leaky, capillaries in brain are known to be much tighter due to astroglial endfeet surrounding capillaries in brain. In studies undertaken with Ole Petter Ottersen and his group at the University of Oslo, and including Søren Nielsen, the protein was localized by immunogold electron microscopy to reside at the blood-brain barrier, the site where vascular space and brain parenchyma are juxtaposed (Nielsen *et al.*, 1997; Nagelhus *et al.*, 1998). Brain capillaries are surrounded by a basement membrane that is surrounded by astroglial endfeet. The AQP4 resides almost entirely in the endfeet membranes directly in contact with the basement membrane and is virtually absent in other astroglial membranes [Fig. 8A, B]. Similarly, at the surface of the brain, AQP4 is abundant at the brain surface where liquid and brain are juxtaposed [Fig. 8C]. These locations strongly predicted that AQP4 would be important in movement of water from blood or cerebrospinal fluid into brain as well as in the reverse direction. AQP4 is extremely heavily expressed at these locations forming microcrystalline forms referred to as "square arrays" with size of the assembly determined by the N-terminus of AQP4 (Furman *et al.*, 2003). Other workers, including Antonio Frigeri in Bari, also documented that movements of water in brain and skeletal muscle may be of pathophysiological significance (2001).

Our student John Neely recognized a protein binding motif in the C-terminus of AQP4 predicting that it would associate with the dystrophin associated proteins. In collaboration with Marv Adams and Stan Froehner in Seattle as well as with our colleagues in Oslo, we evaluated the localization of AQP4 in mice bearing targeted disruption in alpha-syntrophin, a member of the dystrophin-associated proteins (Neely *et al.*, 2001). By immunogold electron microscopy, AQP4 was found to be grossly mislocalized in the syn-null animals. Interestingly, when stressed with brain injury resulting from middle cerebral artery occlusion, the control animals did much worse, having significantly larger areas of infarction and edema, and the syn-null animals did much better (Amiry-Moghaddam *et al.*, 2003). These studies suggest that if AQP4 could be selectively inhibited, patients sustaining brain injuries might be protected from some of the tragic consequences of brain swelling. Protection against hyponatremia-induced brain swelling was observed by Søren Nielsen and coworkers (Vajda *et al.*, 2002). Geoff Manley and colleagues independently established the same conclusions with AQP4-null mice (2000).

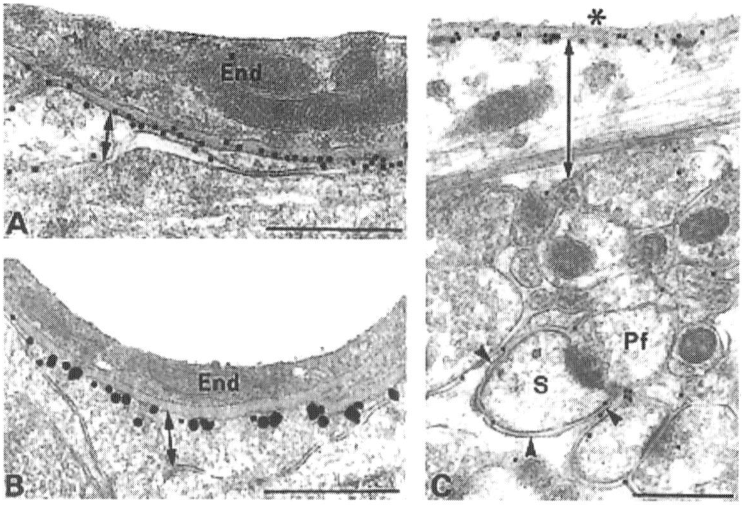

Figure 8. Polarized expression of AQP4 in rat brain.

Panels A and B – Anti-AQP4 immunogold electron microscopy reveals AQP4 in perivascular membranes of astroglial endfeet but not in membranes facing neuropil.

Panel C – Brain surface exhibits anti-AQP4-immunogold labeling of pia. Scale bars: 0.5 μm. Reprinted from *Journal of Neuroscience* with permission (Nielsen *et al.*, 1997).

AQP5 IN SECRETORY GLANDS

Cloned from submandibular salivary glands, AQP5 resides in the apical membrane of acini in salivary, lacrimal, sweat, and other secretory glands (Raina *et al.*, 1995; King *et al.*, 1997). This is the last membrane water crosses during the generation of saliva, tears, and other secretions. It is known that these glands facilitate the release of large amounts of fluid. For example, any individual fasted for several hours will secrete massive amounts of saliva when he senses food is available.

AQP5 may be important in certain clinical disorders. In collaboration with Kazuo Tsubota from Tokyo and Chris Delporte from Brussels, we examined AQP5 distribution in gland biopsies from a small series of patients with Sjögren's Syndrome. This disorder affecting women at mid-life causes dry eye and dry mouth, although the precise biological cause of the disorder is unknown. Using our anti-AQP5 immunohistochemical labeling, the normal apical staining is lost, and the protein appears to be mislocalized throughout the cytoplasm in the Sjögren's biopsies (Tsubota *et al.*, 2001; Steinfeld *et al.*, 2001). Other investigators led by Roland Jonsson in Bergen did not find AQP5 mislocalization in a larger series of Sjögren's patients; however, they noted decreased expression of AQP1 in the myoepithelia surrounding glandular acini (Beroukas *et al.*, 2002). Thus, while Sjögren's may be heterogeneous, aquaporins have been implicated in these two series of patients.

AQP5 is also abundant in sweat glands. In another collaborative study with Søren Nielsen and his group, qualitative analysis of sweating was compared in wild type mice and mice bearing targeted disruption of the the gene encoding AQP5 prepared by Carissa Krane and Anil Menon at the University of Cincinnati. While the morphology of the sweat glands was preserved, the release of amylase in sweat after pilocarpine injection was found to be greatly diminished, indicating that AQP5 must be important for the ability to sweat (Nejsum *et al.*, 2002).

AQUAGLYCEROPORINS AQP7 AND AQP9

AQP7 is expressed in adipocytes. During fasting, energy stored as triglycerides is provided by release of fatty acids, exported by special fatty acid transporters, and glycerol, exported through AQP7. The preferred gluconeogenic substrate, glycerol, passes through AQP9 in the hepatocyte plasma membranes. Glycerol is converted to glucose within the hepatocytes. In the fasted state, AQP9 expression is greatly increased, as demonstrated in the work of Jennifer Carbrey and Dan Gorelick in our group. Multiple research groups have reached the same conclusion that these two proteins, AQP7 and AQP9, form an axis for energy transfer, together maintaining blood glucose levels during starvation (Kuriyama *et al.*, 2002; Carbrey *et al.*, 2003).

AQP7 and AQP9 have a very curious capacity to transport certain heavy metals, a discovery made by Barry Rosen and his colleagues at Wayne State University (Liu *et al.*, 2002). Measurement of arsenic trioxide uptake was performed with yeast expressing AQP7 or AQP9 as well as in oocytes expressing the proteins. Arsenite is uncharged at neutral pH and is very toxic. Clinical features of greatest concern in arsenic poisoning include hepatocellular damage and hepatocellular carcinoma. The World Health Organization has established that an alarming epidemic of arsenic poisoning is now occurring in the Ganges Delta because of arsenic contamination of the groundwater. Since the surface water is infested with *Vibrio cholera*, tube wells were installed, but heavy metal determinations were apparently not performed. While our work explains why these unfortunate people are vulnerable to hepatocellular damage, it does not provide a useful antidote. Prevention of this poisoning requires access to pure drinking water, a basic right deserved by all persons on earth.

NON-HUMAN AQUAPORINS

Aquaporins are also important in plants. A relative of the mustard plant, *Arabidopsis thaliana*, is frequently studied by plant physiologists (also known as "the green people"). Ralf Kaldenhoff at the University of Würzburg demonstrated this dramatically by comparing the gross morphology of a parental Arabidopsis to a plant modified to express the rootlet aquaporin group PIP1b at only 20% of the normal levels. Both plants exhibit similar foliage, but to maintain the same stem turgor, the modified plant sends out more greatly arborized rootlets [Fig. 9] (Kaldenhoff *et al.*, 1998). Thus, the plants are keenly aware of water balance. *Arabidopsis thaliana* has at least 35 different aquaporin genes, and these are presumed to participate in different physiological

Figure 9. Arabidopsis thaliana rootlet aquaporins and stem turgor. Antisense RNA manipulation reduces expression of aquaporin PIP1b proteins by 80%. To maintain stem turgor at normal levels, the Aqp PIP1b deficient plant increases rootlet surface area by arborization. Reprinted with permission from *The Plant Journal* (Kaldenhoff *et al.*, 1998).

processes, including photosynthesis with carbon dioxide passing through leaf aquaporins (Uehlein *et al.*, 2003) and restriction of rootlet water uptake regulated by low pH (Tournaire-Roux *et al.*, 2003). These observations may be relevant to human aquaporins, since similar observations have been reported for AQP1 (Nakhoul *et al.*, 1998) and AQP3 (Zeuthen and Klaerke, 1999).

Aquaporins are also expressed in all forms of micro-organisms. *Escherichia coli* has two – AqpZ, a classic aquaporin, and GlpF, an aquaglyceroporin. The structures of the narrowest section of the channels are somewhat different – narrower and more hydrophilic in aquaporins than in aquaglyceroporins. Archaea, the third kingdom of life, also carry a gene encoding AqpM – homologous to aquaporins and aquaglyceroporins. AqpM may not exist for water transport but may facilitate hydrogen sulfide transport (Kozono *et al.*, 2003). Whatever the function, their presence in all three kingdoms of life reinforces the fundamental importance of this family of proteins. *Sacharomyces cerevisiae* have two aquaporin genes and two aquaglyceroporin genes. Exciting studies by Stefan Hohmann at the University of Göteborg revealed interesting functions – yeast aquaporins confer resistance to freezing, whereas the aquaglyceroporin FPS1 protects the organisms against hypo-osmotic shock (reviewed by Hohmann, 2002).

CONCLUSION

Discovery of the aquaporins has provided the explanation for selective transport of water across the plasma membranes of cells without allowing protons to also cross the membrane. Some homologs have special permeation properties for glycerol, nitrate, or arsenite. The structural models provide remarkable insight into the biophysical functions. Aquaporins have been implicated in multiple clinical disorders. Aquaporins are clearly involved in some forms of renal vascular diseases, including nephrogenic diabetes insipidus. Roles in defense against thermal stress are expected. Last summer during the heat wave in France, 14,000 people died when the temperature reached 40 °C. Many of these individuals were older, and it is likely that they manifested age-dependent diminution of physiological processes involving aquaporins – concentration of urine, sweating, and thirst mechanisms. While these deficiencies may be unimportant during normal life, they may cause severe problems when individuals are thermally stressed. Aquaporins have also been associated with problems of brain edema and loss of vision. Aquaglyceroporins are involved in the defense against starvation. These proteins are expressed throughout the natural world.

A word for the students from Santiago Ramón y Cajal, the father of neuroscience, "In summary, there are no small problems. Problems that appear small are large problems that are not understood" (Ramón y Cajal, 1897). Certainly the issue of water permeation of membranes was initially assumed to be a small problem related to diffusion. Instead it has turned out to be a complex problem with large physiological and pathophysiological significance.

ACKNOWLEDGMENTS

I thank the past and present members of my laboratory and the many individuals with whom we have collaborated in these studies. I particularly wish to thank my former chairmen, Victor McKusick, Department of Medicine, and M. Daniel Lane, Department of Biological Chemistry, for hiring me onto their faculty. I dedicate this lecture to the memory of John C. Parker, a friend and generous mentor.

REFERENCES

Agre, P., Preston, G. M., Smith, B. L., Jung, J. S., Raina, S., Moon, C., Guggino, W. B., & Nielsen, S. (1993a). Aquaporin CHIP: the archetypal molecular water channel. *American Journal of Physiology – Renal Physiology* **265**, F463–F476.

Agre, P., Sasaki, S., & Chrispeels J. (1993b). Aquaporins: a family of water channel proteins. *American Journal of Physiology – Renal Physiology* **265**, F461.

Agre, P., King, L. S., Yasui, M., Guggino, W. B., Ottersen, O. P., Fujiyoshi, Y., Engel, A., and Nielsen, S. (2002). Aquaporin water channels – From atomic struture to clinical medicine. *Journal of Physiology* **542,** 3–16.

Amiry-Moghaddam, M., Otsuka, T., Hurn, P. D., Traystman, R. J., Finn-Mogens, H., Froehner, S. C., Adams, M. E., Neely, J. D., Agre, P., Ottersen, O. P., Bhardwaj, A. (2003) An alpha-syntropin-dependent pool of AQP4 in astroglial end-feet confers bidirectional water flow between blood and brain. *Proceedings of the National Academy of Sciences of the USA* **100**, 2106–2111.

Beroukas, D., Hiscock, J., Gannon, B. J., Jonsson, R., Gordon, T. P., & Waterman, S. A. (2002). Selective down-regulation of aquaporin-1 in salivary glands in primary Sjögren's syndrome. *Laboratory Investigation* **82**, 1547–1552.

Berry, V., Francis, P., Kaushal, S., Moore, A., & Bhattacharya, S. (2000). Missense mutations in MIP underlie autosomal dominant 'polymorphic' and lamellar cataracts linked to 12q. *Nature Genetics* **25**, 15–17.

Calamita, G., Bishai, W. R., Preston, G. M., Guggino, W. B., & Agre, P. (1995). Molecular cloning and characterization of AqpZ, a water channel from *Escherichia coli*. *Journal of Biological Chemistry* **270**, 29063–29066.

Carbrey, J. M., Gorelick-Feldman, D. A., Kozono, D., Praetorius, J., Nielsen, S., & Agre, P. (2003). Aquaglyceroporin AQP9: Solute permeation and metabolic control of expression in liver. *Proceedings of the National Academy of Sciences of the USA* **100**, 2945–2950.

Chou, C. L., Knepper, M. A., van Hoek, A. N., Brown, D., Yang, B., Ma, T., & Verkman, A. S. (1999). Reduced water permeability and altered ultrastructure in thin descending limb of Henle in aquaporin-1 null mice. *Journal of Clinical Investigation* **103**, 491–496.

de Groot, B. L. & Grubmuller, H. (2001). Water permeation across biological membranes: mechanism and dynamics of aquaporin-1 and GlpF. *Science* **294**, 2353–2357.

Deen, P. M., Verdijk, M. A., Knoers, N. V., Wieringa, B., Monnens, L. A., van Os, C. H., & Van Oost, B. A. (1994). Requirement of human renal water channel aquaporin-2 for vasopressin- dependent concentration of urine. *Science* **264**, 92–95.

Denker, B. M., Smith, B. L., Kuhajda, F. P., & Agre, P. (1988). Identification, purification, and partial characterization of a novel Mr 28,000 integral membrane protein from erythrocytes and renal tubules. *Journal of Biological Chemistry* **263**, 15634–15642.

Finkelstein, A. (1987). Water movement through lipid bilayers, pores, and plasma membranes: Theory and reality. New York: Wiley.

Francis, P., Chung, J. J., Yasui, M., Berry, V., Moore, A., Wyatt, M. K., Wistow, G., Bhattacharya, S. S., & Agre, P. (2000). Functional impairment of lens aquaporin in two families with dominantly inherited cataracts. *Human Molecular Genetics* **9**, 2329–2334.

Frigeri, A., Nicchia, G. P., Nico, B., Quondamatteo, F., Herken, R., Roncali, L., & Svelto, M. (2001). Aquaporin-4 deficiency in skeletal muscle and brain of dystrophic mdx mice. *FASEB Journal.* **15**, 90–98.

Fu, D., Libson, A., Miercke, L. J., Weitzman, C., Nollert, P., Krucinski, J., & Stroud, R. M. (2000). Structure of a glycerol-conducting channel and the basis for its selectivity. *Science* **290**, 481–486.

Furman, C. S., Gorelick-Feldman, D. A., Davidson, K. G. V., Yasumura, T., Neely, J. D., Agre, P., & Rash, J. E. (2003). Aquaporin-4 square array assembly: opposing actions of M1 and M23 isoforms. *Proceedings of the National Academy of Sciences of the USA* **100**, 13609–13614.

Fushimi, K., Uchida, S., Hara, Y., Hirata, Y., Marumo, F., & Sasaki, S. (1993). Cloning and expression of apical membrane water channel of rat kidney collecting tubule. *Nature* **361**, 549–552.

Hohmann, S. (2002). Osmotic stress signaling and osmoadaptation in yeasts. *Microbiology and Molecular Biology Reviews* **66**, 300–372.

Jung, J. S., Bhat, R. V., Preston, G. M., Guggino, W. B., Baraban, J. M., & Agre, P. (1994a). Molecular characterization of an aquaporin cDNA from brain: candidate osmoreceptor and regulator of water balance. *Proceedings of the National Academy of Sciences of the USA* **91**, 13052–13056.

Jung, J. S., Preston, G. M., Smith, B. L., Guggino, W. B., & Agre, P. (1994b). Molecular structure of the water channel through aquaporin CHIP. The hourglass model. *Journal of Biological Chemistry* **269**, 14648–14654.

Kaldenhoff, R., Grote, K., Zhu, J. J., Zimmermann, U. (1998). Significance of plasma lemma aquaporins for water-transport in *Arabidopsis thaliana*. *Plant Journal* **14**, 121–128.

King, L. S., Nielsen, S., & Agre, P. (1996). Aquaporin-1 water channel protein in lung: ontogeny, steroid-induced expression, and distribution in rat. *Journal of Clinical Investigation* **97**, 2183–2191.

King, L. S., Nielsen, S., & Agre, P. (1997). Aquaporins in complex tissues. I. Developmental patterns in respiratory and glandular tissues of rat. *American Journal of Physiology – Cell Physiology* **273**, C1541–C1548.

King, L. S., Choi, M., Fernandez, P. C., Cartron, J. P., & Agre, P. (2001). Defective urinary-concentrating ability due to a complete deficiency of aquaporin-1. *New England Journal of Medicine* **345**, 175–179.

King, L. S., Nielsen, S., Agre, P., & Brown, R. H. (2002). Decreased pulmonary vascular permeability in aquaporin-1-null humans. *Proceedings of the National Academy of Sciences of the USA* **99**, 1059–1063.

Kozono, D., Yasui, M., King, L. S., & Agre, P. (2002). Aquaporin water channels: atomic structure and molecular dynamics meet clinical medicine. *Journal of Clinical Investigation* **109**, 1395–1399.

Kozono, D., Ding, X., Iwasaki, I., Meng, X., Kamagata, Y., Agre, P., & Kitagawa, Y. (2003). Functional expression and characterization of an archaeal aquaporin: AqpM from *Methanothermobacter marburgensis. Journal of Biological Chemistry* **278**, 10649–10656.

Kuriyama, H., Shimomura, H., Kishida, K., Kondo, H., Furuyama, N., Nishizawa, H., Maeda, N., Matsuda, M., Nagaretani, H., Kihara, S., Nakamura, T., Tochino, Y., Funahashi, T., Matsuzawa, Y. (2002). Coordinated regulation of fat-specific and liver-specific glycerol channels, aquaporin adipose, and aquaporin 9. *Diabetes* **51**, 2915–2921.

Liu, Z., Shen, J., Carbrey, J.M., Mukhopadhyay, R., Agre, P., & Rosen, B. (2002). Arsenite transport by mammalian aquaglyceroporins AQP7 and AQP9. *Proceedings of the National Academy of Sciences of the USA* **99,** 6053–6058.

Macey, R. I. & Farmer, R. E. L. (1970)Inhibition of water and solute permeability in human red cells. *Biochemica Biophysica Acta* **211**, 104–106.

Manley, G. T., Fujimura, M., Ma, T., Noshita, N., Filiz, F., Bollen, A. W., Chan, P., & Verkman, A. S. (2000). Aquaporin-4 deletion in mice reduces brain edema after acute water intoxication and ischemic stroke. *Nature Medicine* **6**, 159–163.

Moon, C., Preston, G. M., Griffin, C. A., Jabs, E. W., & Agre, P. (1993). The human aquaporin-CHIP gene. Structure, organization, and chromosomal localization. *Journal of Biological Chemistry* **268**, 15772–15778.

Murata, K., Mitsuoka, K., Hirai, T., Walz, T., Agre, P., Heymann, J. B., Engel, A., & Fujiyoshi, Y. (2000). Structural determinants of water permeation through aquaporin-1. *Nature* **407**, 599–605.

Nagelhus, E. A., Veruki, M. L., Torp, R., Haug, F. M., Laake, J. H., Nielsen, S., Agre, P., & Ottersen, O. P. (1998). Aquaporin-4 water channel protein in the rat retina and optic nerve: polarized expression in Muller cells and fibrous astrocytes. *Journal of Neuroscience* **18**, 2506–2519.

Nakhoul, N. L, Davis, B. A., Romero, M. F., & Boron, W F. (1998). Effect of expressing the water channel aquaporin-1 on the CO2 permeability of Xenopus oocytes. *American Journal of Physiology – Cell Physiology* **274**, C297–298.

Neely, J. D., Amiry-Moghaddam, M., Ottersen, O. P., Froehner, S. C., Agre, P., & Adams, M. E. (2001). Syntrophin-dependent expression and localization of Aquaporin-4 water channel protein. *Proceedings of the National Academy of Sciences of the USA* **98**, 14108–14113.

Nejsum, L. N., Kwon, T. H., Jensen, U. B., Fumagalli, O., Frokiaer, J., Krane, C. M., Menon, A. G., King, L. S., Agre, P., & Nielsen, S. (2002). Functional requirement of aquaporin-5 in plasma membranes of sweat glands. *Proceedings of the National Academy of Sciences of the USA* **99**, 511–516.

Nielsen, S., DiGiovanni, S.R., Christensen, E.I., Knepper, M.A., & Harris, H.W. (1993a). Cellular and subcellular immunolocalization of vasopressin-regulated water channel in rat kidney. *Proceedings of the National Academy of Sciences of the USA* **90**, 11663–11667.

Nielsen, S., Smith, B. L., Christensen, E. I., & Agre, P. (1993b). Distribution of the aquaporin CHIP in secretory and resorptive epithelia and capillary endothelia. *Proceedings of the National Academy of Sciences of the USA* **90**, 7275–7279.

Nielsen, S., Smith, B. L., Christensen, E. I., Knepper, M. A., & Agre, P. (1993c). CHIP28 water channels are localized in constitutively water-permeable segments of the nephron. *Journal of Cell Biology* **120**, 371–383.

Nielsen, S. & Agre, P. (1995). The aquaporin family of water channels in kidney. *Kidney International* **48**, 1057–1068.

Nielsen, S., Chou, C. L., Marples, D., Christensen, E. I., Kishore, B. K., & Knepper, M. A. (1995). Vasopressin increases water permeability of kidney collecting duct by inducing translocation of aquaporin-CD water channels to plasma membrane. *Proceedings of the National Academy of Sciences of the USA* **92**, 1013–1017.

Nielsen, S., Nagelhus, E. A., Amiry-Moghaddam, M., Bourque, C., Agre, P., & Ottersen, O. P. (1997). Specialized membrane domains for water transport in glial cells: high- resolution immunogold cytochemistry of aquaporin-4 in rat brain. *Journal of Neuroscience* **17**, 171–180.

Nielsen, S., Frokiaer, J., Marples, D., Kwon, T. H., Agre, P., & Knepper, M. A. (2002). Aquaporins in the kidney: from molecules to medicine. *Physiological Reviews* **82**, 205–244.

Preston, G. M. & Agre, P. (1991). Isolation of the cDNA for erythrocyte integral membrane protein of 28 kilodaltons: member of an ancient channel family. *Proceedings of the National Academy of Sciences of the USA* **88**, 11110–11114.

Preston, G. M., Carroll, T. P., Guggino, W. B., & Agre, P. (1992). Appearance of water channels in Xenopus oocytes expressing red cell CHIP28 protein. *Science* **256**, 385–387.

Preston, G. M., Jung, J. S., Guggino, W. B., & Agre, P. (1993). The mercury-sensitive residue at cysteine 189 in the CHIP28 water channel. *Journal of Biological Chemistry* **268**, 17–20.

Preston, G. M., Smith, B. L., Zeidel, M. L., Moulds, J. J., & Agre, P. (1994). Mutations in aquaporin-1 in phenotypically normal humans without functional CHIP water channels. *Science* **265**, 1585–1587.

Raina, S., Preston, G. M., Guggino, W. B., & Agre, P. (1995). Molecular cloning and characterization of an aquaporin cDNA from salivary, lacrimal, and respiratory tissues. *Journal of Biological Chemistry* **270**, 1908–1912.

Ramon y Cajal, S. (1897). Advice to a Young Investigator (translated by N. Swanson and L. W. Swanson, 1999) Cambridge, MIT Press.

Schnermann, J., Chou, C. L., Ma, T., Traynor, T., Knepper, M. A., & Verkman, A. S. (1998). Defective proximal tubular fluid reabsorption in transgenic aquaporin-1 null mice. *Proceedings of the National Academy of Sciences of the USA* **95**, 9660–9664.

Smith, B. L. & Agre, P. (1991). Erythrocyte Mr 28,000 transmembrane protein exists as a multisubunit oligomer similar to channel proteins. *Journal of Biological Chemistry* **266**, 6407–6415.

Smith, B. L., Preston, G. M., Spring, F. A., Anstee, D. J., & Agre, P. (1994). Human red cell aquaporin CHIP. I. Molecular characterization of ABH and Colton blood group antigens. *Journal of Clinical Investigation* **94**, 1043–1049.

Steinfeld, S., Cogan, E., King, L. S., Agre, P., Kiss, R., & Delporte, C. (2001). Abnormal distribution of aquaporin-5 water channel protein in salivary glands from Sjögren's syndrome patients. *Laboratory Investigation* **81**, 143–148.

Sui, H., Han, B. G., Lee, J. K., Walian, P., & Jap, B. K. (2001). Structural basis of water-specific transport through the AQP1 water channel. *Nature* **414**, 872–878.

Tajkhorshid, E., Nollert, P., Jensen, M. O., Miercke, L. J., O'Connell, H., Stroud, R. M., & Schulten, K., (2002). Control of the selectivity of the aquaporin water channel family by global orientational tuning. *Science* **296**, 525–530.

Tournaire-Roux, C., Sutka, M., Javot, H., Gout, E., Gerbeau, P., Luu, D. T., Bligny, R. & Maurel, C. (2003). Cytosolic pH regulates root water transport during anoxic stress through gating of aquaporins. *Nature* **425**, 393–397.

Tsubota, K., Hirai, S., King, L. S., Agre, P., & Ishida, N. (2001). Defective cellular trafficking of lacrimal gland aquaporin-5 in Sjögren's syndrome. *Lancet* **357**, 688–689.

Uehlein, N., Lovisolo, C., Siefritz, F., & Kaldenhoff, R. (2003). The tobacco aquaporin NtAQP1 is a membrane CO2 pore with physiological functions. *Nature* **425**, 734–737.

Vajda, Z., Pedersen, M., Fuchtbauer, E.-M., Wert, K., Stodkilde-Jorgensen, H., Sulyok, E., Doczi, T., Neely, J. D., Agre, P., Frokiaer, J., & Nielsen, S. (2002). Delayed onset of brain edema and mislocalization of aquaporin-4 in dystrophin-null transgenic mice. *Proceedings of the National Academy of Sciences of the USA* **99**, 13131–13136.

Walz, T., Smith, B. L., Zeidel, M. L., Engel, A., & Agre, P. (1994). Biologically active two-dimensional crystals of aquaporin CHIP. *Journal of Biological Chemistry* **269**, 1583–1586.

Yasui, M., Hazama, A., Kwon, T. H., Nielsen, S., Guggino, W. B., & Agre, P. (1999). Rapid gating and anion permeability of an intracellular aquaporin. *Nature* **402**, 184–187.

Zeidel, M. L., Ambudkar, S. V., Smith, B. L., & Agre, P. (1992). Reconstitution of functional water channels in liposomes containing purified red cell CHIP28 protein. *Biochemistry* **31**, 7436–7440.

Zeidel, M. L., Nielsen, S., Smith, B. L., Ambudkar, S. V., Maunsbach, A. B., & Agre, P. (1994). Ultrastructure, pharmacologic inhibition, and transport selectivity of aquaporin channel-forming integral protein in proteoliposomes. *Biochemistry* **33**, 1606–1615.

Zeuthen, T. & Klaerke, D. A. (1999). Transport of water and glycerol in aquaporin 3 is gated by H(+). *Journal of Biological Chemistry* **274**, 21631–21636.

Roderick MacKinnon

RODERICK MACKINNON

I was born on February 19, 1956 in the middle of a snowstorm. It remains one of those humorous family stories that my mother likes to tell. My father the planner had rehearsed the way to the hospital but apparently things looked a lot different at night in a blizzard. Eventually they made it and so did I, the fourth of seven children. My father was a postal worker when I was very young but studied computers and became a programmer on the big IBM main frames. My mother worked as a part time schoolteacher, but mostly took care of the children at home. Thinking back on it now I know we did not have much money but I never knew that growing up. My parents provided a happy environment and made their expectations clear to us. Television is bad for you, reading is good for you, and you better get an A for effort in school. What you end up doing in life is up to you. Just make sure you enjoy what you do because then you will do it well. We all pursued completely different walks of life. I became the scientist.

I suppose there were some early indications of my tendency to a life of curiosity. Apparently from a very young age I had a habit of asking lots of questions: 'what would happen if…?' was a favorite. And I liked having facts straight and knowing how things work and did not hesitate to give explanations to those around me, apparently to an annoying degree sometimes. I remember one day my father, at the end of his patience, commenting that I was a 'compendium of useless information'. I certainly can understand his plight with one of the seven having way too many questions and answers all the time. On the positive side, I learned a new word that day when I looked up compendium in the dictionary.

There were probably even indications that my curiosity might be scientific. Burlington Massachusetts was rural when I was young and I loved to roam and explore. I had rock collections and read children's books on geology and the history of the earth. I made little volcanoes out of plaster of paris and added baking soda and vinegar to the craters to simulate volcanic eruptions. I had an accident one day that made my mother laugh to my utter frustration: at that young age I failed to appreciate the humor in a little boy telling his mother he had dropped a volcano on his toe! In the summer I collected butterflies, turtles, snakes and other living things. One summer my mother enrolled me in a science enrichment class for elementary school students and I was allowed to take home a microscope. I used it to look at everything I could find: microorganisms from the nearby pond, leaves and blades of grass. I spent hour after hour alone, mesmerized by the tiny little things that I could see.

My scientific curiosity took a back seat to athletics through junior high and high school. Gymnastics was a good match to my small build and to my solitary nature. I was a member of a team but gymnastics is an individual sport. You learn a technique, then a 'move', and then a 'routine'. And then you perfect it through practice, working mostly alone. I had a very good no nonsense teacher, coach Hayes, who really instilled in me the idea of perfection through practice. I was actually not all that bad, particularly at floor exercise and high bar. I even considered pursuing gymnastics in college, but during my final year of high school I began to wonder what I should pursue for a career.

I attended the University of Massachusetts in Boston for one year and then transferred to Brandeis University. Brandeis was an eye opening experience for me. For the first time in my life I was in a seriously intellectual environment. The classes tended to be small, intense, and stimulating. I discovered that I had a passion for science, and that I was very good at it. I chose Biochemistry as a major and a newly arrived assistant professor named Chris Miller for my honors thesis advisor. He had a little laboratory with big windows and lots of light shining in. I studied calcium transport and learned about the cell membrane as an electrode. I could see that Chris Miller was a man having lots of fun in his daily life and it was inspiring to me, and the memory of this stayed with me. But the biggest influence Brandeis had on my life happened in Physics class. There I met my future wife Alice Lee, whose sparkling eyes and sharp mind caught my attention.

Against Chris Miller's advice I went to medical school after Brandeis. I studied at Tufts University School of Medicine and then at Beth Israel Hospital Boston for house officer training in Internal Medicine. I learned a lot but in the end I should have taken Chris' advice to pursue science. Medicine required a lot of memorization and little analytical problem solving. To keep a certain part of my brain active I began to study mathematics, and continue this even today, learning new methods and solving problems with the same disciplined approach I had learned in gymnastics. I started back to science near the end of house officer training working with Jim Morgan studying calcium in cardiac muscle contractility, which was very enjoyable and kept me connected to medicine. But I had a yearning to work on a very basic science problem, which meant I would have to break my medical ties. This was a difficult decision because I had invested so many years in medical education; to abandon it was to admit to myself that I had misspent a big piece of my life. And there were practical considerations as well. It was time finally to get a permanent job; after all, my wife Alice had supported me through years of training. Not to mention I was nearly 30 years old with no real basic science training beyond my Brandeis undergraduate education: would I even be able to make it as a scientist?

Two factors had the greatest influence on my decision. Back in my first year of medical school I lost my sister Elley, an artist only two years my senior. Diagnosed with leukemia during my hematology clerkship as I learned about the dreaded disease, she lasted only two months. This horrifying event im-

Alice Lee in 1986 at the time of my decision to leave medicine for science.

pressed upon me how fragile and precious life is, and how important it is to seize the moment and enjoy what you do while you can. I remember thinking when I look back upon my life at the age of seventy, thirty will seem young: just go for it. And the second factor was Alice. She had complete faith in my ability to succeed. Never mind that postdoctoral studies meant a reduction of my already piddling house officer salary. She simply said you have no choice; we will manage somehow.

Memories of Chris Miller's laboratory beckoned so I returned for postdoctoral studies. Of course I will never out live his reminding me that I should have listened to him in the first place. Feeling far behind in my knowledge I approached my postdoctoral studies with intensity, learning techniques and theory. I felt I should be an expert in electrochemistry, stochastic processes, linear systems theory, and many more subjects. I read books, solved the problem sets, mastered the subjects, and carried out experiments. I had the very good fortune of a coworker Jacques Neyton, a postdoctoral scientist from France. Jacques is a very critical thinker who would brood on a problem. We exchanged ideas often. When I would tell him one of my ideas he had a tendency just to listen quietly. Then, after a while, if his response started with 'Hey Roddy, there's something I don't understand' I knew I was in trouble – my idea was probably no good!

After I completed a series of biophysical studies on K^+ channels it came time to apply for an academic position. During the late 1980s physiology departments were more interested in hiring channel gene cloners than bio-

Recent photograph of Rod and Alice.

physicists. But Peter Hess convinced his colleagues at Harvard that my work showed promise and I was offered an assistant professorship there. My laboratory made good progress on K^+ channels. It was exciting for a while but in just a few years I began to feel that the return on what we could learn from studying the functional effects of mutations was diminishing. We had identified the K^+ channel signature sequence, but without knowing its structure we never would understand the chemical principles of ion selectivity in K^+ channels. I decided at that point to learn X-ray crystallography to someday see a K^+ channel.

I began to learn methods of protein purification and X-ray crystallography while still at Harvard, initially working with channel toxins and a small soluble protein called a PDZ domain. However, I thought it best to move away from my familiar environment at Harvard to pursue channel structure. There were really two reasons motivating me to move. First was the practical issue of obtaining funding to work in an area in which I had no background: start-up funds associated with moving to a new university would be useful for this purpose. The second and far more important reason was that moving would enable me to immerse myself completely in the new endeavor. A change of environment would remove the distractions of everyday life, isolate me from the temptation to fall back on channel physiology studies that I was already good at, and allow me to focus with singular purpose on the structural studies. I needed this to become an expert in membrane protein biochemistry and X-ray crystallography, and to develop a 'feel' for protein structure. When the president of Rockefeller University Torsten Wiesel heard about my scientific plans he suggested that I move to Rockefeller University and I did. Rockefeller provided a wonderful environment for concentrating on a difficult problem.

It has been said that giving up my already successful lab at Harvard in order to pursue the structure of a K^+ channel was a risky thing to do. At the time I was told that my aspirations were altogether unrealistic. From my perspective I had little choice because I wanted to understand K^+ selectivity and I knew that the atomic structure provided the only path to understanding. I would rather fail trying than never try at all. It helped that I was accustomed to making transitions and had become good at teaching myself new subjects. I have to admit that few people working with me at the time wanted much to do with the new endeavor – only one new postdoctoral scientist Declan Doyle was enthusiastic. My wife Alice, an organic chemist, saw that I was going to be pretty lonely and decided to join me in the lab. And to my good fortune she has worked with me since. I have learned that most people do not like change but I do. For me change is challenging, good for creativity, and it definitely keeps life interesting.

I think of the past eight years of my life in New York at Rockefeller University as a personal odyssey. The new laboratory started out very small, with only Declan, Alice and me. But it grew in the first year with the addition of other enthusiastic postdoctoral scientists, including João Morais Cabral and John Imredy. Working with membrane proteins was very difficult as expected. We had our periods of despair, but every time we felt left without options something good happened and despair gave way to excitement. Persistence and dedication eventually paid off. The atomic structure of the K^+ selectivity filter was more informative and more beautiful than I ever could have imagined. My laboratory now is an incredible place, overflowing with excitement and ideas sustained by the continual infusion of bright young scientists who come from around the world to work with me. It gives me great satisfaction to know that these young scientists who are sophisticated in their knowledge of protein chemistry and structure will lead the field of ion channel research into the future. This has been a wonderful adventure.

I owe thanks for the life I have: to Alice, to all my loving family of MacKinnons and Lees, to my scientific family of students, postdocs and colleagues, to senior colleagues who have helped me along my way to pursue my passion, and to the Rockefeller University, the Howard Hughes Medical Institute, and the National Institutes of Health for their support. I am very thankful for my life as a scientist, for the opportunity to understand in some small way the world around me. I hope my best experiment and scientific ideas are yet to come. This hope keeps me going.

POTASSIUM CHANNELS AND THE ATOMIC BASIS OF SELECTIVE ION CONDUCTION

Nobel Lecture, December 8, 2003

by

RODERICK MACKINNON

Howard Hughes Medical Institute, Laboratory of Molecular Neurobiology and Biophysics, Rockefeller University, 1230 York Avenue, New York, NY 10021, USA.

INTRODUCTION

All living cells are surrounded by a thin, approximately 40 Å thick lipid bilayer called the cell membrane. The cell membrane holds the contents of a cell in one place so that the chemistry of life can occur, but it is a barrier to the movement of certain essential ingredients including the ions Na^+, K^+, Ca^{2+} and Cl^-. The barrier to ion flow across the membrane – known as the dielectric barrier – can be understood at an intuitive level: the cell membrane interior is an oily substance and ions are more stable in water than in oil. The energetic preference of an ion for water arises from the electric field around the ion and its interaction with neighboring molecules. Water is an electrically polarizable substance, which means that its molecules rearrange in an ion's electric field, pointing negative oxygen atoms in the direction of cations and positive hydrogen atoms toward anions. These electrically stabilizing interactions are much weaker in a less polarizable substance such as oil. Thus, an ion will tend to stay in the water on either side of a cell membrane rather than enter and cross the membrane. And yet numerous cellular processes, ranging from electrolyte transport across epithelia to electrical signal production in neurons, depend on the flow of ions across the membrane. To mediate the flow, specific protein catalysts known as ion channels exist in the cell membrane. Ion channels exhibit the following three essential properties: (1) they conduct ions rapidly, (2) many ion channels are highly selective, meaning only certain ion species flow while others are excluded, (3) their function is regulated by processes known as gating, that is, ion conduction is turned on and off in response to specific environmental stimuli. Figure 1 summarizes these properties (figure 1).

The modern history of ion channels began in 1952 when Hodgkin and Huxley published their seminal papers on the theory of the action potential in the squid giant axon (Hodgkin and Huxley, 1952a; Hodgkin and Huxley, 1952b; Hodgkin and Huxley, 1952c; Hodgkin and Huxley, 1952d). A fundamental element of their theory was that the axon membrane undergoes changes in its permeability to Na^+ and K^+ ions. The Hodgkin-Huxley theory

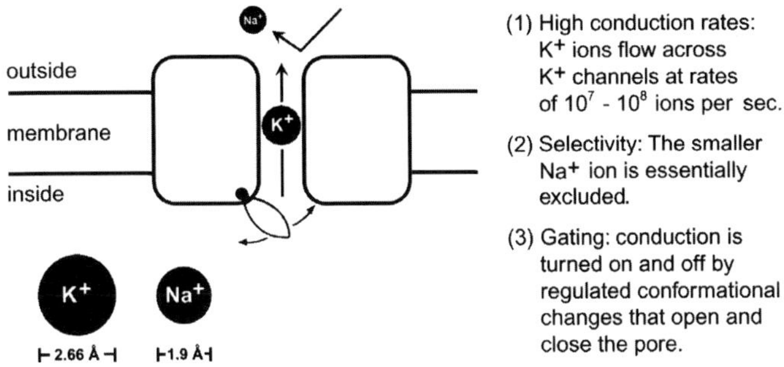

Figure 1. Ion channels exhibit three basic properties depicted in the cartoon. They conduct specific ions (for example K^+) at high rates, they are selective (a K^+ channel essentially excludes Na^+), and conduction is turned on and off by opening and closing a gate, which can be regulated by an external stimulus such as ligand-binding or membrane voltage. The relative size of K^+ and Na^+ ions is shown.

did not address the mechanism by which the membrane permeability changes occur: ions could potentially cross the membrane through channels or by a carrier-mediated mechanism. In their words 'Details of the mechanism will probably not be settled for some time' (Hodgkin and Huxley, 1952a). It is fair to say that the pursuit of this statement has accounted for much ion channel research over the past fifty years.

As early as 1955 experimental evidence for channel mediated ion flow was obtained when Hodgkin and Keynes measured the directional flow of K^+ ions across axon membranes using the isotope $^{42}K^+$ (Hodgkin and Keynes, 1955). They observed that K^+ flow in one direction across the membrane depends on flow in the opposite direction, and suggested that 'the ions should be constrained to move in single file and that there should, on average, be several ions in a channel at any moment'. Over the following two decades Armstrong and Hille used electrophysiological methods to demonstrate that Na^+ and K^+ ions cross cell membranes through unique protein pores – Na^+ channels and K^+ channels – and developed the concepts of selectivity filter for ion discrimination and gate for regulating ion flow (Hille, 1970; Hille, 1971; Hille, 1973; Armstrong, 1971; Armstrong *et al.*, 1973; Armstrong and Bezanilla, 1977; Armstrong, 1981). The patch recording technique invented by Neher and Sakmann then revealed the electrical signals from individual ion channels, as well as the extraordinary diversity of ion channels in living cells throughout nature (Neher and Sakmann, 1976).

The past twenty years have been the era of molecular biology for ion channels. The ability to manipulate amino acid sequences and express ion channels at high levels opened up entirely new possibilities for analysis. The advancement of techniques for protein structure determination and the development of synchrotron facilities also created new possibilities. For me, a scientist who became fascinated with understanding the atomic basis of life's electrical system, there could not have been a more opportune time to enter the field.

MY EARLY STUDIES: THE K^+ CHANNEL SIGNATURE SEQUENCE

The cloning of the *Shaker* K^+ channel gene from *Drosophila melanogaster* by Jan, Tanouye, and Pongs revealed for the first time a K^+ channel amino acid sequence and stimulated efforts by many laboratories to discover which of these amino acids form the pore, selectivity filter, and gate (Tempel *et al.*, 1987; Kamb *et al.*, 1987; Pongs *et al.*, 1988). At Brandeis University in Chris Miller's laboratory I had an approach to find the pore amino acids. Chris and I had just completed a study showing that charybdotoxin, a small protein from scorpion venom, inhibits a K^+ channel isolated from skeletal muscle cells by plugging the pore and obstructing the flow of ions (MacKinnon and Miller, 1988). In one of those late night 'let's see what happens if' experiments while taking a molecular biology course at Cold Spring Harbor I found that the toxin – or what turned out to be a variant of it present in the charybdotoxin preparation – inhibited the Shaker K^+ channel (MacKinnon *et al.*, 1988; Garcia *et al.*, 1994). This observation meant I could use the toxin to find the pore, and it did not take very long to identify the first site-directed mutants of the Shaker K^+ channel with altered binding of toxin (MacKinnon and Miller, 1989). I continued these experiments at Harvard Medical School where I began as assistant professor in 1989. Working with my small group at Harvard, including Tatiana Abramson, Lise Heginbotham, and Zhe Lu, and sometimes with Gary Yellen at Johns Hopkins University, we reached several interesting conclu-

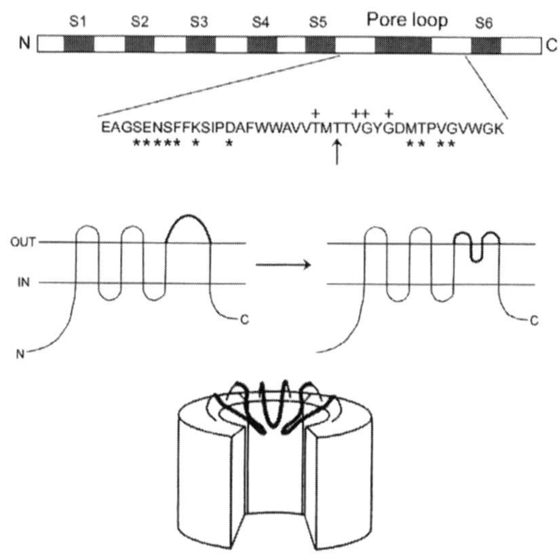

Figure 2. Early picture of a tetramer K^+ channel with a selectivity filter made of pore loops. A linear representation of a Shaker K^+ channel subunit on top shows shaded hydrophobic segments S1 to S6 and a region designated the pore loop. A partial amino acid sequence from the Shaker K^+ channel pore loop highlights amino acids shown to interact with extracellular scorpion toxins (*), intracellular tetraethylammonium (↑) and K^+ ions (+). The pore loop was proposed to reach into the membrane (middle) and form a selectivity filter at the center of four subunits (bottom).

Bacteria	2TM	:TATTVGYG
Archea	6TM	:TATTVGYG
Plant	6TM	:TLTTVGYG
Fruitfly	6TM	:TMTTVGYG
Worm	6TM	:TMTTVGYG
Mouse	6TM	:SMTTVGYG
Human	2TM	:TQTTIGYG
Human	6TM	:TMTTVGYG

Figure 3. The K^+ channel signature sequence shown as single letter amino acid code (blue) is highly conserved in organisms throughout the tree of life. Some K^+ channels contain six membrane-spanning segments per subunit (6TM) while others contain only two (2TM). 2TM K^+ channels correspond to 6TM K^+ channels without the first four membrane-spanning segments (S1-S4 in figure 2).

sions concerning the architecture of K^+ channels. They had to be tetramers in which four subunits encircle a central ion pathway (MacKinnon, 1991). This conclusion was not terribly surprising but the experiments and analysis to reach it gave me great pleasure since they required only simple measurements and clear reasoning with binomial statistics. We also deduced that each subunit presents a 'pore loop' to the central ion pathway (figure 2) (MacKinnon, 1995). This 'loop' formed the binding sites for scorpion toxins (MacKinnon and Miller, 1989; Hidalgo and MacKinnon, 1995; Ranganathan *et al.*, 1996) as well as the small-molecule inhibitor tetraethylammonium ion (MacKinnon and Yellen, 1990; Yellen *et al.*, 1991), which had been used by Armstrong and Hille decades earlier in their pioneering analysis of K^+ channels (Armstrong, 1971; Armstrong and Hille, 1972). Most important to my thinking, mutations of certain amino acids within the 'loop' affected the channel's ability to discriminate between K^+ and Na^+, the selectivity hallmark of K^+ channels (Heginbotham *et al.*, 1992; Heginbotham *et al.*, 1994). Meanwhile, new K^+ channel genes were discovered and they all had one obvious feature in common: the very amino acids that we had found to be important for K^+ selectivity were conserved (figure 3). We called these amino acids the K^+ channel signature sequence, and imagined four pore loops somehow forming a selectivity filter with the signature sequence amino acids inside the pore (Heginbotham *et al.*, 1994; MacKinnon, 1995).

When you consider the single channel conductance of many K^+ channels found in cells you realize just how incredible these molecular devices are. With typical cellular electrochemical gradients, K^+ ions conduct at a rate of 10^7 to 10^8 ions per second. That rate approaches the expected collision frequency of K^+ ions from solution with the entryway to the pore. This means that K^+ ions flow through the pore almost as fast as they diffuse up to it. For this to occur the energetic barriers in the channel have to be very low, something like those encountered by K^+ ions diffusing through water. All the more remarkable, the high rates are achieved in the setting of exquisite selectivity: the K^+ channel conducts K^+, a monovalent cation of Pauling radius 1.33 Å, while essentially excluding Na^+, a monovalent cation of Pauling radius 0.95 Å. And this ion selectivity is critical to the survival of a cell. How does nature accomplish high conduction rates and high selectivity at the same time? The answer to this question would require knowing the atomic structure formed by the signature sequence amino acids, that much was clear. The conservation of the signature sequence amino acids in K^+ channels throughout the tree of life, from bacteria (Milkman, 1994) to higher eukaryotic cells, implied that nature had settled upon a very special solution to achieve rapid, selective K^+ conduction across the cell membrane. For me, this realization provided inspiration to want to directly visualize a K^+ channel and its selectivity filter.

THE KCSA STRUCTURE AND SELECTIVE K^+ CONDUCTION

I began to study crystallography, and although I had no idea how I would obtain funding for this endeavor, I have always believed that if you really want to do something then you will find a way. By happenstance I explained my plan to Torsten Wiesel, then president of Rockefeller University. He suggested that I come to Rockefeller where I would be able to concentrate on the problem. I accepted his offer and moved there in 1996. In the beginning I was joined by Declan Doyle and my wife Alice Lee MacKinnon and within a year others joined including João Morais Cabral, John Imredy, Sabine Mann and Richard Pfuetzner. We had to learn as we went along, and what we may have lacked in size and skill we more than compensated for with enthusiasm. It was a very special time. At first I did not know how we would ever reach the point of obtaining enough K^+ channel protein to attempt crystallization, but the K^+ channel signature sequence continued to appear in a growing number of prokaryotic genes, making expression in *Escherichia coli* possible. We focused our effort on a bacterial K^+ channel called KcsA from *Streptomyces lividans*, discovered by Schrempf (Schrempf *et al.*, 1995). The KcsA channel has a simple topology with only two membrane spanning segments per subunit corresponding to the Shaker K^+ channel without S1 through S4 (figure 2). Despite its prokaryotic origin KcsA closely resembled the Shaker K^+ channel's pore amino acid sequence, and even exhibited many of its pharmacological properties, including inhibition by scorpion toxins (MacKinnon *et al.*, 1998). This surprised us from an evolutionary standpoint, because why should a scorpion want to inhibit a bacterial K^+ channel! But from the utilitarian point of view of protein biophysicists we knew exactly what the scorpion toxin sensitivity meant, that KcsA had to be very similar in structure to the Shaker K^+ channel.

The KcsA channel produced crystals but they were poorly ordered and not very useful in the X-ray beam. After we struggled for quite a while I began to wonder whether some part of the channel was intrinsically disordered and interfering with crystallization. Fortunately my neighbor Brian Chait and his postdoctoral colleague Steve Cohen were experts in the analysis of soluble proteins by limited proteolysis and mass spectrometry, and their techniques

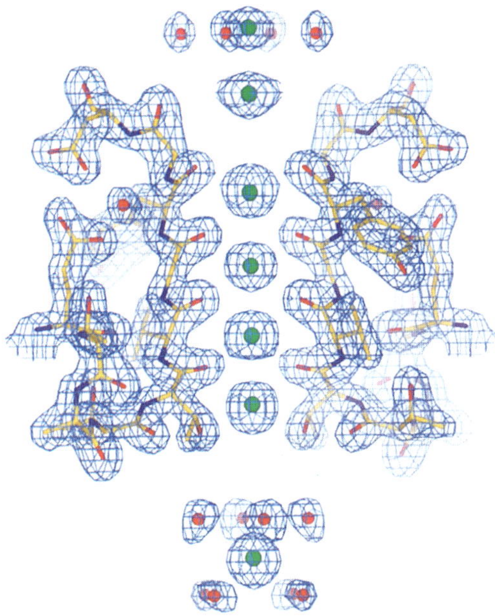

Figure 4. Electron density ($2F_o$-F_c contoured at 2 σ) from a high-resolution structure of the KcsA K^+ channel is shown as blue mesh. This region of the channel features the selectivity filter with K^+ ions and water molecules along the ion pathway. The refined atomic model is shown in the electron density. Adapted from (Zhou *et al.*, 2001b).

applied beautifully to a membrane protein. We found that KcsA was as solid as a rock, except for its C-terminus. After removing disordered amino acids from the c terminus with chymotrypsin the crystals improved dramatically, and we were able to solve an initial structure at a resolution of 3.2 Å (Doyle *et al.*, 1998). We could not clearly see K^+ in the pore at this resolution, but my years of work on K^+ channel function told me that Rb^+ and Cs^+ should be valuable electron dense substitutes for K^+, and they were. Rubidium and Cs^+ difference Fourier maps showed these ions lined up in the pore – as Hodgkin and Keynes might have imagined in 1955 (Hodgkin and Keynes, 1955).

The KcsA structure was altogether illuminating, but before I describe it, I will depart from chronology to explain the next important technical step. A very accurate description of the ion coordination chemistry inside the selectivity filter would require a higher resolution structure. With 3.2 Å data we could infer the positions of the main-chain carbonyl oxygen atoms by applying our knowledge of small molecule structures, that is our chemical intuition, but we needed to see the selectivity filter atoms in detail. A high-resolution structure was actually quite difficult to obtain. After more than three additional years of work by João and then Yufeng (Fenny) Zhou we finally managed to produce high-quality crystals by attaching monoclonal Fab fragments to KcsA. These crystals provided the information we needed, a structure at a resolution of 2.0 Å in which K^+ ions could be visualized in the grasp of selec-

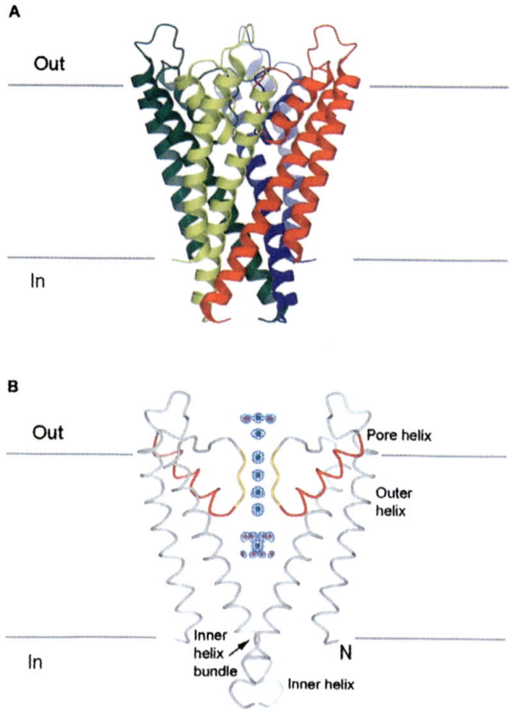

Figure 5. (A) A ribbon representation of the KcsA K^+ channel with its four subunits colored uniquely. The channel is oriented with the extracellular solution on top. (B) The KcsA K^+ channel with front and back subunits removed, colored to highlight the pore-helices (red) and selectivity filter (yellow). Electron density in blue mesh is shown along the ion pathway. Labels identify the pore, outer, and inner helices and the inner helix bundle. The outer and inner helices correspond to S5 and S6 in figure 2.

tivity filter protein atoms (figure 4) (Zhou *et al.*, 2001b). What did the K^+ channel structure tell us and why did nature conserve the K^+ channel signature sequence amino acids?

Not all protein structures speak to you in an understandable language, but the KcsA K^+ channel does. Four subunits surround a central ion pathway that crosses the membrane (figure 5A). Two of the four subunits are shown in figure 5B with electron density from K^+ ions and water along the pore. Near the center of the membrane the ion pathway is very wide, forming a cavity about 10 Å in diameter with a hydrated K^+ ion at its center. Each subunit directs the C-terminal end of a 'pore helix', shown in red, toward the ion. The C-terminal end of an α-helix is associated with a negative 'end charge' due to carbonyl oxygen atoms that do not participate in secondary structure hydrogen bonding, so the pore helices are directed as if to stabilize the K^+ ion in the cavity. At the beginning of this lecture I raised the fundamental issue of the cell membrane being an energetic barrier to ion flow because of its oily interior. KcsA allows us to intuit a simple logic encoded in its structure, and electrostatic calculations support the intuition (Roux and MacKinnon, 1999): the K^+ channel lowers the membrane dielectric barrier by hydrating a K^+ ion deep inside the membrane, and by stabilizing it with α-helix end charges.

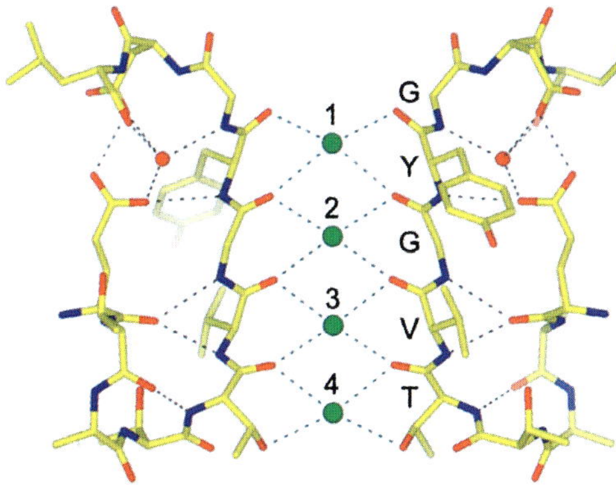

Figure 6. Detailed structure of the K^+ selectivity filter (two subunits). Oxygen atoms coordinate K^+ ions (green spheres) at positions 1 to 4 from the extracellular side. Single letter amino acid code identifies select signature sequence amino acids. Yellow, blue and red correspond to carbon, nitrogen and oxygen atoms, respectively. Green and gray dashed lines show oxygen-K^+ and hydrogen bonding interactions.

How does the K^+ channel distinguish K^+ from Na^+? Our earlier mutagenesis studies had indicated that the signature sequence amino acids would be responsible for this most basic function of a K^+ channel. Figure 6 shows the structure formed by the signature sequence – the selectivity filter – located in the extracellular third of the ion pathway. The glycine amino acids in the sequence TVGYG have dihedral angles in or near the left-handed helical region of the Ramachandran plot, as does the threonine, allowing the main-chain carbonyl oxygen atoms to point in one direction, toward the ions along the pore. It is easy to understand why this sequence is so conserved among K^+ channels: the alternating glycine amino acids permit the required dihedral angles, the threonine hydroxyl oxygen atom coordinates a K^+ ion, and the side-chains of valine and tyrosine are directed into the protein core surrounding the filter to impose geometric constraint. The end result when the subunits come together is a narrow tube consisting of four equal spaced K^+ binding sites, labeled 1 to 4 from the extracellular side. Each binding site is a cage formed by eight oxygen atoms on the vertices of a cube, or a twisted cube called a square antiprism (figure 7). The binding sites are very similar to the single alkali metal site in nonactin, a K^+ selective antibiotic with nearly identical K^+-oxygen distances (Dobler *et al.,* 1969; Dunitz and Dobler, 1977). The principle of K^+ selectivity is implied in a subtle feature of the KcsA crystal structure. The oxygen atoms surrounding K^+ ions in the selectivity filter are arranged quite like the water molecules surrounding the hydrated K^+ ion in the cavity. This comparison conveys a visual impression of binding sites in the filter paying for the energetic cost of K^+ dehydration. The Na^+ ion is apparently too small for these K^+-sized binding sites, so its dehydration energy is not compensated.

The question that compelled us most after seeing the structure was exactly

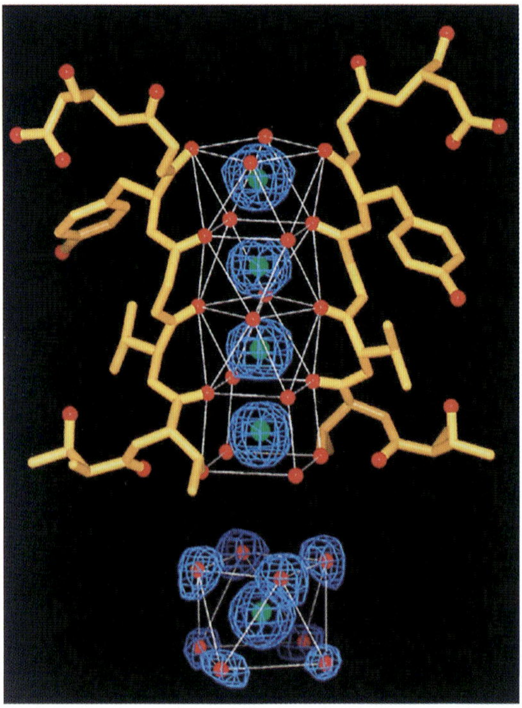

Figure 7. A K^+ channel mimics the hydration shell surrounding a K^+ ion. Electron density (blue mesh) for K^+ ions in the filter and for a K^+ ion and water molecules in the central cavity are shown. White lines highlight the coordination geometry of K^+ in the filter and in water. Adapted from (Zhou *et al.*, 2001b).

how many ions are in the selectivity filter at a given time? To begin to understand how ions move through the filter we needed to know the stoichiometry of the ion conduction reaction, and that meant knowing how many ions can occupy the filter. Four binding sites were apparent, but are they all occupied at once? Four K^+ ions in a row separated by an average center-to-center distance of 3.3 Å seemed unlikely for electrostatic reasons. From an early stage we suspected that the correct number would be closer to two, because two ions more easily explained the electron density we observed for the larger alkali metal cations Rb^+ and Cs^+ (Doyle *et al.*, 1998; Morais-Cabral *et al.*, 2001). Quantitative evidence for the precise number of ions came with the high-resolution structure and with the analysis of Tl^+ (Zhou and MacKinnon, 2003). Thallium is the most ideally suited 'K^+ analog' because it flows through K^+ channels, has a radius and dehydration energy very close to K^+, and has the favorable crystallographic attributes of high electron density and an anomalous signal. The one serious difficulty in working with Tl^+ is its insolubility with Cl^-. Fenny meticulously worked out the experimental conditions and determined that on average there are between two and two and a half conducting ions in the filter at once, with an occupancy at each position around one half.

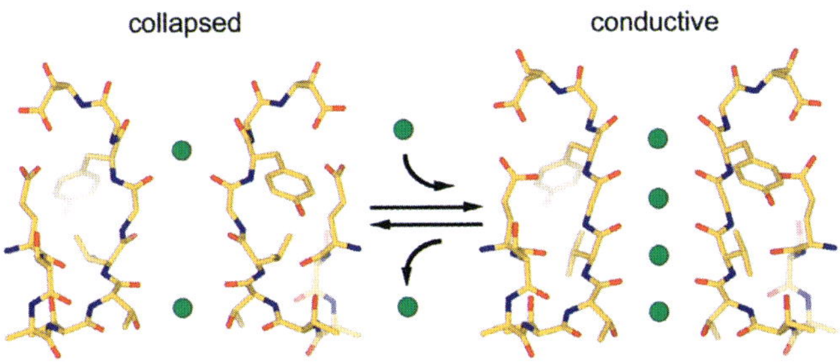

Figure 8. The selectivity filter can adopt two conformations. At low concentrations of K^+ on average one K^+ ion resides at either of two sites near the ends of the filter, which is collapsed in the middle. At high concentrations of K^+ a second ion enters the filter as it changes to a conductive conformation. On average, two K^+ ions in the conductive filter reside at four sites, each with about half occupancy.

We also observed that if the concentration of K^+ (or Tl^+) bathing the crystals is lowered sufficiently (below normal intracellular levels) then a reduction in the number of ions from two to one occurs and is associated with a structural change to a 'collapsed' filter conformation, which is pinched closed in the middle (Zhou *et al.,* 2001b; Zhou and MacKinnon, 2003). At concentrations above 20 mM the entry of a second K^+ ion drives the filter to a 'conductive' conformation, as shown in figure 8. Sodium on the other hand does not drive the filter to a 'conductive' conformation even at concentrations up to 500 mM.

The K^+-induced conformational change has thermodynamic consequences for the affinity of two K^+ ions in the 'conductive' filter. It implies that a fraction of the second ion's binding energy must be expended as work to bring about the filter's conformational change, and as a result the two ions will bind with reduced affinity. To understand this statement at an intuitive level, recognize that for two ions to reside in the filter they must oppose its tendency to collapse and force one of them out, i.e. the two-ion 'conductive' conformation is under some tension, which will tend to lower K^+ affinity. This is a desirable property for an ion channel because weak binding favors high conduction rates. The same principle, referred to as the 'induced fit' hypothesis, had been proposed decades earlier by enzymologists to explain high specificity with low substrate affinity in enzyme catalysis (Jencks, 1987).

In the 'conductive' filter if two K^+ ions were randomly distributed then they would occupy four sites in six possible ways. But several lines of evidence hinted to us that the ion positions are not random. For example Rb^+ and Cs^+ exhibit preferred positions with obviously low occupancy at position 2 (Morais-Cabral *et al.,* 2001; Zhou and MacKinnon, 2003). In K^+ we observed an unusual doublet peak of electron density at the extracellular entryway to the selectivity filter, shown in figure 9 (Zhou *et al.,* 2001b). We could explain this density if K^+ is attracted from solution by the negative protein surface charge near the entryway and at the same time repelled by K^+ ions inside the filter. Two discrete peaks implied two distributions of ions in the filter. If K^+ ions

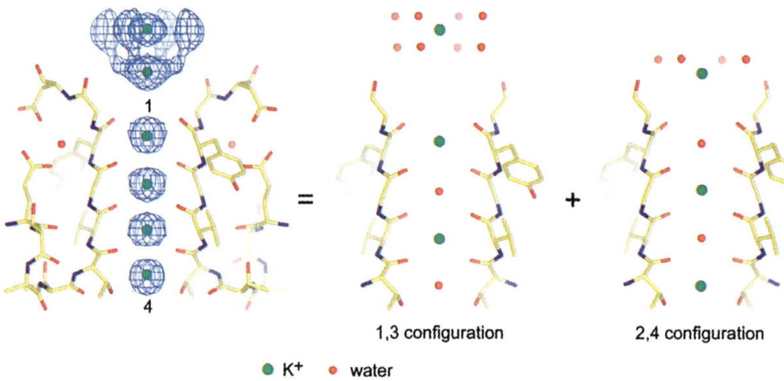

Figure 9. Two K^+ ions in the selectivity filter are hypothesized to exist predominantly in two specific configurations 1,3 and 2,4 as shown. K^+ ions and water molecules are shown as green and red spheres, respectively. Adapted from (Zhou *et al.,* 2001b).

tend to be separated by a water molecule for electrostatic reasons then the two dominant configurations would be 1,3 (K^+ ions in positions 1 and 3 with a water molecule in between) and 2,4 (K^+ ions in positions 2 and 4 with a water molecule in between). A mutation at position 4 (threonine to cysteine) was recently shown to influence K^+ occupancy at positions 2 and 4 but not at 1 and 3, providing strong evidence for specific 1,3 and 2,4 configurations of K^+ ions inside the selectivity filter (Zhou and MacKinnon, 2004).

Discrete configurations of an ion pair suggested a mechanism for ion conduction (figure 10A) (Morais-Cabral *et al.,* 2001). The K^+ ion pair could diffuse back and forth between 1,3 and 2,4 configurations (bottom pathway), or alternatively an ion could enter the filter from one side of the membrane as the ion-water queue moves and a K^+ exits at the opposite side (the top pathway). Movements would have to be concerted because the filter is no wider than a K^+ ion or water molecule. The two paths complete a cycle: in one complete cycle each ion moves only a fraction of the total distance through the filter, but the overall electrical effect is to move one charge all the way. Because two K^+ ions are present in the filter throughout the cycle we expect there should be electrostatic repulsion between them. Together with the filter conformational change that is required to achieve a 'conductive' filter with two K^+ ions in it, electrostatic repulsion should favor high conduction rates by lowering K^+ affinity.

Absolute rates from 10^7 to 10^8 ions per second are truly impressive for a highly selective ion channel. One aspect of the crystallographic data suggests that very high conductance K^+ channels such as KcsA might operate near the maximum rate that the conduction mechanism will allow. All four positions in the filter have a K^+ occupancy close to one half, which implies that the 1,3 and 2,4 configurations are equally probable, or energetically equivalent, but there is no *a priori* reason why this should be. A simulation of ions diffusing around the cycle offers a possible explanation: maximum flux is achieved when the energy difference between the 1,3 and 2,4 configurations is zero be-

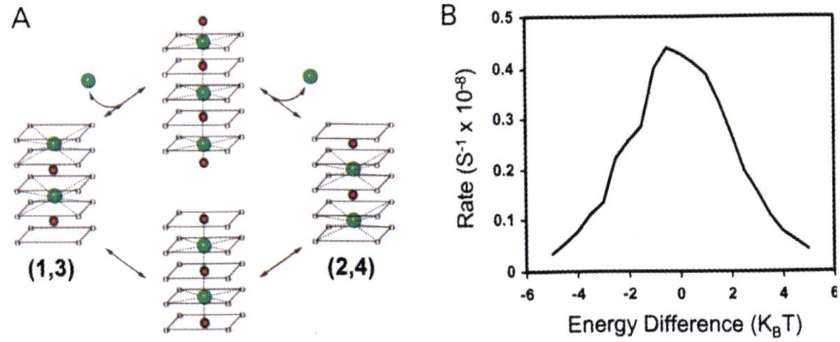

Figure 10. (A) Through-put cycle for K^+ conduction invoking 1,3 and 2,4 configurations. The selectivity filter is represented as five square planes of oxygen atoms. K^+ and water are shown as green and red spheres, respectively. (B) Simulated K^+ flux around the cycle is graphed as a function of the energy difference between the 1,3 and 2,4 configurations. Adapted from (Morais-Cabral *et al.*, 2001).

cause that is the condition under which the 'energy landscape' for the conduction cycle is smoothest (figure 10B). The energetic balance between the configurations therefore might reflect the optimization of conduction rate by natural selection (Morais-Cabral *et al.*, 2001). It is not so easy to demonstrate this point experimentally but it is certainly fascinating to ponder.

COMMON STRUCTURAL PRINCIPLES UNDERLIE K^+ AND Cl^- SELECTIVITY

The focus of this lecture is K^+ channels, but for a brief interlude I would like to show you a Cl^- selective transport protein. By comparing a K^+ channel and a Cl^- 'channel' we can begin to appreciate familiar themes in nature's solutions to different problems: getting cations and anions across the cell membrane. ClC Cl^- channels are found in many different cell types and are associated with a number of physiological processes that require Cl^- ion flow across lipid membranes (Jentsch *et al.*, 1999; Maduke *et al.*, 2000). As is the case for K^+ channels, ClC family genes are abundant in prokaryotes, a fortunate circumstance for protein expression and structural analysis. When Raimund Dutzler joined my laboratory he, Ernest Campbell and I set out to address the structural basis of Cl^- ion selectivity. We determined crystal structures of two bacterial members of the ClC Cl^- channel family, one from *Escherichia coli* (EcClC) and another from *Salmonella typhimurium* (StClC) (Dutzler *et al.*, 2002). Recent studies by Miller on the function of EcClC have shown that it is actually a Cl^- – proton exchanger (Accardi and Miller, 2004). We do not yet know why certain members of this family of Cl^- transport proteins function as channels and others as exchangers, but the crystal structures are fascinating and give us a view of Cl^- selectivity. Architecturally the ClC proteins are unrelated to K^+ channels, but if we focus on the ion pathway certain features are similar (figure 11). As we saw in K^+ channels, the ClC proteins have α-helices pointed at the ion pathway, but the direction is reversed with the positive

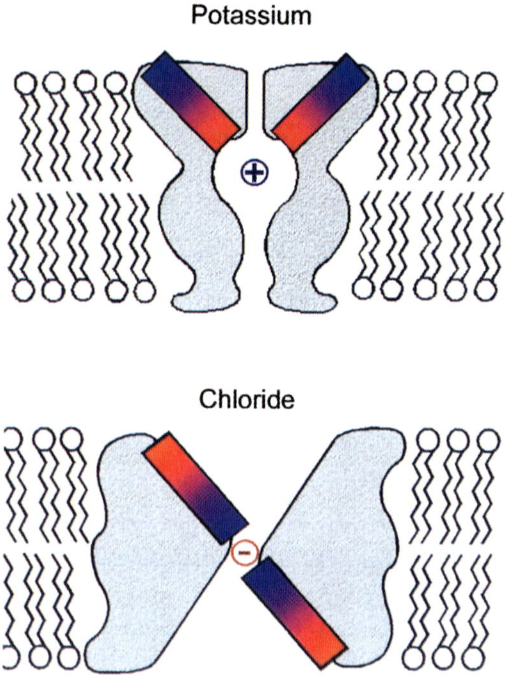

Figure 11. The overall architecture of K^+ channels and ClC Cl^- transport proteins is very different but certain general features are similar. One similarity shown here is the use of α-helix end charges directed toward the ion pathway. The negative C-terminal end charge (red) points to K^+. The positive N-terminal end charge (blue) points to Cl^-.

charge of the N-terminus close to Cl^-. This makes perfect sense for lowering the dielectric barrier for a Cl^- ion. In ClC we see that ions in its selectivity filter tend to be coordinated by main chain protein atoms, with amide nitrogen atoms surrounding Cl^- instead of carbonyl oxygen atoms surrounding K^+ (figure 12). We also see that both the K^+ and Cl^- selectivity filters contain multiple close-spaced binding sites and appear to contain more than one ion, perhaps to exploit electrostatic repulsion between ions in the pore. I find these similarities fascinating. They tell us that certain basic physical principles are important, such as the use of α-helix end charges to lower the dielectric barrier when ions cross the lipid membrane.

TRYING TO SEE A K^+ CHANNEL OPEN AND CLOSE

Most ion channels conduct when called upon by a specific stimulus such as the binding of a ligand or a change in membrane voltage (Hille, 2001). The processes by which ion conduction is turned on are called gating. The conduction of ions occurs on a time scale that is far too rapid to involve very large protein conformational changes. That is undoubtedly one of the reasons why a single KcsA structure could tell us so much about ion selectivity and con-

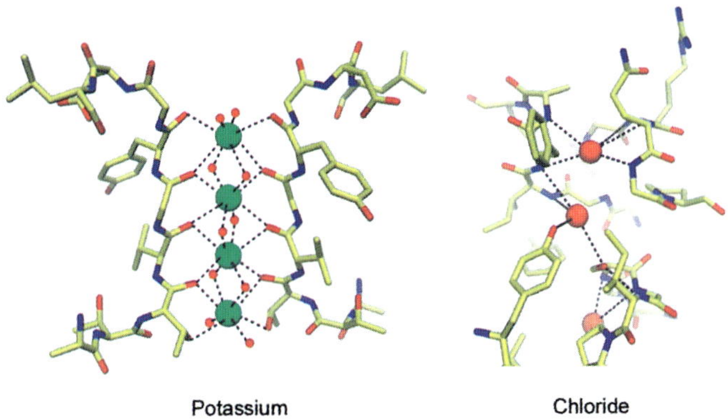

Figure 12. K^+ and Cl^- selectivity filters make use of main chain atoms to coordinate ions: carbonyl oxygen atoms for K^+ ions (green spheres) and amide nitrogen atoms for Cl^- ions (red spheres). Both filters contain multiple close-spaced ion binding sites. The Cl^- selectivity filter is that of a mutant ClC in which a glutamate amino acid was changed to glutamine (Dutzler *et al.*, 2003).

duction. Gating on the other hand occurs on a much slower time scale and can involve large protein conformational changes. The challenge for a structural description of gating is to capture a channel in both opened (on) and closed (off) conformations so that they can be compared.

In the KcsA K^+ channel gating is controlled by intracellular pH and lipid membrane composition, but unfortunately the KcsA channel's open probability reaches a maximum value of only a few percent in functional assays (Cuello *et al.*, 1998; Heginbotham *et al.*, 1998). At first we had no definitive way to know whether a gate was open or closed in the crystal structures. In the 1970s Armstrong had proposed the existence of a gate near the intracellular side of the membrane in voltage dependent K^+ channels because he could 'trap' large organic cations inside the pore between a selectivity filter near the extracellular side and a gate near the intracellular side (Armstrong, 1971; Armstrong, 1974). Following these ideas we crystallized KcsA with a heavy atom version of one of his organic cations, tetrabutyl antimony (TBA), and found that it binds inside the central cavity of KcsA (Zhou *et al.*, 2001a). This was very interesting because the ~10 Å diameter of TBA far exceeds the pore diameter leading up to the cavity: in KcsA the intracellular pore entryway is constricted to about 3.5 Å by the inner helix bundle (figure 5B). Seeing TBA 'trapped' in the cavity behind the inner helix bundle evoked Armstrong's classical view of K^+ channel gating, and implied that the inner helix bundle serves as a gate and is closed in KcsA. Mutational and spectroscopic studies in other laboratories also pointed to the inner helix bundle as a possible gate-forming structural element (Perozo *et al.*, 1999; del Camino *et al.*, 2000).

Youxing Jiang and I hoped we could learn more about K^+ channel gating by determining the structures of new K^+ channels. From gene sequence analysis we noticed that many prokaryotic K^+ channels contain a large C-ter-

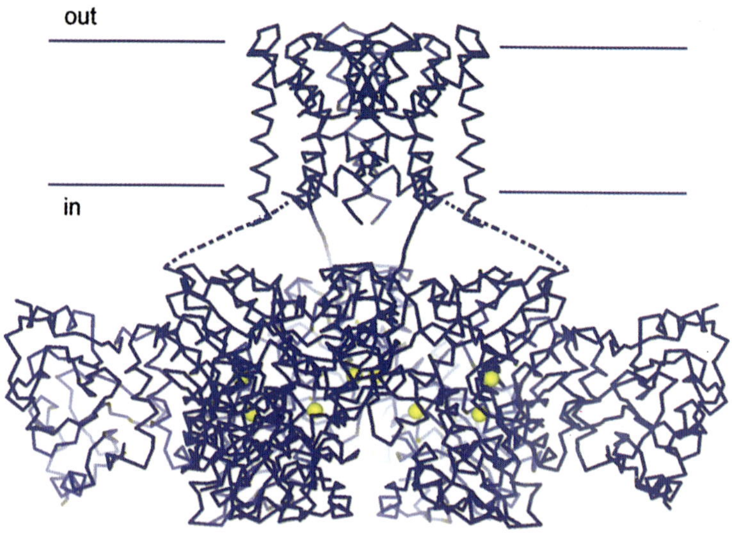

Figure 13. The MthK K^+ channel contains an intracellular gating ring (bottom) attached to its ion conduction pore (top). Ca^{2+} ions (yellow spheres) are bound to the gating ring in clefts in between domains. The connections between the gating ring and the pore, which were poorly ordered in the crystal, are shown as dashed lines.

minus that encodes what we called RCK domains and we suspected that these domains control pore opening, perhaps through binding of an ion or a small-molecule. Initally we determined the structure of isolated RCK domains from an *Escherichia coli* K^+ channel, but by themselves they were not very informative beyond hinting that a similar structure exists on the C-terminus of eukaryotic Ca^{2+}-dependent 'BK' channels (Jiang *et al.*, 2001). We subsequently determined the crystal structure of MthK, complete K^+ channel containing RCK domains, from *Methanobacterium thermoautotrophicus* (figure 13) (Jiang *et al.*, 2002a). This structure was extremely informative. The RCK domains form a 'gating ring' on the intracellular side of the pore. In clefts between domains we could see what appeared to be divalent cation binding sites, and the crystals had been grown in the presence of Ca^{2+}. In functional assays we discovered that the open probability of the MthK channel increased as Ca^{2+} or Mg^{2+} concentration was raised, giving us good reason to believe that the crystal structure should represent the open conformation of a K^+ channel.

In our MthK structure the inner helix bundle is opened like the aperture of a camera (figure 14) (Jiang *et al.*, 2002b). As a result, the pathway leading up to the selectivity filter from the intracellular side is about 10 Å wide, explaining how Armstrong's large organic cations can enter the cavity to block a K^+ channel, and how K^+ ions gain free access to the selectivity filter through aqueous diffusion. By comparing the KcsA and MthK channel structures it seemed that we were looking at examples of closed and opened K^+ channels, and could easily imagine the pore undergoing a conformational change from closed to open. To open, the inner helices would have to bend at a point halfway across

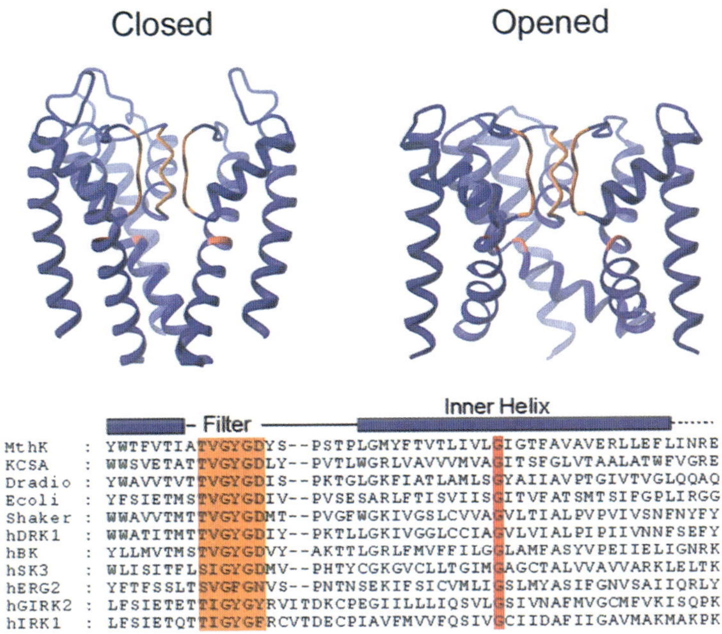

Figure 14. KcsA and MthK represent closed and opened K^+ channels. Three subunits of the closed KcsA K^+ channel (left) and opened MthK K^+ channel (right) are shown. The inner helices of MthK are bent at a glycine gating hinge (red), allowing the inner helix bundle to open. Partial amino acid sequences from a variety of K^+ channels with different gating domains are compared. Colors highlighting the selectivity filter sequence (gold) and inner helix glycine hinge (red) match colors used in the structures. Adapted from (Jiang *et al.*, 2002b).

the membrane as their C terminus is displaced laterally away from the pore axis by conformational changes in the gating ring. A glycine amino acid facilitates the bending in MthK by introducing a hinge point in the middle of the inner helix. Like MthK, KcsA and many other K^+ channels contain a glycine at the very same location; its conservation suggests that the inner helices move in a somewhat similar manner in many different K^+ channels (figure 14).

Gating domains convert a stimulus into pore opening. Further studies are needed to understand how the free energy of Ca^{2+} binding is converted into pore opening in the MthK channel. And the mechanistic details of ligand gating will vary from one channel type to the next because nature is very modular with ion channels, just like with other proteins. Gene sequences show us that a multitude of different domains can be found attached to the inner helices of different K^+ channels, allowing ions such as Ca^{2+} or Na^+, small organic molecules, and even regulatory proteins to control the conformational state of the pore and so gate the ion channel (http://www.ncbi.nlm.nih.gov/BLAST/) (Atkinson *et al.*, 1991; Schumacher *et al.*, 2001; Yuan *et al.*, 2003; Kubo *et al.*, 1993).

A fundamentally different kind of gating domain allows certain K^+, Na^+, Ca^{2+} and nonselective cation channels to open in response to membrane volt-

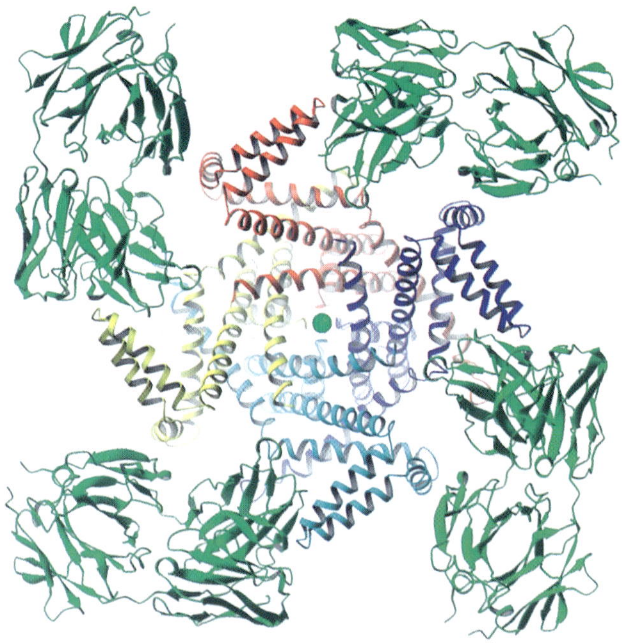

Figure 15. Crystal structure of the KvAP K^+ channel in complex with monoclonal antibody Fab fragments. The channel is viewed along the pore axis from the intracellular side of the membrane, with α-helical subunits colored in blue, yellow, cyan, and red. One Fab fragment (green) is bound to the helix-turn-helix element of the voltage sensor on each subunit. From (Jiang *et al.*, 2003a).

age changes. Referred to as voltage sensors, these domains are connected to the outer helices of the pore and form a structural unit within the membrane. The basic principle of operation for a voltage sensor is the movement of protein charges through the membrane electric field coupled to pore opening (Armstrong and Bezanilla, 1974; Sigworth, 1994; Bezanilla, 2000). Like transistors in an electronic device, voltage-dependent channels are electrical switches. They are a serious challenge for crystallographic analysis because of their conformational flexibility. Youxing Jiang and I working with Alice Lee and Jiayun Chen solved the structure of a voltage-dependent K^+ channel, KvAP, from the thermophilic Archea *Aeropyrum pernix* (Ruta *et al.*, 2003; Jiang *et al.*, 2003a) (figure 15). In the crystal of KvAP the voltage sensors, held by monoclonal Fab fragments, adopted a non-native conformation. This observation in itself is meaningful as it underscores the intrinsic flexibility of voltage sensors: in contrast Fab fragments had little effect on the more rigid KcsA K^+ channel and ClC Cl^- channel homolog, both of which we determined in the presence and absence of Fab fragments (Doyle *et al.*, 1998; Zhou *et al.*, 2001b; Dutzler *et al.*, 2002; Dutzler *et al.*, 2003). KvAP's voltage sensors contain a hydrophobic helix-turn-helix element with arginine residues beside the pore (Jiang *et al.*, 2003a), and functional experiments using tethered biotin and avidin show that this element moves relative to the plane of the membrane (Jiang *et al.*, 2003b). Additional structures revealing different channel conformations will be needed to better understand the mechanistic details of voltage-dependent gating. But the KvAP structure and associated functional studies have provided a conceptual model for voltage-dependent gating – one in which the voltage sensors move at the protein-lipid interface in response to a balance between hydrophobic and electrostatic forces. Rees and colleagues at the California Institute of Technology determined the structure of a voltage regulated mechanosensitive channel called MscS, and although it is unrelated to traditional voltage-dependent channels, it too contains hydrophobic helix-turn-helix elements with arginine residues apparently against the lipid membrane (Bass *et al.*, 2002). MscS and KvAP are fascinating membrane protein structures. They do not fit into the standard category of membrane proteins with rigid hydrophobic walls against the lipid membrane core. I find such proteins intriguing.

We are only just beginning to understand the structural principles of ion channel gating and regulation. Electrophysiological studies have uncovered a multitude of connections between cellular biochemical pathways and ion channel function (Hille, 2001). New protein structures are now beginning to do the same. Beta subunits of certain eukaryotic voltage-dependent channels are structurally related to oxido-reductase enzymes (Gulbis *et al.*, 1999; Gulbis *et al.*, 2000). PAS domains on other K^+ channels belong to a family of sensory molecules (Morais Cabral *et al.*, 1998), and a specialized structure on G protein-gated channels forms a binding site for regulatory G protein subunits (Nishida and MacKinnon, 2002). The interconnectedness of ion channel function with many aspects of cell function is beginning to reveal itself as complex and fascinating.

CONCLUDING REMARKS

I think the most exciting time in ion channel studies is just beginning. So many of the important questions are waiting to be answered and we have the tools in hand to answer them. I am very optimistic about the future, and for the great possibilities awaiting young scientists who are now setting out to study ion channels and other membrane proteins. I consider myself very fortunate to have contributed to some small part of the knowledge we have today. Of course, my contributions would never have been possible without the efforts and enthusiasm of the young scientists who have come from around the world to study ion channels with me (figure 16). I also owe thanks to the Rockefeller University, the Howard Hughes Medical Institute, and the National Institutes of Health for supporting my scientific research.

MACKINNON LABORATORY: 1989–2003

Postdoctoral	*Students*	*Staff Scientists*	*Collaborators*
Laura Escobar	Lise Heginbotham	Tatiana Abramson	Gary Yellen
Zhe Lu	Michael Root	John Lewis	Maria Garcia
Adrian Gross	Patricia Hidalgo	Alice Lee MacKinnon	Gerhard Wagner
Kenton Swartz	Sanjay Aggarwal	Sabine Mann	Andrzej Krezel
Chul-Seung Park	James Morrell	Richard Pfuetzner	Brian Chait
Rama Ranganathan	Alexander Pico	Anling Kuo	Steve Cohen
Chinfei Chen	Vanessa Ruta	Minhui Long	Martine Cadene
Declan Doyle	Ian Berke	Amelia Kaufman	Benoit Roux
John Imredy		Ernest Campbell	Tom Muir
João Morais Cabral		Jiayun Chen	
Youxing Jiang			
Jacqueline Gulbis			
Raimund Dutzler			
Francis Valiyaveetil			
Xiao-Dan Pfenninger-Li			
Ming Zhou			
Ofer Yifrach			
Yufeng Zhou			
Sebastien Poget			
Motohiko Nishida			
Uta-Maria Ohndorf			
Steve Lockless			
Qiu-Xing Jiang			
Seok-Yong Lee			
Stephen Long			

Thanks to Rockefeller University, HHMI, NIH, to the synchrotrons CHESS, NSLS, ALS, APS and ESRF and to my assistant Wendell Chin.

Figure 16. MacKinnon laboratory from 1989 to 2003.

REFERENCE LIST

Accardi, A. and Miller, C. (2004) Proton-coupled chloride transport mediated by ClC-ec1, a bacterial homologue of the ClC chloride channels. Biophys.J. *86[1]*, 286a.

Armstrong, C. M. (1971). Interaction of tetraethylammonium ion derivatives with the potassium channels of giant axons. J. Gen. Physiol. *58*, 413–437.

Armstrong, C. M. (1981). Sodium channels and gating currents. Physiol. Rev. *61*, 645–683.

Armstrong, C. M. and Bezanilla,F. (1974). Charge movement associated with the opening and closing of the activation gates of the Na^+ channels. J. Gen. Physiol. *63*, 533–552.

Armstrong, C. M., Bezanilla, F., and Rojas, E. (1973). Destruction of sodium conductance inactivation in squid axons perfused with pronase. J. Gen. Physiol. *62*, 375–391.

Armstrong, C. M. and Hille, B. (1972). The inner quaternary ammonium ion receptor in potassium channels of the node of Ranvier. J. Gen. Physiol. *59*, 388–400.

Armstrong, C. M. (1974). Ionic pores, gates, and gating currents. Q. Rev. Biophys. *7*, 179–210.

Armstrong, C. M. and Bezanilla, F. (1977). Inactivation of the sodium channel. II. Gating current experiments. J. Gen. Physiol *70*, 567–590.

Atkinson, N. S., Robertson, G. A., and Ganetzky, B. (1991). A component of calcium-activated potassium channels encoded by the Drosophila slo locus. Science *253*, 551–555.

Bass, R. B., Strop, P., Barclay, M., and Rees, D. C. (2002). Crystal structure of *Escherichia coli* MscS, a voltage-modulated and mechanosensitive channel. Science *298*, 1582–1587.

Bezanilla, F. (2000). The voltage sensor in voltage-dependent ion channels. Physiol Rev. *80*, 555–592.

Cuello, L. G., Romero, J. G., Cortes, D .M., and Perozo, E. (1998). pH-dependent gating in the Streptomyces lividans K^+ channel. Biochemistry *37*, 3229–3236.

del Camino, D., Holmgren, M., Liu, Y., and Yellen, G. (2000). Blocker protection in the pore of a voltage-gated K^+ channel and its structural implications. Nature *403*, 321–325.

Dobler,v. M., Dunitz, J. D., and Kilbourn, B. T. (1969). Die struktur des KNCS-Komplexes von nonactin. Helvetica Chimica Acta *52*, 2573–2583.

Doyle, D. A., Morais Cabral, J. H., Pfuetzner, R. A., Kuo, A., Gulbis, J. M., Cohen, S. L., Chait, B. T., and MacKinnon,R. (1998). The structure of the potassium channel: molecular basis of K^+ conduction and selectivity. Science *280*, 69–77.

Dunitz, J. D. and Dobler, M. (1977). Structural studies of ionophores and their ion-complexes. In Biological aspects of inorganic chemistry, A. W. Addison, W. R. Cullen, D. Dolphin, and B. R. James, eds. John Wiley & Sons, Inc.), pp. 113–140.

Dutzler, R., Campbell, E. B., Cadene, M., Chait, B. T., and MacKinnon,R. (2002). X-ray structure of a ClC chloride channel at 3.0 Å reveals the molecular basis of anion selectivity. Nature *415*, 287–294.

Dutzler, R., Campbell, E. B., and MacKinnon, R. (2003). Gating the selectivity filter in ClC chloride channels. Science *300*, 108–112.

Garcia, M. L., Garcia-Calvo, M., Hidalgo, P., Lee, A., and MacKinnon, R. (1994). Purification and charaterization of three inhibitors of voltage-dependent K^+ channels from Leiurus quinquestriatus var. hebraeus venom. Biochemistry *33*, 6834–6839.

Gulbis, J. M., Mann, S., and MacKinnon, R. (1999). Structure of a voltage-dependent K^+ channel beta subunit. Cell *97*, 943–952.

Gulbis, J. M., Zhou, M., Mann, S., and MacKinnon, R. (2000). Structure of the cytoplasmic β subunit-T1 assembly of voltage-dependent K^+ channels. Science *289*, 123–127.

Heginbotham, L., Abramson, T., and MacKinnon, R. (1992). A functional connection between the pores of distantly related ion channels as revealed by mutant K^+ channels. Science *258*, 1152–1155.

Heginbotham, L., Lu, Z., Abramson, T., and MacKinnon, R. (1994). Mutations in the K^+ channel signature sequence. Biophys. J. *66*, 1061–1067.

Heginbotham, L., Kolmakova-Partensky, L., and Miller, C. (1998). Functional reconstitution of a prokaryotic K^+ channel. J. Gen. Physiol. *111*, 741–749.

Hidalgo, P. and MacKinnon, R. (1995). Revealing the architecture of a K^+ channel pore through mutant cycles with a peptide inhibitor. Science *268*, 307–310.

Hille, B. (1970). Ionic channels in nerve membranes. Prog. Biophys. Mol. Biol. *21*, 1–32.

Hille, B. (1973). Potassium channels in myelinated nerve. Selective permeability to small cations. J. Gen. Physiol. *61*, 669–686.

Hille, B. (2001). Ion Channels of Excitable Membranes. (Sunderland, MA: Sinauer Associates, Inc.).

Hille, B. (1971). The permeability of the sodium channel to organic cations in myelinated nerve. J. Gen. Physiol. *58*, 599–619.

Hodgkin, A. L. and Huxley, A. F. (1952a). A quantitative description of membrane current and its application to conduction and excitation in nerve. J. Physiol. *117*, 500–544.

Hodgkin, A. L. and Huxley, A. F. (1952b). Currents carried by sodium and potassium ions through the membrane of the giant axon of Loligo. J. Physiol. *116*, 449–472.

Hodgkin, A. L. and Huxley, A. F. (1952c). The components of membrane conductance in the giant axon of Loligo. J. Physiol. *116*, 473–496.

Hodgkin, A. L. and Huxley, A. F. (1952d). The dual effect of membrane potential on sodium conductance in the giant axon of Loligo. J. Physiol. *116*, 497–506.

Hodgkin, A. L. and Keynes, R. D. (1955). The potassium permeability of a giant nerve fibre. J. Physiol. (Lond) *128*, 61–88.

Jencks, W. P. (1987). Catalysis in Chemistry and Enzymology. (Dover Publications, Inc.).

Jentsch, T. J., Friedrich, T., Schriever, A., and Yamada, H. (1999). The CLC chloride channel family. Pflugers Arch. *437*, 783–795.

Jiang, Y., Lee, A., Chen, J., Ruta, V., Cadene, M., Chait, B., and MacKinnon, R. (2003a). X-ray structure of a voltage-dependent K^+ channel. Nature *423*, 33–41.

Jiang, Y., Ruta, V., Chen, J., Lee, A., and MacKinnon, R. (2003b). The principle of gating charge movement in a voltage-dependent K^+ channel. Nature *423*, 42–48.

Jiang, Y., Lee, A., Chen, J., Cadene, M., Chait, B. T., and MacKinnon, R. (2002a). Crystal structure and mechanism of a calcium-gated potassium channel. Nature *417*, 515–522.

Jiang, Y., Lee, A., Chen, J., Cadene, M., Chait, B. T., and MacKinnon, R. (2002b). The open pore conformation of potassium channels. Nature *417*, 523–526.

Jiang, Y., Pico, A., Cadene, M., Chait, B. T., and MacKinnon, R. (2001). Structure of the RCK domain from the *E. coli* K^+ channel and demonstration of its presence in the human BK channel. Neuron *29*, 593–601.

Kamb, A., Iverson, L. E., and Tanouye, M. A. (1987). Molecular characterization of Shaker, a Drosophila gene that encodes a potassium channel. Cell *50*, 405–413.

Kubo, Y., Reuveny, E., Slesinger, P. A., Jan, Y. N., and Jan, L. Y. (1993). Primary structure and functional expression of a rat G-protein-coupled muscarinic potassium channel. Nature *364*, 802–806.

MacKinnon, R. (1991). Determination of the subunit stoichiometry of a voltage-activated potassium channel. Nature *350*, 232–235.

MacKinnon, R. (1995). Pore loops: an emerging theme in ion channel structure. Neuron *14*, 889–892.

MacKinnon, R., Cohen, S. L., Kuo, A., Lee, A., and Chait, B. T. (1998). Structural conservation in prokaryotic and eukaryotic potassium channels. Science *280*, 106–109.

MacKinnon, R. and Miller, C. (1988). Mechanism of charybdotoxin block of the high-conductance, Ca^{2+}-activated K^+ channel. J. Gen. Physiol. *91*, 335–349.

MacKinnon, R. and Miller, C. (1989). Mutant potassium channels with altered binding of charybdotoxin, a pore-blocking peptide inhibitor. Science *245*, 1382–1385.

MacKinnon, R., Reinhart, P. H., and White, M. M. (1988). Charybdotoxin block of Shaker K^+ channels suggests that different types of K^+ channels share common structural features. Neuron *1*, 997–1001.

MacKinnon, R. and Yellen G. (1990). Mutations affecting TEA blockade and ion permeation in voltage-activated K^+ channels. Science *250*, 276–279.

Maduke, M., Miller, C., and Mindell, J. A. (2000). A decade of CLC chloride channels: structure, mechanism, and many unsettled questions. Annu. Rev. Biophys. Biomol. Struct. *29*, 411–438.

Milkman, R. (1994). An *Escherichia coli* homologue of eukaryotic potassium channel proteins. Proc. Natl. Acad. Sci. 91[9], 3510–3514.

Morais-Cabral, J. H., Zhou, Y., and MacKinnon, R. (2001). Energetic optimization of ion conduction rate by the K^+ selectivity filter. Nature *414*, 37–42.

Morais-Cabral, J. H., Lee, A., Cohen, S. L., Chait, B. T., Li, M., and MacKinnon, R. (1998). Crystal structure and functional analysis of the HERG potassium channel N-terminus: a eukaryotic PAS domain. Cell *95*, 649–655.

Neher, E. and Sakmann, B. (1976). Single-channel currents recorded from membrane of denervated frog muscle fibres. Nature *260*, 799–802.

Nishida, M. and MacKinnon, R. (2002). Structural basis of inward rectification: Cytoplasmic pore of the G protein-gated inward rectifier GIRKI at 1.8 Å resolution. Cell *III*, 957–965.

Perozo, E., Cortes, D. M., and Cuello, L. G. (1999). Structural rearrangements underlying K^+-channel activation gating. Science *285*, 73–78.

Pongs, O., Kecskemethy, N., Muller, R., Krah-Jentgens, I., Baumann, A., Kiltz, H. H., Canal, I., Llamazares, S., and Ferrus, A. (1988). Shaker encodes a family of putative potassium channel proteins in the nervous system of *Drosophila.* EMBO J. *7*, 1087–1096.

Ranganathan, R., Lewis, J. H., and MacKinnon, R. (1996). Spatial localization of the K^+ channel selectivity filter by mutant cycle-based structure analysis. Neuron *16*, 131–139.

Roux, B. and MacKinnon, R. (1999). The cavity and pore helices in the KcsA K^+ channel: electrostatic stabilization of monovalent cations. Science *285*, 100–102.

Ruta, V., Jiang, Y., Lee, A., Chen, J., and MacKinnon, R. (2003). Functional analysis of an archeabacterial voltage-dependent K^+ channel. Nature *422*, 180–185.

Schrempf, H., Schmidt, O., Kummerlen, R., Hinnah, S., Muller, D., Betzler, M., Steinkamp, T., and Wagner, R. (1995). A prokaryotic potassium ion channel with two predicted transmembrane segments from *Streptomyces lividans.* EMBO J. *14*, 5170–5178.

Schumacher, M. A., Rivard, A. F., Bachinger, H. P., and Adelman, J. P. (2001). Structure of the gating domain of a Ca2+-activated K^+ channel complex with Ca^{2+}/calmodulin. Nature *410*, 1120–1124.

Sigworth, F. J. (1994). Voltage gating of ion channels. Q. Rev. Biophys. *27*, 1–40.

Tempel, B. L., Papazian, D. M., Schwarz, T. L., Jan, Y. N., and Jan, L. Y. (1987). Sequence of a probable potassium channel component encoded at Shaker locus of *Drosophila.* Science *237*, 770–775.

Yellen, G. Jurman, M. E., Abramson, T., and MacKinnon, R. (1991). Mutations affecting internal TEA blockade identify the probable pore-forming region of a K^+ channel. Science *251*, 939–942.

Yuan, A., Santi, C. M., Wei, A., Wang, Z. W., Pollak, K., Nonet, M., Kaczmarek, L., Crowder, C. M., and Salkoff, L. (2003). The sodium-activated potassium channel is encoded by a member of the Slo gene family. Neuron *37*, 765–773.

Zhou, M. and MacKinnon, R. (2004). A mutant KcsA K^+ channel with altered conduction properties and selectivity filter ion distribution. J.Mol.Biol. 338, 839–846.

Zhou, M., Morais-Cabral, J. H., Mann, S., and MacKinnon, R. (2001a). Potassium channel receptor site for the inactivation gate and quaternary amine inhibitors. Nature *411*, 657–661.

Zhou, Y. and MacKinnon, R. (2003). The occupancy of ions in the K^+ selectivity filter: Charge balance and coupling of ion binding to a protein conformational change underlie high conduction rates. J. Mol. Biol. *333*, 965–975.

Zhou, Y., Morais-Cabral, J. H., Kaufman, A., and MacKinnon, R. (2001b). Chemistry of ion coordination and hydration revealed by a K^+ channel- Fab complex at 2.0 Å resolution. Nature *414*, 43–48.

Chemistry 2004

AARON CIECHANOVER, AVRAM HERSHKO and IRWIN ROSE

"for the discovery of ubiquitin-mediated protein degradation"

THE NOBEL PRIZE IN CHEMISTRY

Speech by Professor Lars Thelander of the Royal Swedish Academy of Sciences. Translation of the Swedish text.

Your Majesties, Your Royal Highnesses, Ladies and Gentlemen,

This year's Laureates in Chemistry are being rewarded for their discovery of life's own death-labeling system and at the same time for having solved a scientific mystery.

The cells in our bodies contain about one hundred thousand different proteins. Proteins do all the work in the cell, and proteins are directly responsible for its shape and function. One of the things proteins do is to build up the molecular machines that form our muscles; another is to form the enzymes that accelerate and control the various chemical reactions that are necessary for life.

Now, how can a cell possibly keep track of all its proteins? Protein molecules are synthesized and broken down all the time at a high rate. As to the *synthesis* of proteins, we have a good understanding of how this is regulated at molecular level, and the research here has been rewarded with a number of Nobel Prizes. In contrast, the *break down* of proteins in the cell has been considered less interesting, and it has attracted few researchers.

This year's Laureates in Chemistry, Aaron Ciechanover, Avram Hershko and Irwin Rose, went against the stream. They studied precisely how the breakdown of proteins is regulated in the cell. What aroused their interest was reports in the literature that the breaking down of proteins *inside* living cells requires energy. This seemed a paradox since everybody knew that, for example, the degradation of proteins in the intestines – that is, *outside* the cell – takes place with no requirement for added energy. Why is energy needed for degradation *inside* cells?

By studying the mechanisms of energy-dependent protein degradation in cell extracts, this year's three Laureates succeeded at the beginning of the 1980s in identifying a completely new principle for protein degradation. They discovered a system that used a type of "death label" together with three different enzymes to attach it to the proteins to be destroyed. The energy goes to activating the label and enabling the cell to control the process accurately.

The death label itself is a small protein called ubiquitin. The name comes from the Latin *ubique*, which means 'everywhere' and tells us that the protein is found in the cells of almost all organisms. Among all the proteins in the cell, the enzyme system chooses a certain unwanted protein molecule and tags it with the death label. The labeled protein molecule is transported to a type of waste disposer inside the cell called the proteasome, which recognizes the label rather like a key fitting into a lock. Before the labeled protein is sucked into the waste disposer for destruction, the label is removed so that it can be used again to death label more proteins. Inside the waste disposer, the doomed protein is chopped to pieces. These can later be used for synthesizing new proteins.

It was first believed that controlled protein degradation is used only to destroy faulty proteins, which may otherwise damage the cell. Prion diseases and Alzheimer´s disease represent similar cases. Yet constantly growing research has shown something more: that it is at least as important for the cell, by destroying a protein with a certain function, to be able to regulate a biochemical reaction, rather like when you turn off a switch. We now know that controlled protein degradation regulates other very important processes in the cell as well. Examples are the cell cycle, repair of DNA damage and immune defense. In plants, the process is also needed to prevent self-pollination. And a failure in our degradation system may lead to disease.

The discovery of controlled protein degradation explains, at molecular level, the function of a regulation system that is very central for the cell. Among other things, the knowledge can be used to produce new medicines against different diseases.

Professor Ciechanover, Professor Hershko, Professor Rose,

Your discovery of a system for controlled protein degradation in cells has fundamentally changed our way of thinking about protein degradation. We can now understand at molecular level how the cell controls a number of central biochemical processes. More reactions regulated by ubiquitin-mediated protein degradation are being identified every year. On behalf of the Royal Swedish Academy of Sciences, I wish to convey to you our warmest congratulations, and I now ask you to step forward to receive the Nobel Prize in Chemistry from the hands of His Majesty the King.

AARON CIECHANOVER

THE FORMATIVE YEARS – CHILDHOOD IN THE NEWLY BORN STATE OF ISRAEL

I was born in Haifa, a port city in the northern part of Israel, in October 1947, one month before Israel was recognized by the United Nations (UN) as an independent state. It took several additional months to establish the necessary institutions and for the British to leave, and on May 15th 1948, David Ben-Gurion, the founding father of the modern Jewish state and its first Prime Minister made Israel a fact and declared its establishment as a democratic state and a home for every Jew in the world. The neighboring, but even more distant Arab countries, along with powerful Arab parties from within, did not accept the UN resolution and deliberately decided to alter it by force. A bloody and costly war erupted. It lasted a year, and more than 1% of the population of the newly born and defenseless state sacrificed their lives on its defense. I assume that the first two years of my life (1947–1949) were extremely difficult for my parents, Bluma (nee Lubashevsky) and Yitzhak, who immigrated from Poland with their families as adolescents in the mid 1920s. Why did their families leave Poland – their "homeland" – their houses, working places, property, relatives and friends, and decided to make their new home in a place with a vague, if any, clear future, that was part of the British Empire. They were idealists who enthusiastically followed the call of the Zionist movement that was established at the turn of the century by Benjamin Ze'ev Herzel – the seer of the Jewish State – to settle the land and make it – after two thousand years in the Diaspora, since the destruction of the temple in Jerusalem – a home for the Jews. Following the Jewish Congress in Basel (Switzerland) in 1896, Herzel declared: "In Basel I founded the Jewish State". At that time Israel was part of the Ottoman Empire and became in 1917 part of the British Empire. My parents came from religious families, and the move, I believe, had also religious roots: Jews, throughout their lives in the Diaspora, have not stopped dreaming on having their own country, with Jerusalem as its capital, a dream that was driven by a biblical decree and prophecy: *"Thus saith the Lord GOD: Behold, I will take the children of Israel from among the nations, whither they are gone, and will gather them on every side, and bring them into their own land" (Ezekiel 37:21); "And they shall dwell in the land that I have given unto Jacob my servant, wherein your fathers dwelt; and they shall dwell therein, they, and their children, and their children's children, for ever" (Ezekiel 37:25); "And I will rejoice in Jerusalem, and joy in my people; and the voice of weeping shall be no more heard in her, nor the voice of crying" (Isaiah 65:19); "And they shall build houses, and inhabit them; and they shall plant vineyards, and eat the fruit of them" (Isaiah 65:21).*

The question of timing was an important one, as despite centuries of continuous persecution and discrimination in Europe, the initial idea to establish a Jewish State had been the dream of a few. Only small groups of Jews settled in Israel during the 18th, 19th, and the beginning of the 20th century. It was only towards the end of the 19th century, with the ideas of Herzel and the moves that led to the Balfour declaration (the British Minister of Foreign Affairs who declared in 1917 the recognition in the need for a Jewish homeland), that an active Zionist movement and institutions were established, resulting in the translation of the dream into reality. Yet, it took an enormous amount of courage and daring by these European Jews to materialize this dream and try to establish, with almost no resources or support, a homeland in a place they had dreamt of for two thousand years, but that was not theirs at the time. The process was clearly accelerated by the heavy clouds that then covered the skies of Europe and that ended with the Holocaust. Many members of my parents' families immigrated to Israel before the Holocaust, but those who remained in Poland were perished by the murderous German and their loyal Polish collaborators. The conversion of this movement into a State at that particular time (1947–1948) was no doubt the direct historical result of the holocaust, and symbolized the rise of the Jewish Nation from ash.

My father was a clerk in a law firm (later – along with my brother – he studied law and became a lawyer), and my mother was a housewife and English teacher. My brother, Joseph (Yossi), who is 14 years older then me, was already on his national military compulsory service when I was 4 years old, the age from which I remember myself. I grew up in Haifa and enjoyed the wonderful beaches and Mount Carmel that rolls into the Mediterranean Sea. From my early days at home I remember a strong encouragement to study. My father worked hard to make sure we obtain the best possible education, and at the same time he was a member in the "Hagannah" (defense), one of the pre-state military organizations that fought the British for an independent Jewish State. Working in a law firm in the Arab section of the city, he risked his life daily going to work during the pre independence war hostilities and then the war time. My brother told me the family was waiting daily on the balcony to see him returning home safely. At home he used every free minute to delve into classic literature, Jewish religious law (Mishnah and Talmud) and modern law books. An important part of the education at home involved Judaism and Zionism. On the Jewish side we obtained a liberal modern orthodox education. We attended services in the synagogue every Saturday and during holidays, and celebrated at home all Jewish feasts. Needless to say that my mother kept a kosher kitchen. It was extremely important for my parents to educate us as a new breed of proud Israeli Jews in their own independent country. My father inherited me with his love of Jewish studies and cultural life. To this very day, along with several physicians and scientists colleagues, I take regular lessons taught by a rabbinical scholar, on how the Jewish law views moral and ethical problems related to modern medicine and science. Jewish cantorial music reflecting prayers of Jews along many centuries has become my favorite music, and I avidly search for this vanishing vocal expression of Jewish culture in flee markets, used records stores, and auctions all over. Also, different Judaica artifacts decorate my study. In parallel, my parents made sure we should receive an excellent general education. My father spoke fluently several languages, Hebrew, Polish, Arabic, French, English, German and Yiddish, and wanted me to acquire his strong love for books: while our home was not a rich one, we had a huge library. My parents also loved classical music, so we had a great collection of 78 rpm, and later 33 rpm records. I remember that Bizet's Carmen occupied more than twenty, double-sided, RCA (His Master's Voice) 78 rpm bakelite records. The apparently peaceful life of our family in Israel (although under the British Crown) during the years of the Holocaust in Europe (1939–1945) were overshadowed by the murder of their family members and of many families of their relatives and friends that did not escape Europe in time. For my parents, the establishment of the State of Israel as an independent and sovereign Jewish State was a direct historical result of the Holocaust in Europe and a clear statement of *"Never Massadah shall fall again"* (Massadah was one of the last strongholds of Jews during the Roman Empire. It fell in Roman hands after a long curfew during which all its defenders committed suicide in order not to fall as prisoners in Roman hands. While asiring for freedom, they lost their land and lives. They were not ready to live anywhere or under any circumstances, but as free people in their own land). They left us with the idea that the Jewish State will not only protect us as free people, but will allow us to develop our own unique culture in a more general national context, rather than as minorities scattered in different countries in the Diaspora.

FALLING IN LOVE WITH BIOLOGY

From early days I remember my strong inclination towards biology, though it has taken different directions at different times. I remember collecting flowers on Mount Carmel and drying them in the heavy Babylonian Talmud of my brother. I will never forget his rage discovering my love of nature hidden among the pages of the old Jewish tracts. Then came the turtles and the lizards, and extracting chlorophyll from leaves with alcohol, and the first microscope my brother bought me from his trip to England when I was 11 years old. With this microscope I discovered cells (in the thin onion epithelium) and did my first experiment in osmosis, when I followed the alteration in the volume of the cells after immersing the epithelium in salt solutions of different strengths. With friends we tried to launch a self propelled rocket. The flowers collection kept growing, now in special dedicated albums, and with it, a small collection of skeletons of different animals – fish, frog, snake, turtle, and even some human bones I received from an older friend who was a medical student. After several years of amateurish flirting with biology, I decided to formalize my knowledge and love of biology, and to major in Biology in high school. While my years in elementary (1953–1959) and junior high (1959–1963) school were mostly uneventful and passed without any thoughts on my future, the last two years in "Hugim" (circles) high school in Haifa (1963–1965) were not. I had wonderful and inspiring teachers in biology (Naomi Nof), chemistry (Na'ama Greenspon), and physics and mathematics (Harry Amitay) who revealed to me a little of these different and exciting disciplines. Yet, I felt that twice as much was still concealed. Biology at that time was largely a descriptive area. While we studied the mechanism of conversion of glucose to H_2O and CO_2 and production of energy in yeast and mammals (and the opposite process occurring during photosynthesis in plants), and became acquainted with simple graphic descriptions of mitotic and meiotic cell divisions, most of our studies were devoted to detailed descriptions of the flora and fauna in our region, to comparative zoology (I remember well the efforts invested in memorizing the twelve differences between the frog and the toad, or between the circulatory systems and skeletal structure of the cat and dog), and to basic descriptive human anatomy and physiology (e.g. how the human skeleton structure enables posture on two limbs). Pathogenetic mechanisms of diseases had not been taught, and the structure of DNA and the genetic code had entered our textbooks only towards the end of our high school studies, in 1964/5. On the other hand, chemistry and physics appeared to me, maybe naively, strong mechanistic disciplines built on solid mathematical foundations. As a result, I had a deep feeling that the future somehow resided in biology, in deciphering basic mechanisms, as so little was then known. Yet, the complexity of biological and pathological processes looked to me enormous, almost beyond our ability to grasp, and I was intimidated: while I was clearly attracted to the secrets of biology, I was afraid to get lost. Importantly, I had nobody around, close enough, to consult, to clarify my thoughts. While deliberating between the largely unknown in biology and what I naively thought were the already well founded physics and chemistry, medicine emerged as a compromise: it appeared to me as representing a balanced mixture of physics, chemistry, basic biology and physiology, along with interesting pathology and social sciences.

Adding to this complexity was that along these years I lost both of my parents: my mother died in 1958 and my father in 1964. After the death of my mother, I was left with my father who took wonderful care of me. When my father died several years later, my late aunt Miriam (Wishniak; my mother's sister), with the help of my brother and sister-in-law, Atara, took me to her home in Haifa, enabling me to seamlessly complete my high school studies in the same class and along with my friends – without interruption. The other option was to move to Tel Aviv, to my brother's home, but this would have been much more complicated. So I spent the weekdays with my aunt in Haifa, studying, and the weekends and holidays with my brother and sister-in-law, in Tel Aviv. Their help was a true miracle, as thinking of it retrospectively, being left alone without parents at the age of sixteen, the distance to youth delinquency was shorter than the one to the high school class. Yet, with the help of these wonderful family members, I managed to continue.

HOW MY LOVE TO BIOLOGY EVOLVED TO BECOME A CAREER

Towards graduation from high school I had to make a decision. The regular track would have taken me, like most Israelis, to national compulsory service in the Israeli Defense Forces, IDF, a duty we were all eager to fulfill. In addition to the regular service, the army encourages certain high school graduates to postpone their service and first obtain a university education, particularly in areas that are relevant to the military, such as medicine and different disciplines in engineering and sciences. Lacking any financial support, I thought it would be better to acquire a practical profession I could make a living from as soon as I could. As I mentioned, medicine emerged as a compromise between the complexity and mysteries of biological mechanisms to what I thought are the already well founded physics and chemistry. Not less important, medicine has traditionally been the ultimate in "Jewish" professions, the dream of every Jewish mother and family. What also attracted me to medicine is that I was under the impression that diseases can be cured: as children, we may have been influenced by short, self-limiting diseases that affected us, like influenza and measles, and were not directly aware of the major killers that left physicians and scientists alike helpless (much like these days), such as malignancies, vascular diseases and neuro-degenerative disorders: I had not appreciated at the time how far more descriptive medicine is, much more than biology. Practically and not less important (which helped solve my dilemma), was the fact that biology was not an option in this military-supported service postponement program. Last but not least, it was a practical choice, a profession one can make a living on. So, after a fierce competition I was accepted into the only medical school in Israel at that time, that of the Hebrew University and "Hadassah" in Jerusalem (1965). The first four years (1965–1969) were exciting. We studied basic and clinical sciences, and I started to seriously entertain the idea of broadening my knowledge base in biochemistry or pharmacology. Towards the end of the 4th year, once we started to examine patients, serious doubts had begun to arise whether I made the right choice and truly want to become a practicing physician. The imbalance between phenomenology and pathogenetic mechanisms of diseases on one hand, and the lack of any mechanism-based treatment for most of the major killers on the other hand, made me seriously think that I am on the wrong trail. I felt restless and started to realize how little we know, how descriptive is our understanding of disease mechanisms and pathology, and as a consequence how most treatments are symptomatic in nature rather then causative. The statement "with God's help" I heard so frequently from patients that were praying for cure and health received a real meaning. I had a feeling clinical medicine was going to bore me, and decided to take one year off in order to "taste" true and "wet" basic research. The Faculty of Medicine had a special, one year program for the few who elected to broaden their knowledge in basic research, and I decided to major in biochemistry. I had to convince my brother that this was the right thing to do, as I needed his help to further postpone my military service by one year. This was not easy, as he too had a "dream" – to see me independent with a profession from which I could make a living, and which in the traditional Jewish spirit was nothing else but practical medicine. Following our parents' death, he felt he was responsible for my future and well being, and wanted to see me professionally and financially independent as soon as he could. I nevertheless managed to convince him, and during that year (1969–1970), under the guidance of excellent biochemists, Jacob Bar-Tana and Benjamin Shapira, I investigated mechanisms of CCl_4-induced fatty liver in a rat model, and discovered that it may be caused, at least partially, by an increased activity of phosphatidic acid phosphatase, a key enzyme involved in di- and triglycerides biosynthesis. Completing this research year (and obtaining a M.Sc. degree), I knew I had found a new love – biochemistry. Jacob and Benjamin walked me through the exciting maze of biochemical pathways, and I was mystified. Yet, the consummation was still far away. Being loyal to the promise I made to my brother, and also to my commitment to the Israeli army, I completed the clinical years (1970–1972) and graduated Medical School.

To obtain my medical license, I still had to complete one additional year of rotating internship. At that time, colleagues told me that a young talented biochemist, Dr. Avram Hershko, had just completed his post-doctoral training with Gordon Tomkins at the University of California in San Francisco (UCSF) and was recruited by the Dean and founder of the newly established Faculty of Medicine at the Technion in Haifa, the late Professor David Ehrlich, to establish a Unit of Biochemistry. I wrote to Avram, with the intention to relocate to Haifa, to carry out my rotating internship there, and to use this year to complete my M.D. research thesis under his supervision. This was a small thesis I had to submit to the Medical School in partial fulfillment of the requirements for graduation. Typically for this thesis, most medical students are evaluating statistically on-going treatments/procedures, but I decided to return to the laboratory and touch on yet another research project in Biochemistry. Avram agreed to accept me as an M.D. research student, and in October 1972 we started our more than three decades voyage. Avram was still not certain about his own main research direction, and we discussed two possibilities for my M.D. thesis. One was obviously to further dissect the tyrosine aminotransferase (TAT) ATP-dependent proteolytic pathway: Avram started his own trip into the world of intracellular proteolysis with Gordon and discovered that the degradation of the gluconeogenetic enzyme TAT in cells requires energy. This was a corroboration of earlier findings of Simpson who demonstrated in the early 1950s that the degradation of the entire population of cellular proteins in liver slices requires energy. Yet, the mechanism(s) of this thermodynamically paradoxical requirement had remained elusive. The other possibility was to study the mechanism(s) involved in the cell's "pleiotropic response" – the immediate response of serum-starved, G0 synchronized cells to the addition of serum. During his post-doctoral studies with Gordon, Avram found that among the many stimulated processes that follow the addition of serum, are rapid uptake of nucleotides, amino acids, and phosphate. As during my studies on fatty liver I acquired experience in analyzing lipids, and since Avram felt the elucidation of the TAT proteolytic mechanism may be a too difficult undertaking for a limited-in-scope M.D. thesis, we decided to add one additional layer to the study on the "pleiotropic response" and to analyze the effect of serum on synthesis of phospholipids. We assumed that following serum addition, cell membranes undergo major changes that will be reflected in phospholipid metabolism. Indeed, a few minutes after serum addition, we were able to detect a dramatic increase in the turnover of the phospho-inositol moiety on the diglycerol skeleton. A review of the literature revealed a similar effect in different target cells in response to a broad array of stimuli, including parasympathetic secretory cells responding to acetylcholine and thyroid gland cells to their cognate hormones, thyrotropin (TSH). The year (1972–1973) I spent in the laboratory (it was not a real year but rather moon lighting, as a significant part of the time I was busy in the hospital, rotating among the different clinical departments, completing my internship and duties towards graduation. I worked in the laboratory in my free evenings, nights, weekends and holidays) finally convinced me to pursue a career in Biochemistry. But I still had three years of military service ahead of me (1973–1976).

MY MILITARY SERVICE AND PROFESSIONAL CAREER – HAVE THEY COLLIDED WITH ONE ANOTHER?

Following graduation, it was time to repay my national debt and serve in the IDF. I served for three years (1973–1976) and did it gladly. Serving in the army has always been regarded as an integral and important part of Israeli life, and an entry card to its society, giving one the feeling of sharing – everyone takes part in protecting this land and its inhabitants. In addition, the service itself was extremely interesting, technically, but also socially and historically. Technically, since I served in interesting units. Socially, since the military service is a wonderful humane experience, the best melting pot one can go through, generating true friendships during hard times, friendships that are therefore deep, true and lasting. Historically, it spanned an interesting period. Initially I served in the navy, as a physician in the missile boats fleet. The year was 1973, immediately after the October Day of Atonement (Yom Kippur) war, and Israel faced a problem of protecting its southern gates, the Red Sea and the marine entry to its port in Eilat. Marine transportation through the southern gates of the Red Sea, Bab-el-Mandeb strait, and the narrow Tiran (Sharm-a-Sheikh) strait were threatened by the Arab countries that neighbored the water way, mostly Saudi Arabia and Egypt, but also Yemen and Somalia, and Israel had to stretch its marine arm. To do so, it was necessary to transfer missile boats from the main naval bases in the Mediterranean to the Red Sea. At that time Israel did not have diplomatic relationship with Egypt, and the Suez Canal was closed by ships sunken by the Egyptians during the June 1967 Six Day War. Thus, the decision was to bring the boats from Haifa to Eilat, sailing via the Mediterranean Sea, the Gibraltar strait and around the West and then East coasts of Africa. I was the physician on the "Reshef", one of the two modern Israeli missile boats that were built in the Haifa naval shipyard. One can imagine that for small missile boats, such a long, several weeks voyage, a large part of it in the open Oceans, is rather complicated, and for many reasons also risky. Beyond fuelling and provision of supplies and spare parts to the crews and boats, one has to think of sailing in waterways surrounded by hostile countries, many miles away from home and a long flight distance for the Israeli Air Force. Another problem was obviously medical, how one treats emergencies, from possible gunshot wounds through "simple" daily problems like appendicitis, in a small ship, far from any medical facility and with limited diagnostic and treatment capabilities. I was particularly concerned, as I was a young physician with almost no clinical experience. I assume this would have been a challenge for more experienced physicians as well. Luckily, the voyage was smooth. The remaining part of my three years service was also interesting. I spent that time in the Research and Development (R & D) unit of the Medical Corps, developing a broad array of sophisticated devices for the soldier in the battlefield. Because of the broad range of experiences acquired, the military service has been my ever best school for real life "sciences". During all these years (1973–1976) I maintained tight connections with Avram and fulfilled my duties as an "external" department member: during vacations from the military, and along with other members of the department that grew meanwhile, I taught continuously the course in Clinical Biochemistry to 3rd year medical students. I should mention in particular Michael (Mickey; see also below) Fry with whom I have remained a good friend to the very present. Also, in 1975, during the military service, I married Menucha, a physician and a graduate of Tel Aviv University School of Medicine. Menucha was a resident in internal medicine in Tel Aviv Municipal Hospital, and we built our first home in this city. Marrying Menucha brought my wanderings to an end, and I felt I had again a family and a home. During all the years since the death of my father (1963–1975) I did not have a real stable home, and I wandered between the homes of my brother and sister-in-law in Tel Aviv and of my aunt in Haifa. They were truly wonderful, but I needed a base, and Menucha, with her quiet approach and warm acceptance, along with our beautiful apartment, provided me with this, so much needed, shelter.

DISCOVERY OF THE UBIQUITIN SYSTEM – GRADUATE STUDIES

Towards the end of the military service, I had to make what I assume has been the most important decision in my career: to start a residency in clinical medicine, in surgery, which was my favorite choice, or to enroll into graduate school and start a career in scientific research. It was clear to me that I was heading to graduate school. My disillusionment from clinical medicine that diseases can be cured based on understanding of their pathogenetic mechanisms, along with a magical and enchanting attraction to Biochemistry made the decision easy. I received a strong support and encouragement from my wife Menucha, who started to realize she was married to a student in sciences with no clear future rather than to a physician with a bright career and broad financial horizons that she thought she had married. So in November of 1976, after my discharge from the national military service and a two months driving trip across the USA, I started my graduate studies with Avram Hershko. At that time his group focused mostly on studying intracellular proteolysis, and I learnt from him that he had given up on trying to identify the mediator(s) and mechanism(s) involved in serum-induced "pleiotropic response". The model system that was chosen to study proteolysis was degradation of abnormal hemoglobin in the reticulocyte which is the terminally differentiating red blood cell. The reason for the selection of the reticulocyte as a model system was that we were looking for a non-lysosomal and energy-requiring proteolytic system, as from many studies it had become clear that regulated proteolysis of intracellular proteins is non-lysosomal (see the accompanying Nobel Lecture), and the reticulocyte no longer contains lysosomes which are removed during the final stages of its maturation (see below) before its release into the circulation. Interestingly, in the summer of 1978, during a Gordon Conference on Lysosomes, I met Dr. Alex Novikoff from Yeshiva University School of Medicine in New York. Alex, along with Dr. Christian de Duve, was one of the pioneers of the lysosome research field. When I told him we were working on the reticulocyte because this cell does not have lysosomes, he angrily dismissed this argument, telling me that he characterized morphologically acid phosphatase-positive organelles in reticulocytes. He even gave me the relevant paper he published on the subject, though it was not clear that these are proteolytically functional organelles. Another reason for the choice of the reticulocyte as a model for studying intracellular proteolysis was that in its final stages of maturation in the bone marrow and prior to entering the peripheral circulation, a massive proteolytic burst destroys most of its machineries, making it clear that the cell is equipped with an efficient proteolytic system. Earlier studies by Rabinovitz and Fisher demonstrated that the reticulocyte degrades abnormal, amino acid analogue-containing hemoglobin, yet the mechanisms had remained elusive. We assumed that the same mechanism that is involved in differentiation and maturation of the reticulocyte is also involved in the removal of "naturally occurring" mutant abnormal hemoglobins that are synthesized in different hemoglobinopathies, such as thalassemias and sickle cell anemia, and also in the destruction of the amino acid analogs – containing abnormal hemoglobins. We wanted to believe and hoped that this mechanism would turn out to be "universal", and involved in degradation of normal proteins in all cells. Years later this assumption turned out to be correct. Thus, this important piece of information – the existence of a non-lysosomal proteolytic system, made the choice of the reticulocyte an obvious one. It was still necessary to demonstrate that the process requires energy, and indeed, following our initial characterization of degradation of abnormal hemoglobin in the intact cell we showed that the process required energy (was published in 1978 in the proceedings of a proteolysis meeting held in Buffalo, NY), and felt that the time was ripe to break the cell open and isolate and characterize the non-lysosomal and ATP-dependent proteolytic enzyme(s). Shortly before, in 1977, Dr. Alfred Goldberg and his post-doctoral fellow Dr. Joseph Etlinger at Harvard Medical School characterized, for the first time, a cell-free proteolytic system from reticulocyte, which was exactly the point where we wanted to start our own march, so we basically adopted their system.

I will not describe here the detailed history of the discovery of the ubiquitin system, but rather highlight two important points along the five years of my exciting graduate studies (1976–1981) with Avram and Irwin A. Rose (Ernie) that led to the discovery of the system. The more detailed history can be found in several review articles written on the system at that time (most notable is Hershko, A. and Ciechanover, A. (1982). Mechanisms of Intracellullar Protein Breakdown. *Annual Review of Biochemistry*, **51**, 335–364.) and later, and in the accompanying Nobel Lecture.

(1) The first point relates to the multiplicity of enzymatic components of the system: Our first aim along the purification process of the ATP-dependent "protease" was to remove hemoglobin, the major protein in the crude extract. Towards that end, we resolved the extract on an anion exchange resin, where we encountered already the first exciting finding. The proteolytic activity could not be found neither in the non-adsorbed material which we denoted Fraction I, nor in the high salt eluted material (denoted Fraction II). Rather, we recovered the activity following reconstitution of the two Fractions. We learnt two important lessons from this experiment which was published in 1978 in *Biochemical and Biophysical Research Communications* (BBRC; in my opinion the first paper in the long historical trail of the ubiquitin system) and which I regard as one of two or three key publications in the field. We learnt two lessons from this experiment: (i) The first was that the protease we were after was not a "classical" single enzyme that degrades its substrate, but had at least two components. This was already a digression from the paradigm in the field at that time that proteolytic substrates, almost without exception, can be cleared, at least partially, by single proteases with limited, yet defined specificities. Now, following the unraveling of the human genome and the discovery of important common structural domains within several groups of enzymes of the system, we know that the number of components of the ubiquitin system exceeds one thousand, but the first hint was already there; once one is left without a paradigm, all possibilities are open. (ii) The second lesson was a methodological one. Each time we lost activity during purification of any of the components we were characterizing, we "returned" to the chromatographical column fractions and tried to reconstitute it via complementation: "classical" biochemistry at its best was on our side. Standing at a crossroad, we – luckily but thoughtfully – decided to start first with purification and characterization of the active component in Fraction I. We decided so because Fraction I was the hemoglobin-containing fraction that did not adsorb to the resin, and since many proteins do absorb, we thought that this fraction should not contain too many additional proteins beyond hemoglobin, and it would be easy to purify the active component. Ten months after I started my studies

(summer of 1977), Avram started his sabbatical with Ernie at the Fox Chase Cancer Center in Philadelphia, PA, USA, and left me with the task to purify the active component from Fraction I. After many unsuccessful trials along with another graduate student of Avram, Yaacov Hod, my colleague Mickey Fry, who was appointed as my substitute thesis advisor for this year (1977–1978), came with the "crazy" idea to heat Fraction I and see if the active component is heat-stable, and indeed it was. He did so as all our attempts to resolve the activity – despite the large difference in the molecular mass between the active protein (~10 kDa) and hemoglobin, the other major protein in Fraction I (65 kDa) – failed: hemoglobin, that is so abundant, "contaminated" the entire resolution span of each column in every single resolution method we used. Following 5–10 min at 90°C, the hemoglobin in crude Fraction I was "cooked" and precipitated like mud, and the activity remained soluble in the supernatant. It was hard to believe it was a protein, but Mickey remembered several other heat-stable proteins. Immediately after, we showed directly that the activity in Fraction I was also a protein: it was sensitive to trypsin and precipitable with ammonium sulfate. Further characterization revealed that the protein had a molecular mass of ~8,500 Da, and we called it ATP-dependent Proteolysis Factor-1, APF-1, to denote that this was the first component in the system that we characterized. All along the way I corresponded with Avram, sent him my results, and during his sabbatical we wrote the BBRC paper.

(2) The second key finding was also discovered in Haifa during the winter of 1978–1979. We purified APF-1 to homogeneity and labeled it with radioactive iodine. When the radio-labeled protein was incubated in crude reticulocyte Fraction II in the presence of ATP, we observed a dramatic increase in its molecular weight: it now migrated as a sharp peak in the void volume of the gel filtration chromatographical column. For several months we tried to elucidate the mechanism that underlies the change in the molecular weight of APF-1, hypothesizing, for example, that APF-1 could be an activator of a protease that must generate a binary complex with the proteolytic enzyme in order to activate it, but to no avail. An important breakthrough occurred during our 1979 summer stay of several months in the laboratory of Ernie. Through a series of extremely elegant, yet simple, experiments, in the design of which the broad knowledge of Ernie in protein chemistry and enzymology played a critical role, we found that APF-1 is covalently attached to the substrate through a bond that had all the characteristics of a peptide bond. Furthermore, we found that multiple moieties of APF-1 are attached to each substrate molecule, and that the reaction is reversible: APF-1 can be removed from the substrate or it degradation products and recycled, though not via reversal of the conjugation reaction. Accordingly, we hypothesized that covalent attachment of multiple moieties of APF-1 to the target substrate is necessary to render it susceptible to degradation by a downstream protease that recognizes only tagged but not untagged proteins, followed by the release of free and reusable APF-1.

The APF-1 cycle predicted the existence of three, entirely novel activities: (i) APF-1 conjugating enzyme(s), (ii) a protease that recognizes specifically the tagged substrates and degrades them, and (iii) APF-1-recycling enzymes. All these activities were identified later by us (the three conjugating enzymes, E1, E2, and E3) and by others (the conjugates degrading protease known as the 26S proteasome complex, and the ubiquitin recycling enzymes, the isopeptidases; see the accompanying Nobel Lecture). The findings describing the covalent tagging of the target substrate by APF-1 as a degradation signal, along with the first model of the newly discovered proteolytic system, were published in 1980 in two manuscript that appeared in the *Proceedings of the National Academy of Sciences of the USA* (PNAS).

Another important development also occurred during our stay in Ernie's laboratory, and I am not sure whether it was shear luck or serendipity, probably both. We were not aware of any other precedent of a modification of a protein by another protein. The neighboring laboratories of Martin Nemer, Alfred Zweidler, and Leonard Cohen studied dynamics of variants of different histones during sea urchin development. They drew our attention to a protein called A24 (uH2A) which was discovered earlier by Ira Goldknopf and Harris Busch, and that was a covalent conjugate between two proteins – a small, ~8.5 kDa protein called ubiquitin, and histone 2A (H2A). Goldknopf and Busch, and in parallel Margaret Dayhoff, identified the nature of the bond between the two protein moieties in the conjugate. They found that the ubiquitin-histone bond was an isopeptide/bifurcated bond between the C-terminal Gly^{76} residue in the ubiquitin moiety, and the ε-NH_2 group of Lys^{119} in the histone moiety of the conjugate. The role of this conjugate was not clear at the time, though its level was found to be dynamic and change during differentiation, when the histone moiety is subjected to ubiquitination and de-ubiquitination. This information on the ubiquitin-histone adduct along with the striking similarities we found between APF-1 and ubiquitin in their general characteristics such as molecular mass and amino acid composition, led Keith Wilkinson and his colleague Arthur (Art) Haas who were post-doctoral fellows in the laboratory of Ernie, along with Michael Urban from Zweidler's laboratory, to carry out a series of direct experiments, showing unequivocally that APF-1 is indeed ubiquitin. Our study on the characterization of APF-1 and its possible similarity to ubiquitin, and Wilkinson's study (along with Urban and Haas) on the identification of APF-1 as ubiquitin, led to the convergence of two fields, that of histone research and of proteolysis. More important, they suggested that the bond between ubiquitin and the target proteolytic substrate might identical to that between ubiquitin and histone, which turned out later to be true. The two studies were published in tandem in 1980 in the *Journal of Biological Chemistry* (JBC; see the accompanying Nobel Lecture). The identification of the nature and structure of the bond clearly paved the road to the later purification and characterization of the conjugating enzymes and their mode of action.

As for ubiquitin, the protein was identified in the 1970s by Gideon Goldstein (in the Memorial Sloan-Kettering Cancer Center in New York City) as a small, 76 residue thymic polypeptide hormone that stimulates T cell differentiation via activation of adenylate cyclase. Additional studies by

Gideon Goldstein had suggested that it was universally distributed in both prokaryotes and eukaryotes, thus giving rise to its name (coined by Gideon Goldstein). Later studies by Allan Goldstein showed that the thymopoietic activity was due to an endotoxin contamination in the protein preparation, and not to ubiquitin. Using functional assays, it was found in my laboratory (and I believe that in several others as well) that ubiquitin was limited to eukaryotes, and its apparent presence in bacteria was due to contamination of the bacterial extract with the yeast extract in which the bacteria were grown: growing the bacteria in a synthetic medium containing carbon (glucose) and nitrogen (ammonium chloride) sources and vitamins resulted in "disappearance" of ubiquitin from the preparation. The later unraveling of the bacterial genome demonstrated unequivocally that the ubiquitin tagging system does not exist in prokaryotes, though there is some similarity between the proteasome and certain bacterial proteolytic complexes. Thus, in a relatively short period of time, ubiquitin was converted from a ubiquitous thymopoietic hormone to a eukaryotic proteolytic marker. While it appeared that the term ubiquitin was not justified anymore, as it is clearly not ubiquitous, we stopped using the term APF-1 and adopted the term ubiquitin for the modifying protein in the newly discovered proteolytic system. At times habits and tradition are stronger than the scientific validity and/or the logic in nomenclature. Accordingly, we adopted a general policy to use in our terminology the name/term that was first coined by the discoverer of any novel protein.

From that point on, the road was relatively short to the identification and characterization of the conjugation mechanism and the three enzymes involved in this process. En route we followed partially, with great admiration, the footsteps of Dr. Fritz Lipmann, the great biochemist from the Rockefeller University (who was awarded the 1953 Nobel Prize in Physiology or Medicine for the discovery of Coenzyme A). Lipmann continued to contribute enormously to our understanding of basic biochemical processes. Among his many discoveries was the mechanism of non-ribosomal (and hence non-genetically encoded) peptide bond formation that is involved in the biosynthesis of bacterial oligopeptides such as Gramicidin S. We learnt that the basic biochemical principles, such as generation of high-energy intermediates involved in peptide bond formation, were preserved along evolution regardless of whether the bond is encoded genetically or not, or whether it links two amino acids, or two proteins, or an amino acid to the elongoting polypeptide chain. Initially, we identified the general mechanism of activation of ubiquitin in crude extract. Later, using "covalent" affinity chromatography over immobilized ubiquitin and a stepwise elution (that was based on the general mechanism we deciphered earlier), we purified the three conjugating enzymes that act successively, in a cascade-like mechanism, and catalyze this unique process: (1) the ubiquitin-activating enzyme, E1, the first enzyme in the ubiquitin system cascade, (2) the ubiquitin-carrier protein, E2, to which the activated ubiquitin is transferred from E1, and (3) the ubiquitin-protein ligase, E3, the last and critical component in the three step conjugation mechanism that specifically recognizes the target substrate and conjugates it with ubiquitin.

The E3 was also adsorbed to the immobilized ubiquitin, although via an yet unknown mechanism, distinct from that of E1 and E2: the binding of these two enzymes was mediated by the activation mechanism. Later studies by Avram in the late 1980s revealed that the E3 adsorbed by the column was E3α that recognizes substrates via their N-terminal residue. At this point, however, unknowingly and unintentionally, we were extremely lucky when we used as model substrates commercial proteins such as BSA, lysozyme and RNase, that were all recognized (as we learnt later) by this ligase and via a similar targeting motif – their N-terminal residue. Had we used other substrates, such as globin, the protein we used in our initial experiments, the E3 adsorbed to the column would have probably escaped our attention, as E3s do not typically adsorb to ubiquitin. Independently, and in parallel to the later characterization of the enzyme by Avram, I also used this enzyme in order to characterize a distinct subset of proteins recognized via this signal (see below). Last, using antibodies that we raised against ubiquitin with the help of Arthur Haas, we found that the ubiquitin system is involved in degradation of abnormal, short-lived proteins in hepatoma cells, demonstrating that the system is not limited to the terminally differentiating reticulocyte, but is probably distributed "universally" in nucleated mammalian cells, playing an important role in maintaining the cell's quality control, by removing abnormal proteins. During my graduate studies at Avram's laboratory, I collaborated with Hannah Heller, an extremely talented and knowledgeable research associate (who also joined us for some of our summer stays in the laboratory of Ernie in Philadelphia), and with Yaacov Hod who was also a graduate student with Avram at that time. Other colleagues in the laboratory provided me with a lot of help during this period, including Dvorah Ganoth, Sarah Elias, and Esther Eythan who were research associates with Avram, and Clara Segal and Bruria Rosenberg, two dedicated technicians.

THE INTERACTION WITH IRWIN ROSE

As noted, I spent an important part of my graduate studies in Ernie's laboratory. Avram spent a sabbatical in his laboratory in 1977–1978, and I joined him for the first time for several months in the summer of 1978, after I completed the initial characterization of APF-1 in Haifa. I returned to Ernie's laboratory during the summers of 1979, 1980, and 1981. As noted, during our summer stay in 1979, we resolved the problem of the nature of the high molecular mass "compound" generated when APF-1 was incubated with Fraction II in the presence of ATP. This change in the molecular mass of APF-1 was discovered several months earlier in Haifa, however, we were not able to unravel the nature of the "compound"; this had to await the knowledge and wisdom of Ernie. In a breakthrough discovery, we found that the target substrate is covalently modified by multiple moieties of APF-1, a reversible modification that renders the protein substrate susceptible to degradation. This was a novel type of post-translational modification (see however above for the modification of histone H2A by ubiquitin) and clearly a new biological paradigm, that the elucidation of which required – as I feel today in retrospect – a different type of knowledge in biology and enzymology, and an original experimental approach. Elucidation of this modification would not have been possible without Ernie's advice that was based on his immense knowledge in enzymology and protein chemistry, accompanied by his unbiased original thinking and approach to problem resolving. This discovery, along with the discovery in 1980 that APF-1 is ubiquitin, made Ernie and his fellows critically important partners in the historical trail of the discovery of the ubiquitin system. Interestingly, Ernie studied proteolysis before Avram joined him first, but had never published in the field before.

POST-GRADUATE TRAINING AT M.I.T. AND HOW I CONTINUED MY STUDIES ON THE UBIQUITIN SYSTEM INDEPENDENTLY

The five years in graduate school had a significant impact on my future career. Not only because I played an active part in the discovery of such an important pathway, but maybe more important, because I learnt several basic and key principles of how to approach a scientific problem. From my mentors I learnt two principles: first, to select an important biological problem, preferarbly an unobvious one and not in the mainstream, and second, to make sure that there are appropriate research tools to approach it experimentally. I also learnt to become a long books rather than a short stories writer: I learnt not to be opportunistic but rather to adhere to a project, to dig deeply into a problem, to resolve it mechanistically, to unravel complex mazes – peeling them like an onion, not to be tempted to be dragged after fashions. I learnt to pay attention to small details, to carefully examine hints, as the important findings are not always obvious from the apparent results. I learnt to be stubborn, to fight difficulties uphill, but most importantly, to be critical: I believe I developed good senses that enable me to distinguish false from truth, and artifacts from meaningful findings. Interestingly, I learnt all these principles not in frontal lessons or formal presentations, but as an apprentice, following my mentors own attitude and way of thinking. At the same time I also learnt to question, to doubt, to ask, and to discuss, to follow my own gut feeling when it was necessary, not to always take advice and direction for granted, and to trust myself too. It did help in many occasions along the way. Thus, at times I found myself swimming alone against the stream. Altogether, these principles generated an important philosophy and shaped my approach to science, something I try to instill to my own students, as I strongly believe it is the only way one can make an impact, leave an imprint behind.

Towards graduation I had to think of the next step – post-doctoral training and planning of my future career as an independent scientist. I was in a dilemma. On the one hand I knew it was important to obtain training somewhere else, under different mentorship, in a different environment, being exposed to a different culture of science. On the other hand I knew for certain that the ubiquitin system was extremely important and that we were seeing only the tip of its iceberg. I therefore wanted to continue my studies in a related field, learning more on regulated proteolysis, but also to continue my own studies on ubiquitin. I had several ideas in mind of where to go, but the choice was quite narrow and also risky, as I did not have any idea of how much independence I could have as a post-doctoral fellow. Searching for a mentor, and with the advice of my colleague Mickey Fry, I looked for scientists whose work was related to regulated proteolysis. I wrote to Günter Blobel in the Rockefeller University who worked at that time on translocation of proteins across the endoplasmic reticulum (ER) membrane, a process that involves cleavage of the leader peptide by signal peptidase, to Jeffrey Roberts in Cornell University who worked on *E. coli* RecA protein-directed cleavage of phage λ repressor and its requirement for polynucleotide, and to Harvey Lodish at the M.I.T. who worked, among other subjects, on processing of viral polyproteins. I am not sure Harvey was that impressed with the ubiquitin system at that time, but he was the only one to respond positively. Typical of his etiquette (as I learnt later), his response was prompt and direct, and he invited me for an interview after which he accepted me. Günter was kind enough to let me know he did not have space in his laboratory at that time, and Jeffrey never responded.

With two fellowships, one from the Leukemia Society of America and one from the Israel Cancer Research Fund, ICRF, I started a period of three wonderful years (1981–1984) in Harvey's laboratory in the Department of Biology at M.I.T. Harvey gave me complete freedom to choose my research subjects. What I had in mind was to take advantage of the exceptional strength of the laboratory and Harvey's unique expertise in cell biology, but in parallel, to continue my own studies on the ubiquitin system. I realized that Harvey was no longer interested in viral proteins processing, and along with Alan Schwartz who was a visiting scientist (from Harvard Medical School) in the laboratory, we started to characterize the transferrin receptor on a human hepatoma cell line with the aim of disseding the dynamics of the receptor and iron delivery into cells. This collaboration led us, along with another fellow in the laboratory, Alice Dautry-Varsat (from the Pasteur Institute) who joined us later, to the discovery of a fascinating mechanism of how iron is delivered into cells, a process mediated by the transferrin receptor: In the neutral pH of the growth medium, the iron-loaded holo-transferrin binds to its receptor with a high affinity and is endocytosed into the cell. At the low endosomal pH, the affinity between iron and transferrin is weakened dramatically. As a result, the iron cation is released, but the apo-transferrin, which has high affinity for the receptor at acidic pH, remains bound strongly. Along with the receptor, the apo-transferrin recycles to the cell surface. At the neutral pH of the growth medium, the apo-transferrin loses its high affinity to the receptor and is released into the neutral pH extracellular fluid. There, it binds iron with high affinity, rebinds to its receptor and the hololigand is endocytosed again. The transferrin-transferrin receptor pH- and iron loading-dependent cycle has become a "classic" in the field of receptor mediated-endocytosis. Based on this mechanism, other phenomena related to receptor and ligand recycling to the cell surface or targeting to the lysosome could be explained, which are also due to the pH difference between the external environment and the interior of the endocytic pathway vesicles. However, throughout this time, I lived under the strong feeling that the ubiquitin system had barely started to emerge, with only the basic principles unraveled: I felt compelled to get back and work on it. So gradually I started to "crawl" and return to my alma mater research subject.

On one fascinating subject I worked on my own, continuing to explore a mysterious finding I discovered during my graduate training and which I did not pursue at the time: When we purified APF-1/ubiquitin in Haifa, we noticed a large discrepancy between its dry weight and its Lowry or 280 nm protein quantitative measurements. We hypothesized that the protein can be a ribonucleoprotein, RNP, and the remaining mass is that of the nucleic acid component. To test this hypothesis, I added DNase to the crude extract in which we monitored ATP- and ubiquitin-dependent degradation of BSA, that was used as one of our model substrates. The enzyme had no effect. We then added RNase A, and to our surprise proteolysis was completely inhibited, even with an extremely small amount – mere few nanograms – of the enzyme added: it looked as if the enzyme exerted its effect via catalysis – RNA degradation. Avram suggested to test the RNase effect also on lysozyme, a protein that was used as our second model substrate. Here we got no effect, which was kind of a surprise, as proteolysis of the two substrates, BSA and lysozyme, behaved in an identical manner all along the way: ATP as well as all the different factors we resolved from the crude extract, were all required for the degradation of both proteins. Avram suspected that the RNase effect could be an artifact. Meanwhile APF-1 was identified by Keith Wilkinson and his colleagues as ubiquitin (see above), and the amino acid sequence/composition of ubiquitin disclosed the "secret" of the dry weight/protein measurement discrepancy – the molecule has a single tyrosine residue. So we decided not to pursue this subject, and the selective inhibitory effect of RNase A on BSA degradation had remained an unsolved mystery – for the time.

I had not stopped suspecting, however, that the findings must represent some true biological phenomenon, and used the opportunity of my independence at Harvey's laboratory to pull out the late 1970s results from my notebook and start dissecting the RNase effect in a systematic manner. With some advice from Alexander Varshavsky (Alex; M.I.T.), and a lot of help and reagents from Joan Steitz (Yale), Harvey Lodish and Uttam RajBhandary (M.I.T.), I managed to make some progress. I discovered that the degradation of BSA was completely dependent on specific tRNAs, that of Arg and His, and that the destruction of the tRNA led to inhibition of the reaction. The nature of the mechanism of action of the tRNAs and the problem of why the degradation of lysozyme was insensitive to RNase had remained a mystery at that time, which was resolved only when I returned to Israel and established my own laboratory (see below).

The other ubiquitin subject I was studying involved a collaboration with Alex Varshavsky and his then graduate student, Daniel Finley – Dan. At that time Alex was studying the role of mono-ubiquitination of histones (for the histone-ubiquitin adduct H2A known also as protein A24 or uH2A, see above).

Alex noted a series of publications on a temperature sensitive cell cycle arrest mouse mutant cell, ts85, that was generated and described by the group of M. Yamada. These researchers reported that at the non-permissive temperature, the cell lost the histone H2A-ubiquitin adduct. With the ubiquitin cycle unraveled we surmised that this loss could be due to one of two defects: (i) loss of ubiquitination, or (ii) activated de-ubiquitination. Planning our experimental approach, we thought that the defect in these cells is more likely due to loss rather then to gain of function, and we set to dissect the defect. The idea was that the same defect in monoubiquitination of the histone may affect also protein degradation which involves polyubiquitination, though it was clear that the single modification of the histone molecule by ubiquitin does not lead to its targeting to proteolysis. Identification of the biochemical defect in the cells was not too difficult, as we used the isolation technique of the conjugating enzymes developed in Haifa and demonstrated that the defect results from a temperature-sensitive ubiquitin-activating enzyme, E1, the first enzyme in the ubiquitin system cascade (see above). Importantly, inactivation of the enzyme led to inhibition of ubiquitin conjugation to the general population of cellular proteins and was not confined to inhibition of conjugation of histone H2A. Consequently, degradation of both abnormal and normal short-lived proteins was also inhibited, demonstrating that the same enzyme that is involved in ubiquitin activation for histone modification, is also involved in activation of ubiquitin for modification of substrates destined for degradation. We were very lucky in the sense, as if the defect would have been more specific, involving an E3 that targets several substrates, or “worse”, a specific histone E3, we could not have possibly detected an effect on the degradation of the general population of cell proteins: only a defect in a key enzyme such as E1 could have resulted in such a dramatic effect. Identification and characterization of the cell defect further corroborated our earlier general hypothesis that ubiquitination signals proteins for degradation, and that it also occurs in nucleated cells, a finding we had already demonstrated in Haifa, albeit indirectly (using anti-bodies raised against ubiquitin and monitoring the level and dynamics of ubiquitin-protein adducts under conditions of basal and accelerated proteolysis in hepotoma cells; see above). Since the ts85 cell was also a cell cycle arrest mutant, we hypothesized, but did not show at the time experimentally, that the system may be involved in regulating the cell cycle, an hypothesis that later turned out to be correct.

THE RETURN TO ISRAEL – INDEPENDENT RESEARCH CAREER

After three years at M.I.T. (1981–1984), it was time to seek for an independent academic position. After many deliberations and despite attractive offers and a strong temptation to stay in the US, I decided to return home, to Israel. With the help of Avram, I obtained an independent academic position in the Department of Biochemistry at the Faculty of Medicine of the Technion (where I graduated), and returned home towards the end of 1984, after a productive post-doctoral period. Importantly, I already had a research subject

I wanted to pursue, the effect of RNase on ubiquitin-mediated proteolysis.

The years that followed the post-doctoral fellowship (1984–present) have been extremely rewarding. I was happy to return to Israel to my family and friends, to a place I felt I belong. I established my own independent research group and laboratory, obtained extramural competitive funding, and continued my research on the ubiquitin system. I have been lucky to have, along the years, a group of extremely talented graduate students and post-doctoral fellows. In our first series of studies we elucidated the role of tRNA in the proteolytic process, a subject I discovered as a graduate student and continued to study independently while at the M.I.T. (see above). Along with one of my first graduate students, Sarah Ferber, we demonstrated that proteins with acidic N-termini, Asp or Glu, undergo arginylation at the N-terminus, converting the acidic, negatively charged residue at this site to a positively charged residue. The reaction is catalyzed by Arg tRNA-protein transferase, a known protein with an hitherto unknown function. The enzyme uses charged $tRNA^{Arg}$ as a source of activated Arg. Therefore, digestion of the cell extract RNA with RNase A inhibits this reaction. This finding explained the selectivity of the RNase effect to BSA and not to lysozyme: BSA has an Asp residue in the N-terminus, while lysozyme has lysine in this position. Interestingly, the ligase involved in BSA ubiquitination is E3α that was discovered during my graduate studies. As described later by Avram and his gradute student Yuval Reiss, the ligase recognizes several groups of substrates, among them proteins with basic but not with acidic N-termini. Thus, what appeared initially as an artifact turned out to be part of the first specific recognition signal in a target substrate (see below). Parallel to our work on the RNase effect, Avram and Yuval characterized the enzyme and identified on it three distinct substrate binding sites for: (i) basic (the one involved in recognition of basic and Arg-modified acidic N-termini) and (ii) bulky-hydrophobic N-termini, but also for (iii) larger, yet still undefined "body" sites that reside downstream to the N-terminal residue. Because the enzyme recognized certain substrates at their N(α)-terminal residue, it was termed E3α. In parallel and using a systematic genetic approach in the yeast *S. cerevisiae*, Alex Varshavsky and his colleagues formulated a general rule ('N-end rule') for recognition of all 20 different amino acid residues at the N-terminal site.

Research in the laboratory has evolved also in other directions. We have shown that N-α-acetylated proteins are also targeted by the ubiquitin system. This important finding demonstrated that this N-terminally modified "family" of proteins, a group that constitutes a large proportion of cellular proteins, must be targeted by signals that are distinct from the N-terminal residue and reside downstream to it: since they do not have free N-termini, they cannot be recognized by this residue. Along with the discovery of the "body" site in E3α, we felt that N-terminal recognition involves only a small and limited set of proteins, and the mode of recognition of the numerous substrates of the ubiquitin system must be broad and diverse: they must be recognized by multiple and distinct targeting motifs. At that point, the end of the 1980s, we felt it was time to move from studying model substrates to investigating the fate of specific native cellular substrates. We have shown that an important group of cell regulators – tumor suppressors (e.g. p53) and growth promoters (c-Myc) are targeted by the ubiquitin cell free system. We strongly believed that this must be also true for targeting of these substrates *in vivo*, which later, through the work of many others and our own, turned out to be the case. We continued and demonstrated that, unlike the paradigm in the field until that time, that degradation of proteins in the lysosome proceeds independently from the ubiquitin system – the two proteolytic pathways are actually linked to one another, and ubiquitination is required for stress-induced lysosomal degradation of cellular proteins. This area has later evolved in a dramatic manner, and engulfed involvement of the ubiquitin system in receptor-mediated endocytosis and autophagy. Other studies involved elucidation of some of the mechanisms involved in the two step ubiquitin-mediated proteolytic activation of the centrally important transcriptional regulator NF-κB, demonstration of a role for heat shock proteins in targeting certain protein substrates, and identification of a novel site of ubiquitination – the N-terminal residue of the protein substrate. This modification is clearly different and distinct from recognition of the substrate by E3α at the N-terminal residue. In the latter case, the ligase binds to the N-terminal residue while ubiquitination occurs on an internal lysine residue(s). In N-terminal ubiquitination, modification occurs at the N-terminal residue, while the ligase binds, most probably, to an internal sequence in the protein target molecule. This subject has evolved in a surprising manner and changed another paradigm in the field that ubiquitination is limited to internal lysine(s) of the target substrate; we, and later others, have shown that the phenomenon is not limited to the one protein we identified initially – the muscle-specific transcriptional regulator MyoD, and identified a large group of proteins that undergo N-terminal ubiquitination. This group of proteins contain many that have internal lysine(s), but that from some reason(s) cannot be targeted, but interestingly, also a large group of proteins (such as $p16^{INK4a}$ that plays an important role in cell cycle regulation), that are devoid of any lysine residue. To be degraded by the ubiquitin system, these proteins must undergo N-terminal ubiquitination.

These years have not been simple however. The Technion has traditionally been a school of engineering, and life sciences and biomedicine have been foreign to many of its senior leaders, faculty members and policy planners: we were treated in many ways like step children, and thoughts of closing the medical school have been aired at times. This deeply rooted philosophy, which only now starts slowly to change, has severely hampered development in these fields and had left the body of researchers and infrastructure in these areas small and battling for survival. Unlike leaders in other schools of engineering like M.I.T. and Caltech, the Technion's leaders failed to forsee the upcoming revolution in biology and medicine and its huge impact on modern technology. However, through a network of wonderful colleagues all over the world (important among them is my friend Alan Schwartz who is currently at Washington University in St. Louis, but was then at Harvard

Medical School; see above for the beginning of our collaboration at the M.I.T.), and fruitful collaborations, I was able to establish an active research group and carry out what I believe was a good and original research program, even under less than optimal, and at times impossible conditions. This was important in balancing my desire to live in Israel, but at the same time to remain at the forefront of the ubiquitin research field that has grown in its importance to become an extremely exciting, yet a highly competitive, area.

UNPAID DEBTS

Last but not least, I owe a huge debt which I doubt I shall ever be able to repay to several people who helped me cross critical stormy waterways along my life. My aunt, Miriam, who took me to her house after the death of my father and made her home a new home for me, enabling me to complete seamlessly my high school studies without any interruption. My brother Yossi (Joseph) and my sister-in-law Atara, who opened their home for me during the fragile times of my high school and medical studies, and made sure I would not collapse along the way, emotionally, but also economically. And last, my wonderful wife Menucha and my son Tzachi (Yitzhak, Isaac; called after my late father). They engulfed me with love, care, and deep understanding of my needs, and were always there for me when I was flying high on the wings of my dreams, not always seeing or listening to them or being with them, physically and emotionally. Without all these wonderful life partners, I could not have achieved anything.

I also owe special thanks to all my mentors, who each contributed in their own way to my upbringing as a scientist. I owe a big debt to Jacob Bar-Tana and the late Benjamin Shapira from the Hebrew University in Jerusalem who opend for me the gates to the wonderful maze of metabolic pathways, enabling me fall in love with Biochemistry. Their enthusiasm and wisdom convinced me, at a critical stage of my development, to pursue a career in biological sciences. Deep thanks to Avram Hershko, with whom I have come a long way in discovering the ubiquitin system, and from whom I learnt the very basic principles of how to approach a scientific problem. I owe special thanks to Ernie Rose for showing me that methodic thinking is not always necessary in science, and is even interfering at times, and that being erratic and disordered, even absent minded, thinking in a most unconventional manner, can yield wonderful ideas and results. The interaction with Ernie is unique, as it gives one a feeling of instability, casting doubt in one's basic knowledge and beliefs. The real challenge is to select Ernie's correct idea, which then takes you high above any traditional, step wise approach. Last, I owe a huge debt to Harvey Lodish, who is not only a great cell biologist, but a wonderful spiritual mentor in a different way we tend to think of mentors. He gave me complete freedom to choose my own way, but did not let me fall. He always listened carefully and helped me to analyze my results, and with his deep insight was able to find in the ocean of my numbers and graphic data new routes and pathways that

I could have never seen or thought of. He used to gently comment on my approach when he felt I derailed, and helped redirecting me. Yet, he was never imposing: Harvey's active passive educational approach is truly unique. I owe many thanks to all my colleagues, in particular Alan Schwartz, Iasha Sznajder, Yinon Ben-Neriah, and Kazuhiro Iwai, who helped me in many ways along this long voyage. I must also mention my laboratory research associates, initially Sarah Elias (who also helped me in the initial studies) and then Hedva Gonen and Beatrice Bercovich, who have become my eyes and hands since I established my own laboratory. I should mention the major contribution of Hannah Heller, an extremely talented technician of Avram, who was an integral part of our "voyage" and discovery. Dvorah Ganoth and Esther Eythan also helped us along the way, and Clara Segal and Bruvia Rosenberg provided us with skillful technical help. Last but not least, my wonderful graduate students, fellows and visiting scientists, with whom I made new and exciting ways in the rapidly evolving and exciting ubiquitin field.

CURRICULUM VITAE

Birth:	October 1, 1947, Haifa, Israel	
Education:		
Elemetary school	1953–1959	Haifa, Israel ("Hashiloach" and "Hugim" elemetary schools)
High school	1959–1965	Haifa, Israel ("Hugim" high school)
University (undergraduate and graduate studies)	1965–1972	"Hadassah" and the Hebrew University School of Medicine, Jerusalem, Israel (M.Sc., M.D.) M.Sc. thesis advisors: Drs. Benjamin Shapira and Jacob Bar-Tana M.D. thesis advisor: Dr. Avram Hershko
Clinical internship	1972–1973	"Rambam" Medical Center, Haifa, Israel
University (graduate studies)	1976–1981	Faculty of Medicine, Technion-Israel Institute of Technology, Haifa, Israel (D.Sc.) Thesis advisor: Dr. Avram Hershko
Post-doctoral training:	1981–1984	Department of Biology and the Whitehead Institute, Massachusetts Institute of Technology (M.I.T.), Cambridge, MA, USA Laboratory of Dr. Harvey F. Lodish
Faculty position:	1984–present	Faculty of Medicine Technion-Israel Institute of Technology, Haifa, Israel
Degrees:		
M.S.c	1970	Medical Sciences. Faculty of Natural Sciences and the Department of Biochemistry, "Hadassah" and the Hebrew University School of Medicine, Jerusalem, Israel.

M.D.	1975	"Hadassah" and the Hebrew University School of Medicine, Jerusalem, Israel.
D.Sc.	1981	Faculty of Medicine, Technion-Israel Institute of Technology, Haifa, Israel.
Academic Appointments:		
	1977–1979	Research Fellow. Department of Biochemistry, Faculty of Medicine, Technion-Israel Institute of Technology, Haifa, Israel.
	1979–1981	Lecturer. Department of Biochemistry, Faculty of Medicine, Technion-Israel Institute of Technology, Haifa, Israel.
	1984–1987	Senior Lecturer (with tenure). Department of Biochemistry, Faculty of Medicine, Technion-Israel Institute of Technology, Haifa, Israel.
	1987–1992	Associate Professor. Department of Biochemistry, Faculty of Medicine, Technion-Israel Institute of Technology, Haifa, Israel.
	1992–	Full Professor. Department of Biochemistry, Faculty of Medicine, Technion-Israel Institute of Technology, Haifa, Israel.
	2002–	Distinguished Research Professor. Technion-Israel Institute of Technology, Haifa, Israel.
Administrative Appointments:		
	1993–2000	Director. The Rappaport Family Institute for Research in the Medical Sciences. Faculty of Medicine. Technion-Israel Institute of Technology, Haifa, Israel.
Military Service:		
	1974–1977	National Compulsory Service, Israel Defense Forces. Military Physician, Israeli Navy and the Unit for Research and Development, Surgeon General Headquarters. Discharged in the rank of Major.
Visiting Appointments:		
	1978, 1979, 1980, 1981	*Visiting Scientist.* The Institute for Cancer Research, The Fox Chase Cancer Center, Philadelphia, Pennsylvania, USA (Dr. Irwin A. Rose's laboratory).
	1985–present (at different periods)	*Visiting Professor.* The Dana Farber Cancer Institute and Harvard Medical School, Boston, Massachusetts, USA. *American Cancer Society Eleanor Roosevelt Memorial Fellow and Visiting Professor.* Division of Hematology-Oncology, Department of Pediatrics, Children's Hospital, Washington University School of Medicine, St. Louis, Missouri, USA. *Visiting Professor.* University of Kyoto School of Medicine (within the frame of JSPS – Japan Society for promotion of Science Senior Scientist Award).

Visiting Professor. Northwestern University School of Medicine, Chicago, Illinois, USA.
Visiting Professor. STINT Fellow. Microbiology and Tumor Biology Center (MTC). The Karolinska Institute, Stockholm, Sweden.
Visiting Professor. City University of Osaka School of Medicine, Osaka, Japan (within the frame of JSPS – Japan Society for Promotion of Science – Eminent Scientist Award).
Visiting Professor. Rockefeller University, New York, NY, USA.

INTRACELLULAR PROTEIN DEGRADATION: FROM A VAGUE IDEA THRU THE LYSOSOME AND THE UBIQUITIN-PROTEASOME SYSTEM AND ONTO HUMAN DISEASES AND DRUG TARGETING

Nobel Lecture, December 8, 2004

by

Aaron Ciechanover

Cancer and Vascular Biology Research Center, Faculty of Medicine, Technion – Israel Institute of Technology, Haifa, Israel.

ABSTRACT

Between the 1950s and 1980s, scientists were focusing mostly on how the genetic code is transcribed to RNA and translated to proteins, but how proteins are degraded has remained a neglected research area. With the discovery of the lysosome by Christian de Duve it was assumed that cellular proteins are degraded within this organelle. Yet, several independent lines of experimental evidence strongly suggested that intracellular proteolysis is largely non-lysosomal, but the mechanisms involved had remained obscure. The discovery of the ubiquitin-proteasome system resolved this enigma. We now recognize that ubiquitin- and proteasome-mediated degradation of intracellular proteins is involved in regulation of a broad array of cellular processes, such as cell cycle and division, regulation of transcription factors, and assurance of the cellular quality control. Not surprisingly, aberrations in the system have been implicated in the pathogenesis of many human diseases, malignancies and neurodegenerative disorders among them, which led subsequently to an increasing effort to develop mechanism-based drugs, one is already in use.

INTRODUCTION

The concept of protein turnover is hardly 60 years old. Beforehand, body proteins were viewed as essentially stable constituents that were subject to only minor 'wear and tear': dietary proteins were believed to function primarily as energy-providing fuel, traversing metabolic pathways which were completely

Abbreviations used: ODC, ornitihine decarboxylase; G6PD, glucose-6-phosphate dehydrogenase; PEPCK, phosphoenol-pyruvate carboxykinase; TAT, tyrosine aminotransferase; APF-1, ATP-dependent Proteolysis Factor 1 (ubiquitin); UBIP, ubiquitous immunopoietic polypeptide (ubiquitin); MCP, multicatalytic proteinase complex (26S proteasome); CP, 20S core particle (of the 26S proteasome); RP, 19S regulatory particle (of the 26Sproteasome).

distinct from those of the structural and functional proteins of the body. The problem was hard to approach experimentally, as research tools were not available. Important research reagents that were lacking at that time were stable isotopes. While radioactive isotopes were developed earlier by George de Hevesy (de Hevesy G., Chemistry 1943. In: *Nobel Lectures in Chemistry 1942–1962.* World Scientific 1999. pp. 5–41), they were mostly unstable and could not be used to follow metabolic pathways. The concept that the body structural proteins are static and the dietary proteins are used only as a fuel was challenged by Rudolf Schoenheimer from Columbia University in New York city. Schoenheimer escaped from Nazi Germany and joined the Department of Biochemistry in Columbia University founded by Hans T. Clarke (1–3). There he met Harold Urey who worked in the Department of Chemistry and discovered deuterium, the heavy isotope of hydrogen, a discovery that enabled him to prepare heavy water, D_2O. David Rittenberg who had recently received his Ph.D. in Urey's laboratory, joined Schoenheimer, and together they entertained the idea of *"employing a stable isotope as a label in organic compounds, destined for experiments in intermediary metabolism, which should be biochemically indistinguishable from their natural analog"* (1). Urey later succeeded in enriching nitrogen with ^{15}N, which provided Schoenheimer and Rittenberg with a "tag" for amino acids, resulting in a series of studies on protein dynamics. They discovered that following administration of ^{15}N-labled tyrosine to rat, only ~50% was recovered in the urine, *"while most of the remainder is deposited in tissue proteins. An equivalent of protein nitrogen is excreted"* (4). They further discovered that from the half that was incorporated into body proteins *"only a fraction was attached to the original carbon chain, namely to tyrosine, while the bulk was distributed over other nitrogenous groups of the proteins"* (4), mostly as an αNH_2 group in other amino acids. These experiments demonstrated unequivocally that the body structural proteins are in a dynamic state of synthesis and degradation, and that even individual amino acids are in a state of dynamic interconversion. Similar results were obtained using ^{15}N-labeled leucine (5). This series of findings shattered the paradigm in the field at that time that: (i) ingested proteins are completely metabolized and the products are excreted, and (ii) that body structural proteins are stable and static. Schoenheimer was invited to deliver the prestigious Harvey Lecture (1937) and Edward K. Dunham Lecture (1941; at Harvard University) where he presented his revolutionary findings. After his untimely tragic death in 1941, his lecture notes were edited by Hans Clarke, David Rittenberg and Sarah Ratner, and were published in a small book by Harvard University Press. The editors called the book "The Dynamic State of Body Constituents" (6), adopting the title of Schoenheimer's presentation. In the book, the new hypothesis is clearly presented: *"The simile of the combustion engine pictured the steady state flow of fuel into a fixed system, and the conversion of this fuel into waste products. The new results imply that not only the fuel, but the structural materials are in a steady state of flux. The classical picture must thus be replaced by one which takes account of the dynamic state of body structure".* However, the idea that body proteins are turning over was not accepted easily and was challenged as late as the mid-1950s. For example, Hogness and colleagues studied the kinetics of β-galactosidase in *E. coli* and summarized their findings (7): *"To sum up: there seems to be no conclusive evidence that the protein molecules within the cells of mammalian tissues are in a dynamic state. Moreover, our experiments have shown that the proteins of growing E. coli are static. Therefore it seems necessary to conclude that the synthesis and maintenance of proteins within growing cells is not necessarily or inherently associated with a "dynamic state"*. While the experimental study involved the bacterial β-galactosidase, the conclusions were broader, including also the authors' hypothesis on mammalian proteins. The use of the term *'dynamic state'* was not incidental, as they challenged directly Schoenheimer's studies, using his own term. It should be noted, however, that Schoenheimer's result related, as we know now, to the entire population of body proteins, extracellular and intracellular alike. It is now clear that these two classes of proteins are targeted by two different mechanisms (see below). The complex catabolic pathways, along with the lack of recognition of the importance of the process, made progress in the field slow.

Now, after more then six decades of research in the field and with the discovery of the lysosome and later the complex ubiquitin-proteasome system with its numerous tributaries, it is clear that the area has been revolutionized. While the lysosome is involved mostly (but not solely) in targeting extracellular proteins, the ubiquitin-proteasome system degrades intracellular proteins. We now realize that intracellular proteins are turning over extensively, that the process is specific, and that the stability of many proteins is regulated individually and can vary under different pathophyiological conditions. From a scavenger, unregulated and non-specific end process, it has become clear that proteolysis of cellular proteins is a highly complex, temporally controlled and tightly regulated process that plays major roles in a broad array of basic pathways. Among these processes are cell cycle, development, differentiation, regulation of transcription, antigen presentation, signal transduction, receptor-mediated endocytosis, quality control, and modulation of diverse metabolic pathways. Subsequently, this development has changed the paradigm that regulation of cellular processes occurs mostly at the transcriptional and translational levels, and has set regulated protein degradation in an equally important position. With the multitude of substrates targeted and processes involved, it is not surprising that aberrations in the pathway have been implicated in the pathogenesis of many diseases, among them certain malignancies, neurodegeneration, and disorders of the immune and inflammatory system. As a result, the system has become a platform for drug targeting, and mechanism-based drugs are currently developed, one of them is already on the market.

THE LYSOSOME AND INTRACELLULAR PROTEIN DEGRADATION: FACTS

In the mid-1950s, Christian de Duve discovered the lysosome (see, for example, Refs. 8 and 9 and Figure 1). The lysosome was first recognized biochemically

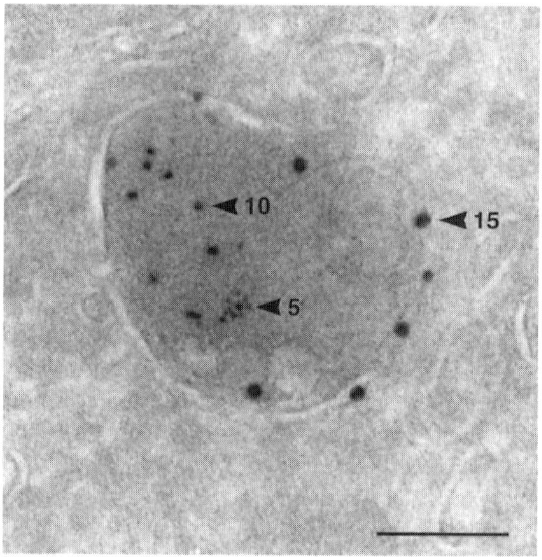

Figure 1: The lysosome: Ultrathin cryosection of a rat PC12 cell that had been loaded for 1 hour with bovine serum albumin (BSA)-gold (5 nm particles) and immunolabeled for the lysosomal enzyme cathepsin B (10-nm particles) and the lysosomal membrane protein LAMP1 (15 nm particles). Lysosomes are recognized also by their typical dense content and multiple internal membranes. Bar, 100 nm. Courtesy of Viola Oorschot and Judith Klumperman, Department of Cell Biology, University Medical Center, Utrecht, The Netherlands.

in the rat liver as a vacuolar structure that contains various hydrolytic enzymes which function optimally at an acidic pH. It is surrounded by a membrane that endows the contained enzymes with latency that is required to protect the cellular contents from their action (see below). The definition of the lysosome has been broadened over the years. This is because it has been recognized that the digestive process is dynamic and involves numerous stages of lysosomal maturation together with the digestion of both exogenous proteins (which are targeted to the lysosome through receptor-mediated endocytosis and pinocytosis) and exogenous particles (which are targeted via phagocytosis; the two processes are known as heterophagy), as well as digestion of endogenous proteins and cellular organelles (which are targeted via micro- and macro-autophagy, in many cases, under stress; see Figure 2). The lysosomal/-vacuolar system as we currently recognize it is a discontinuous and heterogeneous digestive system that also includes structures that are devoid of hydrolases – for example, early endosomes which contain endocytosed receptor-ligand complexes and pinocytosed/phagocytosed extracellular contents. On the other extreme it includes the residual bodies – the end products of the completed digestive processes of heterophagy and autophagy. In between these extremes one can observe: primary/nascent lysosomes that have not been engaged yet in any proteolytic process; early autophagic vacuoles that might contain intracellular organelles; intermediate/late endosomes and phagocytic

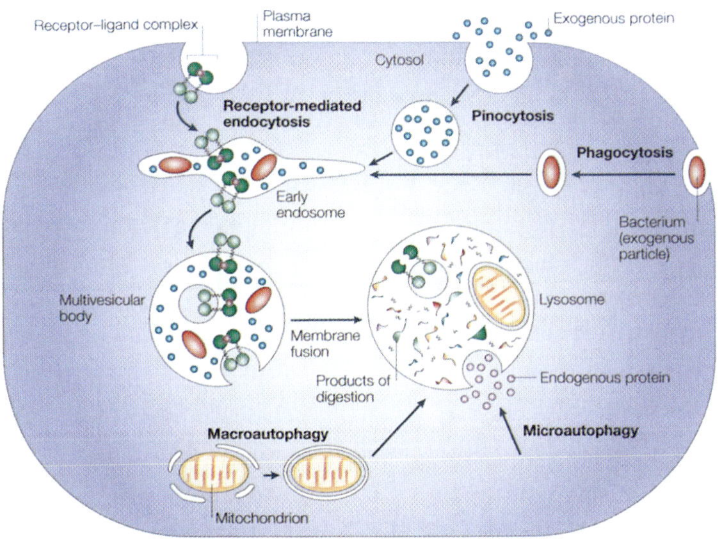

Figure 2: The four digestive processes mediated by the lysosome: (i) specific receptor-mediated endocytosis, (ii) pinocytosis (non-specific engulfment of droplets containing extracellular fluid), (iii) phagocytosis (of extracellular particles), and (iv) autophagy (micro- and macro-; of intracellular proteins and organelles) (with permission from Nature Publishing Group. Published originally in Ref. 83).

vacuoles (heterophagic vacuoles) that contain extracellular contents/particles; and multivesicular bodies (MVBs) which are the transition vacuoles between endosomes/phagocytic vacuoles and the digestive lysosomes.

The discovery of the lysosome along with independent experiments that were carried out at the same time and that have further strengthened the notion that cellular proteins are indeed in a constant state of synthesis and degradation (see, for example, Ref. 10), led scientists to feel, for the first time, that they have at hand an organelle that can potentially mediate degradation of intracellular proteins. The separation of the proteases from their substrates by a membrane provided an explanation for controlled degradation, and the only problem left to be explained (at the time) was how the substrates are translocated into the lysosomal lumen where they are exposed to the activity of the lysosomal proteases and degraded. An important discovery in this respect was the unraveling of the basic mechanism of action of the lysosome – autophagy (reviewed in Ref. 11). Under basal metabolic conditions, portions of the cytoplasm which contain a fraction of the entire population of cellular (cytosolic) proteins, are segregated within a membrane-bound compartment, and are then fused to a primary nascent lysosome, and their contents digested. This process was denoted microautophagy. Under more extreme conditions, starvation for example, mitochondria, endoplasmic reticulum membranes, glycogen bodies and other cytoplasmic entities, can also be engulfed by a process called macroautophagy (see, for example, Ref. 12; the different modes of action of the lysosome in digesting extra- and intracellular proteins particles and organelles, are shown in Fig. 2).

THE LYSOSOME AND INTRACELLULAR PROTEIN DEGRADATION

However, over a period of more than two decades, between the mid-1950s and the late 1970s, it has become gradually more and more difficult to explain several aspects of intracellular protein degradation based on the known mechanisms of lysosomal activity: accumulating lines of independent experimental evidence indicated that the degradation of at least certain classes of cellular proteins must be non-lysosomal. Yet, in the absence of any 'alternative', researchers came with different explanations, some more substantiated and others less, to defend the 'lysosomal' hypothesis.

First was the gradual discovery, coming from several laboratories, that different proteins vary in their stability, and their half-life times can span a range of three orders of magnitude, from a few minutes to many days. Thus, the $t_{1/2}$ of ornitihine decarboxylase (ODC) is ~10 min, while that of glucose-6-phosphate dehydrogenase (G6PD) is 15 hours (for review articles, see, for example, Refs. 13,14). Also, rates of degradation of many proteins was shown to change under different physiological conditions, such as availability of nutrients or hormones. It was conceptually difficult to reconcile the findings of distinct half-lives of different proteins with the mechanism of action of the lysosome, where the microautophagic vesicle contains the entire cohort of cellular (cytosolic) proteins that are therefore expected to be degraded at the same rate. Similarly, changing pathophysiological conditions, such as starvation or re-supplementation of nutrients, were expected to affect the stability of all cellular proteins to the same extent. Clearly, this was not the case.

Another source of concern about the lysosome as the organelle in which intracellular proteins are degraded were the findings that both specific and general inhibitors of lysosomal proteases have different effects on different classes of proteins, making it clear that distinct groups of proteins are targeted by different proteolytic machineries. Thus, the degradation of endocytosed/-pinocytosed extracellular proteins was significantly inhibited by lysosomal inhibitors, a partial effect was observed on the degradation of long-lived cellular proteins, and almost no effect was observed on the degradation of short-lived and abnormal/mutated proteins.

Finally, the thermodynamically paradoxical observation that the degradation of cellular proteins requires metabolic energy, and more importantly, the emerging evidence that the proteolytic machinery uses the energy directly, were in contrast with the known mode of action of lysosomal proteases that under the appropriate acidic conditions, and similar to all known proteases, degrade proteins in an exergonic manner.

The assumption that the degradation of intracellular proteins is mediated by the lysosome was nevertheless logical. Proteolysis results from direct interaction between the target substrates and proteases, and therefore it was clear that active proteases cannot co-excist free and active in the cytosol along with their substrates, which would have resulted in destruction of the cell. Thus, it was recognized that any suggested proteolytic machinery that mediates degradation of intracellular proteins must also be equipped with a mechanism that separates – physically or virtually – between the proteases and their substrates, and enables them to associate only when needed. The lysosomal membrane provided this separating/fencing mechanism. Obviously, nobody could have predicted that a new mode of post-translational modification – ubiquitination – could function as a chemical fence and as a proteolysis signal, and that untagged proteins will remain protected. Thus, while the structure of the lysosome as a distinc organelle could explain the separation necessary between the proteases and their substrates, and autophagy could explain the mechanism of entry of cytosolic proteins into the lysosomal lumen, major problems have remained unsolved. As noted, important among them were: (i) the various half lives of cellular proteins, (ii) the energy requirement for intracellular proteolysis, and (iii) the distinct response of different populations of proteins to lysosomal inhibitors. Thus, according to one model, it was proposed that different proteins have different sensitivities to lysosomal proteases, since their half-lives *in vivo* were in correlation with their sensitivity to the action of lysosomal proteases *in vitro* (15). To explain an extremely long half-life for a protein that was nevertheless sensitive to lysosomal proteases, or alterations in the stability of a single protein under various physiological states, it was suggested that although all cellular proteins are engulfed into the lysosome, only the short-lived proteins are degraded, whereas the long-lived proteins exit back into the cytosol: *"To account for differences in half-life among cell components or of a single component in various physiological states, it was necessary to include in the model the possibility of an exit of native components back to the extralysosomal compartment"* (16). According to a different model, selectivity is determined by the binding affinity of the different proteins for the lysosomal membrane which controls their entry rates into the lysosome, and subsequently their degradation rates (17). For a selected group of proteins, such as the gluconeogenetic enzymes phosphoenol-pyruvate carboxykinase (PEPCK) and fructose-1,6-biphosphatase, it was suggested, though not firmly substantiated, that their degradation in the yeast vacuole is regulated by glucose via a mechanism called 'catabolite inactivation' that possibly involves their phosphorylation. However this regulated mechanism for vacuolar degradation was limited only to a small and specific group of proteins (see for example Ref. 18; reviewed in Ref. 19). More recent studies have shown that at least for stress-induced macroautophagy, a general sequence of amino acids, KFFERQ, that was identified in numerous cellular proteins, directs, via binding to a specific 'receptor' and in cooperation with cytosolic and lysosomal chaperones, the regulated entry of many cytosolic proteins into the lysosomal lumen. While further corroboration of this hypothesis is still required, it can explain the mass, stress-induced, entry into the lysosome of a large population of proteins that contain a homologous sequence, but not the targeting for degradation of a specific protein under defined conditions (reviewed in Refs. 20,21). The energy requirement for protein degradation was described as indirect, and necessary, for example, for protein transport across the lysosomal membrane (22) and/or for the activity of the H^+ pump and the maintenance of the low acidic intralysosomal pH that is necessary for optimal activity of the proteases (23). We now know that both mechanisms require energy. Thus, despite the strained explanations for different stabilities and energy requirement, the lysosome still appeared as the most logical organelle/system that mediates degradation of intracellular proteins. Christian de Duve's view on the subject as summarized in a review article he published in the mid-1960s, saying: *"Just as extracellular digestion is successfully carried out by the concerted action of enzymes with limited individual capacities, so, we believe, is intracellular digestion"* (24) still appeared to be valid even towards the mid 1970s. The problem of different sensitivities of distinct protein groups to lysosomal inhibitors has remained unsolved, and may have served as an important trigger in the future quest for a non-lysosomal proteolytic system that may be involved in at least certain aspects of intracellular protein degradation.

Progress in identifying the elusive, non-lysosomal proteolytic system(s) was hampered by the lack of a cell-free preparation that could faithfully replicate the cellular proteolytic events – degrading proteins in a specific and energy-requiring mode and at neutral pH. An important breakthrough was made by Rabinovitz and Fisher who found that rabbit reticulocytes degrade abnormal, amino acid analogue-containing hemoglobin (25). Their experiments modeled known disease states, the hemoglobinopathies. In these diseases, abnormal mutated hemoglobin chains (such as sickle cell hemoglobin) or excess of unassembled normal hemoglobin chains (generated in thalassemias, diseases in which the pairing chain is not synthesized at all or is mutated and rapidly degraded. Consequently, the bi-heterodimeric hemoglobin complex is not assembled) are rapidly degraded in the reticulocyte (26,27). Reticulocytes are terminally differentiating red blood cells that do not contain lysosomes. Therefore, it was postulated that the degradation of hemoglobin in these cells is mediated by a non-lysosomal machinery. Etlinger and Goldberg (28) were the first to isolate and characterize a cell-free proteolytic preparation from reticulocytes. The crude extract selectively degraded abnormal hemoglobin, required ATP hydrolysis, and acted optimally at a neutral pH, which further corroborated the assumption that the proteolytic activity was non-lysosomal. A similar system was isolated and characterized later by our group (29). Additional studies by the group led subsequently to resolution, characterization, and purification of the major enzymatic components from this extracts, and to the discovery of the ubiquitin signalling system (see below).

THE LYSOSOME HYPOTHESIS IS CHALLENGED

As mentioned above, the unraveled mechanism(s) of action of the lysosome could explain only partially, and at times not satisfactorily, several key emerging characteristics of intracellular protein degradation. Among them were: (i) the heterogeneous stability of individual proteins (ii), the effect of nutrients and hormones on their degradation, and (iii) the dependence of intracellular proteolysis on metabolic energy. The differential effect of selective inhibitors on the degradation of different classes of cellular proteins could not be explained at all.

The development of methods to monitor protein dynamics in cells together with the discovery of specific and general lysosomal inhibitors have resulted in the identification of different classes of cellular proteins, long- and short-lived, and the findings of the differential effects of the inhibitors on these groups (see, for example, Refs. 30,31). An elegant experiment in this respect was carried out by Brian Poole and his colleagues in the Rockefeller University. Poole was studying the effect of lysosomotropic agents – weak bases such as ammonium chloride and chloroquine – that accumulate in the lysosome and dissipate its low acidic pH by neutralizing the H^+ ions: the lysosomal proteases are inactive in neutral pH. It was assumed that this mechanism also underlies the anti-malarial activity of chloroquine and similar drugs that inhibit the activity parasite's lysosome, 'paralyzing' its ability to digest the host's hemoglobin during the intra-erythrocytic stage of its life cycle. Poole and his colleagues labelled metabolically endogenous proteins in living macrophages with ^{3}H-leucine and 'fed' them with broken macrophages that had been labelled metabolically with ^{14}C-leucine prior to their disruption. They assumed, apparently correctly, that the dead macrophages debris and proteins will be phagocytosed by the live macrophages and targeted to the lysosome for degradation. They monitored the effect of lysosomotropic agents on the degradation of these two protein populations, the endogenous and the exogenous. In particular, they studied the effect of the weak bases chloroquine and ammonium chloride, and the acid ionophore X537A which dissipates the H^+ gradient across the lysosomal membrane. They found that these drugs inhibited specifically the degradation of extracellular proteins, but not of intracellular proteins (32). Poole summarized beautifully (I would say also poetically) these experiments and explicitly predicted the existence of a non-lysosomal proteolytic system that degrades intracellular proteins: *"Some of the macrophages labeled with tritium were permitted to endocytise the dead macrophages labeled with ^{14}C. The cells were then washed and replaced in fresh medium. In this way we were able to measure in the same cells the digestion of macrophage proteins from two sources. The exogenous proteins will be broken down in the lysosomes, while the endogenous proteins will be broken down wherever it is that endogenous proteins are broken down during protein turnover"* (33).

The requirement for metabolic energy for the degradation of both prokaryotic (34) and eukaryotic (10,35) proteins was difficult to interpret. Proteolysis is an exergonic process and the thermodynamically paradoxical energy requirement for intracellular proteolysis made researchers believe that energy cannot be consumed directly by proteases or the proteolytic process per se, and is used indirectly. As Simpson summarized his findings (10): *"The data can also be interpreted by postulating that the release of amino acids from protein is itself directly dependent on energy supply. A somewhat similar hypothesis, based on studies on autolysis in tissue minces, has recently been advanced, but the supporting data are very difficult to interpret. However, the fact that protein hydrolysis as catalyzed by the familiar proteases and peptidases occurs exergonically, together with the consideration that autolysis in excised organs or tissue minces continues for weeks, long after phosphorylation or oxidation ceased, renders improbable the hypothesis of the direct energy dependence of the reactions leading to protein breakdown".* Being cautious however, and probably unsure about a single interpretation and conclusion, Simpson still left a narrow slit opened for a proteolytic process that requires energy in a direct manner: *"However, the results do not exclude the existence of two (or more) mechanisms of protein breakdown, one hydrolytic, the other energy-requiring".* Since any proteolytic process must be at one point or another hydrolytic, the statement that makes a distinction between an hydrolytic process and an energy-requiring (yet non-hydrolytic one), is not clear. Judging the statement from a historical perspective and knowing the mechanism of action of the ubiquitin system, where energy is required also in the pre-hydrolytic step (ubiquitin conjugation), Simpson may have thought of a two step mechanism: an initial conditioning step that is energy-requiring yet not hydrolytic, followed by a second hydrolytic step that is energy-independent. However, he did not provide us with a clear explanation. Clearly, he could not have imagined that even the second step will be energy-requiring (as we know is proteasomal activity). At the end of this clearly understandable and apparently difficult deliberation, he left us with a vague explanation linking protein degradation to protein synthesis, a process that was known already at the time to require metabolic energy: *"The fact that a supply of energy seems to be necessary for both the incorporation and the release of amino acids from protein might well mean that the two processes are interrelated. Additional data suggestive of such a view are available from other types of experiments. Early investigations on nitrogen balance by Benedict, Folin, Gamble, Smith, and others point to the fact that the rate of protein catabolism varies with the dietary protein level. Since the protein level of the diet would be expected to exert a direct influence on synthesis rather than breakdown, the altered catabolic rate could well be caused by a change in the rate of synthesis"* (10). With the discovery of lysosomes in eukaryotic cells it could be argued that energy is required for the transport of substrates into the lysosome or for maintenance of the low intralysosomal pH, for example (see above). The observation by Hershko and Tomkins that the activity of tyrosine aminotransferase (TAT) was stabilized following depletion of ATP (36) indicated that energy may be required at an early stage of the proteolytic process, most probably before proteolysis occurs. Yet, it did not provide a clue as for the mechanism involved: the energy could be used, for example, for specific modification of TAT, e.g. phosphorylation, that would sensitize it to degradation by the lysosome or by a yet unknown proteolytic mechanism, or for a modification that activates its putative protease. It could also be used for a more general lysosomal mechanism, one that involves transport of TAT as well as other proteins into the lysosome or dissipation of the intralysosomal acidic pH, for example. The energy inhibitors used by Hershko and Tomkins abolished almost completely degradation of the entire population of cell proteins, confirming previous studies (e.g. 10) and suggesting a general role for energy in protein catabolism. In that respect energy inhibitors had an effect that was distinct from that of protein synthesis inhibitors (e.g. cycloheximide) which affected only enhanced degradation of TAT (induced by steroid hormone depletion), but not basal degradation (36). This finding ruled out, at least partially, a tight linkage between protein synthesis and degradation. In bacteria, that lack lysosomes, an argument involving energy requirement for lysosomal degradation could not have been proposed, but other indirect effects of ATP hydrolysis could have affected proteolysis in *E. coli,* such as phosphorylation of substrates and/or proteolytic enzymes, or maintenance of the 'energized membrane state'. According to this model, proteins could become susceptible to proteolysis by changing their conformation, for example, following association with the cell membrane that maintains a local, energy-dependent gradient of a certain ion. However, such an effect was ruled out (37), and since there was no evidence for a phosphorylation-mediated mechanism (although, similar to the state of affairs in eukaryotes, the proteolytic machinery in prokaryotes had not been identified at that time), it seemed that at least in bacteria, energy is required directly for the proteolytic process. In any event, the requirement for metabolic energy for protein degradation in both prokaryotes and eukaryotes, a process that is exergonic thermodynamically, strongly indicated that proteolysis is a highly regulated process, and that a similar principle/mechanism has been preserved along evolution of the two kingdoms. Implying from the possible direct requirement for ATP in degradation of proteins in bacteria, it was not too unlikely to assume a similar direct mechanism in the degradation of cellular proteins in eukaryotes. Supporting this notion was the description of the cell-free proteolytic system in reticulocytes (28,29), a cell that lacks lysosomes, which suggested that energy is probably required directly for the proteolytic process, although here too, the underlying mechanisms had remained enigmatic at the time. The description of the cell-free system paved however the road for the later detailed dissection of the underlying mechanisms involved and to the discovery of the ubiquitin system.

THE UBIQUITIN-PROTEASOME SYSTEM

The cell-free proteolytic system from reticulocytes (28,29) turned out to be an important and rich source for the purification and characterization of the enzymes that are involved in the ubiquitin-proteasome system. Initial fractionation of the crude reticulocyte cell extract on the anion-exchange resin diethylaminoethyl (DEAE) cellulose yielded two fractions which were both required to reconstitute the energy-dependent proteolytic activity found in the crude extract: the unadsorbed, flow through material was denoted Fraction I, and the high salt eluate of the adsorbed proteins was denoted Fraction II (38).

Table 1: Resolution of the ATP-dependent proteolytic activity from crude reticulocyte extract into two essentially required complementing activities (adapted from Ref. 38; with permission from Elsevier/Biochem. Biophys. Res. Commun.).

Fraction	Degradation of [^{3}H]globin (%)	
	-ATP	+ATP
Lysate	1.5	10
Fraction I	0.0	0.0
Fraction II	1.5	2.7
Fraction I and Fraction II	1.6	10.6

This finding was an important observation and a methodolgical lesson for future dissection of the system. For one, it suggested that the system is not composed of a single 'classical' protease that has evolved evolutionarily to acquire energy dependence [although energy-dependent multi component/-complex proteases, such as the mammalian 26S proteasome (see below) and the prokaryotic Clp family of proteases have been described later], but that it is made of at least two components. This finding of a two component, energy-dependent protease that had no precedent, left us with no paradigm to follow. In an attempt to explain this finding, we suggested, for example, that the two fractions could represent an inhibited protease and its activator. Second, learning from this reconstitution experiment and the essential dependence between the two active components, we continued to reconstitute activity from resolved fractions whenever we encountered a loss of activity along further purification steps. This biochemical 'complementation' approach resulted in the discovery of additional enzymes of the system, all required to be present in the reaction mixture in order to catalyze the multi-step proteolysis of the target substrate. We chose first to purify the active component from Fraction I (see accompanying Biography). It was found to be a small, ~8.5 kDa heat-stable protein that was designated ATP-dependent Proteolysis Factor 1, APF-1. APF-1 was later identified as ubiquitin (see below; I am using the term APF-1 to the point in which it was identified as ubiquitin and then change the terminology accordingly). In retrospect, the decision to start the purification efforts with Fraction I turned out to be strategically important, as Fraction I contained only one single protein – APF-1 – that was necessary to stimulate proteolysis of the model substrate we used at the time, while Fraction II, as we learnt later contained many more components. Starting the purification efforts with Fraction II would have resulted in a much more complicated route. Later studies showed that Fraction I contains additional components necessary for the degradation of other substrates, but these were not necessary for the reconstitution of the system at that time. This choice enabled us not only to purify APF-1, but also to quickly decipher its mechanism of action. A critically important finding that paved the road for future developments in the field was that multiple moieties of APF-1 are covalently conjugated to the target substrate when incubated in the presence of Fraction II, and that the modification requires ATP (39,40; Figures 3 and 4). It was also found that the modifica-

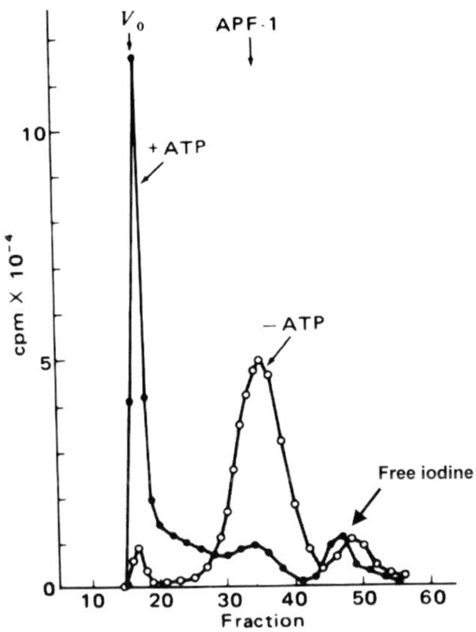

Figure 3: APF-1 is shifted to a high molecular mass "compound(s)" following incubation in ATP-containing crude reticulocyte cell extract. ^{125}I-labelled APF-1 was incubated with reticulocyte crude Fraction II in the absence (open circles) or presence (closed circles) of ATP, and the reaction mixtures were resolved via gel filtration chromatography. Shown is the radioactivity measured in each fraction. As can be seen, following addition of ATP, APF-1 becomes associated with some component(s) in Fraction II, which could be an enzyme(s) or a substrate(s) of the system (with permission from Proceedings of the National Academy of the USA; published originally in Ref. 39).

tion is reversible, and APF-1 can be removed from the substrate or its degradation products and reutilized (40).

The discovery that APF-1 is covalently conjugated to protein substrates and stimulates their proteolysis in the presence of ATP and crude Fraction II, led in 1980 to the proposal of a model according to which protein substrate modification by multiple moieties of APF-1 targets it for degradation by a downstream, at that time an yet unidentified, protease that cannot recognize the unmodified substrate; following degradation, reusable APF-1 is released (40). It was also shown that APF-1 can be released from the intact substrate (40). This can occur, for example, if the substrate was marked eironeously. It was further predicted that the release is mediated by specific protease(s) and not via reversal of the conjugation reaction. Amino-acid analysis of APF-1, along with its known molecular mass and other general characteristics raised the suspicion that APF-1 is ubiquitin (41), a known protein of previously unknown function. Contributing to this suspicion was the previously described conjugate between ubiquitin and histone H2A that raised the possibility that a similar, though obviously not an identical, conjugate is generated between APF-1 and its target substrate (see below). Indeed, in a series of biochemical and functional experiments, Wilkinson and colleagues confirmed unequivo-

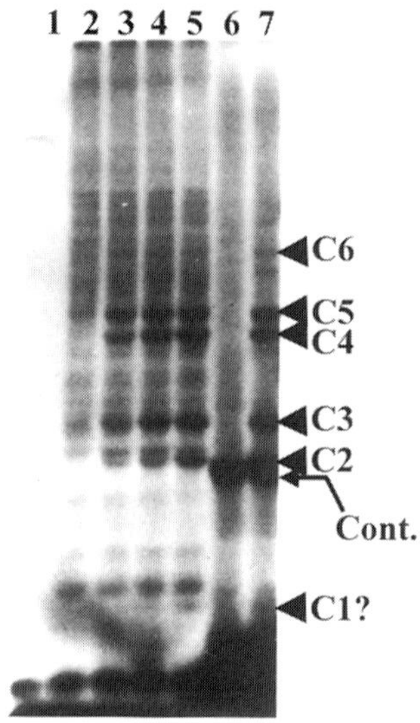

Figure 4: Multiple molecules of APF-1/ubiquitin are conjugated to the proteolytic substrate, probably signalling it for degradation. To interpret the data described in the experiment depicted in Figure 3 and to test the hypothesis that APF-1 is conjugated to the target proteolytic substrate, ^{125}I-APF-1 was incubated along with crude Fraction II (Figure 3 and text) in the absence (lane 1) or presence (lanes 2–5) of ATP and in the absence (lanes 1,2) or presence (lanes 3–5) of increasing concentrations of unlabeled lysozyme. Reaction mixtures resolved in lanes 6 and 7 were incubated in the absence (lane 6) or presence (lane 7) of ATP, and included unlabeled APF-1 and ^{125}I-labeled lysozyme. C1–C6 denote specific APF-1-lysozyme adducts in which the number of APF-1 moieties bound to a the lysozyme moiety of the adduct is probably increasing, from 1 to 6. Reactions mixtures were resolved via sodium dodecyl sulfate-polyacrylamide gel electrophoresis (SDS-PAGE) and visualized following exposure to an X-ray film (autoradiography) (with permission from Proceedings of the National Academy of the USA; published originally in Ref. 40).

cally that APF-1 is ubiquitin (42). Ubiquitin is a small, heat-stable and highly evolutionarily conserved protein of 76 residues. It was first purified during the isolation of thymopoietins (43) and was subsequently found to be ubiquitously expressed in all kingdoms of living cells, including prokaryotes (44). Interestingly, it was initially found to induce lymphocyte differentiation, a characteristic that was attributed to the stimulation of adenylate cyclase (44,45). Accordingly, it was named UBIP for ***ub***iquitous ***i***mmunopoietic ***p***olypeptide (44). However, later studies showed that ubiquitin is not involved in the immune response (46), and that it was a contaminating endotoxin in the preparation that was responsible for the adenylate cyclase and the T-cell differentiating activities. Furthermore, the unraveling of the sequence of several eubacteria and archaebacteria genomes, as well as earlier biochemical analyses in these organisms (unpublished) showed that ubiquitin is restricted only to eukaryotes. The finding of ubiquitin in bacteria (44) was probably due to contamination of the bacterial extract with yeast ubiquitin derived from the yeast extract in which the bacteria were grown. While in retrospect the name ubiquitin is a misnomer as it is restricted to eukaryotes and is not ubiquitous as was previously thought, from historical reasons it has still remained the name of the protein. Accordingly, and in order to avoid confusion, we suggest that names of other novel enzymes and components of the ubiquitin system, but of other systems as well, should remain as were first coined by their discoverers.

An important development in the ubiquitin research field was the discovery that a single ubiquitin moiety can be covalently conjugated to histones (see above), and particularly to histones H2A and H2B. While the function of these adducts has remained elusive until recently, their structure was unraveled in the mid 1970s. The structure of the ubiquitin conjugate with H2A (uH2A; was also designated protein A24) was deciphered by Goldknopf and Busch (47,48) and by Hunt and Dayhoff (49), who found that the two proteins are linked through a fork-like, branched isopeptide bond between the carboxy-terminal glycine of ubiquitin (Gly^{76}) and the ε-NH_2 group of 119 internal lysine (Lys^{119}) of the histone molecule. It was suggested that the isopeptide bond found in the histone-ubiquitin adduct is identical to the bond that was found between ubiquitin and the target proteolytic substrate (50), and between the ubiquitin moieties in the polyubiquitin chain (51,52) that is synthesized on the substrate and that functions, most probably, as a proteolysis recognition signal for the downstream 26S proteasome. In this particular polyubiquitin chain the linkage is between Gly^{76} of one ubiquitin moiety and internal Lys^{48} of the previously conjugated ubiquitin moiety. It appears that only Lys^{48}-based polyubiquitin chain is recognized by the 26S proteasome and serves as a proteolytic signal. In recent years it has been shown that the first ubiquitin moiety can also be attached in a linear mode to the N-terminal residue of the proteolytic target substrate (53). However, even in this case, the subsequent ubiquitin moieties are generating a Lys^{48}-based polyubiquitin chain on the first linearly fused moiety. N-terminal ubiquitination is clearly required for targeting naturally occurring lysine-less proteins for degradation. Yet, several lysine-containing proteins have also been described that traverse this pathway, the muscle-specific transcription factor MyoD, for example. In these proteins the internal lysine residues are probably not accessible to their cognate ligases. Other types of polyubiquitin chains have also been described that are not involved in targeting the conjugated substrates for proteolysis. Thus, a Lys^{63}-based polyubiquitin chain has been described that is probably necessary to activate transcription factors (reviewed recently in Ref. 54). Interestingly, the role of monoubiquitination of histones has also been identified recently. It is involved in regulation of transcription, probably via modulation of the structure of the nucleosomes (for recent reviews, see, for example, Refs. 55,56).

The identification of APF-1 as ubiquitin, and the discovery that a high-energy isopeptide bond, similar to the one that links ubiquitin to histone H2A, links it also to the target proteolytic substrate, resolved at that time the enigma of the energy requirement for intracellular proteolysis (see however below) and paved the road to the untangling of the complex mechanism of isopeptide bond formation. This process turned out to be similar to that of

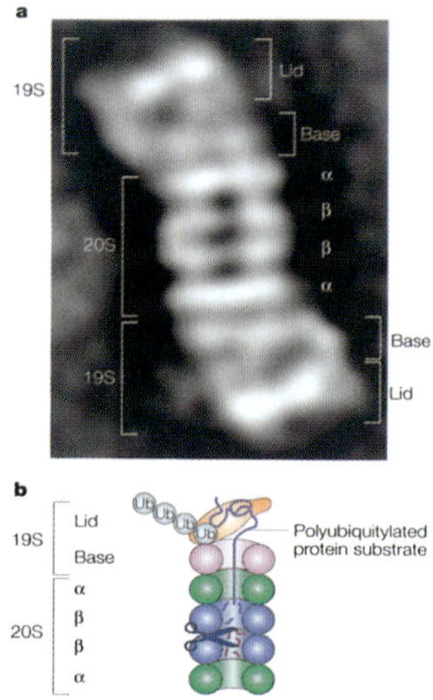

Figure 5: The Proteasome. The proteasome is a large, 26S, multicatalytic protease that degrades polyubiquitinated proteins to small peptides. It is composed of two sub-complexes: a 20S core particle (CP) that carries the catalytic activity, and a regulatory 19S regulatory particle (RP). The 20S CP is a barrel-shaped structure composed of four stacked rings, two identical outer α rings and two identical inner β rings. The eukaryotic α and β rings are composed each of seven distinct subunits, giving the 20S complex the general structure of $\alpha_{1\text{-}7}\beta_{1\text{-}7}\beta_{1\text{-}7}\alpha_{1\text{-}7}$. The catalytic sites are localized to some of the β subunits. Each extremity of the 20S barrel can be capped by a 19S RP, each composed of 17 distinct subunits, 9 in a "base" sub-complex, and 8 in a "lid" sub-complex. One important function of the 19S RP is to recognize ubiquitinated proteins and other potential substrates of the proteasome. Several ubiquitin-binding subunits of the 19S RP have been identified, however, their biological roles and mode of action have not been discerned. A second function of the 19S RP is to open an orifice in the α ring that will allow entry of the substrate into the proteolytic chamber. Also, since a folded protein would not be able to fit through the narrow proteasomal channel, it is assumed that the 19S particle unfolds substrates and inserts them into the 20S CP. Both the channel opening function and the unfolding of the substrate require metabolic energy, and indeed, the 19S RP "base" contains six different ATPase subunits. Following degradation of the substrate, short peptides derived from the substrate are released, as well as reusable ubiquitin (with permission from Nature Publishing Group. Published originally in Ref. 83).
a. Electron microscopy image of the 26S proteasome from the yeast *S. cerevisiae.*
b. Schematic representation of the structure and function of the 26S proteasome.

peptide bond formation that is catalysed by tRNA synthetase or during non-ribosomal synthesis of short peptides (57). In these processes, that occurs during protein synthesis a single amino acid is activated, whereas during activation of ubiquitin, the c-terminal Gly^{76} residue of the protein is activated, but otherwise the three processes appear to be almost identical. Using the unravelled mechanism of ubiquitin activation and immobilized ubiquitin as a 'covalent' affinity bait, the three enzymes that are involved in the cascade reaction of ubiquitin conjugation were purified by Ciechanover, Hershko, and their colleagues. These enzymes are: (i) E1, the ubiquitin-activating enzyme, (ii) E2, the ubiquitin-carrier protein, and (iii) E3, the ubiquitin-protein

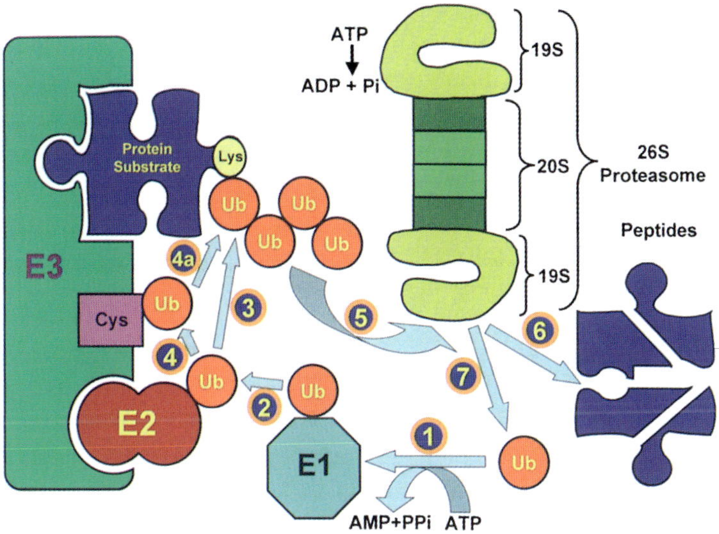

Figure 6: The ubiquitin-proteasome proteolytic system. Ubiquitin is activated by the ubiquitin-activating enzyme, E1 (1) followed by its transfer to a ubiquitin-carrier protein (ubiquitin-conjugating enzyme, UBC), E2 (2). E2 transfers the activated ubiquitin moieties to the protein substrate that is bound specifically to a unique ubiquitin ligase E3. The transfer is either direct [(3) in the case of RING finger ligases and possibly some other smaller groups of E3s such as U-Box ligases], or via an additional thiol-ester intermediate on the ligase [(4, 4a) in the case of HECT domain ligases]. Successive conjugation of ubiquitin moieties to one another generates a polyubiquitin chain that serves as the binding (5) and degradation signal for the downstream 26S proteasome. The substrate is degraded to short peptides (6), and free and reusable ubiquitin is released by de-ubiquitinating enzymes (DUBs) (7).

ligase (58,59). The discovery of the E3 component which is the specific substrate-binding component of the system, pointed to a possible solution to the problem of the heterogeneous stabilities of different proteins – they might be specifically recognized and targeted by different ligases.

Within a short period, the ubiquitin tagging hypothesis received substantial support. For example, Chin and colleagues injected into HeLa cells labeled ubiquitin and hemoglobin and denatured the injected hemoglobin by oxidizing it with phenylhydrazine. They found that ubiquitin conjugation to globin is markedly enhanced by denaturation of hemoglobin, and the level of globin-ubiquitin conjugates was proportional to the rate of hemoglobin degradation (60). Hershko and colleagues observed a similar correlation for abnormal, amino acid analogue-containing short-lived proteins (61). A previously isolated cell cycle arrest mutant that loses the ubiquitin-histone H2A adduct at the permissive temperature (62), was found by Finley, Ciechanover and Varshavsky to harbor a thermolabile E1 (63). Following heat inactivation, the cells fail to degrade normal short-lived proteins (64). Although the cells did not provide direct evidence for substrate ubiquitination as a destruction signal, they still provided a strong and direct linkage between ubiquitin conjugation and intracellular protein degradation.

At this point, the only missing link was the identification of the downstream protease that would specifically recognize ubiquitinated substrates. Tanaka and colleagues identified a second ATP-requiring step in the reticulocyte proteolytic system which was independent of ubiquitination (65), and Hershko and colleagues demonstrated that energy is required for conjugate degradation (66). An important advance in the field was a discovery by Hough and colleagues who partially purified and characterized a high-molecular mass alkaline protease that degraded, in an ATP-dependent mode, ubiquitin adducts of lysozyme but not untagged lysozyme (67). This protease, which was later called the 26S proteasome (see below), provided all the necessary criteria for being the specific proteolytic arm of the ubiquitin system. This finding was confirmed, and the protease was further characterized by Waxman and colleagues who found that it is an unusually large, ~ 1,5 MDa enzyme, unlike any other known protease (68). A further advance in the field was the discovery (69) that a smaller, neutral multi-subunit 20S protease complex that was discovered together with the larger 26S complex, is similar to a "multicatalytic proteinase complex" (MCP) that was described earlier by Wilk and Orlowski in bovine pituitary gland (70). This 20S protease is ATP-independent and has various catalytic activities, cleaving on the carboxy-terminal side of hydrophobic, basic and acidic residues. Hough and colleagues raised the possibility – although they did not show it experimentally – that this 20S protease can be a part of the larger 26S protease that degrades ubiquitin adducts (69). Later studies showed that indeed, the 20S complex is the core catalytic particle of the larger 26S complex (71,72). However, a strong evidence that the active 'mushroom'-shaped 26S protease is generated through the assembly of two distinct sub-complexes – the catalytic 20S cylinder-like MCP and an additional 19S ball-shaped sub-complex (that was predicted to have a regulatory role) – was provided only in the early 1990s by Hoffman and colleagues (73): These researchers mixed the two purified particles and generated the active 26S enzyme.

The proteasome is a large, 26S, multicatalytic protease that degrades poly-ubiquitinated proteins to small peptides. It is composed of two sub-complexes: a 20S core particle (CP) that carries the catalytic activity, and a regulatory 19S regulatory particle (RP) (for the structure of the 26S proteasome, see Figure 5; for a scheme describing the ubiquitin system, see Figure 6).

CONCLUDING REMARKS

The evolvement of proteolysis as a centrally important regulatory mechanism is a remarkable example for the development of a novel biological concept and the accompanying battles to change paradigms. The five decades journey between the early 1940s and early 1990s began with fierce discussions on whether cellular proteins are static as had been thought for a long time, or are in a "dynamic stste" of synthesis and degradation. The discovery of protein dynamics was followed by the discovery of the lysosome, that was believed – between the mid-1950s and mid-1970s – to be the organelle within which in-

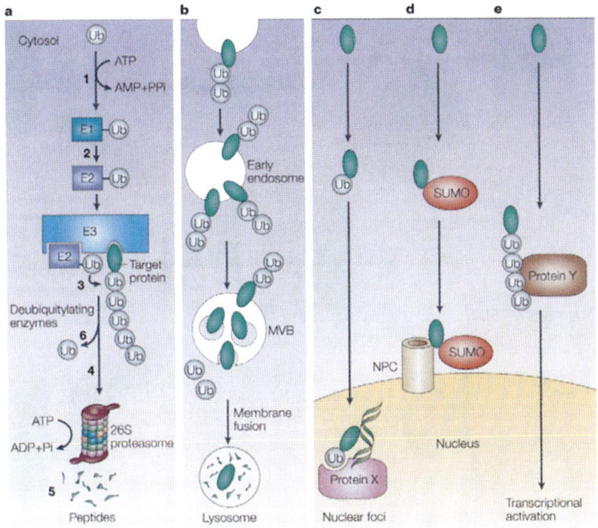

Figure 7: Some of the different functions of modification by ubiquitin and ubiquitin-like proteins. a. Proteasomal-dependent degradation of cellular proteins (see Figure 6). b. Mono⁻ or oligoubiquitination targets membrane proteins to degradation in the lysosome/vacuole. c. Monoubiquitination, or d. a single modification by a ubiquitin-like (UBL) protein, SUMO for example, can target proteins to different subcellular destinations such as nuclear foci or the nuclear pore complex (NPC). Modification by UBLs can serve also other, non-proteolytic functions, such as protecting proteins from ubiquitination or activation of E3 complexes (not shown). e. Generation of a Lys^{63}-based polyubiquitin chain can activate transcriptional regulators, directly or indirectly via recruitment of other proteins (protein Y, for example; shown), or activation of upstream components such as kinases. Ub denotes ubiquitin. (With permission from Nature Publishing Group. Published originally in Ref. 83).

tracellular proteins are destroyed. Independent lines of experimental evidence gradually eroded the lysosomal hypothesis and resulted in the evolvement of new concept, that the bulk of intracellular proteins are degraded – under basal metabolic conditions – via a non-lysosomal machinery. This resulted in the discovery of the ubiquitin system in the late 1970s and early 1980s.

With the identification of the reactions and enzymes that are involved in the ubiquitin-proteasome cascade, a new era in the protein degradation field began at the late 1980s and early 1990s. Studies that showed that the system is involved in targeting of key regulatory proteins – such as light-regulated proteins in plants, transcriptional factors, cell cycle regulators and tumor suppressors and promoters – started to emerge (see for example Refs. 74–78). These studies were accompanied by functional analysis of the system in the yeast *Saccharomyces Cerevisial* (carried out initially mostly by Varshavsky and colleagues), taking advantage of the power of genetics. They were then followed by numerous studies on the underlying mechanisms involved in the degradation of specific proteins, each with its own unique mode of recognition and regulation. The unraveling of the human genome revealed the

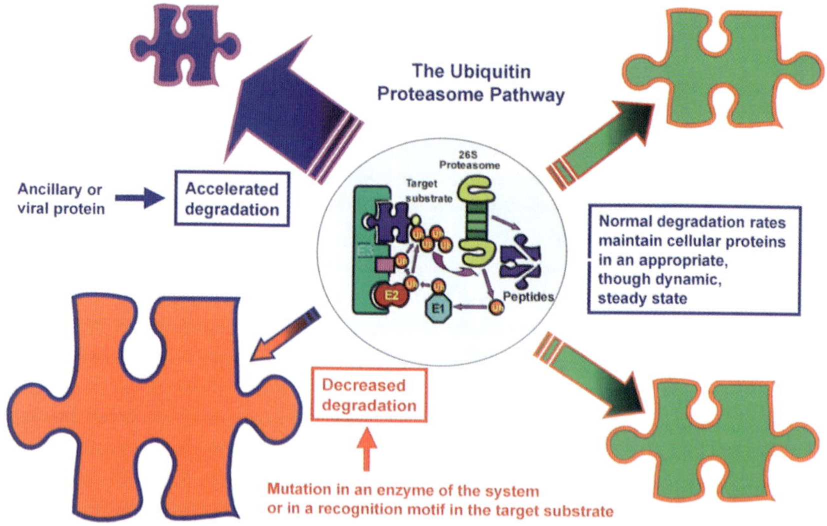

Figure 8: Aberrations in the ubiquitin-proteasome system and pathogenesis of human diseases. Normal degradation of cellular proteins maintains them in a steady-state level, though, this level may change under various physiological conditions (upper and lower right side). When degradation is accelerated due to an increase in the level or activity of an E3 (Skp2 in the case of p27, for example), or overexpression of an ancillary protein that generates a complex with the protein substrate and targets it for degradation (the Human Papillomavirus E6 oncoprotein that associates with p53 and targets it for degradation by the E6-AP ligase, or the cytomegalovirus-encoded ER proteins US2 and US11 that target MHC class I molecules for Endoplasmic Reticulum-Associated Degradation, ERAD), the steady state level of the protein decreases (upper left side). A mutation in a ubiquitin ligase or in a substrate's scaffold/binding protein [such as occurs in *Adenomatous Polyposis Coli* – APC, or in E6-AP (Angelmans' Syndrome)] or in the substrate's recognition motif (such as occurs in β-catenin or in ENaC) will result in decreased degradation and accumulation of the target substrate (lower left side).

existence of hundreds of distinct E3s, attesting to the complexity and the high specificity and selectivity of the system. Two important advances in the field were the discovery of the non-proteolytic functions of ubiquitin such as activation of transcription and routing of proteins to the lysosome/vacuole, and the discovery of modification by ubiquitin-like proteins (UBLs), that is involved in numerous non-proteolytic functions such as directing proteins to their subcellular destination, protecting proteins from ubiquitination, and controlling entire processes such as autophagy (see for example Ref. 79) (for some of the different roles of modifications by ubiquitin and UBLs, see Figure 7). All these studies have led to the emerging realization that this novel mode of post-translational modification/covalent conjugation plays key roles in regulating a broad array of cellular processes – among them cell cycle and division, growth and differentiation, activation and silencing of transcription, apoptosis, the immune and inflammatory response, signal transduction, receptor-mediated endocytosis, various metabolic pathways, and the cell quality control – all that through proteolytic and non-proteolytic mechanisms. The discovery that ubiquitin modification plays a role in routing proteins to the lysosome/vacuole, and that modification by specific and unique ubiquitin-like proteins and modification system controls autophagy closed an exciting historical cycle, since it demonstrated that the two apparently distinct proteolytic systems – the ubiquitin/proteasome and the lysosome – communicate with one another. With the many processes and innumerable substrates targeted by the ubiquitin pathway, it is not surprising to find that aberrations in the system underlie, directly or indirectly, the pathogenesis of many diseases. While inactivation of a major enzyme such as E1 is obviously lethal, mutations in enzymes or in recognition motifs in substrates that do not affect vital pathways or that affect the involved process only partially, may result in a broad array of phenotypes. Likewise, acquired changes in the activity of the system can also evolve into certain pathologies. The pathological states associated with the ubiquitin system can be classified into two groups: (**a**) those that result from loss of function – mutation in a ubiquitin system enzyme or in the recognition motif in the target substrate that result in stabilization of certain proteins, and (**b**) those that result from gain of function – abnormal or accelerated degradation of the protein target (for a general scheme describing aberrations in the ubiquitin system that result in disease states, see Figure 8). Studies that employ targeted inactivation of genes coding for specific ubiquitin system enzymes and substrates in animals can provide a more systematic view into the broad spectrum of pathologies that may result from aberrations in ubiquitination, ubiquitin-mediated proteolysis, and modification by UBLs. Better understanding of the processes and identification of the components involved in the degradation of key regulators will lead to the development of mechanism-based drugs that will target specifically only the involved proteins. While the first drug, a specific proteasome inhibitor is already on the market (80), it appears that one important hallmark of the new era we are entering now will be the discovery of novel drugs based on targeting of specific processes such as inhibiting aberrant Mdm2- or E6-AP-mediated accelerated targeting of the tumor suppressor p53 which will result in regaining its lost function.

Many review articles have been published on different aspects of the ubiquitin system. The purpose of this article was to bring to the reader several milestones along the historical pathway of the evolvement of the ubiquitin system. For additional reading on the ubiquitin system the reader is referred to the many reviews written on the system, among them, for example, are Refs. 81,82. Some parts of this review, including several Figures, are based on another recently published review article (Ref. 83).

ACKNOWLEDGEMENT

Research in the laboratory of Aaron Ciechanover has been supported along the years by grants from the US-Israel Binational Science Foundation (BSF), the Israel Science Foundation (ISF) founded by the Israeli National Academy of Humanities, Arts and Sciences, the German- Israeli Foundation (GIF) for

Scientific Research and Development, the Israel Cancer Research Fund (ICRF) USA, the Deutsche-Israeli Cooperation Program (DIP), the European Commission (EC), the Israel Cancer Society (ICS), the Prostate Cancer Foundation (PCF) – Israel, the Foundation for Promotion of Research in the Technion and various research grants administered by the Vice President of the Technion for Research. Infrastructural equipment for the laboratory of A.C. and for the Cancer and Vascular Biology Research Center has been purchased with the support of the Wolfson Charitable Fund – Center of Excellence for Studies on Turnover of Cellular Proteins and its Implications to Human Diseases.

REFERENCES

1. Clarke, H.T. (1958). Impressions of an organic chemist in biochemistry. *Annu. Rev. Biochem.* **27**, 1–14.
2. Kennedy., E.P. (2001). Hitler's gift and the era of biosynthesis. *J. Biol. Chem.* **276**, 42619–42631.
3. Simoni, R.D., Hill, R.L., and Vaughan, M. (2002). The use of isotope tracers to study intermediary metabolism: Rudolf Schoenheimer. *J. Biol. Chem.* **277** (issue 43), e1-e3 (available on-line at: http:/www.jbc.org).
4. Schoenheimer, R., Ratner, S., and Rittenberg, D. (1939). Studies in protein metabolism. VII: The metabolism of tyrosine. *J. Biol. Chem.* **127**, 333–344.
5. Ratner, S., Rittenberg, D., Keston, A.S., and Schoenheimer, R. (1940). Studies in protein metabolism. XIV: The chemical interaction of dietary glycine and body proteins in rats. *J. Biol. Chem.* **134**, 665–676.
6. Schoenheimer, R. The Dynamic State of Body Constituents (1942). Harvard University Press, Cambridge, Massachusetts, USA.
7. Hogness, D.S., Cohn, M., and Monod, J. (1955). Studies on the induced synthesis of, β-galactosidase in *Escherichia coli:* The kinetics and mechanism of sulfur incorporation. *Biochim. Biophys. Acta* **16**, 99–116.
8. de Duve, C., Gianetto, R., Appelmans, F., and Wattiaux, R. (1953). Enzymic content of the mitochondria fraction. *Nature (London)* **172**, 1143–1144.
9. Gianetto, R., and de Duve, C. Tissue fractionation studies 4. (1955). Comparative study of the binding of acid phosphatase, β-glucoronidase and cathepsin by rat liver particles. *Biochem. J.* **59**, 433–438.
10. Simpson, M.V. The release of labeled amino acids from proteins in liver slices. (1953). *J. Biol. Chem.* **201**, 143–154.
11. Mortimore, G.E., and Poso, A.R. (1987). Intracellular protein catabolism and its control during nutrient deprivation and supply. *Annu. Rev. Nutr.* **7**, 539–564.
12. Ashford, T.P., and Porter, K.R. (1962). Cytoplasmic components in hepatic cell lysosomes. *J. Cell Biol.* **12**, 198–202.
13. Schimke, R.T., and Doyle, D. (1970). Control of enzyme levels in animal tissues. *Annual Rev. Biochem.* **39**, 929–976.
14. Goldberg, A.L., and St. John, A.C. (1976). Intracellular protein degradation in mammalian and bacterial cells: Part 2. *Annu. Rev. Biochem.* **45**, 747–803.
15. Segal, H.L., Winkler, J.R., and Miyagi, M.P. (1974). Relationship between degradation rates of proteins *in vivo* and their susceptibility to lysosomal proteases. *J. Biol. Chem.* **249**, 6364–6365.
16. Haider, M., and Segal, H.L. (1972). Some characteristics of the alanine-aminotransferase and arginase-inactivating system of lysosomes. *Arch. Biochem. Biophys.* **148**, 228–237.
17. Dean, R.T. Lysosomes and protein degradation. (1977). *Acta Biol. Med. Ger.* **36**, 1815–1820.
18. Müller, M., Müller, H, and Holzer, H. (1981). Immunochemical studies on catabolite inactivation of phosphoenolpyruvate carboxykinase in *Saccharomyces cerevisiae. J. Biol. Chem.* **256**, 723–727.

19. Holzer, H. (1989). Proteolytic catabolite inactivation in *Saccharomyces cerevisiae. Revis. Biol. Celular* **21**, 305–319.
20. Majeski, A.E., and Dice, J.F. (2004). Mechanisms of chaperone-mediated autophagy. *Intl. J. Biochem. Cell Biol.* **36**, 2435–2444.
21. Cuervo, A.M., and Dice, J.F. (1998). Lysosomes, a meeting point of proteins, chaperones, and proteases. *J. Mol. Med.* **76**, 6–12.
22. Hayashi, M., Hiroi, Y., and Natori, Y. (1973). Effect of ATP on protein degradation in rat liver lysosomes. *Nature New Biol.* **242**, 163–166.
23. Schneider, D.L. (1981). ATP-dependent acidification of intact and disrupted lysosomes: Evidence for an ATP-driven proton pump. *J. Biol. Chem.* **256**, 3858–3864.
24. de Duve, C., and Wattiaux, R. Functions of lysosomes. (1966). *Annu. Rev. Physiol.* **28**, 435–492.
25. Rabinovitz, M., and Fisher, J.M. (1964). Characteristics of the inhibition of hemoglobin synthesis in rabbit reticulocytes by threo-α-amino-β-chlorobutyric acid. *Biochim. Biophys. Acta.* **91**, 313–322.
26. Carrell, R.W., and Lehmann, H. (1969). The unstable hemoglobin hemolytic anaemias. *Semin. Hematol.* **6**, 116–132.
27. Huehns, E.R., and Bellingham, A.J. (1969). Diseases of function and stability of hemoglobin. *Br. J. Hematol.* **17**, 1–10.
28. Etlinger, J.D., and Goldberg, A.L. (1977). A soluble ATP-dependent proteolytic system responsible for the degradation of abnormal proteins in reticulocytes. *Proc. Natl. Acad. Sci. USA* **74**, 54–58.
29. Hershko, A., Heller, H., Ganoth, D., and Ciechanover, A. (1978). Mode of degradation of abnormal globin chains in rabbit reticulocytes. In: Protein Turnover and Lysosome Function (H.L. Segal & D.J. Doyle, eds.). Academic Press, New York. pp. 149–169.
30. Knowles, S.E., and Ballard, F.J. (1976). Selective control of the degradation of normal and aberrant proteins in Reuber H35 hepatoma cells. *Biochem J.* **156**, 609–617.
31. Neff, N.T., DeMartino, G.N., and Goldberg, A.L. (1979). The effect of protease inhibitors and decreased temperature on the degradation of different classes of proteins in cultured hepatocytes. *J. Cell Physiol.* **101**, 439–457.
32. Poole, B., Ohkuma, S., and Warburton, M.J. (1977). The accumulation of weakly basic substances in lysosomes and the inhibition of intracellular protein degradation. *Acta Biol. Med. Germ.* **36**, 1777–1788.
33. Poole, B., Ohkuma, S. & Warburton, M.J. (1978). Some aspects of the intracellular breakdown of exogenous and endogenous proteins. In: Protein Turnover and Lysosome Function (H.L. Segal and D.J. Doyle, eds.). Academic Press, New York. pp. 43–58.
34. Mandelstam, J. (1958). Turnover of proteins in growing and non-growing populations of *Escherichia coli. Biochem. J.* **69**, 110–119.
35. Steinberg, D., and Vaughan, M. (1956). Observations on intracellular protein catabolism studied *in vitro. Arch. Biochem. Biophys.* **65**, 93–105.
36. Hershko, A., and Tomkins, G.M. (1971). Studies on the degradation of tyrosine aminotransferase in hepatoma cells in culture: Influence of the composition of the medium and adenosine triphosphate dependence. *J. Biol. Chem.* **246**, 710–714.
37. Goldberg, A.L., Kowit, J.D., and Etlinger, J.D. (1976). Studies on the selectivity and mechanisms of intracellular protein degradation. In: Proteolysis and Physiological Regulation (D.W. Ribbons and K. Brew, eds.). Academic Press, New York. pp. 313–337.
38. Ciechanover A., Hod, Y., and Hershko, A. (1978). A heat-stable polypeptide component of an ATP-dependent proteolytic system from reticulocytes. *Biochem. Biophys. Res. Common.* **81**, 1100–1105.
39. Ciechanover, A., Heller, H., Elias, S., Haas, A.L., and Hershko, A. (1980). ATP-dependent conjugation of reticulocyte proteins with the polypeptide required for protein degradation. *Proc. Natl. Acad. Sci. USA.* **77**, 1365–1368.
40. Hershko, A., Ciechanover, A., Heller, H., Haas, A.L., and Rose, I.A. (1980). Proposed role of ATP in protein breakdown: Conjugation of proteins with multiple chains of the polypeptide of ATP-dependent proteolysis. *Proc. Natl. Acad. Sci. USA* **77**, 1783–1786.

41. Ciechanover, A., Elias, S., Heller, H., Ferber, S. and Hershko, A. (1980). Characterization of the heat-stable polypeptide of the ATP-dependent proteolytic system from reticulocytes. *J. Biol. Chem.* **255**, 7525–7528.
42. Wilkinson, K.D., Urban, M.K., and Haas, A.L. (1980). Ubiquitin is the ATP-dependent Proteolysis Factor I of rabbit reticulocytes. *J. Biol. Chem.* **255**, 7529–7532.
43. Goldstein, G. (1974). Isolation of bovine thymin, a polypeptide hormone of the thymus. *Nature (London)* **247**, 11–14.
44. Goldstein, G., Scheid, M., Hammerling, U., Schlesinger, D.H., Niall, H.D., and Boyse, E.A. (1975). Isolation of a polypeptide that has lymphocyte-differentiating properties and is probably represented universally in living cells. *Proc. Natl. Acad. Sci. USA* **72**, 11–15.
45. Schlessinger, D.H., Goldstein, G., and Niall, H.D. (1975). The complete amino acid sequence of ubiquitin, an adenylate cyclase stimulating polypeptide probably universal in living cells. *Biochemistry* **14**, 2214–2218.
46. Low, T.L.K., and Goldstein, A.L. (1979). The chemistry and biology of thymosin: Amino acid analysis of thymosin α1 and polypeptide β1. *J. Biol. Chem.* **254**, 987–995.
47. Goldknopf, I.L., and Busch, H. (1975). Remarkable similarities of peptide fingerprints of histone 2A and non-histone chromosomal protein A24. *Biochem. Biophys. Res. Commun.* **65**, 951–955.
48. Goldknopf, I.L., and Busch, H. (1977). Isopeptide linkage between non-histone and histone 2A polypeptides of chromosome conjugate-protein A24. *Proc. Natl. Acad. Sci. USA* **74**, 864–868.
49. Hunt, L.T., and Dayhoff, M.O. (1977). Amino-terminal sequence identity of ubiquitin and the non-histone component of nuclear protein A24. *Biochim. Biophys. Res. Commun.* **74**, 650–655.
50. Hershko, A., Ciechanover, A., and Rose, I.A. (1981). Identification of the active amino acid residue of the polypeptide of ATP-dependent protein breakdown. *J. Biol. Chem.* **256**, 1525–1528.
51. Hershko, A., and Heller, H. (1985). Occurrence of a polyubiquitin structure in ubiquitin-protein conjugates. *Biochem. Biophys. Res. Common.* **128**, 1079–1086.
52. Chau, V., Tobias, J. W., Bachmair, A., Mariott, D., Ecker, D., Gonda, D. K., and Varshavsky, A. (1989). A multiubiquitin chain is confined to specific lysine in a targeted short lived protein. *Science* **243**, 1576–1583.
53. Ciechanover, A., and Ben-Saadon R. (2004). N-terminal ubiquitination: More protein substrates join in. *Trends Cell Biol.* **14**, 103–106.
54. Muratani, M., and Tansey, W.P. (2003). How the ubiquitin-proteasome system controls transcription. *Nat. Rev. Mol. Cell Biol.* **4**, 192–201.
55. Zhang, Y. (2003). Transcriptional regulation by histone ubiquitination and deubiquitination. *Genes & Dev.* **17**, 2733–2740.
56. Osley, M.A. (2004). H2B ubiquitylation: The end is in sight. *Biochim. Biophys. Acta.* **1677**, 74–78.
57. Lipman, F. (1971). Attempts to map a process evolution of peptide biosynthesis. *Science* **173**, 875–884.
58. Ciechanover, A., Elias, S., Heller, H. & Hershko, A. (1982). "Covalent affinity" purification of ubiquitin-activating enzyme. *J. Biol. Chem.* **257**, 2537–2542.
59. Hershko, A., Heller, H., Elias, S., and Ciechanover, A. (1983). Components of ubiquitin-protein ligase system: Resolution, affinity purification and role in protein breakdown. *J. Biol. Chem.* **258**, 8206–8214.
60. Chin, D.T., Kuehl, L., and Rechsteiner, M. (1982). Conjugation of ubiquitin to denatured hemoglobin is proportional to the rate of hemoglobin degradation in HeLa cells. *Proc. Natl. Acad. Sci. USA* **79**, 5857–5861.
61. Hershko, A., Eytan, E., Ciechanover, A. and Haas, A.L. (1982). Immunochemical Analysis of the turnover of ubiquitin-protein conjugates in intact cells: Relationship to the breakdown of abnormal proteins. *J. Biol. Chem.* **257**, 13964–13970.
62. Matsumoto, Y., Yasuda, H., Marunouchi, T., and Yamada, M. (1983). Decrease in uH2A (protein A24) of a mouse temperature-sensitive mutant. *FEBS Lett.* **151**, 139–142.

63. Finley, D., Ciechanover, A., and Varshavsky, A. (1984). Thermolability of ubiquitin-activating enzyme from the mammalian cell cycle mutant ts85. *Cell* **37**, 43–55.
64. Ciechanover, A., Finley D., and Varshavsky, A. (1984). Ubiquitin dependence of selective protein degradation demonstrated in the mammalian cell cycle mutant ts85. *Cell* **37**, 57–66.
65. Tanaka, K., Waxman, L., and Goldberg, A.L. (1983). ATP serves two distinct roles in protein degradation in reticulocytes, one requiring and one independent of ubiquitin. *J. Cell Biol.* **96**, 1580–1585 (1983).
66. Hershko, A., Leshinsky, E., Ganoth, D., and Heller, H. (1984). ATP-dependent degradation of ubiquitin-protein conjugates. *Proc. Natl. Acad. Sci. USA* **81**, 1619–1623.
67. Hough, R., Pratt, G., and Rechsteiner, M. (1986). Ubiquitin-lysozyme conjugates: Identification and characterization of an ATP-dependent protease from rabbit reticulocyte lysates. *J. Biol. Chem.* **261**, 2400–2408.
68. Waxman, L., Fagan, J., and Goldberg, A.L. (1987). Demonstration of two distinct high molecular weight proteases in rabbit reticulocytes, one of which degrades ubiquitin conjugates. *J. Biol. Chem.* **262**, 2451–2457.
69. Hough, R., Pratt, G., and Rechsteiner M. (1987). Purification of two high molecular weight proteases from rabbit reticulocyte lysate. *J. Biol. Chem.* **262**, 8303–8313.
70. Wilk, S., and Orlowski, M. (1980). Cation-sensitive neutral endopeptidase: Isolation and specificity of the bovine pituitary enzyme. *J. Neurochem.* **35**, 1172–1182.
71. Eytan, E., Ganoth, D., Armon, T., and Hershko, A. (1989). ATP-dependent incorporation of 20S protease into the 26S complex that degrades proteins conjugated to ubiquitin. *Proc. Natl. Acad. Sci. USA.* **86**, 7751–7755.
72. Driscoll, J., and Goldberg, A.L. (1990). The proteasome (multicatalytic protease) is a component of the 1500-kDa proteolytic complex which degrades ubiquitin-conjugated proteins. *J. Biol. Chem.* **265**, 4789–4792.
73. Hoffman, L., Pratt, G., and Rechsteiner, M. (1992). Multiple forms of the 20S multicatalytic and the 26S ubiquitin/ATP-dependent proteases from rabbit reticulocyte lysate. *J. Biol. Chem.* **267**, 22362–22368.
74. Shanklin, J., Jaben, M., and Vierstra, R.D. (1987). Red light-induced formation of ubiquitin-phytochrome conjugates: Identification of possible intermediates of phytochrome degradation. *Proc. Natl. Acad. Sci. USA* **84**, 359–363.
75. Hochstrasser, M., and Varshavsky, A. (1990). *In vivo* degradation of a transcriptional regulator: the yeast α2 repressor. *Cell* **61**, 697 708.
76. Scheffner, M., Werness, B.A., Huibregtse, J.M., Levine, A.J., and Howley, P.M. (1990). The E6 oncoprotein encoded by human papillomavirus types 16 and 18 promotes the degradation of p53. *Cell* **63**, 1129–1136.
77. Glotzer, M., Murray, A.W., and Kirschner, M.W. (1991). Cyclin is degraded by the ubiquitin pathway. *Nature* **349**, 132–138.
78. Ciechanover, A., DiGiuseppe, J.A., Bercovich, B., Orian, A., Richter, J.D., Schwartz, A.L., and Brodeur, G.M. (1991). Degradation of nuclear oncoproteins by the ubiquitin system *in vitro*. *Proc. Natl. Acad. Sci. USA* **88**, 139–143.
79. Mizushima, N., Noda, T., Yoshimori, T., Tanaka, Y., Ishii, T., George, M.D., Klionsky, D.J., Ohsumi, M., and Ohsumi, Y. (1998). A protein conjugation system essential for autophagy. *Nature* **395**, 395–398.
80. Adams J. (2003). Potential for proteasome inhibition in the treatment of cancer. *Drug Discov. Today.* **8**, 307–315.
81. Glickman, M.H., and Ciechanover, A. (2002). The ubiquitin-proteasome pathway: Destruction for the sake of construction. *Physiological Reviews* **82**, 373–428.
82. Pickart, C.M., and Cohen, R.E. (2004). Proteasomes and their kin: Proteases in the machine age. *Nature Rev. Mol. Cell Biol.* **5**, 177–187.
83. Ciechanover, A. (2005). From the lysosome to ubiquitin and the proteasome. *Nature Rev. Mol. Cell Biol.* **6**, 79–86.

Portrait photo of Professor Ciechanover by photographer Dan Porges.

Avram Hershko אברהם הרשקו

AVRAM HERSHKO

I was born on December 31, 1937, in Karcag, Hungary. Karcag is a small town of around 25,000 inhabitants, about 150 kilometers east of Budapest. It had a Jewish community of nearly one thousand people. My father, Moshe Hershko, was a schoolteacher in the Jewish elementary school in Karcag; most of the Jewish children in that town were his students. His former students from Hungary, and later from Israel, described him with admiration as an inspiring teacher and a role model educator. My mother Shoshana/ Margit ("Manci") was an educated and musically gifted woman. She gave some English and piano lessons to children in Karcag. My older brother, Chaim, was born in 1936, less than two years before me. My mother wanted very much to have also a girl baby, but the times were at the eve of World War II, Hitler's screams could be frequently heard on the radio, my parents became apprehensive of the future and thus did not try to have more children. Still, my recollections of my early childhood are of very happy times, with loving and supporting parents, growing up in a nice house with a beautiful garden, created by my father who was also an amateur (but avid) gardener. A family picture from these times, with my parents, my brother and I as an infant, shows well the warmth of my family.

This early paradise was lost rapidly and brutally. World War II broke out, and soon Hungary joined in as an ally of Nazi Germany. In 1942, my father was taken by the Hungarian Army to serve as a forced laborer, with companies of other Jewish men. They were sent to the Russian front, where most of them perished. Luckily for my father, the Soviet Army advanced so rapidly after Stalingrad that he was captured by the Soviets before the Nazis could kill him. Then, he was used by the Soviets as a forced laborer. He was released only in 1946, so we did not know for four years whether or not he was alive.

In the spring of 1944, Hungary's dictator Horthy understood that Germany was loosing the war, and planned to desert. The Germans sensed this and quickly occupied Hungary. This was followed by the rapid extermination of much of the Jewish population of Hungary. In May – June 1944, most Jewish people were concentrated in ghettos and then transported to death camps in Poland. I was six years old at that time. We were in a ghetto at the outskirts of Karcag for a couple of weeks and then were transferred to a terribly crowded ghetto in Szolnok, which is a larger city in the same district. From there, Jews from the entire district were transported further on freight-trains. They were told that they were sent to work, but after the war we learned that most of the trains were headed for Auschwitz. By some random event, my family and I were put on one of the few trains that headed for Austria, where

My parents, my brother (middle, up) and I (middle, down) at around the end of 1938.

Jews were actually used for labor. This group included my mother with us two children, my paternal grandparents and my three aunts. In Austria we were in a small village near Vienna, where adults worked in the fields and in a factory. We were liberated by the Soviet Army on the spring of 1945. My maternal grandparents perished in the Holocaust, along with 360,000 Hungarian Jews and almost two-thirds of the Jewish people of Karcag. Following our reunion with my father in 1946, the family lived for three years in Budapest, where my father found job as a schoolteacher. The family emigrated to Israel in 1950.

In Israel we settled in Jerusalem and I started a new and very different life. Of course, there were initial difficulties of being new immigrants. We had to learn a new language, Hebrew. This was not too difficult for children (I was less than 13 at that time), but it was more difficult for my parents. Still, my father studied Hebrew and soon started to work, again as a schoolteacher. (Later he taught in at a teachers' seminary and authored mathematics textbooks, which were very popular in Israel). As always, education of their children was my parents' highest priority. Although we were quite poor immigrants at that time, my brother and I were sent to an expensive private school in Jerusalem. I suspect that most of the salary of my father was spent on our tuition fees.

At school I was received well by the other children. These were times of massive immigration to Israel, so a new immigrant child with a Hungarian accent did not stand out too much (I am told that I still have some Hungarian accent, especially in English, though my Hungarian language is quite poor now). I was a good student, and learned easily different subjects, such as mathematics, physics, literature, history and even Talmud! That became a problem when I finished high school, I was interested in too many subjects, so it was difficult for me to decide how to continue. I chose to study medicine, probably by default, because my brother Chaim was already a medical student, so I could inherit his books for free! Chaim always wanted to be a physician, and he is now a very well-known hematologist and an authority on iron metabolism.

In 1956, I started to study at the Hebrew University – Hadassah Medical School in Jerusalem, which was the only medical school in Israel at that time (there are now four). In the basic science part of my medical studies, I fell in love with biochemistry. I studied biochemistry in three different courses: organic chemistry, basic biochemistry and a course called "physiological chemistry", which was medically oriented biochemistry. I was very fortunate to have outstanding teachers in all three courses. Organic chemistry was taught by Yeshayahu Leibowitz, a legendary person in Israel, a highly original thinker whose knowledge encompassed philosophy, political science, Bible, Talmud, medicine, chemistry and more. He was probably my best teacher, it was an intellectual feast to listen to him. Leibowitz loved biochemistry, and he sneaked biochemistry into his lectures on organic chemistry whenever he could, which was often. Basic biochemistry was taught by Shlomo Hestrin, also an inspiring teacher who had a special talent to transfer his enthusiasm of science to the students. Physiological chemistry was taught by Ernst Wertheimer, a professor of German Jewish origin whom we had some difficulty to understand because of his heavy German accent, but who had an excellent perspective of integration of metabolism at the level of the total body and of physiological contexts of biochemistry. Another part of the same course was taught by Jacob Mager. Mager was an outstanding biochemist and a man of encyclopedic knowledge. However, he was very shy and quite a bad classroom teacher (though an excellent teacher in the laboratory, as I learned later). Most of his lectures were delivered while he was writing whole metabolic pathways on the blackboard, without any notes, with his face to the blackboard and his back directed to the class. Still, I was so much impressed by the depth and breadth of his knowledge of biochemistry that I decided to ask Mager to do some research in his laboratory.

I started to work in Mager's laboratory in 1960. At that time, there was no formal M.D. - Ph. D. program at the Hebrew University, but it was possible to do a year of research between the preclinical and the clinical years of medical studies. I did that, and although I completed medical studies later on, I already knew by the end of that year that I was going to do research, rather than clinical practice. I was very fortunate to have had Jacob Mager as my mentor and tutor of biochemical research. He was a scientist of incredible scope of interests and knowledge. He was interested in every subject in biomedicine, he knew almost everything about every subject and he worked simultaneously on 3–4 completely different research projects. This undoubtedly caused fragmentation of his contributions to science, but provided his students with a

Jacob Mager.

Judy and Avram Hershko in 1977.

broad experience in different areas of biochemistry in a single, relatively small laboratory. In a period of a few years I worked with Mager on subjects as different as the effects of polyamines on protein synthesis *in vitro*, glucose-6-hosphate dehydrogenase deficiency and a variety of aspects of purine nucleotide metabolism, including enzymology and regulation. During this time, I also finished my medical studies, did my military service as a physician (1965–1967) and then returned for two more years to Mager's laboratory to finish my Ph. D. thesis (1967–69). I received not just a broad view of biochemistry from Mager, but also a very solid basis. He was a very rigorous experimentalist, every experiment had to be done with all possible positive and negative controls, all experiments were carried out in the duplicate, and every significant new finding had to be repeated several times to make it sufficiently credible. I owe a lot to Jacob Mager for a strong background of rigorous biochemistry.

I met Judith (née Leibowitz) in 1963, and we married at the end of the same year. Judy was born and raised in Switzerland. After her studies in biology, she decided to spend a year in Israel. During this year, she worked in the hematology laboratory of the Hadassah hospital in Jerusalem. One day, I walked over to the hematology laboratory to get a blood sample that I needed for my research, and we literarily bumped into each other. This collision caused her to stay in Israel for more than one year, and now we are married for over 41 years. We have three sons: Dan (1964), Yair (1968) and Oded (1975). Dan is a surgeon, Yair is a computer engineer and Oded is a medical student. We have

Judy, Avram and family in 2003 (Judy's 60th birthday party). Left to right ; standing: Vardit, Yair, Avram, Judy; middle row: Dan, Ori, Sharon, Oded; sitting: Lee, Maya, Roni, Ela.

now six grandchildren: Maya (1994), Lee (1997), Roni (1998), Ela (2000) Ori (2002) and Shahar (2004). Needless to say, both Judy and I are crazy about all our grandchildren. During all our years together, I got tremendous support from Judy. Although she came from one of world's most peaceful countries to one of the least, and from a very comfortable and pampering environment to quite primitive surroundings, she stood ground with a lot of energy, courage and cheerful optimism. She always took care of all my possible needs, as well as the needs of our children and grandchildren. Judy is not only a very beautiful woman, but she also irradiates a lot of caring, love and compassion. In addition to providing so much support at home, she also helped me a lot in the laboratory, over a period of more than 15 years. The ubiquitin system was helped by Judy in more than one way.

In 1969-71 I was a post-doctoral fellow with Gordon Tomkins at the Department of Biochemistry and Biophysics of the University of California in San Francisco. I met Gordon the previous year, when he gave some lectures in Israel. He was very different from Mager: outgoing, vivacious, bursting with original ideas. Unlike Mager, Gordon did not care much about controls or experimental detail, but he was a volcano of a man, constantly erupting with great ideas and he was a wonderful stimulator of many other researchers' work as well. Many distinguished scientists who knew Gordon Tomkins at that time (unfortunately, he died at an early age) are still speaking of him with great admiration. He exuded a great personal charm and I liked him instantly. I

Gordon Tomkins.

thought that Gordon may add some new dimensions to my development in science and this indeed was the case. I got a lot of stimulation and biological perspective from Gordon, while I continued to use what I learned from Mager about rigorous controls. As described in the accompanying lecture, I learned about protein degradation and got fascinated with this process while I was working with Gordon Tomkins.

I returned from San Francisco to Israel in 1971. Originally, I planned to return to the Hebrew University in Jerusalem, but a new medical school opened in Haifa and I was offered to be its Chairman of Biochemistry. This sounded very challenging and I agreed, but later it turned out to be a very minute Unit of Biochemistry in a very small Faculty of Medicine of the Technion, so at the beginning I chaired mainly myself. One initial reason for its being so small was that there was not enough space to house much faculty. The whole Faculty of Medicine was housed, on a temporary basis, in an old two-floor monastery. This "temporary" situation lasted for more than 15 years, until the new building of the Faculty of Medicine was completed in 1987. However, I had great times in that old monastery, and much of the discovery of the ubiquitin system was done right there. Isolation may at times lead to creativity, since one is not bothered by what others are doing and does not feel compelled to work on currently popular, "fashionable" subjects. I was very fortunate to assemble there a highly devoted research team, that included at the beginning Hanna Heller and Dvora Ganoth, and later, at different times, Ety Eytan, my

Old and new buildings of the Faculty of Medicine, Technion, Haifa. A. Old building. My laboratory was at the right corner of the upper floor. B. New building.

wife Judy, Sarah Elias and Clara Segal. Dvora and Ety still work with me. My first graduate students were David Epstein, Yaacov Hod and Michael Aviram. For a number of years, we tried to establish a cell-free system that reproduces energy dependent protein degradation in the test tube, essential for the biochemical analysis of this system. For this purpose, we tried different sources, such as liver homogenates and extracts from cultured cells and even from bacteria. We did not have any success in all these attempts. I remember that a biochemist friend from Jerusalem visited my laboratory and at the end of the visit she told me that I should not have most of my laboratory working on a hopeless subject. However, I was very obstinate and was obsessed with the idea that it would be possible to find out how proteins are degraded only with a biochemically analyzable cell-free system. Maybe I was lucky to work in such a remote and small place; in a larger institution, my graduate students and research assistants may have deserted me for some less frustrating research. Finally we used for biochemical fractionation the reticulocyte cell-free system established in the Goldberg laboratory (see Lecture). At that time, Aaron Ciechanover joined my laboratory for a D. Sc. thesis, after completing his medical studies and Army service. Aaron was the most incredibly hard-working graduate student that I ever had. With his huge energies, he contributed a lot to the discovery of the ubiquitin system. He was also a natural manager, already as a graduate student. I recall that at the end of my sabbatical year in Philadelphia in 1978 (see below), after telling Ernie Rose how small were

Israeli research grants, Ernie suggested that I should apply for a foreign research grant from the NIH to support my work in Israel. I was inclined to do a couple of more experiments instead of writing a grant application, but Aaron pushed me into a chair and commanded: "now write the NIH grant application!". I wrote it and got the grant, the first of five consecutive grant periods supported by the NIH. It saved the situation in the Haifa lab at a very critical time. I am very grateful to the NIH for supporting my work and also to Aaron for forcing me to write the initial grant application.

The story of the discovery of the ubiquitin system is described in my lecture, and here I add only some anecdotal episodes from these times. The fractionation of reticulocyte lysates to Fractions 1 and 2 was based on a trick that I learned from Mager in the purification of enzymes of purine nucleotide metabolism from erythrocytes. Hemoglobin constitutes about 80–90% of total protein of erythrocytes and reticulocytes, and therefore the first task in the purification of any enzyme from these cells is to get rid of the great mass of hemoglobin. This is most conveniently done by using the anion exchange resin DEAE-cellulose, which binds most non-hemoglobin proteins, but not hemoglobin. In our case this procedure resulted in loss of activity, which could be recovered by adding back Fraction 1 that contained not only hemoglobin but also ubiquitin. In fact, in our laboratory jargon we called ubiquitin for some time "Red", because of the red color of hemoglobin in this fraction. After we found that the factor in this fraction (i. e., ubiquitin) remains active after boiling for 30 minutes, we consulted a protein expert at the Technion who told us that our factor cannot be a protein. We found, however, that it is a protein, based on its sensitivity to the action of proteinases. Maybe the lesson from this story is that it is dangerous to consult experts.

After working for 6 years at the Technion, I had a sabbatical year due in 1977–78. I had a problem in choosing a person with whom I would spend my sabbatical year. I knew the people in the (then) small protein degradation field, and I was not very enthusiastic. Many people in the field had their pet theories about the cause for the high selectivity of intracellular protein degradation, without much (or any) experimental evidence. Once again, I was lucky. In 1976, I attended a Fogarty meeting on a quite general subject at the National Institutes of Health. Irwin Rose also attended this meeting, and one morning I joined him at the breakfast table. Ernie was well known for his work on enzyme mechanisms. In the course of our conversation I asked Ernie in what else was he interested, and his reply was: "protein degradation". I was a bit taken aback and told him that I never saw anything published by him on protein degradation. His reply was:" there is nothing worth publishing on protein degradation"! I liked his critical attitude and Ernie being such a character and therefore I asked him to spend my sabbatical year in his laboratory. It turned out that Ernie Rose was really interested in protein degradation. When he had been a young faculty in the fifties at the Department of Biochemistry of Yale University, he talked to Melvin Simpson, another young faculty there, and Simpson told him about his experiments on the energy-dependence of the liberation of amino acids from proteins in liver slices (see Lecture). This

Irwin Rose at Fox Chase Cancer Center in 1992.

aroused Ernie's interest, and from time to time he did experiments trying to understand the energy-dependence of protein degradation. He did not make any significant progress in these experiments, therefore he did not publish anything on protein degradation.

Ernie Rose is the third person, in addition to Mager and Tomkins, who had a great influence on my scientific life. He is very different from both Mager and from Tomkins. He likes problem solving, and his attitude to science is highly analytical. I am more intuitive, so we complemented each other very well. He is so brilliant that people do not always understand his ideas and are a little afraid of him. People are also often apprehensive of him because he can be very critical, and does not hesitate to voice his criticisms. We got along very well over a period of 20 years, which included several sabbaticals and many summer visits in his laboratory at Fox Chase Cancer Center in Philadelphia. Our only disputes were when he refused to be co-author of work to which he actually made significant contributions. In the case of the few papers on which he is co-author, I had to force him to agree. He was most unselfish in our joint work, a rare phenomenon in today's science. I asked him once why does he keep inviting me back to his laboratory, and his answer was: "I like the excitement". Ernie always downplayed his contributions to the ubiquitin field. He wrote an autobiographical article for *Protein Science* in 1995, and the word "ubiquitin" is not mentioned in this recollections paper. In our conversations he always described his role in the ubiquitin story as being merely supportive, but this is certainly not true. Although on occasions when I worked in his laboratory and he was adsorbed with some problem in enzyme mechanisms, he would forget about my existence for a week or two, but then suddenly he would come up with a bright suggestion about my current work. I can state that Ernie's input of ideas, inspiration and helpful criticism were essential for the discovery of the ubiquitin system and for the delineation of some of the main enzymatic reactions in this pathway.

The rest of my story is a lot of more work, but also a lot of more scientific excitement and fun. I continued to be obstinate, and continued to do what many considered to be old-fashioned biochemistry in the eighties, when the powerful technologies of molecular biology became available. This biochemical work resulted in the discovery of the three types of enzymes involved in ubiquitin-protein ligation (E1, E2 and E3), and of some further enzymes of this system. Subsequently, I became interested in the roles of ubiquitin-mediated protein degradation in the cell division cycle. This led me to the Marine Biological Laboratory (MBL) at Woods Hole, due to the availability of a clam oocycte cell-free system, which faithfully reproduces cell cycle related events in the test tube. This system was important for the discovery of the Cyclosome/Anaphase-Promoting Complex, as described in the Lecture. In the past decade, I am spending my summers at the MBL for the same reason that I spent my summers previously at Fox Chase Cancer Center – to be able to devote almost all my time to do experiments in a tranquil environment. Benchwork is my great hobby; I also do benchwork in Haifa, but on a more part-time basis. I have always loved to do experiments with my own hands, both for peace of mind and for excitement. Also, my own experiments were important for almost every significant progress made in my laboratory. One cannot have a more beautiful place than the MBL for doing experiments: the great natural beauty of the surroundings, the tranquility and outstanding scientific environment- all combine to make the MBL a great place for doing summer research.

When I look back at my life until now, I am amazed how fortunate I have been in both my personal and my scientific life. After escaping the Holocaust, both my parents lived in Israel to a good old age. I am very happy with my wife, children and grandchildren. I was very fortunate to have outstanding mentors in science, and then to be able to use the knowledge gained for a significant contribution. If only there were some peace in the world, including between Israel and its neighbors – I would be completely satisfied.

THE UBIQUITIN SYSTEM FOR PROTEIN DEGRADATION AND SOME OF ITS ROLES IN THE CONTROL OF THE CELL DIVISION CYCLE

Nobel Lecture, December 8, 2004

by

AVRAM HERSHKO

Unit of Biochemistry, the Rappaport Faculty of Medicine and Research Institute, Technion – Israel Institute for Technology, Haifa, Israel.

INTRODUCTION

All living cells contain many thousands of different proteins, each of which carries out a specific chemical or physical process. Due to the importance of proteins in basic cellular functions, there has been a great interest in the problem of how proteins are synthesized. In the fifties and sixties of the 20th century, the discovery of the double helical structure of DNA and the cracking of the genetic code focused attention on the mechanisms by which the order of bases in DNA determines the sequence of amino acids in proteins, and on further molecular mechanisms that regulate the expression of specific genes. Because of the intensive research activity on protein synthesis, little attention was paid to at that time to the fact that many proteins are rapidly degraded to amino acids. This dynamic turnover of cellular proteins had been previously known by the pioneering work of Schoenheimer and co-workers, who were among the first to introduce the use of isotopically labeled compounds to biological studies. They administered ^{15}N-labeled L-leucine to adult rats, and the distribution of the isotope in body tissues and in excreta was examined. It was observed that less than one-third of the isotope was excreted in the urine, and most of it was incorporated into tissue proteins (1). Since the weight of the animals did not change during the experiment, it could be assumed that the mass and composition of body proteins also did not change. It was concluded that newly incorporated amino acids must have replaced those in tissue proteins in a process of dynamic protein turnover (1). Schoenheimer's studies on the dynamic state of proteins and of some other body constituents were published in a small booklet in 1942, soon after his untimely death (ref. 2, see Fig. 1).

In the subsequent decades research on protein degradation was neglected, mainly because of the great interest in the mechanisms of protein synthesis, as described above. However, gradually experimental evidence accumulated which indicated that intracellular protein degradation is extensive, selective and has basically important cellular functions. It was observed that abnormal

THE DYNAMIC STATE
OF BODY CONSTITUENTS

BY

RUDOLF SCHOENHEIMER, M.D.

Late Associate Professor of Biological Chemistry, Columbia University

CAMBRIDGE, MASSACHUSETTS
HARVARD UNIVERSITY PRESS
1942

Figure 1. Front page of Schoenheimer's collected lectures, edited by his colleagues soon after his death.

proteins produced by the incorporation of some amino acid analogues are selectively recognized and are rapidly degraded in cells (3). However, intracellular protein degradation was not thought to be merely a "garbage disposal" system for the elimination of abnormal or damaged proteins. By the late sixties, it became apparent that normal proteins are also degraded in a highly selective fashion. The half-life times of different proteins ranged from several minutes to many days, and rapidly degraded proteins usually had important regulatory functions. These properties of intracellular protein degradation and the role of this process in the regulation of the levels of specific proteins were summarized in an excellent review by Schimke and Doyle in 1970 (4). Thus, it was known at that time that protein degradation has important functions, but it was not known what is the biochemical system that carries out this process at such a high degree of selectivity and sophistication.

MY FIRST ENCOUNTER WITH PROTEIN DEGRADATION

I became interested in the problem of how proteins are degraded in cells when I was a post-doctoral fellow in the laboratory of Gordon Tomkins in 1969–71 at the University of California, San Francisco. At that time, Gordon was mainly interested in the mechanisms by which steroid hormones induce the synthesis of specific proteins. His model system for this purpose was the synthesis of the enzyme tyrosine aminotransferase (TAT) in cultured hepato-

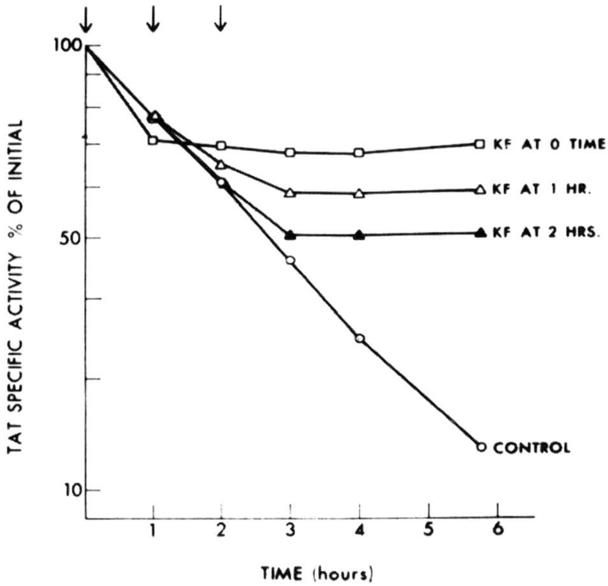

Figure 2. Energy-dependence of the degradation of tyrosine aminotransferase. From ref. 5.

ma cells. When I arrived there I saw that it was a large laboratory, with many post-doctoral fellows working on different aspects of the synthesis of TAT. I thought that this was a bit too crowded and I asked Gordon for a different project. He suggested that I should study the degradation of TAT, a process that also regulates the level of this enzyme. This was how I became involved in protein degradation, a problem on which I have been working ever since.

Fig. 2 shows one of the first experiments that I did as a post-doctoral fellow in the Tomkins lab. It was quite easy to follow the degradation of TAT: first, hepatoma cells were incubated with a steroid hormone, which caused a great increase in the level of this protein. Then the hormone was removed by changing the culture medium, and a rapid decline in the level of this protein, due to its degradation, could be observed. As with other regulatory proteins, this protein also had a relatively rapid rate of degradation, with a half-life time of about 2-3 hours. I found then that the degradation of TAT was completely arrested by potassium fluoride, an inhibitor of cellular energy production (ref. 5 and Fig. 2). The effect was not specific to fluoride, because I got similar results with other inhibitors of cellular energy production. These results confirmed and extended earlier findings of Simpson (6) on the energy-dependence of the liberation of amino acids from proteins in liver slices. This observation was later dismissed as being indirect, and that energy is needed to keep the acidic pH inside the lysosomes (described in ref. 7). However, in the case of TAT energy was needed for the selective degradation of a specific enzyme, and it did not seem reasonable to assume that engulfment into lysosomes can be responsible for the highly selective degradation of specific cellular proteins. Since ATP depletion also prevented the inactivation of the enzymatic activity of TAT, it was concluded that energy is required at an early step in the process of protein degradation (5).

I was very much intrigued by the energy-dependence of intracellular protein degradation. Energy is usually needed to synthesize a chemical bond, and not to break a chemical bond. Thus, the action of extracellular proteinases of the digestive system is an exergonic process, *i. e.*, it actually releases energy. This suggested that within cells a novel, as yet unknown proteolytic system exists, that presumably uses energy to attain the high selectivity of the degradation of cellular proteins.

DISCOVERY OF THE ROLE OF UBIQUITIN IN PROTEIN DEGRADATION

Parts of the story of the discovery of the ubiquitin system have been described previously (7–9). Following my return to Israel and setting up my own laboratory at the Technion in Haifa, I continued to pursue this problem of how proteins are degraded in cells, and why is energy required for this process. It was clear to me that the only way to find out how a completely novel system works is that of classical biochemistry. This consists of using a cell-free system that faithfully reproduces the process in the test tube, fractionation to separate its different components, purification and characterization of each component and reconstitution of the system from isolated and purified components. A cell-free ATP-dependent proteolytic system from reticulocyte lysates was first established by Etlinger and Goldberg (10). Subsequently, my laboratory subjected this system to biochemical fractionation, with the aim of the isolation of its components and the characterization of their mode of action. A great part of this work was done by Aaron Ciechanover, who was my graduate student at that time (1976–1981). This work has also received a lot of support, great advice and helpful criticism from Irwin Rose, in whose laboratory at Fox Chase Cancer Center I worked in a sabbatical year in 1977–78 and in many summers afterwards.

In the initial experiments, we have resolved reticulocyte lysates on DEAE-cellulose into two crude fractions: Fraction 1, which contained proteins not adsorbed to the resin, and Fraction 2, which contained all proteins adsorbed to the resin and eluted with high salt. The original aim of this fractionation was to get rid of hemoglobin, which was known to be in Fraction 1, while most non-hemoglobin proteins of reticulocytes were known to be in Fraction 2. We found that neither fraction was active by itself, but ATP-dependent protein degradation could be reconstituted by the combination of the two fractions (11). The active component in Fraction 1 was a small, heat stable protein; we have exploited its stability to heat treatment for its purification to near homogeneity. We termed this protein at that time APF-1, for ATP-dependent Proteolysis Factor 1. The identity of APF-1 with ubiquitin was established later by Wilkinson *et al.* (12), subsequent to our discovery of its covalent ligation to

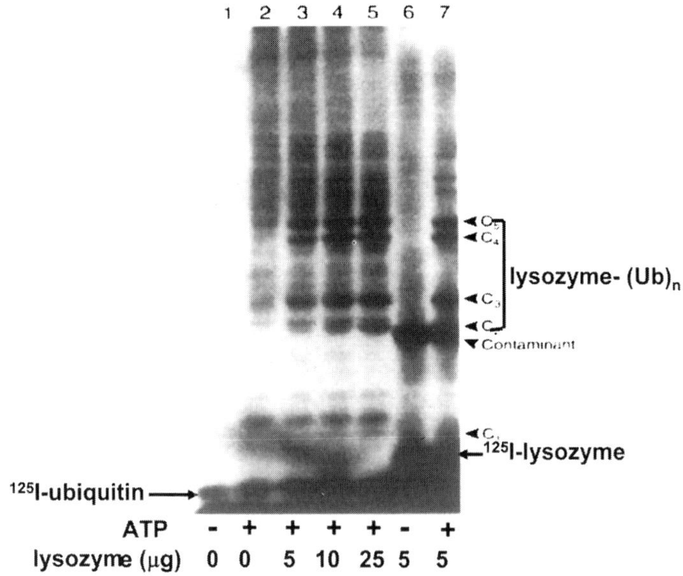

Figure 3. Discovery of covalent ligation of ubiquitin to substrate protein. See the text. From ref. 16.

protein substrates, as described below. Ubiquitin was originally isolated by Goldstein and co-workers in a search for hormones of the thymus, but was subsequently found to be present in all tissues and eukaryotic organisms, hence its name (13). The functions of ubiquitin were not known, though it was discovered by Goldknopf and Busch that ubiquitin was conjugated to histone 2A by an isopeptide linkage (14).

The purification of APF-1/ubiquitin from Fraction 1 was the key to the elucidation of the mode of its action in the proteolytic system. It looked smaller than most enzymes, so at first we thought that it might be a regulatory subunit of some enzyme (such as a protein kinase or an ATP-dependent protease) present in Faction 2. To test this notion, we looked for the association of APF-1/ubiquitin with some protein in Fraction 2. For this purpose, purified radiolabeled APF-1/ubiquitin was incubated with Fraction 2 in the presence or absence of ATP, and was subjected to gel filtration chromatography. A marked ATP-dependent association of APF-1/ubiquitin with high molecular weight material was observed (15). It was very surprising to find, however, that ubiquitin was bound by covalent amide linkage, as indicated by the resistance of high molecular weight derivative to alkali, hydroxylamine and boiling with SDS in the presence of mercaptoethanol (15). The analysis of reaction products by SDS-polyacrylamide gel electrophoresis showed that ubiquitin was ligated to a great number of endogenous proteins. Since crude Fraction 2 from reticulo-

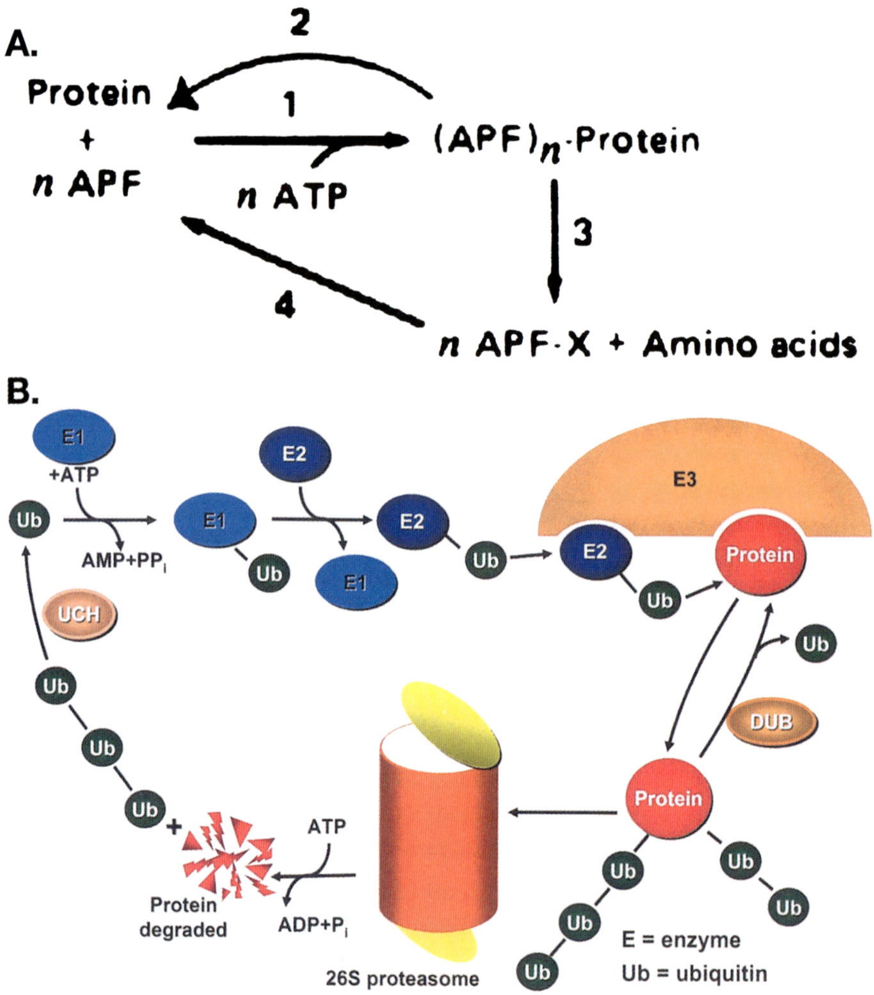

Figure 4. The ubiquitin system then and now. A. Original proposal of the sequence of events in protein degradation. See the text; from ref. 16. B. Our current view of the main enzymatic reactions in ubiquitin-mediated protein degradation. See the text. Ub, ubiquitin; DUB, deubiquitylating enzyme; UCH, ubiquitin carboxyl-terminal hydrolase.

cytes contained not only enzymes, but also endogenous substrates of the proteolytic system, we began to suspect that ubiquitin may be linked to protein substrates, rather than to an enzyme. We indeed found that proteins that are good (though artificial) substrates of the ATP-dependent proteolytic system, such as lysozyme, are conjugated to ubiquitin (16). The original experiment is shown in Fig. 3. We found that similar high-molecular weight derivatives were formed when ^{125}I-labeled ubiquitin was incubated with unlabeled lysozyme

(Fig. 3, lanes 3–5), or when ^{125}I-labeled lysozyme was incubated with unlabeled ubiquitin in the presence of ATP (Fig. 3, lane 7). Based on these observations, we proposed that proteins are targeted for degradation by covalent ligation to APF-1/ubiquitin (16). The original hypothesis from 1980, formulated jointly with Irwin Rose, is shown in Fig. 4A. We proposed that a putative enzyme, which we called "APF-1-protein amide synthetase", ligates multiple molecules of ubiquitin to the protein substrate (Step 1) and then some other enzyme degrades proteins which are linked to several molecules of ubiquitin (Step 3), and finally free and reutilizable ubiquitin is released (Step 4). According to this proposal, ubiquitin is essentially a tag, which when attached to a protein, dooms this protein to be degraded.

IDENTIFICATION OF ENZYMES OF THE UBIQUITIN-MEDIATED PROTEOLYTIC SYSTEM

In subsequent work, we tried to isolate and characterize enzymes of the ubiquitin pathway, using the same biochemical fractionation-reconstitution approach. The original proposal (Fig. 4A) was found to be essentially correct, but much important detail was added. Our present knowledge of the main enzymatic steps in the ubiquitin-mediated proteolytic pathway is shown in Fig. 4B (reviewed in ref. 17). This scheme summarizes about 10 years of our work (from 1980 to 1990), as well as that of some other researchers. Thus, we found that ubiquitin is ligated to proteins not by one enzyme, but by the sequential action of three enzymes. These are the ubiquitin-activating enzyme E1 (18), a ubiquitin-carrier protein E2 (19) and a ubiquitin-protein ligase E3 (19). E1 carries out the ATP dependent activation of the carboxy-terminal glycine residue of ubiquitin (20) by the formation of ubiquitin adenylate, followed by the transfer of activated ubiquitin to a thiol site of E1 with the formation of thiolester linkage (18, 21). Activated ubiquitin is transferred to a thiol site of E2 by transacylation, and is then further transferred to amino group of the protein substrate in a reaction that requires E3 (19). We found that the role of E3 is to bind specific protein substrates (22). Based on this observation, it was proposed that the selectivity of ubiquitin-mediated protein degradation is mainly determined by the substrate specificity of different E3 enzymes (23). This notion was verified by subsequent work in many laboratories on the selective action of a large number of different E3 enzymes on their specific protein substrates. Proteins ligated to polyubiquitin chains are degraded by a large 26S proteasome complex (discovered by other investigators) and free ubiquitin is released by the action of ubiquitin-C-terminal hydrolases or isopeptidases (reviewed in 17).

MECHANISMS OF THE DEGRADATION OF CYCLIN B: DISCOVERY OF THE CYCLOSOME/ANAPHASE-PROMOTING COMPLEX

All our studies on the basic biochemistry of the ubiquitin pathway were carried out in the reticulocyte system, using artificial model protein substrates. Though many gaps remained in our understanding of the basic biochemical

Figure 5. The north Atlantic surf clam, *Spisula solidissima.*

processes of the ubiquitin system, at around 1990 I thought that it was important to turn to the question of how is the degradation of specific cellular proteins is carried out by the ubiquitin system in a selective and regulated fashion. This is how I became interested in the roles of the ubiquitin system in the cell division cycle, because the levels of many cell cycle regulatory proteins oscillate in the cell cycle. I first worked on the biochemical mechanisms of the degradation of cyclin B in the early embryonic cell cycle. Cyclin B was discovered by Hunt and co-workers as a protein that is degraded at the end of each mitosis (24). It was subsequently found that it is a positive regulatory subunit of protein kinase Cdc2/Cdk1 (Cyclin-dependent kinase 1) (reviewed in ref. 25). In the early embryonic cell cycles, cyclin B is synthesized during the interphase and then is rapidly degraded in the metaphase-anaphase transition. The active protein kinase Cdk1-cyclin B (also called MPF or M phase-promoting factor) is formed at the beginning of mitosis and promotes entry of cells into mitosis. The inactivation of MPF, caused by the degradation of cyclin B, is required for exit from mitosis. Our question was: what is the system that degrades cyclin B and why does it act only at the end of mitosis?

I have approached this problem again by biochemistry, and here the quest for a cell-free system led me to marine biology and to the surf clam, *Spisula solidissima* (Fig. 5). This is a large clam that produces large numbers of oocytes. Luca and Ruderman (26) established a cell-free system from fertilized clam oocytes that faithfully reproduced cell cycle stage-specific degradation of mitotic cyclins. In this work, I was greatly helped first by Robert Palazzo and Leonard Cohen, and then by collaboration with Joan Ruderman. Initial fractionation of the system (27) showed that in addition to E1, two novel components were required for the ligation of cyclin B to ubiquitin: these were a specific E2 called E2-C and an E3-like activity, which in clam extracts, was associated with particulate material. We have solubilized the E3-like activity and found it to be a large (~1,500 kDa) complex that has ubiquitin ligase activity on mitotic cyclins. The activity of this enzyme is regulated in the cell cycle: it is inactive in the interphase and becomes active at the end of mitosis, an event that requires the action of Cdk1/cyclin B (28). We called this ubiquitin ligase complex the cyclosome, to denote its large size and important roles in cell cycle regulation (28). A similar complex was isolated at about the same time from extracts of *Xenopus* eggs by the Kirschner laboratory and was called the Anaphase-Promoting Complex (29). Parallel genetic work in yeast by the Nasmyth group identified several subunits of the Anaphase-Promoting Complex/cyclosome (or APC/C, as it is now called) as products of genes required for exit from mitosis (30). Thus, the discovery of APC/C was due to the convergence of biochemical with genetic work. Subsequent work by other investigators showed that the APC/C is also involved in the degradation of several other important cell cycle regulators, such as securin, an inhibitor of anaphase onset (reviewed in 31, 32). In addition, the APC/C is the target of the spindle assembly checkpoint system, an important surveillance mechanism that allows the separation of sister chromatids only after they are all properly attached to the mitotic spindle (reviewed in 33).

ROLE OF SCF^{SKP2} UBIQUITIN LIGASE IN THE DEGRADATION OF THE CDK INHIBITOR P27^{KIP1}

Another problem on which I have been working recently, in collaboration with Michele Pagano, is on the mode of the degradation of the mammalian Cdk inhibitor p27^{Kip1}. This inhibitor is present at high levels in G0/G1, preventing the action of Cdk2/cyclin E and Cdk2/cyclin A to drive cells into the S-phase. Following growth stimulation by mitogenic agents p27 is rapidly degraded, allowing the action of these kinases to promote entry into the S-phase (reviewed in 34). It has been shown that p27 is degraded by the ubiquitin system (35). We have tried to identify the ubiquitin ligase system that targets p27 for degradation. It was first found that the process of p27-ubiquitin ligation can be faithfully reproduced *in vitro* in extracts of HeLa cells. Thus, the rate of ligation of p27 to ubiquitin was much greater in extracts from growing cells than in extracts from G1-arrested cells. It was also found that the phosphorylation of p27 on T187 by Cdk2/cyclin E is required for p27-ubiquitin ligation *in vitro* (36), as is the case *in vivo* (37). Having established that the cell-free system accurately reflects the characteristics of p27 ubiquitylation in cells, we then proceeded to utilize this cell-free system to identify the ubiquitin ligase (E3 enzyme) involved in this process. Because of the requirement for the phosphorylation of the p27 substrate, we suspected that an SCF (Skp1-Cullin1-F-box protein) type ubiquitin ligase might be involved. SCF complexes comprise a large family of ubiquitin-protein ligases, whose variable F-box protein subunits recognize a variety of phosphorylated protein substrates (reviewed in 38). We have identified Skp2 (S-phase kinase-associated protein 2) as the specific F-box protein component of an SCF complex that ubiquitylates p27, based on the following biochemical evidence: (a) Immunodepletion of extracts from proliferating cells with an antibody directed against Skp2 abolished p27-ubiquitin ligation activity; (b) Addition of recombinant, purified Skp2 to such immunodepleted extracts completely restored p27-ubiquitin ligation; (c) Specific binding of p27 to Skp2, dependent upon phosphorylation of p27 on T187, could be demonstrated *in vitro*. Combined with further *in vivo* evidence from the Pagano lab, Skp2 was identified as the specific and rate-limiting component of an SCF complex that targets p27 for degradation (39). It is notable that levels of Skp2 also oscillate in the cell cycle, being very low in G1, increasing upon entry of cells into the S-phase and declining again later on (40). These fluctuations in Skp2 levels provide an important mechanism for cell cycle stage-specific regulation of p27 degradation.

We have next tried to reconstitute the SCF^{Skp2} system that ligates p27 to ubiquitin from purified components. We found that in addition to the known components (Cullin 1, Skp1, Skp2, Roc1, Cdk2/cyclin E, E1 and the E2 enzyme Cdc34), an additional protein factor is required for this reaction. We have purified the missing factor from extracts of HeLa cells and have identified it as Cks1 (cyclin kinase subunit 1), both by mass spectrometry sequencing and by functional reconstitution with recombinant Cks1 protein (41). Cks1 belongs to the highly conserved Suc1/Cks family of proteins, which bind to some cyclin-dependent kinases and to phosphorylated proteins, and are essential for several cell cycle transitions (42). Human Cks1, but not other members of this protein family, reconstituted p27-ubiquitin ligation in a completely purified system. While all members of the Suc1/Cks protein family have Cdk-binding and anion-binding sites, only mammalian Cks1 binds to Skp2 and promotes the association of Skp2 with p27 phosphorylated on T187 (41). Similar results were independently obtained by another research group (43). More recently, we have mapped the Skp2-binding site of Cks1 by site-directed mutagenesis and found that it is located on a region that includes the $\alpha 2$ and $\alpha 1$ helices, well separated from the other two binding sites of Cks1. All three binding sites of Cks1 are required for its action to promote p27-ubiquitin ligation and for the association of Skp2 with T-187-phosphorylated p27 (44). Based on these and on further observations, a model was proposed according to which Cks1 serves as an adaptor necessary for enzyme-substrate interaction: the Skp2-Cks1 complex binds to phosphorylated p27, a process which requires the anion-binding site of Cks1. The affinity of Skp2 to the substrate is then further strengthened by the association of the Cdk-binding site of Cks1 with Cdk2/cyclin E, to which phosphorylated p27 is tightly bound (44). It is notable that the expression of Cks1 also oscillates in the cell cycle (45, 46), providing an additional mechanism for the regulation of p27 degradation.

CONCLUDING REMARKS

The ubiquitin system has come a long way since its humble beginnings described in this paper. Ubiquitin-mediated degradation of positively or negatively acting regulatory proteins is involved in a variety of cellular processes such as the control of cell division, signal transduction, transcriptional regulation, immune and inflammatory responses, embryonic development, apoptosis and circadian clocks, to mention but a few. The involvement of malfunction of ubiquitin-mediated processes in diseases such as certain cancers, and the therapeutic implications of this knowledge, are also beginning to emerge. I am quite certain that we are still seeing only the tip of the iceberg of the multitude of functions of the ubiquitin system in health and disease. The main lesson from the story of the discovery of the ubiquitin system that I would like to convey, mainly to young researchers, is the continued importance of biochemistry in modern biomedical research. In his book *"For the Love of Enzymes"*, Arthur Kornberg divided the history of biomedical research into four main periods. First were the "microbe hunters", the great microbiologists of the 19th century. They were followed by the "vitamin hunters", the discoverers of the vitamins. Next were the "enzyme hunters" – the biochemists, followed by the "gene hunters" – the molecular geneticists. However, the times of enzyme (or protein) hunting are far from being over. With the completion of the human genome project, all genes have been "hunted", but we know the functions of only about one-third of our genes. If we want to know what are the roles of the rest of our genes in health and in disease, we shall have to continue to use biochemistry, in combination with functional genetics, well into the future. Our story shows that the ubiquitin system could not have been discovered without biochemical approaches. We could not have a clue to the ubiquitin tagging mechanisms by genetics alone, or by the sequence of genes of the ubiquitin system. On the other hand, once the basic biochemistry was known, molecular genetic approaches were essential for the discovery of the multitude of functions of the ubiquitin proteolytic pathway. So my advice to young investigators in biomedical sciences is: if you have a problem that cannot be solved by molecular genetics alone, do not be afraid to use biochemistry, do not hesitate to enter the cold room, and do not be wary of approaching the FPLC machine!

ACKNOWLEDGEMENTS

In experimental sciences, including biochemistry, discoveries are not made by a single person, but require the assistance of dedicated research teams and the help of friends, colleagues and collaborators. In my laboratory at the Technion, Haifa, I was very fortunate to receive devoted help, at different times over a period of more than 30 years, from Dvora Ganoth, Hanna Heller, Esther Eytan, Sarah Elias, Clara Segal and from my wife, Judith Hershko. Among my former graduate students, Aaron Ciechanover did tremendous work in the exciting times of the discovery of ubiquitin-protein ligation 25 years ago. Subsequently, many other graduate students (too many to list here) did very important work on the basic biochemistry of the ubiquitin system and

Figure 6. At the end of summer of 1979 in Fox Chase Cancer Center, Philadelphia. Seated left to right: Avram Hershko, Sandy Goldman, Jessie Warms, Hanna Heller. Standing left to right: Zelda Rose, Arthur Haas, Aaron Ciechanover, Mary Williamson, Irwin Rose, Keith Wilkinson and Leonard Cohen (last three people standing on the right side not identified).

more recently, on some roles of this system in cell cycle control. Out of my several friends-collaborators, mentioned in this paper, Irwin Rose had a very special role. My association with Ernie started with a sabbatical year in his laboratory in Fox Chase Cancer Center, Philadelphia, in 1977–78 (see also accompanying biography). During this year, I continued to work on the initial fractionation of the reticulocyte system and the purification of ubiquitin, which we started in Haifa. In the following summer of 1979, Ernie invited me back to his laboratory, together with my graduate student Aaron Ciechanover and research assistant Hanna Heller. When we got there we already knew, from work done in the Haifa lab, that ubiquitin becomes bound to proteins in an ATP-dependent process. However, the discovery that a covalent amide bond is formed between ubiquitin and the substrate protein was made together with Ernie Rose in that summer in Philadelphia. A group picture, taken at the end of this memorable summer of 1979 at Fox Chase Center, included the people involved (Fig. 6). The results of this summer's work were reported in ref. 16.

REFERENCES

1. Schoenheimer, R., Ratner, S. and Rittenberg, D. (1939) *J. Biol. Chem.* **130**, 703–732.
2. Schoenheimer, R. (1942). *The Dynamic State of Body Constituents.* Harvard University Press, Cambridge, Mass.
3. Rabinowitz, M. and Fisher, J. M. (1964) *Biochim. Biophys. Acta* **91**, 313–322.
4. Schimke, R. T. and Doyle, D. (1970) *Annu. Rev. Biochem.* **39**, 929–976.
5. Hershko, A. and Tomkins, G. M. (1971) *J. Biol. Chem.* **246**, 710–714.
6. Simpson, M. V. (1953) *J. Biol. Chem.* **201**, 143-154.
7. Hershko, A., Ciechanover, A. and Varshavsky, A. (2000). *Nature Med.* **6**, 1073–1081.
8. Hershko, A. (1996). *Trends Biochem. Sci.*, **21**, 445–449.
9. Wilkinson, K. D. (2004) *Cell* **119**, 741–745.
10. Etlinger, J. D. and Goldberg, A. L. (1977) *Proc. Natl. Acad. Sci. USA* **74**, 54–58.
11. Ciechanover, A., Hod, Y. and Hershko, A.(1978) *Biochem. Biophys. Res. Commun.* **81**, 1100–1105.
12. Wilkinson, K. D., Urban, M. K. and Haas, A. L.(1980) *J. Biol. Chem.* **255**, 7529–7532.
13. Goldstein, G., Scheid, M., Hammerling, U., Boyse, E. A., Schlesinger, D. H. and Niall, H. D. (1975) *Proc. Natl. Acad. Sci. USA* **72**, 11–15.
14. Goldknopf, I. L. and Busch, H. (1977) *Proc. Natl. Acad. Sci. USA* **74**, 864–868.
15. Ciechanover, A., Heller, H., Elias, S., Haas, A. L. and Hershko, A. (1980) *Proc. Natl. Acad. Sci. USA* **77**, 1365–1368.
16. Hershko, A., Ciechanover, A., Heller, H., Haas, A. L., and Rose, I. A. (1980) *Proc. Natl. Acad. Sci. USA* **77**, 1783–1786.
17. Hershko, A. and Ciechanover, A. (1998). *Annu. Rev. Biochem.* **67**, 425–479.
18. Ciechanover, A., Heller, H., Katz-Etzion, R. and Hershko, A. (1981) *Proc. Natl. Acad. Sci. USA* **78**, 761–765.
19. Hershko, A., Heller, H., Elias, S. and Ciechanover, A. (1983) *J. Biol. Chem.* **258**, 8206–8214.
20. Hershko, A., Ciechanover, A. and Rose, I. A. (1981) *J. Biol. Chem.* **256**, 1525–1528.
21. Haas, A. L., Warms, J. V., Hershko, A. and Rose, I. A. (1982) *J. Biol. Chem.* **257**, 2543–2548.
22. Hershko, A., Heller, H., Eytan, E. and Reiss, Y. (1986) *J. Biol. Chem.* **261**, 11992–11999.
23. Hershko, A. (1988) *J. Biol. Chem.* **263**, 15237–15240.
24. Evans, T., Rosenthal, E. T., Youngbloom, J., Distel, D. and Hunt, T. (1983) *Cell* **33**, 289–396.
25. Dorée, M. and Hunt, T. (2002) *J. Cell Sci.* **115**, 2461–2464.
26. Luca, F. C. and Ruderman, J. V. (1989) *J. Cell Biol.* **109**, 1895–1909.
27. Hershko, A., Ganoth, D., Sudakin, V., Dahan, A., Cohen, L. H., Luca, F. C., Ruderman, J. and Eytan, E. (1994) *J. Biol. Chem.* **269**, 4940–4946.
28. Sudakin, V., Ganoth, D., Dahan, A., Heller, H., Hershko, J., Luca, F. C., Ruderman, J. V. and Hershko, A. (1995) *Mol. Biol. Cell* **6**, 185-198.
29. King, R. W., Peters, J. M., Tugendreich, S., Rolfe, M., Hieter, P. and Kirschner, M. W. (1995) *Cell* **81**, 279–288.
30. Irniger, S., Piatti, S., Michaelis, C. and Nasmyth, K. (1995) *Cell* **81**, 269–277.
31. Zachariae, W. and Nasmyth, K. (1999). *Genes Dev.* **13**, 2039–2058.
32. Peters, J. M. (2002) *Mol. Cell* **9**, 931–943.
33. Bharadwaj, R. and Yu, H. (2004) Oncogene **23**, 2016–2027.
34. Slingerland, J. and Pagano, M. (2000) *J. Cell Physiol.* **183**, 10–17.
35. Pagano, M., Tam, S. W., Theodoras, A. M., Beer-Romano, P., Del Sal, G., Chau, V., Yew, P. R., Draetta, G. F. and Rolfe, M. (1995). *Science* **269**, 682–685.
36. Montagnoli, A., Fiore, F., Eytan, E., Carrano, A. C., Draetta, G. F., Hershko, A. and Pagano, M. (1999). *Genes Dev.* **13**, 1181–1189.
37. Vlach, J., Hennecke, S. and Amati, B. (1997). *EMBO J.* **16**, 5334–5344.
38. Deshaies, R. J. (1999). *Annu. Rev. Cell Dev. Biol.* **15**, 435–467.
39. Carrano, A., Eytan, E., Hershko, A. and Pagano, M. (1999).*Nature Cell Biol.* **1**, 193–199.
40. Lisztwan, J., Marti, A., Sutterluti, H., Gstaiger, M., Wirbelauer, C. and Krek, W. (1998). *EMBO J.* **17**, 368–383.

41. Ganoth, D., Bornstein, G., Ko. T. K., Larsen, B., Tyers, M., Pagano, M. and Hershko, A. (2001). *Nature Cell Biol.* **3**, 321–324.
42. Harper, J. W. (2001). *Curr. Biol.* **11**, R431–R435.
43. Spruck, C., Strohmaier, H., Watson, M., Smith, A. P. L., Ryan, A., Krek, W. and Reed, S. I. (2001). *Mol. Cell* **7**, 639–650.
44. Sitry, D., Seeliger, M. A., Ko, T. K., Ganoth, D., Breward, S. E., Itzhaki, L. S., Pagano, M. and Hershko, A. (2002). *J. Biol. Chem.* **277**, 42233–42240.
45. Richardson, H. E., Stueland, C. S., Thomas, J., Russel, P. and Reed, S. I. (1990). *Genes Dev.* **4**, 1332–1344.
46. Bashir, T., Dorrello, N. V., Amador, V., Guardavaccaro, D. and Pagano, M. (2004) *Nature* **428**, 190–193.

Portrait photo of Professor Hershko by photographer Dan Porges.

Irwin A. Rose

IRWIN ROSE

We left my birthplace, Brooklyn, New York in 1939 when I was 13. I enjoyed the ethnic variety and the interesting students in my public school, P.S. 134. The kids in my neighborhood were only competitive in games although unfriendly gangs tended to define the limits of our neighborhood. The major extracurricular activities that I can remember were a Victory Garden on school grounds, our contribution to the war effort, and a favorite sport, handball, played between the walls of our apartment house.

Mother, Ella Greenwald, was American born into a family that included one sister and four brothers, all born in Hungary. Father, Harry Royze, had two brothers and a sister from the Odessa region of Russia. The Greenwalds and the Roses were secular Jews and the children more so although my younger brother and I spent some time in Hebrew school to please Grandfather Rose.

Due to my brother having rheumatic fever the family was advised to go to a high and dry climate, Spokane, Washington, where my mother's sister had a comfortable home that could accommodate us. This left my father behind tending his flooring business, an arrangement that I never understood and felt conflicted about. Father's visits were few and far between. The war was going on. Mother did secretarial work in the Navy Supply Depot in Spokane while we kids were making our way through the Spokane school system.

I worked during the summers at a local hospital, chiefly helping out in the psychiatric ward. In time I came to see myself following some career that involved solving medical problems. No one in my family had followed a career in research. Uncle Arthur G. was an excellent violinist and artist, and taught cabinet making at a trade school in Brooklyn. Uncle Dave R. would have become a lawyer had the economic depression not led him into the U.S. Internal Revenue Service. There was no one in my circle from whom I could expect to get advice.

Initially I thought problems on how the brain works to be the most interesting. But it was necessary to be practical, and concentrate on less obscure matters when I entered Washington State College. Besides, there were no courses given in neurobiology. However, I was strongly influence by Prof. Herbert Eastlick who urged his zoology students to set high standards for themselves, and then to the University of Chicago after a brief period in the Navy. My PhD thesis problem was to determine if the DNA content of rat tissues increased if there was B12 in the diet. This problem was suggested by my adviser based on the observation that thymine could replace vitamin B12 in a lactobacillus. I analyzed the DNA of tissues of rats fed diets that varied in

B12. This project was doomed to failure when the genetic nature of DNA was revealed, and I found that the DNA content per cell of liver was independent of diet[1].

PhD Work. I had to think of a new thesis project. Anxious to make up for lost time I picked a problem out of my freshman biochemistry lecture notes. The Putnam/Evans group was interested in determining the origin of the nucleic acid components of bacteriophage synthesized in *E. coli* and Frank Putnam's lectures described experiments of Hammarsten, Reichard, and Saluste[2] as background information. ^{15}N-cytosine, the free base, had been found not to be incorporated into DNA although ^{15}N-cytidine was incorporated into rat liver DNA. It was obvious for me to ask if there might be direct utilization of the whole of cytidine, ribose and all, in the biosynthesis of deoxycytidine. That would be a shock. I learned from Peter Reichard, during a 2004 meeting in Stockholm that the export to Sweden of ^{14}C-compounds was forbidden by the U.S. Atomic Energy Commission at that time, otherwise they certainly would have done the obvious follow-up experiment, using uniformly U-^{14}C labeled cytidine themselves.

I made RNA from *Euglena gracilis* grown on $^{14}CO_2$. I had to work out the determination of the independent specific activities of the sugars and bases which I did by treating the nucleosides with nucleoside phosphorylase and hypoxanthine to exchange for the base to be analyzed. Then by paper chromatography, using a medium containing borate to retard the migration of ribosides, I could also isolate deoxyinosine and cytosine. Although U-^{14}C cytidine did not label the deoxyribose of *E. coli* DNA, I found the deoxycytidine of DNA of rat organs to be almost uniformly labeled. The ^{14}C content was far in excess of the negligible radioactivity in the purine deoxynucleotides[3]. Therefore by both criteria it appeared certain that the ^{14}C reached the deoxyribose directly from the cytidine. Reichard repeated and extended this experiment with U-^{14}C uridine in 1957 with much the same result for the deoxycytidine and thymidine[4].

It would have been reasonable for me to try to work out the enzymology of ribonucleotide reduction after graduating. Peter Reichard at Yale on a post-doc from Sweden, asked me about my intentions. But I was not anxious to take on a heroic problem at this early point in my career. I was interested in learning more about the principals of enzymology.

Stereochemistry at Chicago. Ogston's 1948 paper proposing, in effect, that the ability of an enzyme-substrate complex to distinguish between identical groups on a tetrahedral carbon was a consequence of the asymmetry of the complex[5], was a matter of hot debate in chemistry/biochemistry circles at Chicago in 1950 where the enzyme was still a black box and the emphasis was on the chemistry of changes in the substrate. In particular the Ogston idea could justify the conclusion in the experiments of Myron Bender, done in Chemistry at University of Chicago, that the absence of back labeling of an ester in ^{18}O- water during enzymatic hydrolysis could not rule out a tetrahedral intermediate. Bender had already shown that back labeling occurred during

1950 at Chicago.

1997 at Fox Chase.

ester hydolysis in alkali[6]. In the case of the enzymatic reaction based on Ogston one would expect to lose all the ^{18}O on stereospecific return of such an intermediate to the ester. These thoughts morphed into the positional isotope exchange idea in 1976.

Thus I became challenged to establish the absolute stereochemistry of enzymatic reactions and determine its mechanistic significance, if any. This did not seem such a formidable task although it was not until 1963 that Kenneth Hanson and I solved the historically important problem of the prochirality of citric acid[7] which was necessary for me to gain a proper perspective on the aconitate hydratase reaction.

Yale. In 1955, after post-doctorals at Western Reserve University with C. E. Carter and at New York University with Severo Ochoa, I was fortunate to be invited by Joseph Fruton to become an Instructor in Biochemistry at Yale University Medical School. The first year at Yale was notable for the following three developments. Not willing to spend the time it would take to get the Department's mass spectrometer working, I turned to the scintillation counter that was available in the medical school lab of Seymour Lipsky, an M.D. with a passion for exploring and exploiting new methods. One of the pioneer instruments to become available came from a small start-up company in New Haven that Lipsky had been encouraging, the Technical Instrument Company. Lipsky also had a sample of tritiated water which together with his counter got me started on experiments I wanted to do.

A second important event of my first year at Yale was to learn from Mel

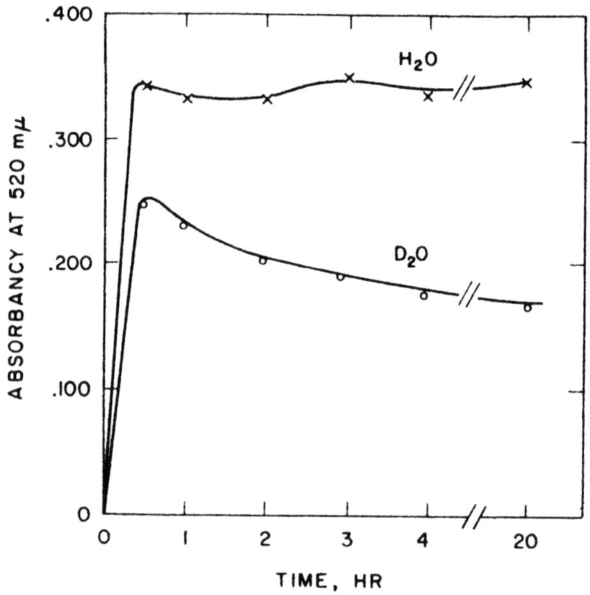

Figure 1. Fructose-6-P formation from glucose-6-P in H_2O and D_2O, as measured colorimetrically by the Roe test for ketoses.

Simpson of his paper showing an apparent ATP requirement for protein breakdown in a liver slice system. This observation required further study which I attempted on the side for the next twenty years.

But the crowning event of 1954–1955 was my proposal of marriage to Zelda Budenstein, a graduate student in the Department. Fortunately, I caught up with her before she graduated. Her mother, widowed since Zelda was age five, came to live with us. She was much loved and a great help with the four children that were in our onrushing future. She enabled Zelda to have a research career, often paralleling mine, which she continued until 1987 when she retired to devote full time to her peace and social interests.

Aldose-ketose isomerases. Probably the most interesting experiment of my nine years at Yale, 1954–1963, was interesting from the way it developed and the confidence it gave me that I might be able to do research after all. I had been looking for evidence of proton transfer in enzymes that catalyzed aldose-ketose interconversions. We had been mistakenly unsuccessful in not finding the small amount of transfer that was later detected in the triose P isomerase reaction. Y. J. Topper had reported that glucose 6-P isomerase in D_2O formed glucose-6-P (G6P) containing about one deuterium using crystallization of the barium salt as a G6P trap[8] suggesting that there might be no proton transfer between reactants of this enzyme. The importance of showing some transfer would be that it would provide a clue to the catalytic process. Complete transfer

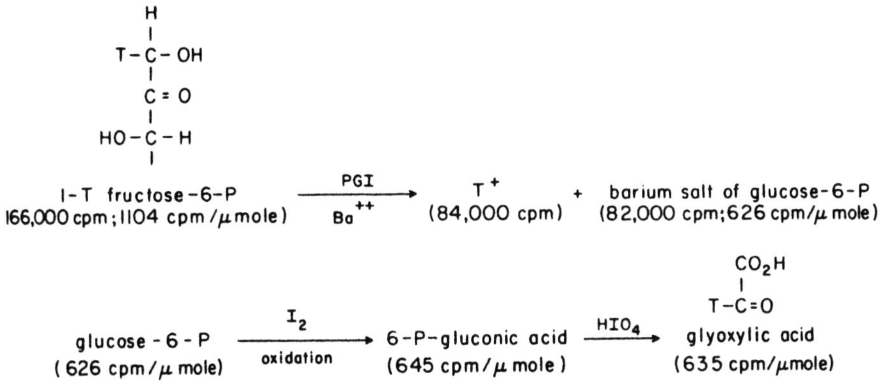

Figure 2. Demonstration of tritium transfer in F-6-P→G-6-P.

would suggest a hydride transfer. Complete exchange would suggest a carbanion intermediate but would not implicate the enzyme as the base. However, the occurrence of both transfer and exchange would result if the abstracted proton were to exchange to some extent before a second proton transfer. Such a result would imply a single base mechanism. No such result had yet been reported. In unrelated experiments I observed a puzzling phenomenon with G6P isomerase. When G6P was used with isomerase in D_2O the colorimetric analysis for fructose-6-P passed through a maximum before reaching a final value. This overshoot of equilibrium that occurred only in D_2O, Figure 1, was very puzzling.

Professor Julian Sturtevant of the Chemistry Department sagely asked me if I was sure of my assumptions. I soon figured out that Topper's experiment might have been misleading. Perhaps his barium trap of G6P was not good enough. If there were both transfer and exchange in D_2O then as the product returns to the enzyme there would be another opportunity for exchange until the product becomes fully exchanged. The fructose-P would go from a partly H- form initially, to an all D- form in the exchanged position at C-1. Now the only thing necessary for my strange result to make sense was to find an isotope effect in the color reaction for ketoses in acid. This was shown by finding the equal amounts of fructose-6-P at 20 hrs when assayed by a different method[9].

We could show transfer using 1-T fructose-6-P with the isomerase using Ba to form the crystalline salt of G6P as it formed in the presence in a small amount of seed crystals, Figure 2. The 6Pgluconate made from the G6P was

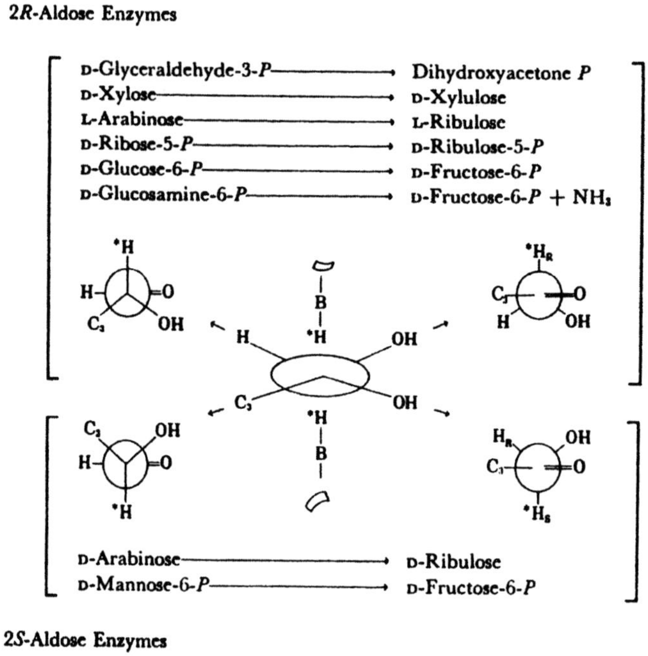

Table 1. Stereochemical course of 2R and 2S isomerase.

only labeled at C-2, the site to which transfer had occurred. We also showed that transfer was between carbons of the same molecule and that the extent of transfer was greatest at low temperature, indicating that the greatest effect of heat was on the dissociation of the intermediate E-T. The proposed intermediate, EH.enediol-P was found to partition about equally in the forward and back direction as shown by the equal incorporation of T-water into product and substrate at early times. A low specific activity of the product was consistent with the slow exchange of the intermediate which would also be expected to show some discrimination against T of the solvent. Since only one T entered the C-1 position of the fructose-6P at equilibrium in tritiated water the abstraction and exchange with solvent were stereospecific. The absolute stereochemistry of 1-^{2}H-fructose-6P was rigorously established by neutron diffraction crystallography as suggested by Lindo Patterson of Fox Chase using the ^{6}Li salt of monodeuteroglycolate[10]. This was the first example of the use of neutron diffraction to establish the absolute configuration of a molecule made asymmetric by isotopic substitution.

All isomerases use cis-enediols intermediates. As shown in Table 1 isomerases that are specific for 2R-aldoses were found to activate the 1-R proton of the ketose produced. Isomerases specific for 2S-aldoses produce ketoses that exchange their 1S proton[11]. For these proton transfers to have been intramolecular, they would both have come from the same face of the enediol intermediates. This

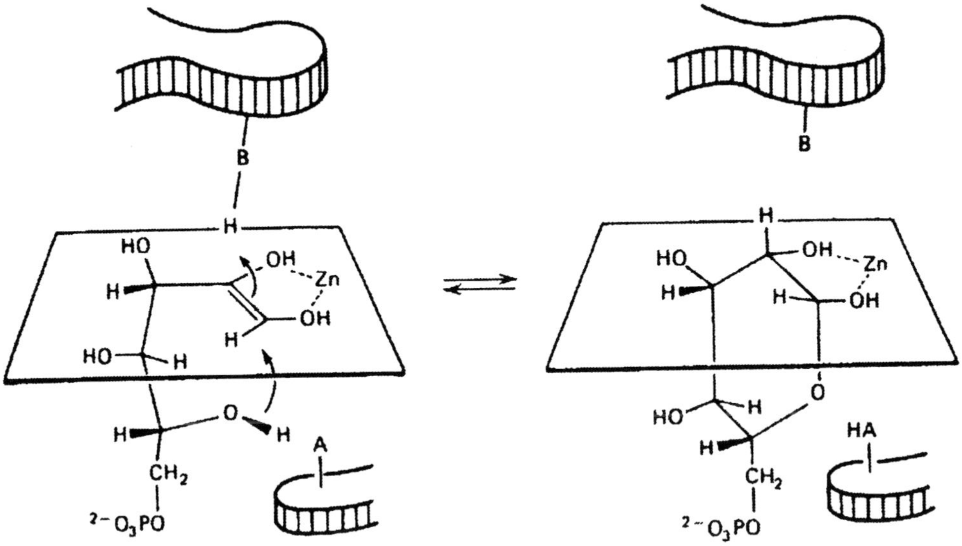

Figure 3. Ring closing to give β-mannose-6-P.

implies a cis-enediol intermediate. cis-Enediol intermediates have the advantage of requiring only a single electrophile to polarize the carbonyl of either substrate. The requirement for one base and one electrophile may be a major factor in preserving the cis-enediol mechanism in evolution.

A relation between the cis-enediol mechanism and the anomeric specificity of isomerases. The open chain forms of sugars are minor species in aqueous equilibria. Therefore a ring opening step will be the first step catalyzed by isomerases for any pentose or hexose substrate. Since ring closing is the reverse of ring opening, we wondered if enzymes with a cis-enediol intermediate would be defined in the ring closing step. Ring closing should occur from the face opposite that used by the proton donor. For example, if the proton approaches C-3 from above the plane, the C2-C3 bond will become oriented below the plane and the C5OH must approach the C-1 carbonyl that is being generated to form the cyclic product from below, see figure 3[11–13].

The anomeric specificities of 5 isomerases were determined in the course of our studies on the isomerases that continued until 1973. The results were as predicted, Table 2. Thus α-xylose were the substrates for the 2R- aldose enzymes in their production of fructose 6P+NH_3 and xylulose, respectively, and β –L-arabinose and B-D-mannose-6-P were the specific substrate for the 2S-aldose isomerase.

The specificity of glucose-6-P isomerase is unusual. Consistent with the above considerations α-G6P and α-F6P are the favored substrates. However,

Anomeric Specificities

ISOMERASE SUBSTRATE	SPECIFICITY	
	Predicted	Found
D-Glucose-6-P	α	α >β
D-Glucosamine-6P	α	α
D-Xylose	α	α
L-Arabinose	β	β
D-Mannose-6-P	β	β

Table 2. Anomeric specificties.

isomerization of *β*-G6P occurs at a significant rate. It has been proposed that the open chain aldose while on the enzyme undergoes torsional inversion at the C2-C3 bond before ring closing. This kind of motion, if it were to occur in the ene-diol of G6P could add the C-2 proton to the flip face and produce mannose-6-P. Seeholzer has reported the activation of the C-2 proton of mannose-6-P by G6P isomerase[14].

A nomenclature for ring face designation[15]. In looking for a way to designate the ring face of a simple ring compound such as G6P or of the two rings of a nucleoside, or of fused ring compounds such as sterols we realized that a universal method could be devised using the established rules for numbering the atoms of rings as contained in standard chemical handbooks[15]. A face is considered *α* if the numbered ring atoms are seen to increase in a clockwise direction, otherwise this face is designated the, *β*-face. The method is not subject to changes in ring substituents as is the D/L naming of sugars. One can readily communicate structural information without resort to pictures by this method. For example, noting that in Watson-Crick base pairing the paired bases have the same face-orientation in forming the antiparallel double helix tells us much of what we need to know to build the correct structure.

Fox Chase. In 1963 we moved to The Institute for Cancer Research of the Fox Chase Cancer Center in Philadelphia. The Institute was a unique place, in my experience. Since its founding it was governed by the notion that to understand cancer one needed a broader understanding of biological science. In its first expansion from center city in 1949 its faculty included crystallographers, embryologists, chemists, biochemists and medical biologists without departmental restrictions. I was attracted by the chance to learn from a wide range of researchers attending each other's seminars as well as freedom from teaching. Research support came from our own competitive grants but the Institute had a generous history of tiding you over. We were not required to go after our own salaries which came from an NIH Core Grant, the first of its kind, which also supported institutional facilities. Decisions were made by the

Director, Timothy R. Talbot, M.D. with the advice of a group of the staff and eminent outsiders. Zelda had a lab and grant of her own. She quit science to be more active in the Nuclear Freeze and other peace efforts in 1987. In 1977 when Avram Hershko requested to share our lab space and facilities he and his group were welcomed for the 22 years he was with us either on sabbatical or during summers.

Whole cell systems. In the next few years I started a series of investigations using isotopes to examine problems of metabolism. It had been observed by Harland Wood's group at Western Reserve that when ^{14}C-lactate or ^{14}C-glycerol are fed to a fasted animal the labeling of the glucose units of liver glycogen was such as to suggest that triose-P isomerase had failed to equilibrate its two triose phosphate substrates during synthesis or that the condensation reaction of FDP aldolase did not give equal labeling of the two halves of FDP. We had earlier investigated this latter using isotope exchange at equilibrium (possibly for the first time) showing that the C456 of FDP exchanged more rapidly than C123[16].

Asymmetric labeling due to incomplete equilibration at the triose-P isomerase step seemed unlikely given the enzyme's high efficiency. This question was settled by showing, with Wood's group, that the asymmetry obtained with ^{14}C-labeled glycerol in which a deuterium was present in the position abstracted in the isomerase reaction was greater in the direction predicted by incomplete equilibration of that enzyme[17], an expression of the primary isotope effect of the isomerase reaction that we had shown previously.

Yale graduate student Bob Kemp determined the fates of reduced pyridine nucleotides, NAPDH and NADH in *Leuconostoc mesenteroides* fermentation and growth using 1-T glucose, 1-T 2 deoxy-glucose and 3-T glucose. Preferential use of 1-T for ethanol production and 3-T for lactate production indicated partial separation of dehydrogenase products[18]. NADH derived from glucose 6P dehydrogenase which also oxidizes 2 deoxyglucose-6-P was the primary source of reducing equivalents for lipid brosynthesis.

Control of glycolysis. In the following years we reported studies on the regulation of glucose degradation in human red blood cells. Glucose transport is rapid so that the first irreversible step, hexokinase, must determine the rate of net flux. ^{14}C- glucose utilization was shown to be inversely related to the G6P level over a 40 fold range using methylene blue and inosine to vary the steady state G6P, even as many other metabolites varied greatly without showing effects on this linearity[19]. Thus any condition that affects the rate of glucose utilization of red cells must do so by either having an effect on G6P or on the activity of the hexokinase itself. A remarkable effect of Pi in stimulating red cell glycolysis was traced to its interference with the binding of G6P to the enzyme without itself having an effect on the hexokinase rate[20].

Using measurements of isotope exchange rates we were led to conclude that the accumulation of triose-P intermediates in human red cells that were incubated with high levels of orthophoshate was due to equilibrium and not a rate limiting step. The low levels of NAD and pyruvate caused by net synthesis of 2, 3-bisPglycerate shifted the equilibrium of the glycolytic intermediates toward high triose-Ps[21].

Glucose-1, 6-P_2 of the brain. Lowry *et al.* reported the rapid depletion of the normally significant amount of G16P_2 during mouse brain ischemia. We have been particularly interested in the role this compound might play in brain function. We purified a enzyme that used G1P with glycerate-1,3-P_2, as the phosphate donor, not ATP or FDP[22, 23]. Mg was a required confactor but Zn was equally active and much more tightly held. About 65% of the activity of a fresh brain extract, resistant to EDTA, is probably the enzyme in the Zn-form. The synthetase is strongly inhibited by concentrations of citrate and FDP that are found in the brain suggesting a regulatory role of G16P_2 and that its synthetic rate may be different in different regions of the brain just as its distribution was shown to be localized[24].

Zelda discovered a brain-specific G16P_2 phosphatase with an absolute requirement for inosine-5-P[25]. Since IMP is the major product of ATP break-down in the ischemic brain this strongly points to activation of the phosphatase as the cause of the rapid fall in G-1,6-P_2 in ischemia.

Mitochondrial hexokinase. A problem that warrants further study derives from our observation in 1967 that instead of being in the cytosol most of the hexokinase of tumor and animal cells is associated with the outer membrane of the mitochondrial fraction[26]. It has recently been reported that the release of mitochondrial hexokinase and some associated proteins may play a role in the onset of the apoptosis cascade[27].

The mechanism of synthetases. When I felt that we understood control of the red cell glycolytic system my attention returned to mechanism questions: ATP driven synthetases, reactions such as glutamine synthetase (gluamate+ NH_3 +ATP–> glutamine +ADP +Pi) may be written to go in three steps: ATP + E–> ADP + E~P, E~P + glutamate –> E. glutamyl~P, E. glutamyl~P + NH_3–>-glutamine + Pi[28]. However, no ADP/ATP isotope exchange could be found unless all of the reaction components were present. On the other hand Meister had performed a pulse/chase experiment that seemed to suggest the formation of an activated intermediate. The enzyme, pulsed with ATP and ^{14}C-glutamate was chased with unlabeled glutamate plus hydroxylamine. ^{14}C-glutamyl-hydroxamate was found in the chase. However, one could argue that a non-covalent complex of E.ATP.glutamate may not have lost the labeled glutamate more rapidly than the chemical reaction would have occurred when the NH_2OH entered the complex. A similar assumption was made by Jacob bar Tana who I met at the Hadassah Medical School in Jerusalem on our sabbatical in 1972. He was having trouble with irreproducible results with phosphofructokinase testing for an E~P mechanism by preincubating the enzyme with ATP[32] and chasing with a mixture of F-6-P and unlabeled ATP. I suggested that he might be capturing E.ATP rather than E~P.ADP. We agreed to examine the assumption that dissociation of a binary complex must be faster than chemistry on bar Tana's next visit to the U.S. We used yeast hexokinase with ^{14}C-glucose (the pulse) followed by ATP plus a large excess of unlabeled glucose (the chase). Labeled G6P was indeed found in Michaelis proportion to the ATP in the chase up to the full measure of enzyme occupied by glucose in the pulse[29]. Dissociation of glucose was indeed slower than its phosphorylation contrary to the generalization that chemistry is slow. Thus was demonstrated the first pulse/chase experiment from which could be determined the functionality of binary complexes, their rate of dissociation and the extent of dissociation of glucose from the ternary complex.

Pulse/chase variants. Subsequently our lab used the pulse/chase method to study hexokinase in the steady state by comparing the ^{14}C- G6P formed in an acid quench with that formed in a high substrate chase with time of quenching, a procedure that gave the position of the central equilibrium, ~1, and the rate constants for all interconversions[30]. We did pulse/chase experiments in which the labeled glucose came from hexokinase in the crystalline state either alone or together with ADP[31]. Although the crystals dissolve slowly the results suggest that their form at that time reflects the form in the crystal. Crystals with glucose alone required the same amount of ATP for 50% trapping as did E.glucose in solution. On the other hand E.glucose.ADP grown crystals from which the ADP had been removed by washing required much less ATP. We also used standard pulse/chase to find that 4 atoms of tritium, with the specific activity of water in the pulse have positions on the enzyme aconitate hydratase, probably an arginine-threonine pair in the active site that could be trapped by the cis-aconitate in the conversion to citrate[32].

Positional isotope exchange (PIX). Although traditional ATP/ADP exchange as a test of E~PO_3 or E-glutamyl~PO_3 intermediates in the glutamine synthetase reaction would fail if the ADP did not leave the enzyme a different kind of isotope exchange might be observed, one in which the beta-gamma bridge oxygen and the beta-nonbridge oxygens of reisolated ATP had mixed with each other. This would only require the phosphoryl oxygens of the bound ADP intermediate be capable of undergoing positional isotope exchange, depending on a symmetrical torsional motion before returning to ATP. This new exchange process was observed in a cleverly designed experiment done by post doc, Fred Midelfort[33]. The PIX rate in the presence of glutamate was consistent with the reverse rate of the enzyme. The exchange rate was decreased by NH_4^+ consistent with a decrease in the concentration of E.glut~P.ADP complex.

Recycling. More recently we have become interested in how an enzyme recovers after product formation. Changes in the active site of an enzyme that occur when a substrate is converted to product must be reversed before the reaction can occur again. With rare exception this recycling of the active site occurs before the product has left the enzyme and is therefore counted as part of the product-off step. Are recycling steps ordered or random? Does the sequence replicate the order of steps during the reaction itself? Fumarase has several properties that mark it as a rare example in which, when salt is present to accelerate the release of product, a further sequence of steps remains that is rate limiting for the whole reaction cycle. Distinctive enzyme isoforms become identifiable as shown by the occurrence of noncompetitive inhibition by product analogues[34, 35]. Added labeled reaction product rebounds to substrate when unlabeled substrate is added showing that the product form of the enzyme is slow to recycle, the Britton counterflow effect[36, 37]. D_2O and glycerol show uncompetitive inhibition patterns indicating that their effects occur after an irreversible step, product release. Noncompetitive inhibition is rarely seen in enzymes that catalyze reactions requiring only one substrate. When it is seen, it is an indication of slow recycling.

The chemical interconversion of fumarase reaction, a carbanion mechanism, is the most rapid part of the reaction cycle. Thus at equilibrium, ^{18}O- water exchanges into malate more rapidly than does fumarate. Perhaps the details of the reaction chemistry can be determined from the details of the recycling. That this seems to be the case is indicated by the pattern of inhibition. The carbanion intermediate analog, 2-hydroxy, 3-nitropropionate is competitive with respect to both malate and fumarate and therefore represents the intermediate species in the recycling whereas all other inhibitors are competitive with only one or the other of the substrates and therefore represent product species.

The ability of salt to increase the rate of product release is lost when the basic amino acid residues that reside between the active site and the solvent are mutated to neutral forms[38]. The mutated enzyme is now rate limited by the product-off step, not recycling. We speculated that the negatively charged products are released through a positively charged channel that is made more rapid by salt.

Enol-pyruvate. Heavy atom NMR spectroscopy of the hydrogens of P-enolpyruvate led to Mildred Cohn's solution of the reaction stereochemistry of the reaction of enolase using (3R)-phosphoglyceric acid-3-d[39]. This led us to the solution of many reaction stereochemistries with enzymes using specifically labeled 3D, T–PEP[40].

Our longstanding belief that enolpyruvate is the intermediate in the pyruvate kinase reaction was confirmed by direct chemical analysis of pyruvate kinase in the steady state[41]. We could also show that enolpyruvate synthesized by treating PEP with a phosphatase was used by pyruvate kinase at an appropriate rate although its affinity was less than expected for an intermediate[42].

Other problems. In a surprise result, the Schiffs bases of dihydroxyacetone-P and of FDP were found to precipitate with the protein of muscle aldolase in cold acid[43]. This property due to the stability of the Schiffs base intermediates in

Zelda and Irwin Rose, 1957.

acid and the acid lability of the eneamin-Ps were used to determine their concentrations as intermediates in the aldolase reaction[44]. Ring formation of the adducts was indicated. Thus 5-deoxyFDP formed a much less stable acid precipitate than did FDP.

The interconversion of citrate and isocitrate by aconitate hydratase was shown to occur with complete retention of the proton transferred between C_2 and C_3 but with no retention of the OH- between C_3 and C_2. The bound cis-aconitate intermediate must flip over, probably around a CH_2COO- pivoting on the ferrous iron of the active site so that the enzyme bound proton can approach from the opposite faces of the intermediate in generating either of the products[45].

Aconitate isomerase catalyzes the interconversion of cis- and trans-aconitate by a 1, 3 allylic rearrangement that uses the (pro-S) hydrogens of the two substrates. The stereochemistry and the transfer of tritium are consistent with a single base carbanion mechanism[46].

The physical separation of H- from D-labeled molecules made possible by their different rates of reaction was used to establish that the transfer of hydrogen in G6P isomerase was intramolecular[9] and that the site of attachment of CO_2 to ribulose-1, 5-P_2 to produce two molecules of phosphoglyceric acid had to be to C-2 rather than C-4 in the RUDP carboxylase reaction[47].

In this chapter I have mentioned most of the problems that have engaged me and my coworkers until the time of my retirement in 1997 with the exception of ubiquitin related studies that are summarized in the next chapter, "Ubiquitin at Fox Chase". Following my retirement, Zelda and I moved to

Laguna Woods in Southern California where I was able to continue lab work using space and facilities shared by Ralph Bradshaw at nearby U.C. Irvine.

During this period Dr. James Nowick of Chemistry and I examined the mechanism of methylglyoxal synthetase (dihydroxyacetone-P –> methylglyoxal) trying to understand our observation[48] that the methyl group of the product was formed nonstereospecifically. We could show spectroscopically that the enol-aldehyde of methylglyoxal was really the enzyme's product[49]. Ketonization occurs off the enzyme and therefore is nonstereospecific. Ketonization to methylglyoxal is very slow but is assisted by thiols which interrupt the double bond conjugation by formation of a thiolhemiacetal. It was perhaps for this reason that the next enzyme, of the sequence, glyoxalase I, has the adduct of methylglyoxal with glutathione as its substrate and D-lactyl – glutathione as its product.

We concluded our paper with the observation that organisms that had no glutathione would have to evolve another mechanism to dispose of methyglyoxal which is toxic to cells due to its action in cross linking macromolecules. Our most recent study shows that the mycobacteria that have mycothiol, (N-acetyl cysteine-glucosamine-inositol), as its major reducing compound have indeed evolved an enzyme that carries out the same kind of sequence with a "glyoxalase" that produces lactyl-mycothiol (unpublished). This is therefore a second example in which an enzyme uses the non-enzymatic product of a previous step as its substrate.

Looking back on my 50 year eclectic journey in research, I am grateful that it has gone as well as it has although still not clever enough to open the black box of enzyme structure. The approach I have taken was successful, in the least, in attracting outstanding postdocs some of whom were on hand in 1975–1980 when with Avram Hershko, we pursued the ubiquitin/protein breakdown work.

I am especially grateful to those who have helped me find solutions to the many problems that troubled us in the early days of biochemistry. Among them were Jacob bar Tana, Steve Benkovic, J. F. Biellmann, Mildred Cohn, Aaron Ciechanover, Don Creighton, Raj Gupta, Arthur Haas, Kenneth Hanson, Avram Hershko, Anthony Jaworowsky, G. Kaklij, Robert Kemp, Judith P. Klinman, David Kosow, Donald Kuo, Gustav Lienhard, Oliver Lowry, Alton Meister, C. Fred Midelfort, Gerd Mullhofer, Koko Murakami, James Nowick, Edward O'Connell, Frank Oski, Lindo Patterson, Cecile Pickart, Sidney Rieder, John Richard, James Robinson, Zelda Rose, Keith Schray, Steve Seeholzer, Frank Solomon, Michael Summers, Jessie Warms, Mary Wimmer, Keith Wilkinson, James Willard, Harland Wood and Alvin Zipursky.

REFERENCES TO BIOGRAPHY

1. Rose, I. A. and Schweigert, B. S. (1952) *Proc. Soc. Exptl. Biol.* 79, 541–544.
2. Hammarsten, E., Reichard, P., and Saluste, E. (1950) *J. Biol. Chem.* 183, 105–109.
3. Rose, I. A. and Schweigert, B. S. (1953) *J. Biol. Chem.* 202. 635–645–645.
4. Reichard, P. (1957) *Acta Chem. Scand.* (1957) 11, 11–16.
5. Ogston, A. G. (1948) *Nature* 162, 963.

6. Bender, M. L. (1951) *J. Am. Chem. Soc.* 73, 1626–1629 and Bender and Heck, H. (1967) *J. Am. Chem. Soc.* 89,1211.
7. Hanson, K. R. and Rose, I. A. (1963) *Proc. Natl. Acad. Sci.* USA 50, 981–988.
8. Topper, Y. J. (1957) *J. Biol. Chem.* 225, 519–425.
9. Rose, I. A. and O'Connell, E. L. (1961) *J. Biol. Chem.* 236, 3086–3092 and Brookhaven Symposium in Biology (1962), 15, 293–309.
10. Johnson, C. K., Gabe, E. J. Tayler, M. R., and Rose. I. A, (1965) *J. Am. Chem. Soc.* 87, 1802–1804.
11. Rose, I. A. (1975) *Advances in Enzymology,* 43, 491–517.
12. Schray, K. J. and Rose, I.A, (1971) *Biochemistry,* 10, 1058–1061.
13. Rose, I. A., O'Connell, E. L., and Schray, K. J. (1973) *J. Biol. Chem.* 248, 2232–2234.
14. Seeholzer, S. H., (1993) *Proc. Natl. Acad. Sci.* USA. 90, 1237–1241.
15. Rose, I.A., Hanson, K., Wilkinson, K.D., and Wimmer, M.J. (1980) *Proc. Natl. Acad. Sci. USA.* 77, 2439–2441.
16. Rose, I.A, (1958) *Proc. Natl. Acad. Sci. USA* 44, 10–15.
17. Rose, I. A., Kellermeyer, R., Stjernholm, R. and Wood, H. G. (1962) *J. Biol. Chem.* 237, 3325–3331.
18. Kemp, R. G. and Rose, I. A. (1964) *J. Biol. Chem.* 239, 2998–3006.
19. Rose, I. A. and O'Connell, E. L. (1964) *J. Biol. Chem.* 239, 12–17.
20. Rose, I. A. and Warms, J. V. B. and O'Connell, E. J. (1964) *Biochem. Biophys. Res. Commun.* 15, 33–37.
21. Rose, I. A and Warms, J. V. B. (1966) *J. Biol. Chem.* 241, 4848–4854.
22. Rose, I. A., Warms, J. V. B., and Kaklij, G. (1975) *J. Biol. Chem.* 250, 3466–3470.
23. Rose, I. A., Warms, J. V. B., and Wong, L.-J. (1977) *J. Biol. Chem.* 252, 4262–4268.
24. Yip, V., Pusateri, M. E., Carter, J. Rose, I. A. and Lowry, O. H. (1988) *J. Neurochem.* 50, 594–602.
25. Guha, S. K. and Rose, Z. B. (1982) *J. Biol. Chem.* 257, 6634–6637.
26. Rose, I. A. and Warms, J. V. B. (1967) *J. Biol. Chem.* 242, 1635–1645.
27. Majewski, N., Nogueira, V., Robey, R. B. and Hay, N. (2004) *Mol. Cell. Biol.* 24, 730–740.
28. Meister, A. (1974) *The Enzymes,* 10, 699–714.
29. Rose, I. A., O'Connell, E. J., Litwin, S. and Bar Tana, J. (1974) *J. Biol. Chem.* 249, 5163–5168.
30. Wilkinson, K. D. and Rose, I. A. (1979) *J. Biol. Chem.* 254,12567–12572.
31. Wilkinson, K. D. and Rose, I. A. (1981) *J. Biol. Chem.* 256, 9890–9894.
32. Kuo, D. J. and Rose, I. A. (1987) *Biochemistry* 26, 7589–7596.
33. Midelfort, C. F. and Rose, I. A. (1976) *J. Biol. Chem.* 251, 5881–5887.
34. Rose, I. A., (1997) *Biochemistry* 36, 12346–12354.
35. Rose, I. A. (1998) *Biochemistry* 37, 17651–17658.
36. Rose, I. A., Warms, J. V. B., and Yuan, R. G. (1993) *Biochemistry* 32, 8504–8511.
37. Britton, H. G. (1973) *Biochem. J.* 133, 255–261.
38. Rose, I. A. and Weaver, T. M. (2004) *Proc. Natl. Acad. Sci.* 101, 3393–3397.
39. Cohn, M., Pearson, J. E., O'Connell, E. L. , and Rose, I. A. (1970) *J. Am. Chem. Soc.* 94, 4095–4098.
40. Rose, I. A., O'Connell, E. L., Noce, P., Utter, M., F., Wood, H. G., Willard, J. M., Cooper, T. G., and Benziman, M. (1969) *J. Biol. Chem.* 244, 6130–6133.
41. Seeholzer, S. H., Jawarowski, A., and Rose, I. A. (1991) *Biochemistry* 30, 722–726.
42. Kuo, D. J. and Rose, I. A. (1978) *J. Am. Chem. Soc.* 100, 6288–6289.
43. Rose, I. A. and Warms, J. V. B. (1985) *Biochemistry* 24, 3952–3957.
44. Rose, I. A., Warms, J. V. B. , and Kuo, D. J. (1987) *J. Biol. Chem.* 262, 692–701.
45. Rose, I. A. and O'Connell,E. L. (1967) *J. Biol. Chem.* 242, 1870–1879.
46. Klinman, J. P. and Rose, I. A. (1971) *Biochemistry* 10, 2259–2266.
47. Mullhofer, G. and Rose, I. A. (1965) *J. Biol. Chem.* 240, 1341–1344.
48. Summers, M. C. And Rose, I.A. (1977) *J. Am. Chem. Soc.* 99, 4475–4478.
49. Rose, I. A. and Nowick, J. S. (2002) *J. Am. Chem. Soc.* 124, 13047–13052.

UBIQUITIN AT FOX CHASE

Nobel Lecture, December 8, 2004

by

IRWIN ROSE

University of California, Department of Physiology and Biophysics, College of Medicine, Irvine, CA 92697–4560, USA.

My interest in protein breakdown as a research problem began in 1955 at about the time I joined the Biochemistry Department of Yale University. It was known that proteins break down intracellularly in the mature animal[1, 2]. In 1955 I learned from Melvin Simpson about experiments he had published 2 years earlier[3] showing that a variety of conditions that should lower the ATP level of liver slices (anaerobiosis, cyanide, 2, 4-dinitrophenol) decreased the rate of liberation of labeled methionine from protein of rat liver slices. Simpson and I had just joined the Biochemistry Department and we had down the hall labs from each other. Simpson's research goal at Yale was to look for an *in vitro* system that performed protein synthesis rather than protein breakdown. During my Yale years I concentrated on enzyme mechanism questions relating to the keto-aldose isomerases using tritium to establish the occurrence of proton transfer[4] but I kept an eye out for reports in the literature on progress in the protein breakdown field.

Nothing new was being reported by 1963 when we relocated to the Institute for Cancer Research of Fox Chase, Philadelphia. At that time I began looking for a cell free system that would show an ATP dependence using the Ehrlich ascites cells that were available from our animal colony. My usual procedure was to label cells in suspension with an essential amino acid, wash the cells, lyse them by homogenization, and look for acid soluble counts adding either ATP or 2-deoxyglucose plus hexokinase to deplete the endogenous ATP.

During 1972 on a half year sabbatical divided between Oxford (J. Knowles) and Jerusalem (Y. Stein), I was able to do further experiments with tissues supplied by Hans Krebs and by Jacob Mager, who had been Avram Hershko's Ph.D research director at the Hadassah Medical School. Both men were highly supportive. But I was not able to find a cell free system. At the same time I was talking to Jacob Bar Tana of the Biochemistry Department and we came up with a Pulse/Chase method that could be used to determine the functionality, binding constant and rate of dissociation of potential enzyme-substrate complexes[5].

GETTING IT ALL TOGETHER

Hershko had been with Gordon Tompkins in San Francisco on a post doctoral fellowship essentially confirming Simpson's findings but with the tyrosine aminotransferase of cells in culture. When he returned to Israel to set up his own lab in Haifa he also looked for a cell free system. I did not meet Avram Hershko until 1975 at a Fogerty Conference on Regulation in Bethesda where we learned of each others interest in protein breakdown. Meanwhile Tompkins died tragically during a brain tumor operation. And Hershko was looking for a U.S. lab in which to expand his further work during sabbatical years and summers. It is not clear to me why he chose our lab in suburban Philadelphia for this purpose. We had no reputation in protein breakdown having never published in the field. Our limited expertise was expected to be in mechanistic enzymology.

In 1977, when we joined forces at Fox Chase, Etlinger and Goldberg[6] had done the first successful experiment using a cell free system with lysates of rabbit reticulocytes. Hershko and his student Aaron Ciechanover[7] had already started fractionating the reticulocyte extract and recognized the heat stable factor (APF-1) before they came to Fox Chase that summer. Their subsequent work, much of which was reported in collaboration with members of the Rose lab, is to be found in the Hershko chapter. The key observation that 8 kDa APF-1 formed units of a chain linked to the protein targeted for breakdown was the most unexpected and unique aspect of the process[8].

The identification of APF-1 with ubiquitin was made by three post docs at Fox Chase: Keith Wilkinson and Arthur Haas from the Rose/Hershko lab and Mike Urban from the next-lab over. Urban was working with chromatin and especially knew about their interesting components, the histones[9]. The pivotal question asked of Urban was: Do you know any examples of two proteins that are linked covalently. This recalled the small protein of unknown function, ubiquitin, a covalent ligand of histone H2A. The size and amino acid composition of APF-1 reported by Hershko and known for ubiquitin were in agreement.

The failure of many cell extracts to give ATP dependent protein breakdown is probably due to a lysosomal trypsin-like protease that destroys the ubiquitin of the preparation. This was discovered by Haas *et al.*[10] who observed the loss of enzyme binding capacity of an ubiquitin affinity column when liver extracts were put through the column. Haas could show ATP/ubiquitin dependent protein breakdown with liver extracts that were preincubated to inactivate the trypsin-like activity. Ubiquitin was already known to be trypsin sensitive at the arg74-gly75 bond.

THE UBIQUITIN ACTIVATING ENZYME

Enzymologists usually study the initial rates of reactions measuring product formation as a function of substrate concentration or other variable. Cell biologists are more likely to want to know the effect of a change on the steady

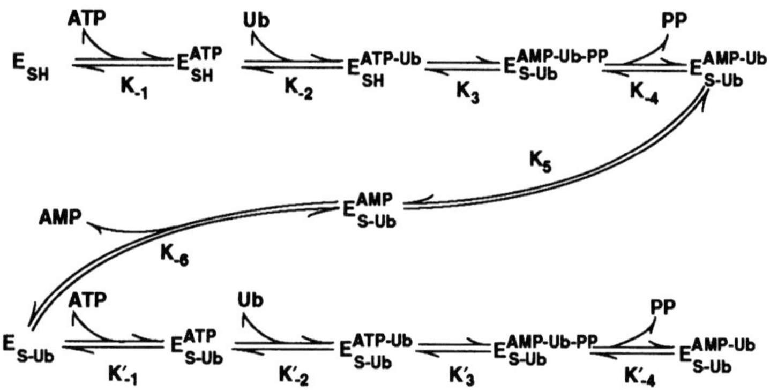

Scheme 1. The sequence and distribution of enzyme intermediates (Haas & Rose, 1982).

state behavior of a complex system. When the ubiquitin activating enzyme, E1, was discovered by Hershko it could not be studied by rate of product formation because the enzyme produced a covalently linked end-product. In 1982, Art Haas used isotope exchange at equilibrium to establish the reaction sequence and a number of equilibrium and rate constants of E1, the only enzyme of the cell that uses Ubiquitin, *per se,* as a substrate.

The Ub activation process is defined by the formation of two equivalents of PPi, one equivalent of bound AMP-Ub and two exchange reactions, ATP with PPi and ATP with AMP:

E1 + ATP + Ub <– > E1.AMP-Ub + PPi
E1.AMP-Ub <–> E1-S-Ub + AMP
E1-S-Ub + ATP + Ub <–> E1-S-Ub.AMP-Ub + PPi

AMP-Ub, prepared from E1, ATP, and Ub and eluted by denaturing the enzyme is sensitive to both DTT and hydroxylamine indicating an acyl-P anhydride linkage. The AMP-Ub could be converted back to ATP upon addition of PPi and Mg^{2+} or to E-S-Ub which did not require PPi or Mg^{2+}[11-13]. Only the formation of enzyme-bound Ub was inhibited by iodoacetamide indicating transfer to cysteine of the enzyme, (Ub being cysteine free). The linkage to Ub had already been established to involve the C-terminal glycine of Ub[14].

ATP was shown to precede Ub in combining with E1 and with E1-S-Ub (see Scheme 1) since Ub above 10 μM caused inhibition of ATP/PPi exchange and at 400 μM inhibition was complete. Addition was not random[12]. Equilibrium constants of the expanded Scheme I could be estimated from

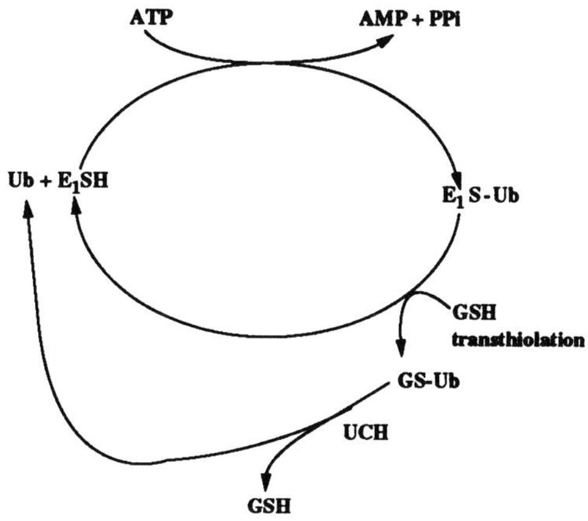

Scheme 2.

the influence of varying AMP and PPi on the concentrations of EAMP-Ub and ES-Ub which were determined with T-ATP by acid precipitation and labeled Ub by electrophoresis. The affinity for ATP, ~40 μM , tighter than the level of ATP in the cell assures that in the cell's E1 will be in the E1-ATP or E1-S-Ub-(ATP) form ready to pick up any free Ub (K_m=0.58 μM). The sensitivity to AMP as an inhibitor, K_6= 0.027 μM, much tighter than ATP as a substrate, suggests that AMP may be acting as a feedback inhibitor at an allosteric site as well as with E-S-Ub.

UB-CARBOXY TERMINAL HYDROLASE: DISCOVERY AND MECHANISM

It was observed that the usual one turnover assay of E1 used ATP well beyond expectation when glutathione was included in the assay. The explanation: The Ub of E1-S-Ub is readily transferred nonenzymatically to mild nucleophiles such as glutathione, DTT, and hydroxylamine. By itself this would increase the consumption of ATP to the amount of Ub present in the assay. But much more was taken up. Were the Ub derivatives unstable? None of the expected Ub-S-DTT from an incubation of AMP-Ub and DTT could be isolated on a covalent-Hg^+ column unless great care was taken in the conversion. Most revealingly the yield was increased if urea was added immediately after the reaction. This indicated that an enzyme carried over with the AMP-Ub from the E1 preparation was regenerating the Ub. These two observations: the transfer of Ub from E1-S-Ub to mild nucleophiles and an enzyme contaminating E1 that would restore free Ub were needed to explain why much more ATP was consumed than could have been expected in the E1 assay[15]. The combined action of E1, glutathione, and the hydrolase results in a futile cycle converting ATP to AMP+PPi, Scheme 2. AMP-Ub is normally too tightly bound to E1 to lead to a futile cycle of its own in the presence of an active nucleophile.

Amides of Ub were not available to test as substrates of the new Ubiquitin-thioesterase until their synthesis was made possible by Cecile Pickart by action of E1+E2 on primary amines[16]. This was of great interest because the enzymes required for recovery of Ub from conjugation in the newly emerging Ub system were believed to be isopeptidases. An E1.E2-S-Ub complex normally transfers Ub to the ε-NH2 of lysine of proteins. To determine the substrate specificity of this system Dr. Pickart found that a variety of small primary amines at much higher concentration are also very good Ub acceptors. The Ub-amides were good substrates for the previously purified Ub-thiolesterase which henceforth has been known as Ub-Carboxy-terminal hydrolase, UCH. The pure enzyme, m.w. 29kDa, was readily prepared from mature human red cells (~0.1 Units/ml packed cells). Its turnover rate is quite high, almost diffusion limited and its V_{max} is ~10 s^{-1}. Ub conjugated to glutathione or to a polyamine in cells with such activity should be negligible.

The rate of the E1.E2, UCH, amine system can be determined by measuring ATP consumption in a futile cycle similar to that of Scheme 2 where the rate can be used to test the activity or the specificity of the component present in rate determining amount. UCH enzymes are limited in the size of the substrate they will act upon. Any role in the deubiquitination of polyubiquitin chains is doubtful. This activity is given over to much larger enzymes, DUBs.

Considerations of the mechanism of UCH start with the observation that these enzymes are inactivated by iodoacetamide, protected by Ub, and therefore should have an active site thiol group and possibly a Ub-thiolester intermediate. We found that inactivation of UCH was brought about by borohydride if ubiquitin was also present[17]. Both T from borohydride and label from Ub were tightly fixed to the inactive enzyme. Both isotopes were released upon mild acid denaturation. T traveled with the Ub. The released product was ~1000x more inhibitory than Ub in an assay using T-butanol-4-NH_2-Ub as a substrate. This effect was lost with borohydride addition to the inhibitor which was protected from reduction if enzyme were added first. We concluded that the acid-liberated inhibitor must be the C-terminal aldehyde form of Ub. This was shown with T-$NaBH_4$, and isolation of T-ethanolamine among the acid hydrolysis products. The basis for the protection of the Ubal by the enzyme must be the formation of a strong complex that shields the aldehyde function. A combination of chemical and physical forces would result from the addition of the active site -SH to the carbony, a thiolhemiacetal on the one hand and multiple protein-protein interactions between the active site and the remainder of the Ubal. Additional stabilization may come from the resemblance of the thiohemiacetal to the tetrahedral intermediates and transition states in the amidase reaction as would be drawn for papain and cathepsin B, Scheme 3.

This interpretation of the mechanism of action of Ubal has been confirmed and extended by x-ray crystallography of Ubal complexed with the UCH

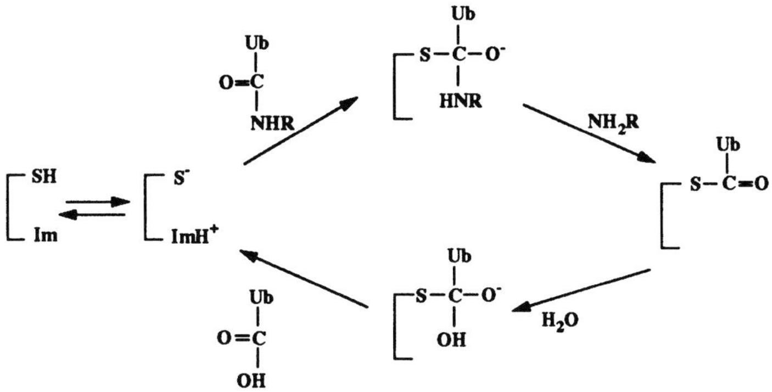

Scheme 3.

of yeast, Yuh 1 by Johnston *et al.*[18] and with the 352 residue UCH domain that was cut out of the 1102 residue long HAUSP deubiquitinating enzyme (DUB) by Hu *et al.*[19]. In both cases Ubal caused significant distance and angle rearrangements in the catalytic triad regions compared with structures done without Ubal[19, 20], respectively as well as placement of H-bonding residues to accommodate the expected oxyanion hole of a thiohemiacetal Ub adduct.

The large DUBs, about 80 of which have been identified, serve to reverse the lengthening of the polyubiquitin chains that leads to the destruction of the targeted protein at the proteasome. HUASP is also known as a co-tumor suppressor protein because, by deubiquitinating p53, the tumor suppressor transcription factor of tissues, it should raise its steady state level. The large size of DUBs is consistent with high specificity and signal control. The roles of the UCHs are not yet clear. The size of the Ub-extension that can be removed by these smaller enzymes is smaller than a Ub so that Ub-Ub cleavage intermediates, if they arise in the deubiquitination process will not be further degraded by a UCH.

UNRESOLVED ISSUES

1. Current discussions of the E1 E2 E3 system often ignore the potential of E1-AMP-Ub to transport Ub units between the solvent and E1-SH to E2 without dissociation of the E1E2 complex. This possibility should be easily evaluated by pulse chase using heavy isotope labeled Ub in the pulse, unlabeled Ub in the chase and analyzing the product by mass spec sequencing.

2. This approach may fail if the E2.E3. protein interactions are weak. In

addition the experiment should give information about the processivity of the system, how many Ubs can be added in succession.

3. This approach should also answer the question of whether when Ubs are added to the growing chain they are added to the distally located end of the chain as is often stated but which seems unreasonable as the chain length increases.

4. An interesting problem stems from the observation by Cecile Pickart[16,17] that hydroxylamine at K_m ~1 mM inactivates the C-terminal hydrolase of erythrocytes, and possibly all UCHs, in the required presence of Ub. Using the hydroxamate of Ub as a substrate, complete inactivation requires ~2000 turnovers of the enzyme. Labeled Ub is not found on the reisolated enzyme, nor is activity recovered. Unless hydroxylamine has some unexplored way of reacting with the S- of the Ub-thiolester- enzyme intermediate, the classical reaction products should be Ub-hydroxamate and active enzyme. In view of the important role of the deubiquitinating enzmes and a practical interest in their inactivation it will be useful to know how hydroxylamine works.

CREDITS

It is a pleasure to pay tribute to Dr. Avram Hershko whose analytic intuition was always productive and whose generous spirit made our interactions always harmonious. The post docs: Aaaron Ciechanover, Art Haas, Cecile Pickart, and Keith Wilkinson were all major contributors who have continued their careers in the Ub field with notable success. Thanks are also due to Hannah Heller and Jesse Warms for their contributions in the lab.

Finally, we wish to pay tribute to Melvin V. Simpson who in 1953 demonstrated that protein breakdown in whole cells was energy dependent (3) and to Etlinger and Goldberg who in 1977 opened the way to *in vitro* studies with the demonstration of ATP dependent protein breakdown in rabbit reticulocyte lycates (6).

REFERENCES

1. Schoenheimer, R. (1942) *The Dynamic State of Body Constituents*, Harvard Press, Cambridge, MA.
2. Schimke, R. T. and Doyle, D. (1970) *Annu. Rev. Biochem.* 39, 929–976.
3. Simpson, M. V. (1953) *J. Biol. Chem.* 20, 143–154.
4. See previous autobiographical chapter.
5. Rose, I. A., *et.al.* (1974) *J. Biol. Chem.* 249, 5163–5168.
6. Etlinger, J. D. and Goldberg, A. L. (1977) *Proc. Natl. Acad. Sci. USA*, 86, 7751–7755.
7. Ciechanover, A., Hod, Y., and Hershko, A. (1978) *Biochem. Biophys. Res. Commun.* 81, 1100–1105.
8. Hershko, A., Ciechanover, A., Heller, H., Haas, A. L., and Rose, I. A. (1980) *Proc. Natl. Acad. Sci. USA* 77, 1783–1786.
9. Wilkinson, K. D., Urban, M. K., and Haas, A. L., (1980) *J. Biol. Chem.* 255, 7529–7532.
10. Haas, A. L., Murphy, K. E., and Bright, P. M. (1985) *J. Biol. Chem.* 260, 4694–4703.
11. Haas, A. L., Warms, J.V.B., Hershko, A. and Rose, I. A.(1982) *J. Biol. Chem.* 257, 2543–2548.
12. Haas, A. L. and Rose, I. A. (1982) *J. Biol. Chem.* 257, 10329–10337.

13. Haas, A. L., Warms, J.V.B. and Rose, I.A. (1983) *Biochemistry* 22, 4388–4394.
14. Hershko, A., Ciechanover, A., and Rose, I.A. (1981) *J. Biol. Chem.* 256, 1525–1528.
15. Rose, I.A, and Warms, J.V.B. (1983) *Biochemistry* 22, 4288–4294.
16. Pickart, C.M. and Rose, I.A. (1985) *J. Biol. Chem.* 260, 7903–7910.
17. Pickart, C.M. and Rose, I.A. (1986) *J. Biol. Chem.* 261, 10210–10217.
18. Johnston, S.C. Riddle, S.M., Cohen, R.E. and Hill, C.P. (1999) *The EMBO J.* 18, 3877–3887.
19. Hu, M., Li, P., Li, M., Li, W. Yao, T., Wu, J-W., Gu, W., Cohen, R, E. an Shi, Y. (2002) *Cell,* 111, 1041–1054.
20. Johnston, S.C., Larsen, C.N., Cook, W.J., Wilkinson, K.D. and Hill, C.P. *The EMBO J.* 16, 3787–3796.

Chemistry 2005

YVES CHAUVIN, ROBERT H. GRUBBS and RICHARD R. SCHROCK

"for the development of the metathesis method in organic synthesis"

THE NOBEL PRIZE IN CHEMISTRY

Speech by Professor Per Ahlberg of the Royal Swedish Academy of Sciences. Translation of the Swedish text.

Your Majesties, Your Royal Highnesses, Ladies and Gentlemen,

This year's Nobel Laureates in Chemistry have created fantastic opportunities to build new organic molecules. It is easy to see why this is important. It is among such molecules that we are finding new materials and new medicines, and developing methods to do old things better.

Organic compounds contain the element carbon. Carbon atoms can bind to each other to form long chains and rings, but they can also bind to other elements such as hydrogen, oxygen and nitrogen to form very complex compounds. All life on earth is based on nature's own carbon compounds but these and other organic compounds can be made artificially using organic synthesis. The diversity of organic molecules is enormous. So far only a small proportion of all the possible organic molecules have been investigated. Nevertheless we have already obtained new medicines, new materials and new surface coatings undreamed of a few years ago.

The achievements of this year's Laureates have given us one of the most important methods for building new organic molecules – metathesis. The word metathesis means change places. In metathesis double bonds between carbon atoms are broken and the parts are paired together in a new way. This does not happen by itself. It takes place with the help of the special molecular tools – catalyst molecules – that the Laureates have developed. The phenomenon of metathesis was discovered in the nineteen-fifties in the polymer industry, but nobody knew what the catalyst looked like, or how it worked: just that it *did* work. However, awareness of the opportunities offered by metathesis for constructing new organic compounds was developing.

A number of researchers made proposals about how metathesis works. The breakthrough came in 1971 when Yves Chauvin presented new experiments and suggested that the catalyst is a carbon/metal compound in which the metal is bonded to carbon with a double bond. Chauvin also presented a novel mechanism for how the catalyst works in metathesis.

The mechanism resembles a dance. The catalyst, in which the metal holds its carbon partner with both hands, "dances" among carbon/carbon pairs, which are molecules with carbon-carbon double bonds. When the "catalyst pair" meets a carbon/carbon-pair, the two dancing pairs unite in a round dance. After a while they let go of each other's hands, leave their old dance partners and dance along with their new ones. The new "catalyst pair" is now ready to catch another dancing carbon/carbon pair for a new round dance. In other words it continues to act as a catalyst for metathesis.

By combining parts of different molecules in this way new molecules with new properties are created. One example of this is biologically active molecules that can function as medicines.

Chauvin's results opened the way for the design and construction of catalysts. Now, the big question concerned the detailed catalyst design required to obtain efficient catalysts for metathesis. The critical progress was made by Richard Schrock and Robert Grubbs.

Schrock, who initiated his basic research in the early nineteen-seventies, used a number of different metals in his investigations. Gradually he discovered which metals worked the best and how the other parts of the catalyst should be constructed. A breakthrough came in 1990 when Schrock presented a group of very active catalysts containing the metal molybdenum.

A new breakthrough came in 1992 when Grubbs published his discovery of a catalyst based on the metal ruthenium. Grubbs' catalysts are stable in air and show higher selectivity but have lower reactivity than Schrock's catalysts. They have become the first well-defined catalysts for general use in metathesis in ordinary laboratories, and new visions of the opportunities of organic synthesis have emerged from their use. Today metathesis is used in the pharmaceuticals, foodstuffs, chemical, biotechnical, polymer and paper industries. Metathesis also saves energy and material, and is kind to the environment. It takes us a step towards a "greener" future.

Dr. Chauvin, Dr. Grubbs, Dr. Schrock,

I have briefly described your basic research. It has led to the efficient metathesis catalysts that have become such revolutionary general tools in organic synthesis, and that have changed our thinking about how to make molecules. Your achievements make it easy to answer the question: to what use can science be put? They are also entirely in the spirit of Alfred Nobel.

On behalf of the Royal Swedish Academy of Sciences I wish to convey our warmest congratulations to you, and I ask you to step forward to receive the Nobel Prize from the hands of His Majesty the King.

YVES CHAUVIN

I was born on 10 October 1930 in Menin (Menen in Flemish) in western Flanders, on the border between Belgium and France. My parents were French. To be more precise, they were not only both from the Tours region, but also descended from long-established families in the little village of Beaumont-la-Ronce. I used to spend my holidays there in my grandparents' large family house, with my numerous cousins. When I die, I am going to be buried in the village cemetery. My grandmother was fond of painting and playing the piano. She had been given lessons by Emmanuel Chabrier, who used to spend the summer months in nearby Membrolle. He said in his correspondence that he did not much care for his pupils on the whole and my grandmother found him very strict. Most of my ancestors were small farmers. My father was an electrical engineer. After three years of military service and quite a difficult time in the First World War, he was sent by the company that had taken him on after his demobilisation to Ypres and then Menin, to work on rebuilding (building) the electricity network in this war-ravaged province. I remember him always being very motivated and working very long hours. He was sent to war again in 1939 and taken prisoner. There were 5 brothers and sisters in the family and we had quite a strict upbringing. From my bedroom, I looked out over our large garden (roses and vegetables) and the Lys river, which at that particular point separates France from Belgium and was where the flax was retted in tanks then dried in little bundles in a field (what a foul smell!). I watched the barges go past, towed by horses or even by men. It was a fascinating sight: the man pulling the rope remained bent over and unmoving for several minutes before the barge started to move. I also remember Vauban's fortifications in this land of invasions, and the smell of roasted chicory... I still have many paintings of Flanders that my father bought from contemporary painters – somewhat classical, but not devoid of charm. I went to pre-school in Flanders and then the French primary school, which meant that I crossed the border every day. I then continued my secondary and higher education in various towns. During the war, I managed to come through the bombings unscathed, though sometimes only just. The war taught me to eat what there was; I am still not a fussy eater, although I do enjoy good food. To be perfectly truthful, I was not a very brilliant student, even at chemistry school. I chose chemistry rather by chance, because I firmly believed (and still do) that you can become passionately involved in your work whatever it is. Various circumstances, mainly to do with my military service, prevented me from doing a PhD and I have often regretted it, though you do

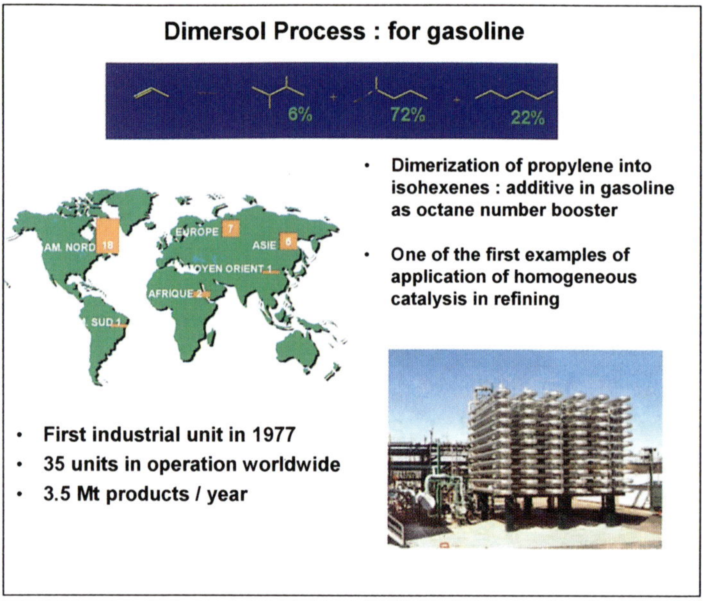

Figure 1.

need to choose the "right" supervisor in the "right" discipline – no easy task when you are totally inexperienced.

So I took a job in industry, but the fact that process development consisted primarily of copying what already existed, with no possibility of exploring other fields, prompted me to resign. Furthermore, I discovered that this was a very common attitude among managers. They are afraid of anything new: "Do what everyone else does and change as little as possible: at least we know it will work." It is the opposite of my way of thinking, which, I must admit, is a bit of an obsession! I have often got into arguments about it. My motto is more, "If you want to find something new, look for something new!" There is a certain amount of risk in this attitude, as even the slightest failure tends to be resounding, but you are so happy when you succeed that it is worth taking the risk.

The whole contradiction of research (whether applied or fundamental) generally lies in the fact that we have to start out with the knowledge handed down by our predecessors, but be able to depart from it "at the right time."

I joined Institut Français du Pétrole in 1960 and managed to focus my work on what I thought would be most interesting. I got married the same year and over the course of time we had two sons and five grandsons.

The oil industry essentially uses heterogeneous catalysis: cracking, reforming, hydrodesulphurization, hydrogenation, etc., but that was not what interested me. I have always tried to avoid areas that have been perfected with time. At the time, nothing much was being done in France on coordination chemistry, organometallics or homogeneous catalysis by transition metals and I was fascinated by the achievements in Italy (G. Natta), Great Britain (J. Chatt), Germany (at the Max-Planck-Institute in Mülheim) and the United

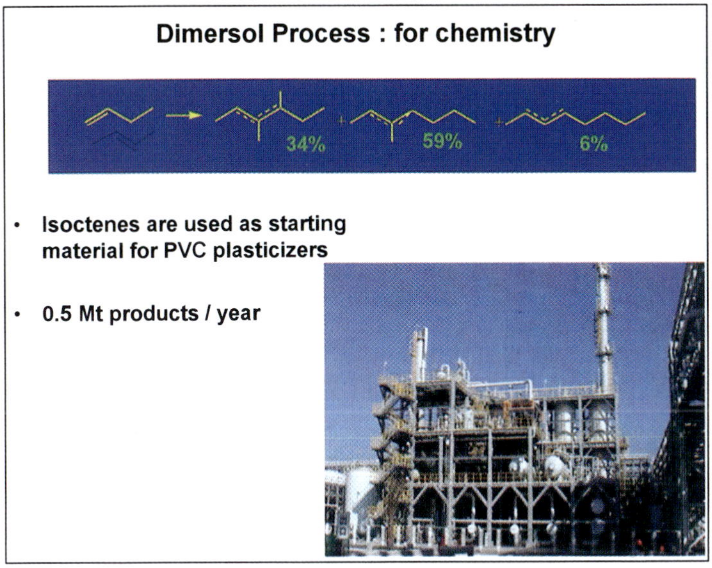

Figure 2.

States. As a result, I unwittingly became the French specialist in these disciplines, which brought me into contact with both the positive and the unwieldy aspects of the various commissions at the CNRS. I spent the best part of my time on applied chemistry, which was what I had been employed for and which I was quite happy about. This was how I came to develop two homogeneous catalysis processes. The first one, which uses a nickel-based catalyst, was called "Dimersol" and exists in 2 basic versions.

The "gasoline" version (Figure 1) consists of dimerising propene to high-octane isohexenes. There is, quite often, an excess of propene, especially in oil refineries that do not have petrochemicals, as in the United States. There are currently 35 plants in operation (including 18 in the USA), with a combined annual output of 3.5 million tonnes. It was the first and only time that coordination catalysis had been used in refining.

The "chemical" version of the process (Figure 2) consists of dimerising n-butenes to isooctenes, basic inputs for plasticizers, using the "oxo" reaction. Current production levels stand at 400,000 tonnes a year.

The second process I developed, and which uses a titanium-based catalyst, was called "Alphabutol." It consists of dimerising ethylene to 1-butene (Figure 3), the co-monomer of low-density linear polyethylene. The benefits of such a process were not evident to begin with and stem from a number of causes. There are currently 20 plants operating worldwide, with a combined output of 400,000 tonnes a year. However, others are under construction, which will take total output to over 0.5 million tonnes a year.

While there are obvious drawbacks to not having done a PhD (especially when you supervise them!), the advantage is that at least your mind is free to focus on whatever presents itself. At the time, I was working on batteries and,

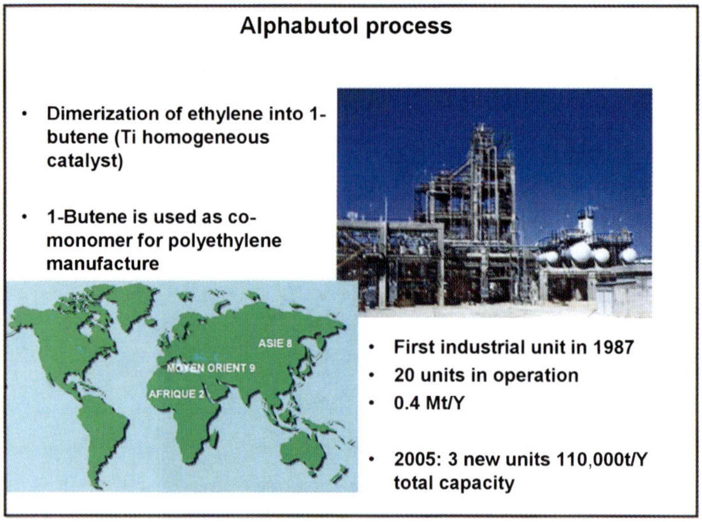

Figure 3.

in particular, the non-aqueous electrolytes used to extend their electrochemical window. I thought it would be a good idea to try to use these electrolytes, which belong to the class of ionic liquids, as catalyst solvents. These liquids feature very low vapour pressure and virtual non-solubility in hydrocarbons, paving the way for a biphasic catalysis. The mixture of alkylimidazolium chloride and aluminium chloride forms a liquid with a very low melting point (below ambient temperature) (Figure 4). It proved to be a first-rate solvent for nickel-based dimerisation catalysts ("Dimersol" catalysts). The diagram

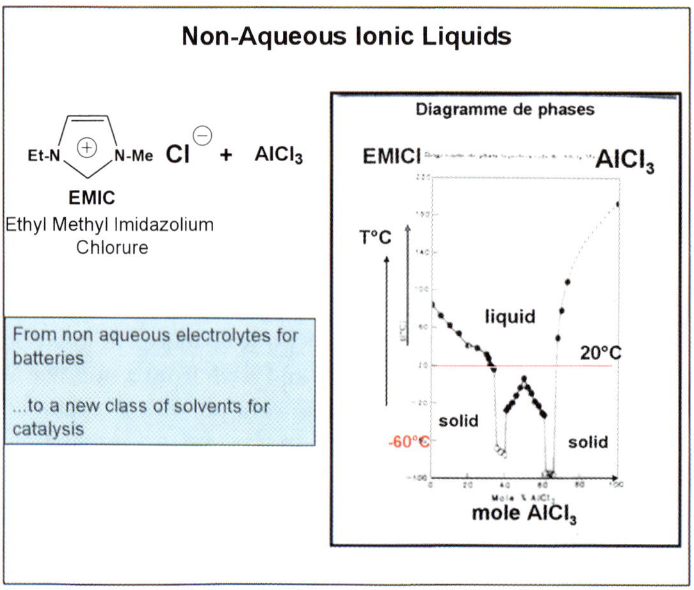

Figure 4.

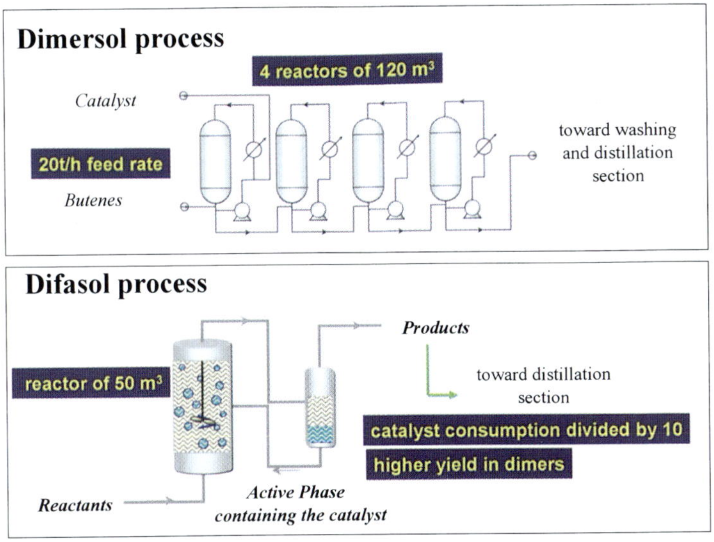

Figure 5.

for this process, called "Difasol," is shown in Figure 5. The reaction volume required for a biphasic system is 10 times smaller than for a homogeneous system (important for safety: refineries do not like to have large volumes in reaction because they are potential "bombs," especially at start-up); likewise for nickel consumption. This new process, dealt with in a PhD project in 1990, will see the light of day thanks to the inventiveness and determination of Hélène Olivier Bourbigou, who took over from me in the laboratory.

What applied chemistry has taught me is the need for absolute solidarity between the research laboratory and the "downstream" side (pilot testing, marketing, setting up industrial plant): same enthusiasm, same determination, especially when everything goes wrong!

There is no difference between fundamental research and applied research. Although this is my view, based on personal taste and the areas I have worked in, it is not necessarily true for others. The PhD either led to, or were derived from processes. I have spoken so much about "processes" because they took up about three-quarters of my working time. However, I also took an interest in other aspects of coordination chemistry, such as palladium catalysis, rhodium catalysis, asymmetric amino-acid synthesis, and so on. After retiring in 1995, I was invited to work in J.-M. Basset's laboratory in Lyon, which allows me to pursue a reasonable level of activity.

OLEFIN METATHESIS: THE EARLY DAYS*

Nobel Lecture, December 8, 2005

by

YVES CHAUVIN

Honorary Director of the Institut Français du Pétrole, Rueil-Malmaison, and of the Centre National de la Recherche Scientifique, Lyon, France.

At no time in my life did I dream of such a Prize. Indeed, as I explained in my biography, I had no training in research as such and as a consequence I am in a sense self-taught. What I owe to the Institut Français du Pétrole is some freedom to choose my research area. I have always been an avid reader of chemical literature, eager for what is new. Like all sciences, chemistry is marked by magic moments. For someone fortunate enough to live such a moment, it is an instant of intense emotion: an immense field of investigation suddenly opens up before you. There were very many of these moments in the course of my career. For example, the discovery of ferrocene, the stereospecific polymerization of olefins by G. Natta (I never failed to read any of his articles!), the homogeneous catalysis of hydrogenation by rhodium complexes (G. Wilkinson), the homogeneous catalysis of dimerization of olefins by nickel complexes (G. Wilke and B. Bogdanovic) and the catalysis of asymmetric hydrogenation. I experienced the latter with an especially keen intensity because we were in very close touch with Henri Kagan, who came every month to tell us about his chemistry; it was outstanding! And 1964 was an especially magical year (Figure 1): a revelation for me, by R.L. Banks of Phillips Petroleum, of the disproportionation of olefins catalyzed by a molybdenum- or tungsten-based heterogeneous catalyst deposited on alumina, the homogeneous catalysis of polymerization of cyclopentene by ring opening, published by G. Natta, the existence of a new metal-carbon bond, the carbenes of E.O. Fischer. *A priori* these three had nothing in common.

The disproportionation of linear olefins, since called *metathesis*, is a equilibrated reaction, governed basically by entropy. This is doubtless why, from the start, "*pairwise mechanisms*" was favored (Figure 2), which assume a perfect symmetry of the reaction intermediates. In addition, these intermediates are based on the only bonds known at the time: π bonds and σ bonds between hydrocarbons and transition metals. Other mechanisms were proposed (Figure 3).

The polymerization of cyclopentene into polypentenylene, for its part, is

* The title is taken from the excellent article by A. Maureen Rouhi, a very competent scientific journalist, published in *Chemical & Engineering News* of 23 December 2002, a special issue on metathesis.

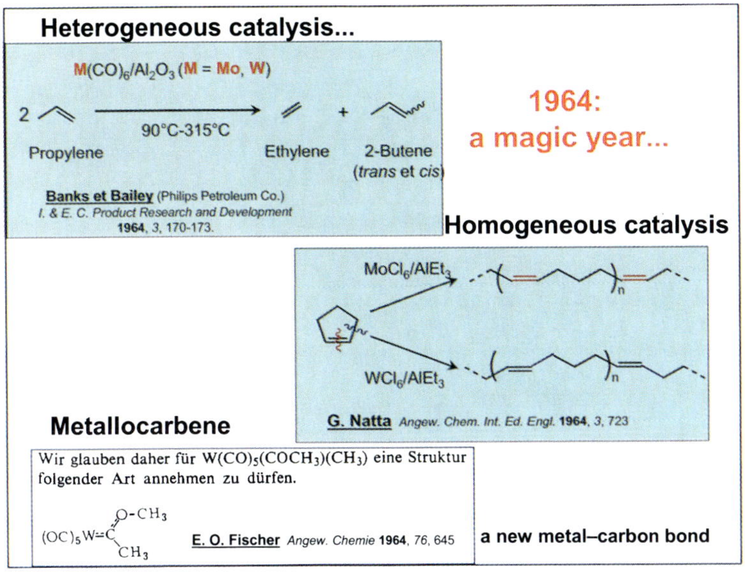

Figure 1.

governed mainly by the enthalpy linked to the release of the ring strain when the ring is cleaved, but entropy also plays a part there due to the multiple conformation of the chain and, like the metathesis of acyclic olefins, it is equilibrated. These factors are no longer the same in the case of the higher cycles, as we shall see later. To explain the high molecular weights observed from the start of the reaction, G. Natta assumed the existence of a σ bond between the transition metal and the growing chain, the proposed mechanism being an α-cleavage of the ring such as is often invoked in classic organic chemistry.

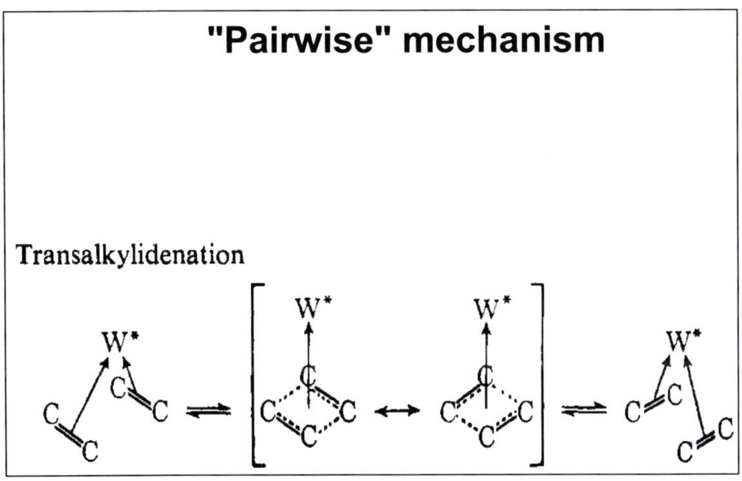

Figure 2.

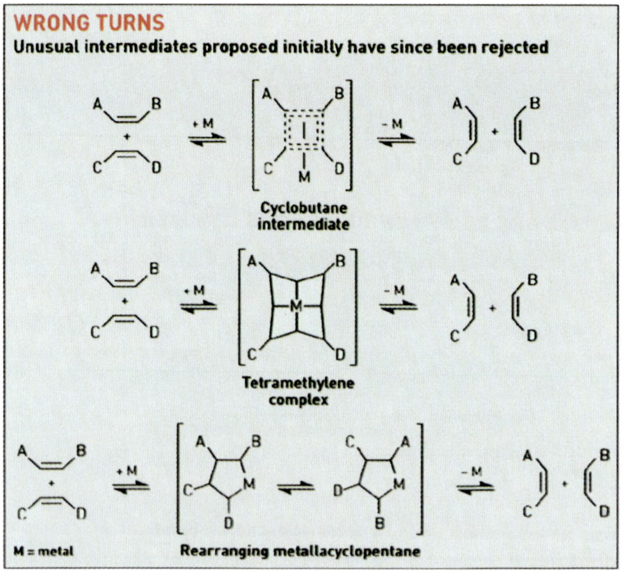

Figure 3.

My immediate reaction back in 1964 was the thought that disproportionation and ring opening were part of the same reaction. In addition to the fact that both reactions preserved the number and type of double bonds (only the molecular weight of the products was changed), they used the same transition metals, molybdenum and tungsten. In fact, at the time, in a way that may be surprising today, there was a gap – a total lack of understanding – between homogeneous catalysis and heterogeneous catalysis (some people even attributed the fundamental role in the transformation to the catalyst support!). Since I was familiar with both types of catalysis in our Institut, that was not a real problem for me. But how could the statistical aspect of the one be reconciled with the chain growth of the other? Meanwhile, N. Calderon had also started building a bridge between homogeneous catalysis and heterogeneous catalysis [1]. It then seemed obvious to associate the two reactions, where the enthalpy of ring opening would in a sense "freeze" the statistical reactivity of the acyclic olefin [2]. Accordingly cyclopentene was associated with 2-pentene in the molar ratio 1:1 with the homogeneous catalyst $WOCl_4$ and $SnBu_4$ or $AlEt_2Cl$. The result was as follows (Figure 4): in addition to the di-unsaturated product of the expected coupling of cyclopentene with 2-pentene (5+5), there were other products, also di-unsaturated (5+4) and (5+6), in the ratio 1:2:1 for the condensations in carbon 9/10/11; then higher tri-unsaturated products containing (5+5+5), (5+5+4), and (5+5+6) carbon atoms; and so on, in decreasing quantities, thereby forming "triads". This was, then, a "telomerization" reaction in which the telogene is n-pentene and the taxogene is cyclopentene. Since the participation of the 2-pentene in the telomerization reaction is at each instant equal to its participation in the metathesis reaction, the telomerization products are as a consequence not derived from the metathesis

Telomerization of cyclopentene with 2-pentene

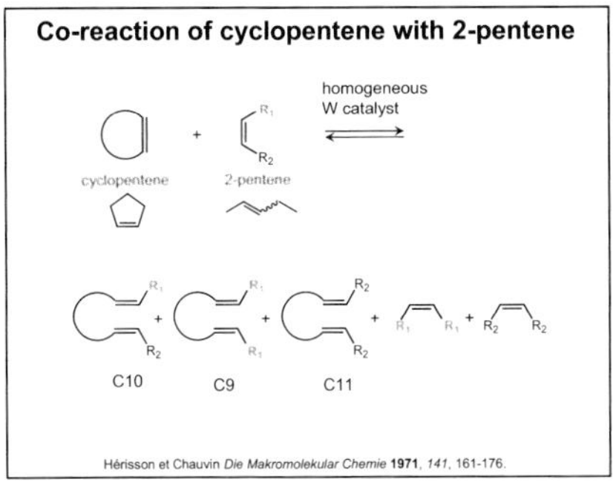

Catalyst Reaction time	$WOCl_4/Sn(C_4H_9)_4$ 7 min.		14 min.		$WOCl_4/Al(C_2H_5)_2Cl$ 24 h.	
Relative Molar quantity						
Cyclopentene	32		35			108
Butenes	9,3		18,7			
Pentenes	58		31,3			159
Hexenes	8,8		17,7			
C_9H_{16}	4,5		8,1		16,8	
$C_{10}H_{18}$	9,5	18,3	18,1	34,4	36,8	70,4
$C_{11}H_{20}$	4,3		8,2		16,8	
$C_{14}H_{24}$	2,0		3,6		8,9	
$C_{15}H_{26}$	4,3	8,3	6,9	13,9	18,5	36,0
$C_{16}H_{28}$	2,0		3,4		8,6	
$C_{19}H_{32}$	1,0		1,5		5,0	
$C_{20}H_{34}$	2,8	4,8	3,0	6,0	8,7	18,6
$C_{21}H_{36}$	1		1,5		4,9	
$C_{24}H_{40}$			0,8		n.d	
$C_{25}H_{42}$			1,6	3,1	n.d	
$C_{26}H_{44}$			0,7		n.d	
Chlorobenzene	80		90		-	
D % (conversion of disproportionation reaction)	24		54		-	
C % (conversion of cross-reaction)	29		46			

Figure 4. Telomerization of cyclopentene with 2-pentene.

products. The reaction, like the metathesis and polymerization of cyclopentene, is equilibrated. The general characters of the telomerization reaction have been verified for many pairs of olefins, such as 2-pentene/cyclooctene, 2-hexene/cyclopentene, and 3-heptene/cyclopentene. α-olefins such as propene and 1-pentene are, as in the case of metathesis, less reactive, but the telomers form in ratios from 1:10:1 to 1:20:1. 1,5-cyclooctadiene (cod) and 1,5,9-cyclododecatriene (cdt) lead to telomers of the same type, but in this case the successive "triads" are separated by 4 atoms of carbon and not by an integral multiple of the number of carbon atoms forming the ring.

This therefore means that alcoylidene "residues" are left on the transition metal. The most obvious hypothesis is the formation of a **metallocarbene** sequestering an olefin to form a metallocyclobutane intermediate (Figure 5).

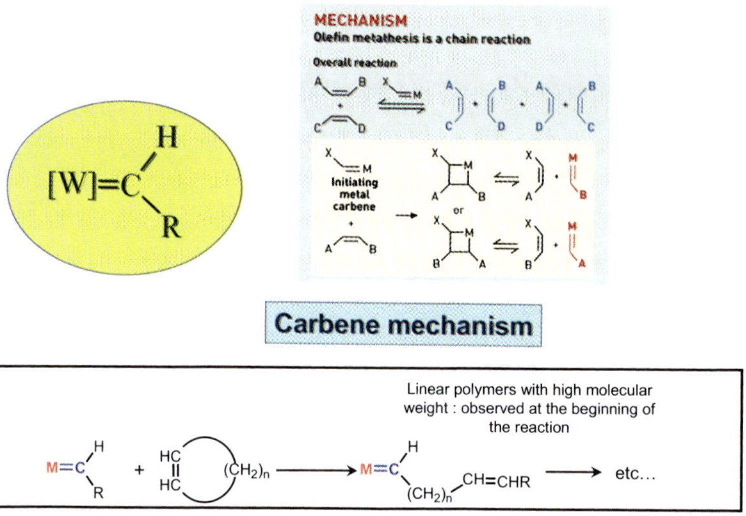

Figure 5.

The mechanism may be a little more complex in reality: the reactivity of cod and cdt as "butadienylidene"

$$(=CH\text{-}CH_2\text{-}CH_2\text{-}CH=)$$

even at the very start of the reaction, so that it cannot be explained as a *back-biting* reaction (reaction of the metallocarbene with one of the double bonds of the growing chain), remains to be explained.

While the cyclopentene is in equilibrium with its polymer (polypentenylene), this is not true of the higher cycles such as cyclooctene and cyclooctadiene: their polymers are in equilibrium with the cyclooligomers [3]. This is due to the relative stability of the cycles, as P. J. Flory has shown in the case of the lactones (Figure 6) [4]. These macrocycles are formed by *back-biting*.

All three reactions can be represented as shown in Figure 7.

It remains to be shown how the initial metallocarbene forms when a tung-

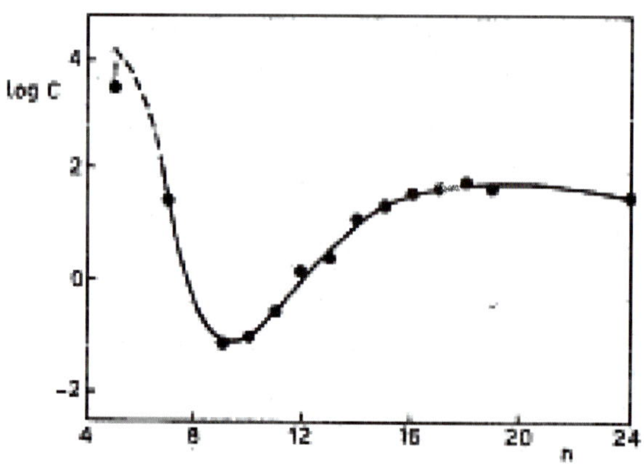

Figure 6. Log of the cyclization constant for the lactonization of ω-hydroxyacids vs. ring size (P. J. Flory).

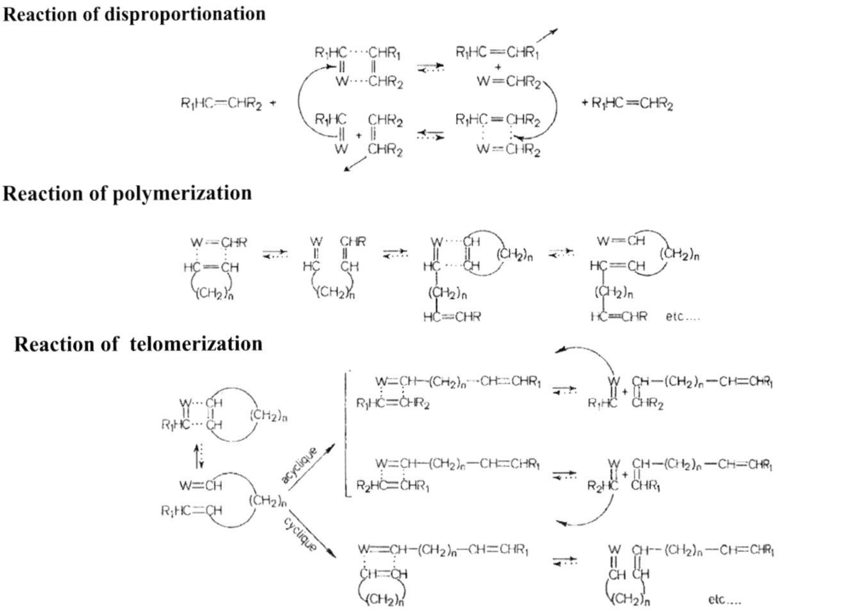

Hérisson et Chauvin *Die Makromolekular Chemie* **1971**, *141*, 161-176.

Figure 7.

sten halide and an organometallic of the principal series are mixed. For this we chose methylated derivatives that cannot give rise to β-H elimination, methyllithium and tetramethyltin. All that could be hoped for was to favor α-H elimination with the formation of a **methylidene** species that could simply be "trapped" by means of olefins with a di-substituted double bond: 2-butene, 2-pentene, 3-hexene. Here is what can be observed (Figure 8) [5]. The formation of propene and of 1-butene clearly demonstrates the existence on the tungsten of a **methylidene** group. The formation of methane is explained by an α-H elimination, doubtless followed by a reducing elimination resulting from a double alkylation of the tungsten.

E.O. Fischer's carbene is itself a very active catalyst for the polymerization of cyclopentene in the presence of $TiCl_4$ activated, probably photochemically, by chlorination of the tungsten (it is well known that $TiCl_4$ is photochemically reduced by alkanes) [6].

It is also possible to characterize and count the metallocarbenes formed on rhenium peroxide impregnated on alumina, by means of a 1,2-disubstituted olefin, then a reaction with ethylene to lead to propene [7] by a sequence of reactions that might be:

$$re^{**}{=}O + CH_3CH{=}CHCH_3 \blacktriangleright re{=}CHCH_3 + CH_3CH{=}O$$
$$re{=}CHCH_3 + CH_2{=}CH_2 \blacktriangleright CH_3CH{=}H_2 + re{=}CH_2$$

** re: symbolic writing formerly used by K. Ziegler.

Formation of the metallocarbene :
WCl_6 + $Sn(CH_3)_4$
Products of reaction of WCl_6, $M(CH_3)_n$ and an olefin

molar % relative to W

Olefin	WCl_6 (g)	Co-reactants (***)	Solvent	T (°C)	Olefin / WCl_6	Methane + éthane	Propy-lene	Butene-1
Butene-2	0,5	$LiCH_3$	Benzene (15 ml)	4	10	22	6,3	-
Butene-2	0,5	$Sn(CH_3)_4$	Benzene (15 ml)	4	10	3,5	3,6	-
Hexene-3	0,2	$Sn(CH_3)_4$	Benzene (30 ml)	4	10	n.d.	-	2,6
	0,5	$LiCH_3$	Benzene (15 ml)	4	10	22	4,4	4,7
	0,2	$Sn(CH_3)_4$	Benzene (30 ml)	4	10	10	9,6	10,4
Pentene-2 (*)	0,2	$Sn(CH_3)_4$	Chloro-benzene (30 ml)	-20	10	n.d.	8,5	6,4
	0,4	$Sn(CH_3)_4$	Chloro-benzene (30 ml)	-20	2,5	n.d.	6,8	6,9

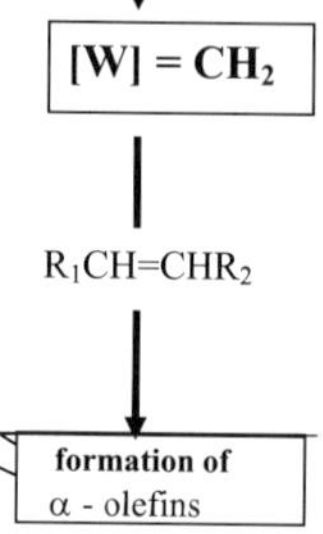

(*) we observe the formation of butene-2et hexene-3 from the disproportion of pentene-2 in excess.
(**) Sn/W = 2 ; Li/W = 2.
Soufflet, Commereuc et Chauvin C. R. Acad. Sc. 1973, 276, 169-171.

Figure 8.

CONCLUDING REMARKS

Obviously, not everyone has the opportunity to witness the birth of a new bond like the one discovered by E.O. Fischer, and the event is likely to become increasingly rare! Nor that of a unified concept encompassing two types of catalysis that until then had not been on speaking terms. It was an opportunity that had to be seized!

ACKNOWLEDGEMENT

I owe a great deal to everyone who took part in what I can rightly call an "adventure", because at the start we could not be sure of reaching a significant result. And more particularly to the talented students directly involved – Jean-Louis Hérisson, J.-P. Soufflet, and G. Zaborowski – without whom this work (and therefore the Prize!) would not have existed.

REFERENCES

1. N. Calderon, H.Y. Chen, K.W. Scott, *Tetrahedron Lett.* **1967**, 3327.
2. J.-L. Hérisson, Y. Chauvin, *Makromol. Chem.* **1971**, 141, 161–176 ; J.-L. Hérisson, *Thèse*, Paris 1970.
3. Y. Chauvin, D. Commereuc, G. Zaborowsky, *Makromol. Chem.* **1978**, 179, 1285–1290.
4. P.J. Flory, *Principle of Polymer Chemistry*, Cornell University Press, New-York 1953, 96, 497, 514.
5. J.-P. Soufflet, D. Commereuc, Y. Chauvin, C. R. Acad. Sc. Paris, **1973**, 276, 169–171.

6. Y. Chauvin, D. Commereuc, D. Cruypelinck, *Makromol. Chem.* **1976**, 177, 2637–2646.
7. Y. Chauvin, D. Commereuc, *J. Chem. Soc. Chem. Commun.*, **1992**, 462–464.

There have been many works on metathesis; the last and without doubt the best documented of them is the book by R.H. Grubbs, "Handbook of Metathesis", 2003, Wiley-VCH.

Portrait photo of Yves Chauvin by photographer U. Montan.

ROBERT H. GRUBBS

I was born at home in rural Kentucky in 1942, in a house that my father, Howard, had built. He did most of the construction himself and built it on land that his father had given him when he married my mother Faye. In some places, my birthplace is listed as Calvert City and in others Possum Trot. I was actually born between the two, so either one really is correct. Both of my parents were from small farm families in western Kentucky where most people stayed close to their homes. My many aunts, uncles and cousins formed a very supportive environment during my childhood. Although my grandmother spent her years working on the farm, she was very well educated and set a high standard that has resulted in many of her grandchildren becoming teachers and educators. She always read, told us stories and maintained an intellectual level that was unusual in rural Kentucky at that time. My mother was one of six surviving children of my grandmother and grandfather, a strawberry farmer. Because she was often sick when she was young and of little help on the farm, she was sent to college, where she received her teaching certificate after two years. She subsequently taught school for over 35 years as she moved from a one room school house teaching all grades, to a two room school teaching lower grades and finally to an elementary school where she taught first grade. Some of my early memories are of going to school with her when the baby sitter was not available and going to night and weekend classes with her while she finished her BA degree. It took her 28 years, but she finished her degree. She and my grandmother provided great role models for the value of education. Although my father did not attend college, he was a very gifted mechanic and a practical engineer. I grew up helping him to rebuild car engines, to install plumbing, to build houses and to work on my uncles' farms. After my father returned from serving two years in the US army in Europe during WW II, we moved into Paducah where he took night classes to become a diesel mechanic. He then spent many years working for the Tennessee Valley Authority, (TVA), operating and maintaining heavy equipment that was used in the construction of the extensive series of dams in Kentucky and Tennessee. The academic model of my mother and grandmother and the very practical, mechanical training from my father turned out to be perfect training for organic chemical research. I was the middle child between two sisters. My older sister became a teacher and has been heavily involved in art and painting. My younger sister became the first female journeyman electrician in western Kentucky and has spent a number of years working for the TVA.

As a child I was always interested in building things. Instead of buying candy, I would purchase nails which I used to construct things out of scrap wood. My mother always claimed that my spending my money on nails instead of on candy was why I was so skinny as a kid. As I grew older, farms in Kentucky provided me with many jobs in hauling hay and in cutting tobacco. In addition to helping fund my college years, these jobs helped me to meet an array of very interesting and amazing men and women.

My interest in science started in Junior High School where an outstanding science teacher, Mrs. Baumgardner, introduced me to the joys of science. In the 50's there were many excellent scientists in Paducah, Kentucky, who had settled there to become teachers after working in what we called the "Atomic Plant". This was a major uranium refining facility that was located in western Kentucky to take advantage of the power and production facilities resulting from the TVA construction in the area.

Since I had grown up with parents who had started life on farms in rural Kentucky and I had spent many hours working on farms, it was natural that I started college at the University of Florida as an Agricultural Chemistry major. This field combined my interests in science and agriculture. One of my summer jobs in college was in an animal nutrition laboratory where I analyzed steer feces all summer long. Fortunately, one of my friends was working in an organic laboratory for the summer and invited me to help him at night. He was working for Merle Battiste, a new faculty member at the University of Florida. I found that organic chemicals smelled much better than steer feces and that there was great joy in making new molecules. Battiste started me on the chemical journey and saved me from a life of analyzing animal matter. Shortly after joining his group, I read an organic text, *Mechanisms and Structure in Organic Chemistry* by E. S. Gould on reaction mechanisms which explained how chemical reactions take place. I was fascinated by being able to do rather simple chemical transformations to learn about the details of how organic compounds reacted at the molecular level. This direct coupling of simple observations with fundamental chemical reactions is the power of organic chemistry. Also, building new molecules was even more fun than building houses. The reactions of cyclopropenes served as the basis for most of my work in the Battiste group, and many years later a similar reaction of cyclopropenes opened the way to the synthesis of our first generation of catalysts. Battiste not only introduced me to organic research, but after he had trained me to be a productive researcher, he told me that I would have to leave his group. Although it would have been best for him for me to stay in his laboratory to finish my degree, he encouraged me to move on to a different group in a different part of the US to receive broader training. He encouraged me to go to Columbia University in New York City where I worked for Ron Breslow. Battiste had been Breslow's first PhD student. Just before going to Columbia, I heard a young Australian chemist, Rolli Pettit, who worked at the University of Texas, lecture at Florida. He talked about the stabilization of unstable molecules by coordination to transition metals. Rolli, who unfortunately died much too young, was an inspirational scientist who helped to direct me toward the use of metals in organic chemistry. For my graduate work,

I initially chose a project related to the Pettit work and finished studying the anti-aromaticity of cyclobutadiene. It was an exciting time in the Breslow group. A number of students and postdoctoral fellows who have since gone on to stellar academic careers worked in the group at that time and, along with Breslow, provided an amazingly stimulating environment. Working on projects involving metals confirmed my desire to turn my research toward transition metal organometallic chemistry and I obtained an NIH fellowship to work with Jim Collman who had just moved to Stanford. At that time organometallic chemistry was in its infancy and it provided a fertile field for a mechanistic chemist. There appeared to be an incredible array of important catalytic processes emerging in the field while little was known about the fundamental transformation involved. Collman had developed a systematic method of discussing reaction types that provided a basis for starting to understand the steps in Catalytic processes. One of the most exciting of these processes was olefin metathesis. It turned out that completely new fundamental steps were required to understand this reaction.

In 1969, as I finished my fellowship with Collman, Michigan State University, (MSU), was the only school that offered me a position in which to start my independent academic career. Harold Hart, Mike Karabatsos, Gene LeGoff, Don Farnum, Bill Reusch and Pete Wagner served as my early mentors at MSU and provided a very supportive environment for a starting faculty member. After nine very productive years at MSU where I started my work in olefin metathesis and a number of other areas of catalysis, I was offered a position at Caltech in 1978. Numerous outstanding graduate students contributed to my program at MSU. One postdoctoral fellow, however, played a major role in developing my research program. Dr. Akira Miyashita, who died in 2004, brought a number of new techniques to the group as well as an exemplary work ethic. He moved with me to Caltech in 1978 where he helped to reestablish the group in California.

The faculty at Caltech had created a rich environment for the training of outstanding graduate students, and these students, along with creative postdoctoral fellows and new faculty colleagues allowed my program to grow in many directions to take advantage of the growing field of organometallic catalysts. Bob Ireland and Dave Evans helped me to return to organic synthesis and to the development of new synthetic processes based on transition metals, and Peter Dervan and Dennis Dongherty, who provided guidance in physical organic chemistry as well as friendship over many years. John Bercaw and Harry Gray kept me honest in inorganic chemistry. Shortly after I arrived at Caltech, Fred Tebbe, a DuPont chemist, reported the structure and reactions of the complex that we later named the "Tebbe Reagent." Working with Dave Evans and Stan Pine on sabbatical from CSULA, we demonstrated that this reagent provided the first general route for the conversion of esters to vinyl ethers. This work also provided the basis for our mechanistic work that resulted in the synthesis of the first metathesis active metallacyclobutane. Many of the lessons learned with the Tebbe reagent were important in later developments. We discovered that it was important to develop a commercial source of this organometallic reagent so that other researchers could try the reaction without making a major investment of time in developing the techniques for making the reagent. The utility of the reagent was always limited by its instability under normal laboratory conditions. It required special air and water free conditions for its use. The desire for ease of use drove the early development of the ruthenium catalysts. We pursued the ruthenium based catalysts since they did not suffer many of the instability problems that had limited the use of the earlier tungsten and molybdenum based catalysts. We also developed early commercial sources and later commercially viable methods for the production of the well defined ruthenium based catalysts. In my lecture I discussed many of the developments from this point that let to the discovery of the family of ruthenium based metathesis catalysts and I named key members of the group who contributed directly to their development. Surrounding the group who made the key discoveries were others who made equally significant contributions that served as the basis for the specifically listed advances.

The olefin metathesis mechanism was one of my first projects at MSU and my group explored various aspects of this reaction throughout my career. In addition, we explored other mechanistic questions. An early project returned me to my interest in cyclobutadiene and we designed a mechanistic study to determine if cyclobutadiene was really formed from the decomposition of the Pettit iron tricarbonyl complexes. Other studies explored the mechanism of the Ziegler catalysts in the polymerization of ethylene and propylene. Using isotopes and stereochemistry, while working along with Mike Steigerwald, we developed a method that determined the critical steps in this reaction. Other work involved the study of metallacyclopentanes, a class of complexes we had become interested in from our early mechanistic studies of metathesis. Although metallacyclopentanes are not important in olefin metathesis, they have been found to be important in a variety of other olefin dimerization reactions. In a similar way, the discoveries in olefin metathesis have led to new processes that can benefit from detailed mechanistic explorations. Many of our advances in ruthenium catalyst development were the result of applying the techniques developed earlier in our group in the study of other complex catalytic processes.

One of the important developments in the group was the growth of research in polymer chemistry. Many of the students and postdoctoral fellows who helped to develop our polymer program have moved on to outstanding careers and have helped to establish polymer chemistry in other universities. The polymer program has opened many new opportunities and has led the group in a number of directions from biomedical applications to the synthesis of new membranes and barrier films. Although ring opening olefin metathesis polymerization, (ROMP), provided the path into polymer science, the techniques learned have opened possibilities outside of olefin metathesis. For example, they have resulted in the development of a new material that has found applications in light adjustable interocular lenses.

In the present commercial environment, it is difficult for fundamental discoveries to be transitioned into large companies. After a number of early frustrations that resulted from attempts to move our work directly into major companies, we were involved in the creation of a company to aid in the transition of technology from discovery to product. Mike Giardello, in addition to developing the first commercially viable route to a ruthenium metathesis catalyst, has played a major role in establishing Materia, Inc., where many of the commercial applications of the ruthenium technology – and other metathesis based products – were developed.

It has been very pleasing that a process that we started to study from pure intellectual curiosity has resulted in new processes which have very practical applications. I hope the success of the process has been a sufficient reward for the patience of the granting agencies who have allowed me to follow the metathesis reaction on the wandering path that has resulted from many chance and unanticipated discoveries. There are still many turns left on the metathesis path.

During my career, over 200 students and postdoctoral fellows have worked in my research group. They have all left their mark. I thank them all for their hard work, for their creative contributions and for making chemistry fun.

During my second year at Columbia, I met a wonderful lady from Brooklyn, Helen O'Kane. She has been my companion and best friend since that time. I thank her and our three children, Barney, now a Professor of Chemistry at Dartmouth, Brendan, an MD resident at USC and Kathleen, a PhD in Clinical Psychology program at University of Hawaii, who have supported me professionally as well as personally during this chemical quest.

OLEFIN METATHESIS CATALYSTS FOR THE PREPARATION OF MOLECULES AND MATERIALS

Nobel Lecture, December 8, 2005

by

ROBERT H. GRUBBS

California Institute of Technology (Caltech), The Division of Chemistry and Chemical Engineering, Pasadena, CA, 91125, USA.

This is a story of our exploration of the olefin metathesis reaction, a reaction that has been the major emphasis of my independent research. As with all stories of scientific discovery, there are three components: the discoveries, the resulting applications, and, perhaps the most important of all, the people involved. Starting from observations made from seemingly unrelated work, our investigations into the fundamental chemistry of this transformation have been an exciting journey, with major advances often resulting from complete surprises, mistakes, and simple intuition. Ultimately, these efforts have contributed to olefin metathesis becoming the indispensable synthetic tool that it is today.[1]

Much of modern organometallic and polymer chemistry, as we know it, started with a chance observation made in the early 1950's by reseachers in the Ziegler laboratory in Mulheim, Germany.[2] During this time, Ziegler was continuing work that had been initiated during WW II in exploring the use of alkyl aluminum complexes for the oligomerization of ethylene to produce lubricating oils. On one occasion, it was noted that this reaction produced 1-butene from ethylene instead of the C_{10}–C_{20} hydrocarbons normally observed. Subsequent analysis of the reaction autoclave found the presence of nickel. When nickel salts were deliberately added to this reaction, 1-butene was again observed. This discovery has since served as the foundation for an amazing array of nickel chemistry and catalysis. In addition, as nickel had been found to possess unexpected reactivity, other metal salts were also investigated. In particular, when titanium and zirconium halides were used in combination with alkyl aluminums, a new form of polyethylene was obtained. Natta further demonstrated that similar catalysts could promote the formation of stereoregular polymers from propylene. The 1963 Nobel Prize in Chemistry was awarded to Ziegler and Natta for this work.

The application of Ziegler-Natta-type systems toward the polymerization of cyclic olefins afforded additional unexpected results. A group at DuPont observed that the polymerization of norbornene did not produce a saturated polymer as expected, but instead afforded an unsaturated polymer in which one of the rings had been opened. While surprising, the origin of this unexpected polymeric structure was not to be pursued until later. Natta subsequently observed a similar result when he attempted to polymerize cyclopentene using tungsten and molybdenum halides – he, too, obtained a ring-opened, unsaturated polymer. Finally, Banks and Bailey of Phillips Petroleum Co., while investigating the possible polymerization of propylene over cobalt molybdate, found instead the formation of ethylene and 2-butene. These three observations, seemingly unrelated, were beginning to indicate a fundamentally new olefin transformation.[3]

My involvement with this puzzle started in 1968–69 while I was a postdoctoral fellow at Stanford. Jim Collman, my postdoctoral mentor, had just returned from a trip to Phillips Petroleum, where he had learned of an amazing reaction that converted propylene into ethylene and 2-butene. During a group meeting (Barry Sharpless, Nobel Prize 2001 was also a member of the Collman group at that time), we began to discuss possible mechanisms for this transformation. With my earlier training in mechanistic organic chemistry as an undergraduate and master's student in the laboratories of Merle Battiste, and as a doctoral student with Ron Breslow, I felt this unknown reaction mechanism would be an ideal problem to study.

Around this time, N. Calderon and his group at Goodyear had developed a soluble catalyst system that could not only induce polymerization via ring-opening, but also convert propylene into ethylene and 2-butene. Calderon's observations established that the polymerization of cyclic olefins observed by DuPont and Natta and the scrambling of acyclic olefins observed by Banks and Bailey were similar reactions. He also established that the mechanism of this reaction involved the cleavage of carbon-carbon double bonds rather than a transfer of alkyl groups between olefins via single-bond cleavage. This first critical mechanistic observation set the stage for subsequent studies. Banks and Bailey went on to propose a "quasicyclobutane" mechanism to account for their interchange reaction that was also consistent with Calderon's results. Other mechanisms were also proposed to account for this transformation, now termed olefin metathesis. What all these mechanisms had in common was the pairwise involvement of carbons in reaction intermediates. Theoretical studies of this reaction carried out by Frank Mango at Shell supported this pairwise exchange of two-carbon fragments.[4]

My independent career started at Michigan State University in 1969, where I proposed an alternative pairwise mechanism and initiated experimental work. Shortly thereafter, however, a growing body of evidence against such a mechanism began to appear. While investigating the effect of acyclic olefins in determining the molecular weight of polypentenamers, Chauvin and Hérrison observed products that were not consistent with the diolefin metal complexes proposed in the "quasicyclobutane"-type mechanisms. They instead proposed that the reaction did not go through a diolefin metal complex, but rather via a one-carbon metal carbene complex and metallocyclobutanes, intermediates that contain an odd number of carbons. In a similar study, Katz proposed an analogous mechanism where he was able to rationalize the observations that Chauvin had found to be inconsistent with his non-

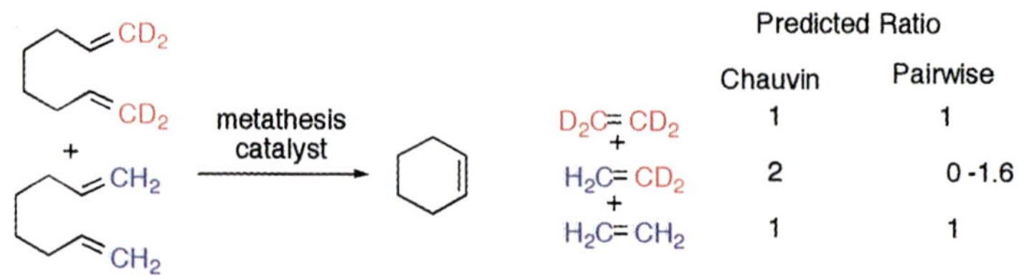

Scheme 1. RCM of 1,7-octadiene.

pairwise mechanism. Chuck Casey subsequently went on to find a model for this transformation utilizing preformed metal carbene complexes, and Schrock was able to determine that metal carbene complexes could, in fact, be generated under conditions similar to those used to prepare Ziegler-Natta-type catalysts.[5] Incorporating this new information into our mechanistic consideration, we designed an experiment that would allow most of the ambiguities of the Chauvin experiment to be addressed in detail. Rather than using a cross-metathesis reaction that would require us to analyze the role of alkyl groups on the intermediates, we instead selected to study the simplest system possible: isotopic substitution on a ring-closing metathesis (RCM) reaction.

The RCM of 1,7-octadiene generates cyclohexene and ethylene (Scheme 1). Cyclohexene is one of the few simple cyclic olefins that will not undergo subsequent metathesis reactions. Consequently, the fate of the two olefins in 1,7-octadiene can be determined via examination of the metathesis products. If the pairwise mechanisms involving even numbers of carbons were involved, the other two carbons would be required to couple together. In contrast, the Chauvin mechanism would be expected to scramble the two terminal carbons. A mixture of 1,1,8,8-tetradeutero-1,7-octadiene and 1,7-octadiene was used to investigate this. In the simplest case, the even carbon mechanisms should produce only ethylene and tetradeuteroethylene, while the odd carbon mechanisms should afford a statistical mixture of ethylene, dideuteroethylene, and tetradeuteroethylene in a 1:2:1 ratio. Careful analysis of all of our experiments found that a statistical mixture of ethylenes was kinetically formed in this reaction, a result consistent with the prediction made using the Chauvin mechanism. Subsequent studies by Katz on related systems provided additional support for this conclusion.[6]

By determining the key reaction intermediates in olefin metathesis, these combined mechanistic studies now enabled the use of rational design for further catalyst optimization. Prior to this time, metathesis catalysts were produced utilizing inconsistent, ill-defined systems. By confirming the non-pairwise mechanism of Chauvin, the identification of new metal alkylidene (metal carbene) complexes capable of promoting olefin metathesis now became our target for further improving reaction efficiency.

Fred Tebbe, a coworker of Schrock during his time at DuPont, developed one of the first well-defined metathesis systems. Now known as the Tebbe reagent, titanium- complex **1,** a metal carbene precursor, was found to exhibit metathesis activity in addition to its ability to promote Wittig-type

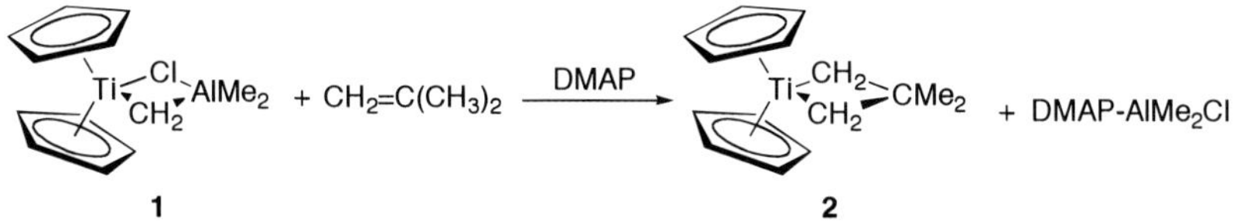

Scheme 2. Preparation of an isolable metallacycle.

reactions on esters. As the Tebbe catalyst was well-defined, both the starting and propagating carbene species could now be observed during a metathesis reaction, making it an ideal system for mechanistic study. In particular, we became interested in investigating the stereochemistry of the intermediate metallacycle. Our initial efforts to isolate a metallacycle complex proved unsuccessful until Tom Howard was able to serendipitously isolate metallacycle **2** by employing DMAP, which facilitated aluminum removal (Scheme 2). Tom was later able to demonstrate this metallacycle to be the low-energy state for this catalyst system. In addition, **2** was found to be an active metathesis catalyst. This observation provided the first example of a metallacycle intermediate and demonstrated that metallacycles could function as kinetically competent catalysts for olefin metathesis.[7]

Utilizing the well-defined Tebbe system, additional mechanistic and synthetic studies were carried out. Laura Gilliam found that metathesis conducted using this system was inherently "living," allowing for the preparation of polymers with defined structures. Polymer preparation and analysis consequently became a major focus within my group to further explore these observations. It was during this period that Schrock spent a 4-month sabbatical at Caltech, where we coauthored a paper demonstrating that one of his tantalum catalysts could also serve as a living initiator. In fact, it is now well established that virtually all defined metathesis initiators can be used to make polymer structures via living polymerization.[8]

Subsequent studies in the Schrock laboratory resulted in the design and synthesis of a number of very efficient molybdenum and tungsten metathesis catalysts, the historical development of which will be discussed in his lecture. Despite the activity of Schrock systems, the use of these early-metal catalysts had a major limitation: the oxophilicity of the metal center afforded poor functional group tolerance and necessitated catalyst preparation and handling under an inert atmosphere. To fully exploit the potential of metathesis, we believed that the development of new, functional group-tolerant catalysts would be crucial. Once again, serendipity would play a role in achieving this objective.

The development of ruthenium-based metathesis catalysts started with the goal of preparing interesting polymeric structures. Polymer chemistry provides an excellent means of studying metathesis catalysts: miniscule catalyst loadings have the capacity to generate large amounts of polymeric material, the structure of which can provide a historical record of catalyst activity. The polymer program within my group was focused on understanding the basic

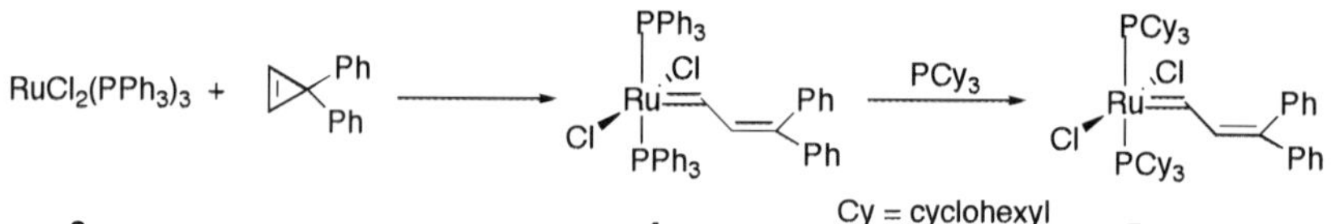

3 4 5

Scheme 3. Preparation of first well-defined ruthenium carbene catalyst.

principles of living metathesis polymerization. From this knowledge, we hoped to develop novel polymers that possessed interesting structures and physical properties. During the course of this work, model building had suggested that the ring-opened polymer from 7-oxo-norbornenes could be used to produce ionophoric membranes for selective ion transport. Bruce Novak took on the challenge of making these desired polymers. Much to our dismay, however, none of the available catalysts were found to promote this reaction. Examining the literature, Bruce came upon reports by Michelotti[9] and later Natta, where late-transition metal catalysts had been used to polymerize strained olefins in protic media. Returning to the lab, Bruce was able to demonstrate that these systems could also effectively promote the formation poly-7-oxo-norbornenes. More importantly, he went on to discover that ruthenium (II) complexes could serve as much more active polymerization catalysts. The outcome of Bruce's investigations resulted in the development of a robust catalyst system that tolerated most functional groups and aqueous media. Despite the activity of these ruthenium salt catalysts, however, only a small percentage of the ruthenium added to a reaction was found to produce an active catalytic center. In addition, as the structure of the active catalysts was totally undefined, it was impossible to make rational changes for further improvements.[10] However, this work defined the path forward; a ruthenium (II) complex and a strained olefin were found to be required to form an active catalyst system.

Assuming that olefin metathesis must occur via the formation of metal carbenes, we felt that the development of well-defined ruthenium carbene complexes was vital for further catalyst optimization. It took a few years, but eventually graduate student Son Binh Nguyen was able to prepare air-stable carbene complex **4** via the reaction of ruthenium(II) phosphine complex **3** with cyclopropene (Scheme 3). To our surprise, complex **4** was found to be an effective catalyst for the polymerization of norbornenes in protic media. Although **4** was not very active, its structure was well-defined, finally providing us with a scaffold by which to make structural changes for catalyst optimization. While many modifications have since been made (e.g. ligand exchange using tricyclohexylphosphine to generate the more active catalyst **5**), the basic structure of our catalyst systems still resembles that of our original "Son Binh Catalyst." This well-defined family of catalysts is distinguished by its amazing tolerance to oxygen and water, making it ideal for application to organic synthesis.[11]

In the early 1990's, Greg Fu joined our group as a postdoctoral scholar

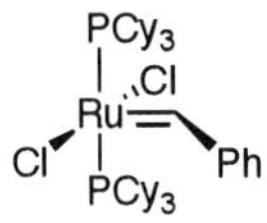

6

Figure 1. First-generation commercial catalyst.

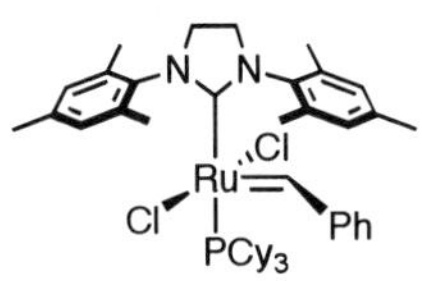

7

Figure 2. Second-generation commercial catalyst.

from Dave Evan's group at Harvard and began investigating the activity of various metathesis polymerization catalysts in organic synthesis. Relatively quickly, he was able to demonstrate that the Schrock molybdenum catalysts were effective for a number of important organic transformations, particularly those involving ring-closing reactions. Just prior to leaving the group to join the faculty at MIT, he was able to demonstrate that our newly prepared ruthenium systems could perform the same transformations on the benchtop without the use of a glovebox. In addition, many previously unreactive substrates were found to readily undergo ruthenium-catalyzed olefin metathesis in high yields.[12]

In our earlier work with the Tebbe reagent, we had found that its general use had been limited due to its air sensitivity and difficult preparation. In contrast, our user-friendly ruthenium catalysts were in high demand, and efforts now had to be made to make them readily available. Unfortunately, the Son Binh synthesis was difficult to run on large scale. Eventually, building off of work performed by Marcia France, Peter Schwab was able to develop a method based on the use of a diazo precursor. Although this reagent is potentially explosive, Mike Giardello developed a procedure to tame this reagent, allowing complex **6** to be commercially made on multikilogram scales (Figure 1).[13] With commercial sources, the organic synthetic community found a wide array of creative uses of olefin metathesis catalysts in the synthesis of complex structures.

Ruthenium's preference for soft Lewis bases and Lewis-acids, such as olefins, over hard bases, such as oxygen-based ligands, is responsible for its high tolerance to air and water. It also makes ruthenium catalysts fundamentally different from titanium-, tungsten- and molybdenum-based systems. Consequently, the rules learned in the development of early-metal catalysts no longer applied: for example, as ruthenium possesses a metal center rich in d electrons, strongly electron-donating ligands, rather than electron-poor ones, are required for high catalyst activity. Detailed mechanistic studies to obtain an understanding of the catalyst's structure/activity profile were therefore required. During the course of such investigations, we found that five-coordinate complexes, such as **5** and **6**, are catalytic precursors. In order to generate the metathesis-active species, one of the neutral ligands must be lost. The remaining neutral ligand of the resulting 14-electron complex is responsible for catalyst turnover. This knowledge was used to rationalize why bulky, basic tricyclohexylphosphine ligands are more effective than

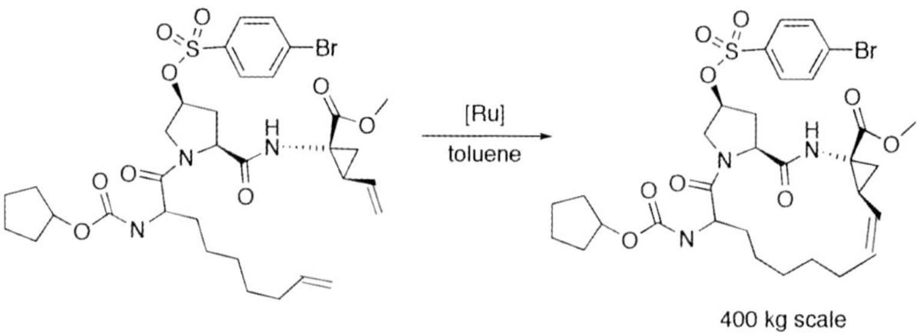

Scheme 4. Commercial application of RCM.

triphenylphosphine ligands in the ruthenium-catalyzed metathesis of cyclooctene.[14]

Further catalyst tuning via the substitution of one of the phosphine ligands on **6** with an *N*-heterocyclic carbene (NHC) ligand resulted in a series of highly active catalysts (e.g. catalyst **7**; Figure 2). This increase in metathesis activity was attributed to an increased rate of catalyst turnover due to the favorable electron donation and steric bulk of the NHC ligand. Additional modification by replacing the phosphine in **7** with a weaker ligand, such as pyridine, resulted in an increase in activity up to a factor of 10^4 relative to catalyst **6**.[15] A number of other research groups have played a role in further optimizing these catalytic systems.

One of the most rewarding aspects of my research has been the opportunity to study the potential applications of a new catalyst. The facile preparation of our ruthenium catalysts, coupled with their functional group tolerance and environmental stability, enabled olefin metathesis to finally realize its broad potential. Although this reaction was already commercially utilized in the processing of hydrocarbons, the application of olefin metathesis in organic synthesis and in the preparation of highly functionalized polymers had yet to be fully developed.

Most commercial products that contain organic molecules possess at least one carbon-carbon double bond or, if one is not present, it is likely that an olefin was used in its preparation. This being the case, the potential applications of olefin metathesis are endless. To date, commercial products range from a wooden baseball bat, treated with metathesis-based polymers to improve durability, to the preparation of highly functionalized pharmaceutical agents, many of which are in the advanced stages of testing. For example, Boehringer Ingelheim recently reported the use of olefin metathesis for the commercial preparation of 400 kg of a compound under investigation to treat hepatitis C (Scheme 4).[16]

Some of the earliest commercial applications of olefin metathesis involved the ring-opening metathesis polymerization (ROMP) of monomers containing strained, unsaturated rings. The polymerization of dicyclopentadiene (DCPD) is one of the best-known examples of this (Scheme 5). DCPD is an attractive monomer for polymer production, as it is inexpensive, and the re-

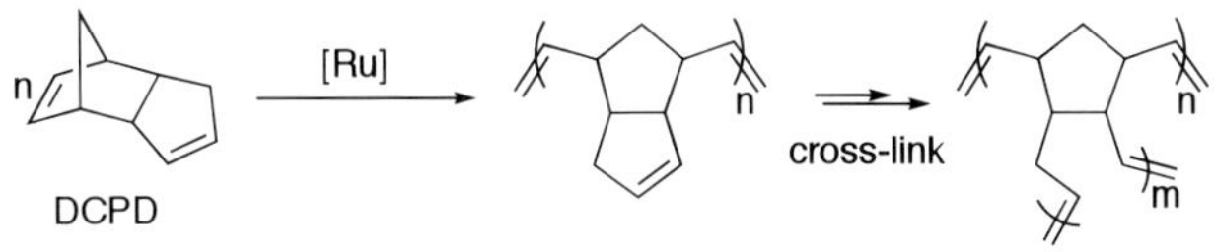

Scheme 5. ROMP for the preparation of DCPD polymers.

sulting polymer products are useful for a variety of applications. Hercules and Goodrich developed one of the first commercial processes for this reaction, where early metal complexes were combined with alkylaluminum to form catalysts to produce DCPD polymers via injection molding. Unfortunately, the extreme sensitivity of these catalysts toward air and water mandated the use of rigorous processing conditions and highly purified monomers.

In contrast, the functional and environmental tolerance of the ruthenium catalysts enables DCPD polymerization to be carried out in open molds in the presence of numerous additives. In addition, DCPD monomers bearing functional groups for further synthetic elaboration can also be utilized. A wide variety of products, such as the composite baseball bats discussed previously, as well as bathroom fixtures, ballistic panels, and large equipment panels, are currently in/or being developed for production utilizing ruthenium-catalyzed ROMP.[17]

The ruthenium systems can also be used to prepare polymers with well defined structures. Like the Tebbe complex, ruthenium olefin metathesis catalysts can be used for the production of living polymers. In these cases, strained olefins are ring open polymerized by the metal center to produce a polymer that contains an active ruthenium alkylidene at the end of the polymer chain. After the first monomer added to the catalyst is consumed, a second monomer can be added and the polymers start to grow again. If the monomers contain different functional groups, the different segments of the polymer will have different properties and functions. Since the systems are living, if all the chains are initiated at the same time, the length of the chain and therefore the molecular weight is controlled by statistics and produces a very narrow distribution of chain lengths. Consequently, both the functionality and molecular weight of polymers produced using the ruthenium systems can be controlled. Since the ruthenium systems have been demonstrated to tolerate many functional groups on other applications, the polymers prepared using ruthenium initiators can contain many functionalities for use in a wide variety of applications. For example, if one of the blocks of the polymer is water soluble and the other is insoluble, the block polymers will form micelles that are nanoparticles that are being explored for applications in medical diagnostics. If the two blocks are incompatible, the solids formed from the block polymers will have microphase separated structures that are important for a number of applications in materials science.

Ring-closing metathesis (RCM) and, to a lesser extent, cross metathesis (CM) have now become standard transformations in organic synthesis. As ruthenium catalysts will tolerate most functional groups, protection-deprotec-

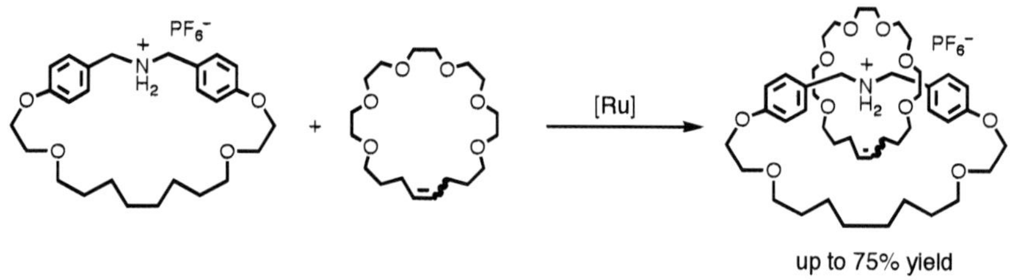

Scheme 6. Olefin metathesis for the preparation of "magic" rings.

tion strategies – which often hinder the application of other reaction methodologies – are seldom required. In addition, ruthenium-catalyzed olefin metathesis is ideal for use during the late stages of a total synthesis due the chemoselectivity exhibited by the catalyst and the mild reaction conditions required. Ruthenium metathesis catalysts are also ideally suited for use in tandem reactions involving acetylenes and ring systems, as they are stable toward a variety of reaction conditions and reagents and, often, their presence will not impede subsequent reaction transformations. This favorable combination of reactivity and functional group compatibility has resulted in the extensive use of olefin metathesis by the chemical community, which is reflected in large number of citations to the use of ruthenium based metathesis catalysts.

A useful feature of olefin metathesis lies in the reaction's reversibility. Olefin metathesis is a thermodynamically controlled reaction; therefore, the self-assembly of the most stable product of a metathesis reaction can be attained by simply allowing the reaction to reach equilibrium. This feature has proven to be advantageous for a variety of applications, particularly in the use of RCM to produce interlocked, "magic" ring systems in high yield (Scheme 6).[18]

In a modern industrial setting, the use of efficient chemical processes, preferably those with minimal impact on the environment, is essential. The term "green chemistry" has evolved as an umbrella concept to represent this general approach to conducting reactions. Metathesis attains the goals of green chemistry in three ways: (1) by providing a more efficient route over traditional methods to carbon-carbon bond formation and avoiding by-product formation; (2) through enabling the use of renewable resources; and (3) by providing a means to attain environmentally friendly products. For instance, the processing of seed oils represents an excellent example of the first two goals. Using solvent-free conditions, olefin metathesis enables vegetable oils to be efficiently processed into compounds that can serve as renewable sources of petroleum product alternatives.[19]

The solvent-free CM of 1-hexene with hexenylacetate (Scheme 7) represents an example of the use of metathesis to generate environmentally-friendly products. The product of this reaction can be readily converted into a peach twig borer pheromone, which can be used as an environmentally-friendly means of insect control in lieu of the use of broad-spectrum pesticides. CM is currently under investigation for the commercial preparation of

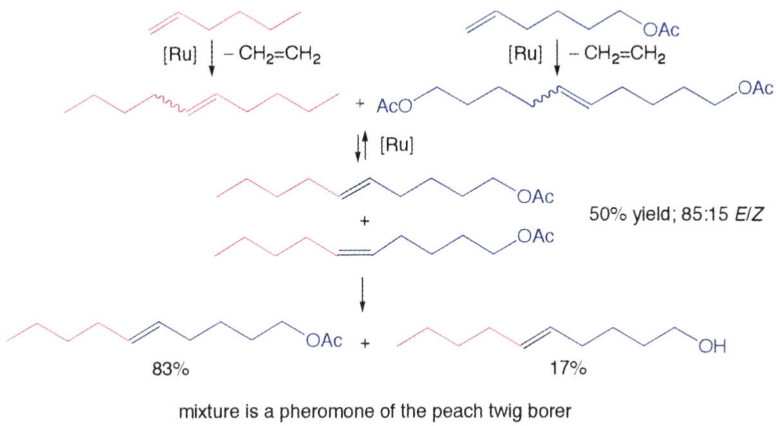

Scheme 7. Commercial application of CM in the production of "green" compounds.

other pheromones, as this route represents an efficient means to generate these compounds starting from inexpensive starting materials.[20]

Beginning as merely an interesting anomaly over forty years ago, the olefin metathesis reaction has transitioned into one of chemistry's most valuable reactions for carbon-carbon bond formation. In the early days of trying to understand this fundamental process, we never envisioned that this reaction would one day achieve the import that it has today. Our exploration of this reaction has been a fascinating journey, and it is one that is ongoing. As work on olefin metathesis continues in both industry and academia, we can look forward to exciting future developments.

I would like to thank the over 200 coworkers who have contributed to my work, as well as all the other interesting people I have interacted with during my career in chemistry. I would also like to thank the group at Materia Inc. who has played a significant role in making this technology available for many commercial applications. You have all made the journey very interesting. I especially want to thank my wife Helen and our kids (Barney, Brendan and Katy) who have provided the support that has made it all possible and worthwhile.

REFERENCES

1. R. H. Grubbs, *Tetrahedron* **2004**, *60*, 7117–7140 *Handbook of Metathesis, 3 Volumes.*; R. H. Grubbs, Ed., Wiley-VCH Verlag GmbH & Co., Weinheim, 2003.
2. "The Chain Straighteners: Fruitful Innovation: The Discovery of Linear and Steroregular Synthetic Polymers." Frank M. McMillan, London, MacMillan, **1979**.
3. W. L.Truett, D. R. Johnson, I. M. Robinson *J. Am. Chem. Soc*, **1960**, 2337.
4. N. Calderon, E. A. Olfsead, J. P. Ward, W. A. Judy, K. W. Scott, *J. Am. Chem. Soc.*, **1968**, *90*, 4133.; R. L.. Banks and G. C. Bailey, *I & EC Product Research and Development*, **1964**, 170.
5. J. L. Herisson, Y. Chauvin *Makromol. Chemie*, **1971**, *141*, 162.; T. J. Katz, J. McGinnis, *J. Am. Chem. Soc.*,**1975**, *97*, 1592.; C. P. Casey, T. J. Burkhardt, *J. Am. Chem. Soc.*, **1974**, *96*, 7808.

6. R. H. Grubbs, D. D. Carr, C. Hoppin and P. L. Burk, *J. Am. Chem. Soc.* **1976**, *98*, 3478.; R. H. Grubbs, P. L. Burk and D. D. Carr, *J. Am. Chem. Soc.* **1975**, *97*, 3265.; T. J. Katz and R. Rothchild, *J. Am. Chem. Soc.* **1976**, *98*, 2519.
7. T. R. Howard, J. B. Lee and R. H. Grubbs, *J. Am. Chem. Soc.,* **1980**,*102*, 6876.; S. H. Pine, R. Zahler, D. A. Evans and R. H. Grubbs, *J. Am. Chem. Soc.,* **1980**,*102*, 3270.; J. R. Stille, R. H. Grubbs, *J. Am. Chem. Soc.* **1986**, *108*, 855.; F.N. Tebbe, G. W. Parshall, G. S. Reddy, *J. Am. Chem. Soc.* **1978,** *100*, 3611.
8. L. R. Gilliom and R. H. Grubbs, *J. Am. Chem. Soc.* **1986**, *108*, 733.; R. R. Schrock, J. Feldman, L. F. Cannizzo and R. H. Grubbs, *Macromolecules* **1987**, *20*, 1169.
9. F. W. Michelotti, W. P. Keaveney, *J. Poly. Sci., Part A,* **1965,** 895.
10. B. M. Novak and R. H. Grubbs, *J. Am. Chem. Soc.* **1988**, *110*, 960.
11. S. T. Nguyen, L. K. Johnson, R. H. Grubbs, and J. W. Ziller, *J. Am. Chem. Soc.* **1992**, *114,* 3974.; S. T. Nguyen, R. H. Grubbs, and J. W. Ziller, *J. Am. Chem. Soc.* **1993**, *115*, 9858.
12. G. C. Fu, S. T. Nguyen, and R. H. Grubbs, *J. Am. Chem. Soc.* **1993**, *115*, 9856–9857; G. C. Fu and R. H. Grubbs, *J. Am. Chem. Soc.* **1992**, *114 (18)*, 7324–7325.
13. P. Schwab, M. B. France, J. W. Ziller, and R. H. Grubbs, *Angew. Chem. Int. Ed. Engl.* **1995**, *34,* 2039. (*Angew. Chem.* **1995**, *107*, 2179).; T. E. Welhelm, T. R. Belderrain, S. N. Brown, R. H. Grubbs, *Organometallics*, **1997**, *16*, 3867.
14. E. L. Dias, S. T. Nguyen, and R. H. Grubbs, *J. Am. Chem. Soc.* **1997**, *119*, 3887–3897.; M. S. Sanford, M. Ulman, and R. H. Grubbs, *J. Am. Chem. Soc.* **2001**, *123,* 749–750.; M. S. Sanford, J. A. Love, and R. H. Grubbs, *J. Am. Chem. Soc.* **2001**, *123*, 6543–6554.
15. M. Scholl, S. Ding. C. W. Lee, R. H. Grubbs, *Org. Lett.* **1999**, *1*, 953.; C. W. Bielawski, R. H. Grubbs. *Angew. Chem., Int. Ed.* **2000**, *39*, 2903.; T. M. Trnka, J. P. Morgan, M. S. Sanford, T. E. Wilhelm, M. Scholl, T-L. Choi, S. Ding, M. W. Day, R. H. Grubbs, *J. Am. Chem. Soc.* **2003**, *125 (9)*, 2546.; J. A. Love, M. S. Sanford, M. W. Day, R. H. Grubbs, *J. Am. Chem. Soc.* **2003**, *125*, 10103.
16. T. Nicola, M. Brenner, K. Donsbach, P. Kreye, *Organic Process and Development*, **2005,** 27.; I. Kodota, H. Takamura, K. Sato, A. Ohno, K. Matusuda, Y. Yamamoto, *J. Am. Chem. Soc.* **2003**, *125*, 46.
17. Patent No. 6,310,121 B1, issued 10/30/01. C. S. Woodson, Jr. and R. H. Grubbs, Patent No. 5,939,504, issued 8/17/99. C. S. Woodson, Jr., R. H. Grubbs.
18. J.-P. Sauvage, B. Mohr, R. H. Grubbs, and M. Weck. *Angew. Chem.* **1997**, *36 (12)*, 1308–1310.; A. F. M. Kilbinger, S. J. Cantrill, A. W. Waltman, M. W. Day, and R. H. Grubbs, *Angew. Chem. Int. Ed. Engl.* **2003**, *42*, 3281.; E. N. Guidry, S. J. Cantrill, J. F. Stoddart, R. H. Grubbs, *Org. Let.* **2005**, *7*, 2129.
19. Frost &Sullivan's Industrial Bioprocessing Alert, Sept. 9, 2005; http://www.epa.gov/greenchemistry/whats_gc.html.
20. Patent No. 6,696,597, issued 2/24/04. R. L. Pederson, R. H. Grubbs,; Patent No. 6,215,019 B1, issued 4/3/01. R. L. Pederson, R. H. Grubbs.

Portrait photo of Robert H. Grubbs by photographer R. Paz.

Richard R. Schrock

RICHARD R. SCHROCK

I was born in Berne, a northeast Indiana farming community proud of its Swiss heritage. Life was not easy for my parents, Noah J. Schrock, the second of six children, and Martha A. Habegger, the second of ten children. They married in 1933 during the Depression. My oldest brother, Luther, was born in 1934, Theodore in 1939. A few months after I appeared in1945, the family moved to Decatur, about 13 miles north of Berne, where we lived until the summer after my fifth birthday. My most lasting memory of our first home is its proximity to the city swimming pool where I spent many happy summer days.

We moved into an old house on the west side of South 13th Street in 1950. The house required a good deal of work, but my father, who had been a carpenter for fifteen years, accomplished the renovation over a period of several years. The house was located on what seemed to me to be an enormous plot of land (one acre); the back yard took forever to mow on a steamy summer day and the vegetable garden produced quantities of corn, strawberries, melons, tomatoes, and raspberries. A large vacant lot extended south toward a small railroad that dove beneath the highway on its way east to a mill where tiles were made from the clay dug on the west side of the road. The process of mining the clay created ponds, which I explored along with the surrounding woods at length. In the summer I would fish, catch snakes and frogs, and build simple huts in the woods, which probably were the first indications of my love for designing and building, and for the outdoors. In the winter I learned how to ice skate and often (so it seemed) would come close to actually freezing toes and fingers. We never had much money, but the house was comfortable (except during the first, and especially cold, winter) and the food (fresh in the summer, canned in the winter) was plentiful for the five of us.

My father built a woodworking shop in the two car garage at the rear of the house where I also spent much time discovering, among other things, that it is not easy to drive nails into maple. I was not allowed to operate power tools at a young age, but my father introduced me to several as I became older, except his prize 1941 Delta table saw, which I eventually inherited when he died of leukemia in 1980 at age 69. He was a patient teacher even though I did not fully appreciate the difference between a chisel and a screwdriver or clear pine and a prized piece of birds eye maple when I was young. I learned many things by trial and error but nevertheless grew to appreciate what one can do with wood and the right tools used the right way. I maintain that interest to

The Schrock family in 1946.

this day. Like most young Hoosier men at that time, I played on the school basketball, baseball, and football teams. Although I did not continue to play organized sports beyond age ten or eleven, sports provided me with a respect and need for physical exercise that I still have.

My curiosity and love of building things played into the hand of my older brother Theodore who presented me with the proverbial chemistry set on my eighth birthday. He was thirteen at the time and beginning to appreciate the beauties of science. His interest in chemistry continued through his college years but he eventually studied medicine instead and became a highly successful surgeon. I was hooked. I created a small laboratory at the end of a storage area for canned goods and used my budding woodworking skills to build shelves for the ever expanding collection of test tubes, beakers, and flasks. I obtained most of my equipment through a mail order supply house with money earned from an early morning paper route. I carried out simple experiments (combining acids and bases to make salts, making pleasant smelling esters, etc.) following the directions in chemistry laboratory texts handed down to me. When I reached the age of thirteen, Harry Dailey, the high school chemistry teacher, stoked my interest in chemistry with more textbooks and discarded equipment. I thought all equipment wonderful to behold. As I became aware of the power of burning natural gas, the lowly alcohol burner was replaced by the common Bunsen burner, and the Bunsen burner ultimately by a high tech broad-headed model capable of putting out a good deal of heat, enough to melt metals in a porcelain crucible and even sodium chloride.

After we moved to a house on Jackson Street in 1958, my laboratory grew in size, diversity, and complexity. I now had at my disposal relatively sophisticated and, if misused, dangerous substances in a small room in the basement. I was often interested in testing recipes for mixtures of oxidizing agents and oxidizeable materials as well as nitrating common household substances. Thankfully, there were no serious mishaps, although my mother tells stories that belie that statement, including one in which the local fire department was called to our home; fortunately only a small rug was burning, not the house. At least I also was quick to think and act.

My father went to San Diego, California, in the fall of 1958 to work in the construction industry with his brother, Clarence, and to explore the feasibility of moving west. In 1959 my mother and I joined him. She drove and I navigated cross country with my laboratory carefully packed in the trunk of the car. By the time I finished Mission Bay High School in 1963, both my parents had also fulfilled their dreams of finishing high school, something they had not been allowed to do in their youth. My interest in chemistry expanded in San Diego. I found a laboratory supply house where I could buy classic equipment (a 250 mL retort was coveted and purchased) and a drugstore where I could buy basic chemicals with some adult help of course. I discovered many wonderful things such as how to make bromine from KBr and sulfuric acid, how to make sodium (and chlorine) by electrolizing molten sodium chloride, and how to analyze for metals by making sulfides with brilliant and characteristic colors. I took up surfing and skin diving, and continued my interest in woodworking by designing, making, and selling fins for surfboards. I entered a regional science fair with a project that concerned osmotic processes in sea urchin eggs and managed to win a prize for it. I collected the sea urchins at low tide and harvested the eggs myself. That, and dissection of a sheep's brain (which I greatly enjoyed) in physiological psychology later in college, was the closest I would come to following my brother into medicine.

I always assumed I would attend college and study chemistry. The only financially viable option was the University of California. I was accepted at Berkeley but chose to attend Riverside, a relatively new campus about 90 miles north of San Diego, because I thought that a smaller school might allow me to do more independent research earlier in my career. That proved to be the case. After the first exam in my first chemistry course at UCR, I was ap-

Research as an undergraduate at the University of California at Riverside in the group of James N. Pitts in the mid sixties.

proached by Professor James Pitts who asked if I wanted a summer job. I agreed and began research in what broadly could be called atmospheric chemistry, a hot topic in the smog ridden Los Angeles basin and surrounding area at that time. In actuality, I spent my time learning to blow glass and construct vacuum lines, and to measure low concentrations of photolysis products using a temperamental, delicate, almost impossible to align, multi-pass Perkin-Elmer IR machine connected to a vacuum line. (Fourier Transform machines were not yet known.) A paper entitled "The Detection of Ethylketen and enol-Crotonaldehyde in the Vapour-phase Photolysis of *trans*-Crotonaldehyde" reported some of my work in 1968 after I had moved on to graduate school. I also worked "up the hill" with Dr. E. A. (Ed) Schuck and in 1966 found my name on a paper entitled "Rate Constant Ratios During Nitrogen Dioxide Photolysis." I learned many things, scientific and otherwise, that need not be detailed here. One that I might mention was the joy (and sometimes discomfort) of hiking in the Sierra Nevada Mountains. I still enjoy mountain hiking, although the frequency has decreased considerably. I capitalized on my knowledge of IR spectroscopy during a summer of research at Dow in Midland, Michigan, where my oldest brother Luther was an engineer.

At Riverside I was influenced by a talented and enthusiastic teacher in physical chemistry named Jerry Bell. Jerry decided that I had enough ability to attend Harvard University for graduate study where he had received his Ph.D. I liked the idea, applied, and was accepted. I celebrated by listening to Rachmaninoff's second piano concerto played by Byron Janis, loudly, through a Fischer amplifier and large home built speakers, each with a volume approaching 12 cubic feet; an arts in western civilization course at UCR had boosted my interest in music that I had acquired in high school. During the last semester at UCR I took an inorganic chemistry course taught by Fred Hawthorne who appropriately spent a good deal of time discussing boron compounds. Although I enjoyed organic chemistry, the possibility of exploring the chemistry of all elements in the periodic table was fascinating to me. Yet for some reason I still regarded myself as a physical chemist.

When I arrived at Harvard, I was scheduled to live in Perkins Hall, a graduate student dormitory. I was offered several possible roommates, among them David Swickard, a childhood friend from Decatur, who unknown to me was beginning graduate school in political science. Needless to say, he was surprised to find me in the dorm room when he arrived. I lived in Perkins for the first year and then moved to an apartment with several other friends near Central Square in Cambridge. The apartment was adequately furnished, convenient to Harvard, and cheap. I spent most of my time in the lab anyway. I was one of the constant residents in the apartment for the next three years, although from time to time I was uncertain how many roommates I actually had.

I had not visited Harvard or researched the faculty before arriving and had no plan concerning the kind of physical chemistry I wanted to pursue. My desire to permute compositions of matter did not seem to fit with the physical chemistry being done at Harvard at that time. Perhaps I had the wrong impression of what I wanted to do or what "physical chemistry" was. Therefore picking a research supervisor proved problematic. One day while walking dejectedly down the main hall of Mallinckrodt Laboratory I passed a long office/lab (in fashion in years past) where a recently arrived assistant professor sat at the desk. I entered and talked with John Osborn, who told me about transition metal chemistry, about the excitement of creating new, colorful, crystalline compounds, and about catalysis by transition metal species such as the rhodium catalysts he had developed as part of his doctoral study with Geoffrey Wilkinson at Imperial College. That sounded like what I wanted to do, make new compounds with potential uses, so I signed on. I am not sure John was totally pleased that I chose to work on rhodium since he may have felt that it was too close to his own doctoral work, although in the end the cationic complexes that I discovered turned out to be quite useful for asymmetric hydrogenation, at least those that contained the appropriate enantiomerically pure phosphine ligand. I also met Ryoji Noyori who was exploring catalysis by transition metal species in E. J. Corey's group, and Mike Strem, who had started Strem Chemicals and for whom I made a sample of deuterated Zeise's salt for the fantastic wage of $100, half a month's pay at that time.

The late sixties were wild in Cambridge as in other university areas, with the Vietnam war escalating, and rebellions seething, or actually erupting. I decided I needed a break and had to see the world. I informed John that I would be traveling abroad for fifteen weeks during the summer of 1968 on a grand European tour. The mantra was $5 a day. I think I almost achieved that goal. Hitchhiking was still acceptable and helped the budget greatly, as did youth hostels and a diet of bread and cheese. I could not believe I actually was in Europe, visiting all the famous museums and practicing my high school Spanish and college German. My success with French was more modest. While I am eager to learn other languages, I do not have the special talent required to achieve that goal without actually having lived abroad.

In November of 1969, the 16th I believe, I meandered to a party near Davis Square where I met a tall blond school teacher named Nancy Carlson. She not only was beautiful but also bright and seemed attracted to me, which I found refreshing. We saw each other regularly that year and most of the next, and married in August of 1971, shortly after I finished and defended my Ph.D. thesis and she completed a master's degree in library science. Only one job was available in 1971, for which there were several hundred applicants, so I considered myself fortunate to obtain a postdoctoral fellowship from the National Science Foundation to work at Cambridge University in the laboratory of Jack Lewis, later Lord Lewis. The year in England was the beginning of a long relationship with England that continues today. Our present house is graced by a brass rubbing of Roger de Trumpington (gold and silver wax on black paper), one of the oldest brasses in England. It took some time to rub a brass that size, and the church in the village of Trumpington south of Cambridge around Christmas time was unrelentingly cold. I discovered a great deal about the relative heat capacity of my body versus a stone floor. We rubbed other brasses, but that particular event remains in my memory because of the cold and the boy's choir practicing their Christmas program in the background. I returned to the U.S. in the winter to look for permanent employment. Job offerings were still relatively meager, although I garnered a couple of offers. Upon returning to Cambridge I met Earl Muetterties, an associate director at the duPont Experimental Station in Wilmington, Delaware, who was on a sabbatical at Cambridge, which was unusual for a nonacademic. He interviewed me and ultimately gave me an offer to work at the Central Research Department, which I accepted. At that time CRD was an academic department in an industrial setting; nothing close to it exists today. Nancy and I moved to Wilmington in August of 1972 with two bicycles purchased in England. I still have mine and ride it occasionally.

At duPont I shared a lab with the remarkable Fred Tebbe, who was studying, among other things, reactions of ethyl aluminum reagents with titanocene dichloride. His most important and surprising discovery was what has become known as Tebbe's reagent, a species made through the reaction of titanocene dichloride with trimethylaluminum. The most remarkable feature of Tebbe's reagent is a methylene group bridging between titanium and aluminum, which could replace the oxygen in a carbonyl group in a Wittig-like reaction. I dabbled in cyclooctatetraene chemistry and synthesized compounds such as the anion of triscyclooctatetraene niobium, and blue $Ta(C_8H_8)Me_3$, prepared through addition of the dianion of cyclooctatetraene to tantalum trimethyldichloride, the only known alkyl of tantalum, which had been reported by G. L.

Juvinall in 1964. I found $TaMe_3Cl_2$ to be easy to prepare on a large scale from $TaCl_5$ and dimethylzinc, which at that time could be purchased neat in large quantities. I became interested in the possibility of preparing other tantalum alkyl complexes. I was heavily influenced by G. Wilkinson's synthesis of hexamethyltungsten. Wilkinson was a consultant for duPont at the time. On a trip to duPont in the spring of 1973 he told us about the synthesis and properties of that remarkable compound. I subsequently concentrated on the synthesis of tantalum pentaalkyl complexes and soon had made highly volatile and highly unstable pentamethyltantalum by treating $TaMe_3Cl_2$ with methyllithium. By July the attempted synthesis of pentaneopentyltantalum led to trineopentylneopentylidenetantalum, $Ta[CHC(CH_3)_3][CH_2C(CH_3)_3]_3$ instead. That compound marked the beginning of high oxidation state carbene, or alkylidene chemistry, as stated in a book by Bill Nugent and Jim Mayer (*Metal-Ligand Multiple Bonds*, 1988) in which the actual DuPont notebook page which describes that discovery was reproduced. (Fortunately, that particular page also was an example of how to keep a proper notebook.) Shortly thereafter I prepared the first isolable terminal methylene complex, the structure of which was solved by Lloyd Guggenberger. George Parshall, my group leader, allowed me to do what I thought interesting and to work after normal hours, which greatly increased my productivity. In the fall of 1972 I first heard the term "olefin metathesis" from Earl Muetterties, who remained interested in the subject after he moved to Cornell in the summer of 1973. I started to follow the metathesis literature and began to suspect that the new alkylidene complexes that I discovered in 1973 might be relevant to that process, even though tantalum was not a known catalyst for olefin metathesis.

I was impressed by the high oxidation state chemistry of osmium imido complexes being carried out by K. Barry Sharpless at MIT and invited him to duPont to give a talk, probably in late 1974. He was excited about the potential of the new alkylidene compounds I had discovered. He and several others at MIT pushed for an offer from MIT, which came in the spring of 1975. I accepted, even though I took a cut in salary. I do not recall salary being much of an issue. MIT fulfilled my dream of obtaining an academic position at a top institution. I looked forward to exploring tantalum chemistry further and teaching. Nancy and I moved to Brighton (part of Boston) in August of 1975. While I pushed ahead with research at MIT, Nancy added a master's degree in art history to her graduate degree in library science and learned bookbinding through an apprenticeship.

I was fortunate to obtain support from the National Science Foundation early in my career at MIT and to build a group of ten students by the end of my third year. I could not have been happier. High oxidation state organometallic chemistry developed rapidly, thanks to the talented students and support for them from the National Science Foundation. By 1980 we had transferred the principles behind tantalum chemistry to tungsten, molybdenum, and rhenium, and had shown what type of tungsten species would metathesize olefins. By the mid 1980's we had developed what I call well-defined catalysts for both the olefin and acetylene metathesis reactions that con-

Me with Nancy, Andrew and Eric (right) in 1989.

tained sterically protecting imido and/or alkoxide ligands. I became interested in dinitrogen chemistry around 1980, and polymer chemistry (first ring-opening metathesis polymerization and later olefin and acetylene polymerization) a few years after that. In the mid 1990's I began a fruitful collaboration with Amir H. Hoveyda on asymmetric metathesis reactions and their applications to organic chemistry, which was and still is supported by the National Institutes of Health. With support of the NIH I also was able to achieve in 2003 a long sought goal for hundreds of researchers over a period of 40 years, the catalytic reduction of dinitrogen with protons and electrons at room temperature and pressure. It turns out that a single molybdenum center within a properly designed protective ligand is sufficient. Time will tell whether these findings are relevant to how dinitrogen is reduced to ammonia in nature on a huge scale by nitrogenase enzymes.

My son Andrew was born in 1978 and Eric in 1981. The family became too large for the house in Brighton so in 1983 we moved to Winchester, Massachusetts, where we slowly renovated a 1904 Arts and Crafts style house. A bindery on the third floor allowed Nancy to practice book conservation while our sons were growing up. Half the basement became my woodworking shop; the other half became at various stages a cub scout den, a teenage getaway, and later a finishing room for items built in my shop.

In 1997 Nancy accepted a position as the Chief Collections Conservator of

Karin and John Osborn with Nancy in 1996.

the Harvard College Libraries in Widener Library at Harvard University; in the summer of 2006 she will return to the MIT Libraries as the Thomas F. Peterson, Jr., Conservator of the Special Collections. With the children on their own we enlarged the kitchen/living area in 2000 and in the process expanded the basement, adding another room for woodworking and a small wine cellar. We enjoy cooking together in the new kitchen and gardening on our manageable 1/4 acre lot.

As of 2006 I have published approximately 425 papers and have trained ~65 graduate students and ~75 postdoctoral students. I still find the process of unlocking nature's secrets an enormously satisfying profession and hope to be fortunate enough to continue to practice it for some time.

MULTIPLE METAL-CARBON BONDS FOR CATALYTIC METATHESIS REACTIONS

Nobel Lecture, December 8, 2005

by

RICHARD R. SCHROCK

Department of Chemistry 6-331, Massachusetts Institute of Technology, 77 Massachusetts Avenue, Cambridge, Massachusetts 02139, USA.

It's my great priviledge to be here today, in a position I never thought possible. I hope the story that I will tell you will give you some idea what I have contributed to the area for which the Nobel Prize in Chemistry was awarded this year.

The story begins thirty two years ago in 1973, the year the Nobel Prize was shared by G. Wilkinson and E. O. Fischer. Wilkinson's Nobel Lecture[1] concerned the nature of a single bond between a transition metal and a carbon atom in an alkyl group, and emphasized the fact that the metal-carbon bond is not inherently weak. E. O. Fischer in his Nobel Lecture[2] summarized the extensive chemistry of transition metal "carbene" complexes[3,4] that contain a metal-carbon double bond discovered by him and his group in 1964 (Fig 1).[5] He also reported new "carbyne" complexes that contain a metal-carbon triple bond.[6] It was clear that metal-carbon single bonds were of great importance in the emerging area of homogeneous catalysis. However, no catalytic reactions involving species that contain metal-carbon double or triple bonds were known. When I went to the Central Research Department of E. I. DuPont de Nemours and Company in 1972, transition metal organometallic chemistry and homogeneous catalysis were of great interest as a consequence of their huge potential in organic chemistry and therefore in industry.

In the early 1970's inorganic chemists knew that many transition metal species containing a metal-carbon bond are subject to various modes of decomposition that are much more rapid than in a non-transition metal species such as $Zn(CH_2CH_3)_2$ or $Al(CH_2CH_3)_3$. The most common of these involves transfer of a β hydrogen, from an ethyl group (MCH_2CH_3) for example, to the metal to

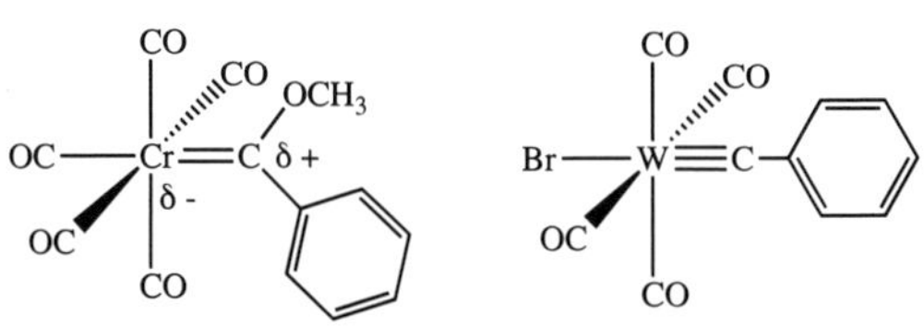

Figure 1. "Low oxidation state" carbene (left) and carbyne complexes (right).

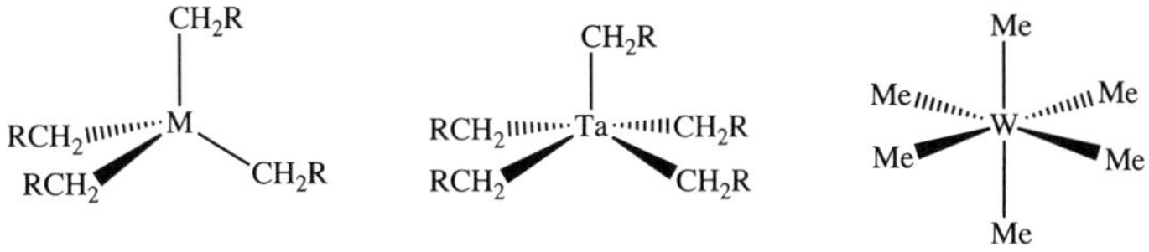

M = Ti, Zr, Hf (?)

Figure 2. "Peralkyl" complexes of Group 4, 5, and 6, metals in their highest possible oxidaton states.

yield a metal-hydride and an alkene. The relative stabilities of high oxidation state "homoleptic" or "peralkyl" compounds such as $M[CH_2Si(CH_3)_3]_4$, $M(CH_2C_6H_5)_4$, and $M[CH_2C(CH_3)_3]_4$ (M = Ti, Zr, or Hf; Fig 2), were rationalized on the basis of the fact that unlike a compound having an ethyl ligand, the alkyl ligands in these species lack β hydrogens and so of course cannot undergo decomposition processes that involve β hydrogens.[7] In 1973 Wilkinson published the synthesis of $W(CH_3)_6$.[8] Unlike $M(CH_3)_4$ species (where M = Ti, Zr, or Hf) $W(CH_3)_6$ is stable at 22°C. The methyl carbon is α with respect to the metal; there is no β carbon and so no β hydrogens. However, methyl species *are not sterically protected* against bimolecular reactions that involve *α hydrogens.* As you can see from the note added in the proof of Wilkinson's Nobel address,[9] I was intrigued by high oxidation state peralkyl complexes and had chosen to explore the organometallic chemistry of tantalum soon after my arrival at DuPont. Little alkyl chemistry was known of the metals in group 5 (V, Nb, Ta) at that time; I chose tantalum because it is next to tungsten in group 6 (Cr, Mo, W), and like tungsten, tantalum is relatively stable in its highest possible oxidation state, Ta(5+). Peralkyl tantalum complexes therefore must be pentaalkyls.

My starting point for tantalum alkyl chemistry was a paper by G. L. Juvinall[10] in which he described the low yield synthesis of niobium and tantalum trimethyl dichloride species through addition of dimethylzinc to the metal pentachlorides. I found that crystalline $TaMe_3Cl_2$ not only could be prepared quantitatively in pentane on a large scale (Eq 1), but it could be stored for

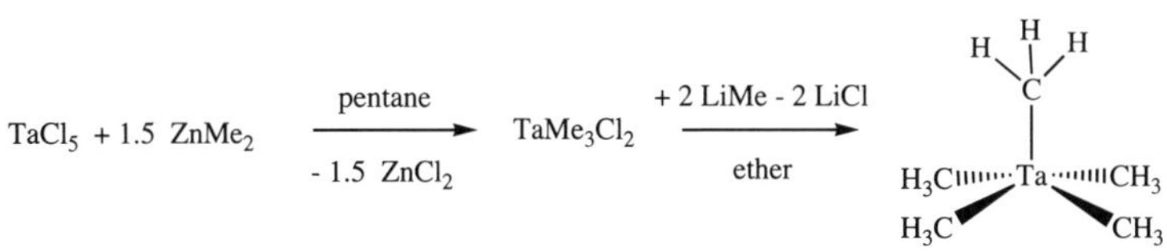

(1)

long periods at –40 °C in the solid state. Moreover, it reacted with two equivalents of LiMe to generate volatile, yellow, crystalline $TaMe_5$.[11] This species is much less stable than $W(CH_3)_6$, but much more stable than $Hf(CH_3)_4$. Pentamethyltantalum decomposes above 0 °C to yield ~3.7 equivalents of methane, and it does so in a bimolecular fashion. Pentamethyltantalum has 15 sterically unprotected α C-H bonds and the metal is highly electron-deficient (10 electrons, 8 shy of the preferred 18 electron count). Interactions between one metal center and CH bonds in another metal complex thereby are encouraged, and are easier in $TaMe_5$ than in WMe_6 for simple steric reasons.

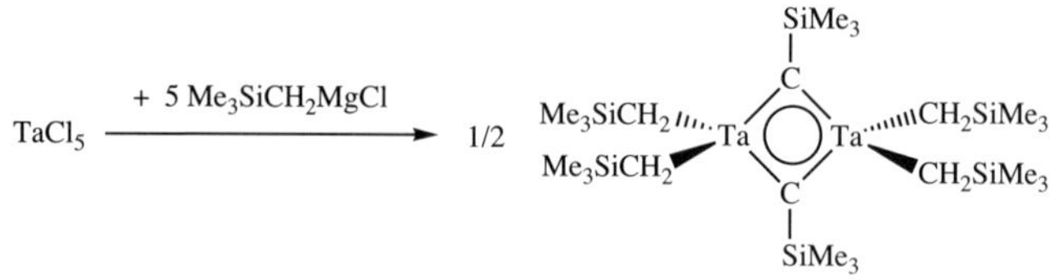

(2)

Wilkinson had just published a reaction that yielded an unusual dimeric species containing what he called "carbene bridge," μ-$CSiMe_3$ (actually a carbyne bridge; Eq 2). With respect to $Ta(CH_2SiMe_3)_5$ he said "It is assumed that a penta-alkyl complex cannot exist for steric reasons."[12] Therefore, I turned to experiments involving the neopentyl (CH_2CMe_3) ligand in order to probe the limit of steric crowding in homoleptic d^0 tantalum pentaalkyls and the nature of the decomposition pathways that might be observable in such circumstances. The crucial experiment (Eq 3) consisted of an attempt to prepare

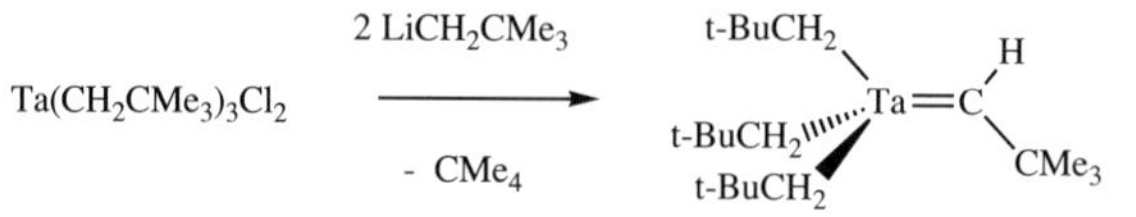

(3)

$Ta(CH_2CMe_3)_5$ through addition of two equivalents of $LiCH_2CMe_3$ to $Ta(CH_2CMe_3)_3Cl_2$. Instead of $Ta(CH_2CMe_3)_5$, orange, crystalline, and thermally stable $(Me_3CCH_2)_3Ta{=}CHCMe_3$ was formed in quantitative yield.[13] The exact mechanism is still not known, but it is likely that an α hydrogen is activated by the metal and subsequently removed as a proton by a neopentyl group, possibly in an *intramolecular* manner in pentaneopentyltantalum itself (Eq 4).

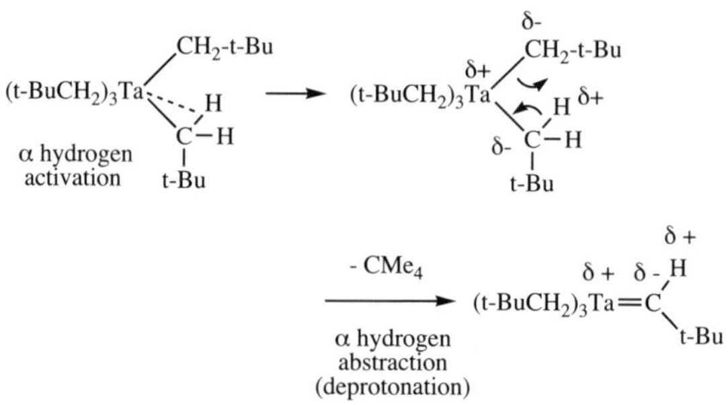

(4)

$(Me_3SiCH_2)_3Ta{=}CHSiMe_3$ may be an intermediate in the reaction shown in equation 2, but it must decompose intermolecularly to give tetramethylsilane and the observed dimeric species; $(Me_3CCH_2)_3Ta{=}CHCMe_3$ does not.

The tantalum neopentylidene complex, $(Me_3CCH_2)_3Ta{=}CHCMe_3$, was new in several respects. It was the first example of a stable transition metal M=CHR species. Second, the terminal alkylidene was formed through a new type of reaction, an intramolecular α hydrogen abstraction from an alkyl. Third, unlike "carbene" complexes discovered by Fischer and his group, $(Me_3CCH_2)_3Ta{=}CHCMe_3$ was highly electron-deficient (10 electrons in metal-based bonding orbitals). Fourth, $(Me_3CCH_2)_3Ta{=}CHCMe_3$ appeared to behave as if the metal were in its highest possible oxidation state with the Ta=C bond being polarized so that the metal is relatively positive and the carbon relatively negative, opposite to what is found in Fischer-type carbene complexes. Fifth, the high thermal stability and resistance of $(Me_3CCH_2)_3Ta{=}CHCMe_3$ toward intermolecular decomposition reactions, *as well as toward further intramolecular reactions*, suggested that other high oxidation state, four-coordinate species that contain four sterically demanding covalently bound ligands might be viable. This simple but important principle pervaded much of the high oxidation state chemistry that was developed over the next 30 years in my group.

"Deprotonation" of a neopentyl ligand to give a neopentylidene ligand suggested that a neopentylidene ligand might be deprotonated further by an external base. Indeed, addition of *n*-butyllithium to $(Me_3CCH_2)_3Ta{=}CHCMe_3$ led to formation of the related high oxidation state, anionic, *neopentylidyne* species, $\{(Me_3CCH_2)_3Ta{\equiv}CCMe_3\}Li$[14] (Eq 5).

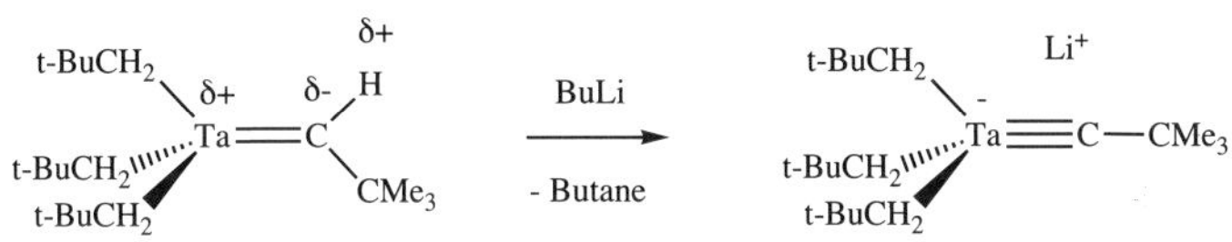

(5)

Finally, it was demonstrated that even a methylene ($Ta{=}CH_2$) species could be prepared through deprotonation of $[TaCp_2(CH_3)_2]^+$.[15,16] Even though $TaCp_2(CH_3)(CH_2)$ contains 18 electrons in metal-based orbitals, it is unstable toward bimolecular decomposition to yield an ethylene complex in 50% yield, plus related species that are formed in the presence of some ligand L (Eq 6). Bimolecular decomposition of alkylidenes, especially methylenes, is a

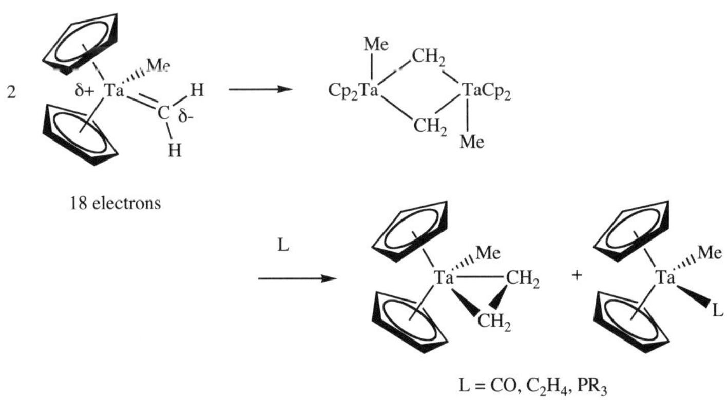

(6)

battle that continues to be fought today. The main point is that between 1973 and 1975 *high oxidation state* tantalum-carbon double and triple bonds had been prepared, and they appeared to be significantly different in several important respects from the *low oxidation state* species prepared by Fischer.

At some point in the early 1970's I became aware of the "olefin metathesis" reaction (Eq 7), a startling and mysterious reaction catalyzed by homogeneous Mo and W catalysts (and heterogeneous Re catalysts), the identities of which were not known.[17-20] Physical organic experiments by (*inter alia*) R. H.

$$2\ RCH{=}CHR' \rightleftharpoons RCH{=}CHR + R'CH{=}CHR' \quad (7)$$

(8)

$$2\ RC{\equiv}CR' \rightleftharpoons R'C{\equiv}CR' + RC{\equiv}CR \quad (9)$$

Grubbs,[21-23] T. J. Katz,[24,25] C. P. Casey,[26] and Y. Chauvin[27] were designed to determine whether the reaction proceeded by a "pair-wise" or "non-pair-wise" mechanism.[20] It was shown that the reaction proceeded in a "non-pair-wise" fashion, and Chauvin was the first to propose the correct "carbene/metalacyclobutane" mechanism (Eq 8) that would account for all observations.[27] However, at that time no "carbene" complexes were known that would react efficiently with ordinary olefins as proposed by Chauvin.

Alkyne metathesis (Eq 9), both heterogeous[28,29] and homogeneous,[30] catalyzed by Mo and W compounds of unknown type, also had been discovered. T. J. Katz[24] proposed a mechanism of alkyne metathesis analogous to that for olefin metathesis (Eq 10), i.e., one that involves formation of *metalacyclobutadiene* intermediates. Yet no "carbyne" complexes were known that would catalyze alkyne metathesis.

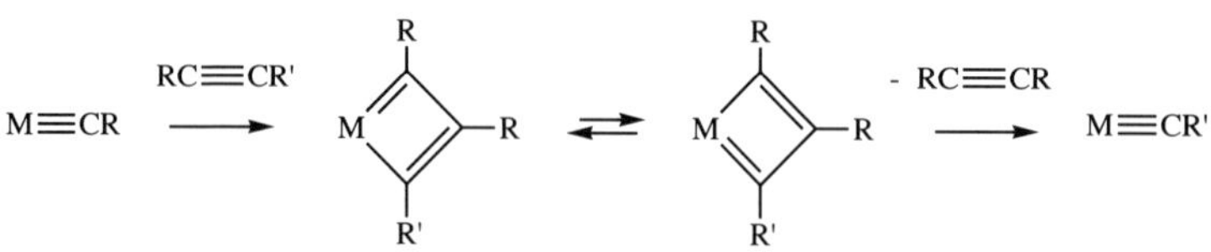

(10)

I asked myself whether the new types of tantalum "alkylidene" and "alkylidyne" complexes might not be at least the type of species that metathesize olefins and acetylenes, respectively, even though tantalum was not known to be a catalyst for either of these reactions at that time.

After my move to MIT in 1975, I began to explore reactions between various tantalum alkylidene complexes and olefins in some detail. It became clear (Eq 11) that electron deficient tantalum and niobium alkylidenes would

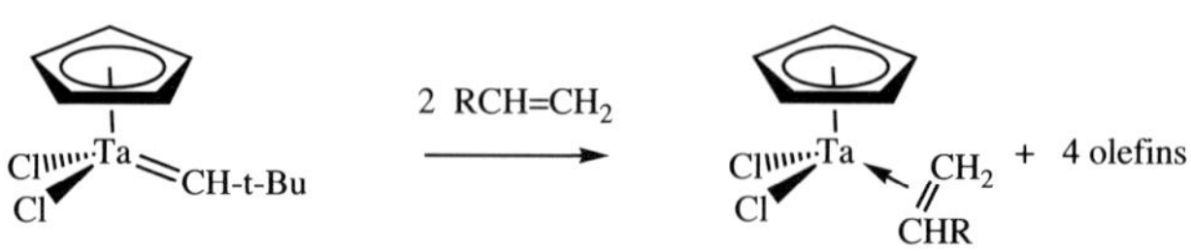

(11)

react with olefins readily to give metalacyclobutane intermediates, but these species would rearrange via a β hydride process to give as many as four of the possible rearrangement products instead of losing an olefin to reform a new alkylidene (Eqs 12 and 13); the alkylidene chain reaction never started and

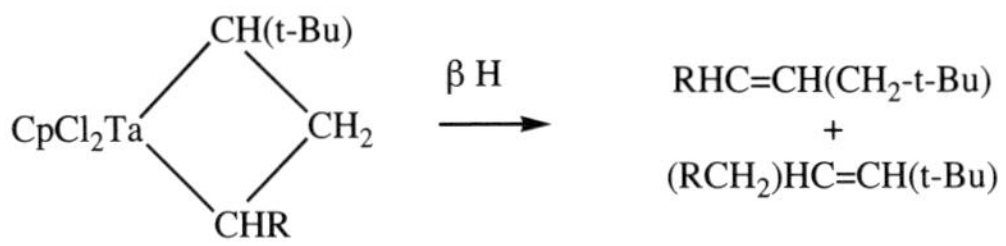

(12)

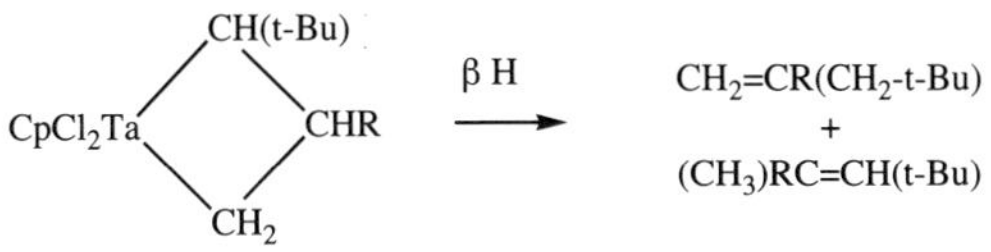

(13)

no metathesis products therefore were observed.[31] At an international symposium on olefin metathesis (ISOM III) in 1979 in Lyon I reported that niobium and tantalum complexes that contain *t-butoxide* ligands in place of chloride ligands, such as M(CH-t-Bu)(O-t-Bu)$_2$Cl(PMe$_3$), could be coaxed into metathesizing olefins. Several dozen catalytic metathesis cycles could be observed for an olefin such as cis-2-pentene.[32] It was proposed that the t-butoxide ligands "prevent reduction" of the metal and "promote metathesis." At that same conference I reported that an attempt to carry out a "Wittig-like" reaction between a tantalum alkylidene and a tungsten oxo complex to yield W(CH-t-Bu)(O-t-Bu)$_4$ and Ta(O)L$_2$Cl$_3$, instead produced an unanticipated 18 electron oxo neopentylidene complex of tungsten (Eq 14).[33] This oxo

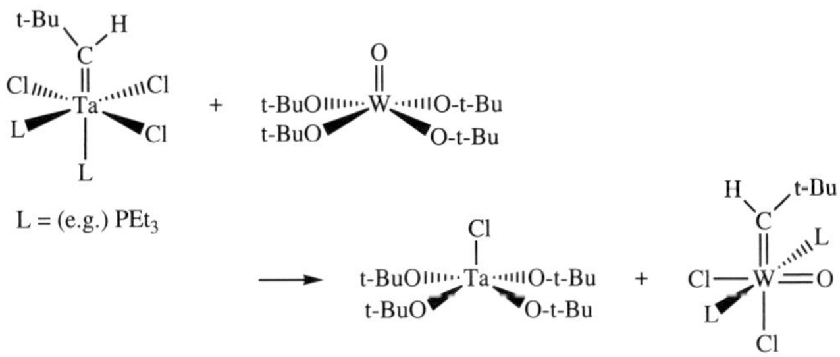

(14)

neopentylidene complex would react with olefins to yield metathesis products and new alkylidenes that one would expect from the metathesis reaction, including a methylene complex, and especially rapidly in the presence of a trace of AlCl$_3$. This was a convincing demonstration that the oxo alkylidene species react with olefins readily and reform an alkylidene complex of the same type from half an olefin that was present. The role of AlCl$_3$, it was proposed, was to remove either chloride or phosphine (reversibly) and thereby create an empty coordination site at the electronically saturated metal center.

A four-coordinate neopentylidene species containing large, covalently bound ligands was believed to be the most likely isolable, but reactive, electron deficient species. A neopentylidene ligand clearly was required. We also knew that alkoxides were desirable, but what should the fourth, and necessarily dianionic, ligand X (Fig 3) be? An oxo ligand was unsuitable because it almost certainly would encourage bimolecular decomposition reactions. We focussed on an imido ([NR]$^{2-}$) ligand, an isoelectronic analog of an oxo ligand, since the imido ligand could be sterically protected by a large R group. After consid-

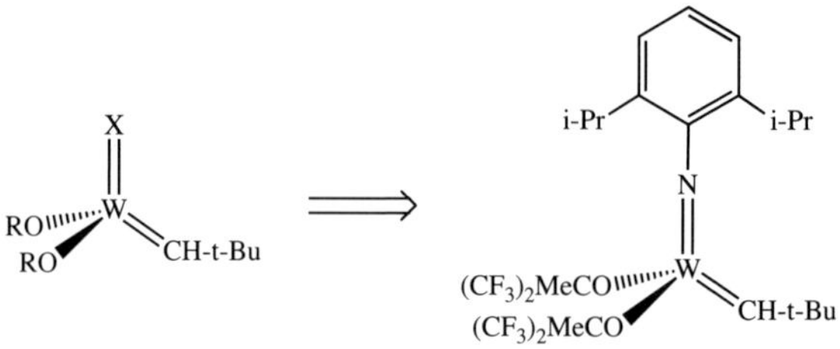

Figure 3. Design of a stable tungsten imido alkylidene bisalkoxide complex.

ering the likely synthetic difficulties associated with R being a 2,6-di-t-butylphenyl group, we settled on the 2,6-diisopropylphenyl group, a choice that in part was in response to a comment by K. B. Sharpless concerning the value of isopropyl groups in general versus t-butyl groups. (The 2,6-diisopropylphenyl group is found in many nitrogen-based ligands in a large variety of catalysts today.) Several large and sterically demanding tertiary alcohols and phenols were available, including fluorinated t-butanols such as $(CF_3)_2MeCOH$ whose pKa in water (~9) is significantly less than that of t-butanol itself (~19). We felt that the electron withdrawing ability of the hexafluoro-t-butoxide ligand should increase the electrophilicity of the metal dramatically and therefore also increase the rate of the reaction of an alkylidene complex with an olefin. Therefore, we aimed to prepare $W(NAr)(CHCMe_3)[OCMe(CF_3)_2]_2$ (Fig 3) with the expectation that this 14 electron species not only could be isolated, but that it would metathesize olefins readily.

Being aware of the almost magical properties of the neopentyl ligand (as well as neopentylidene and neopentylidyne ligands), we also were curious about simple reactions between W(VI) compounds (initially WCl_6) and neopentyllithium or neopentylmagnesium chloride, and had begun exploring such reactions in 1977. We discovered that a volatile, yellow, crystalline alkylidyne complex could be isolated from such reactions,[34] and that the reaction between $W(OMe)_3Cl_3$ and six equivalents of Me_3CCH_2MgCl in diethyl ether produced that species in a yield of ~50% (Eq 15).[35] $(Me_3CCH_2)_3W{\equiv}CCMe_3$ is

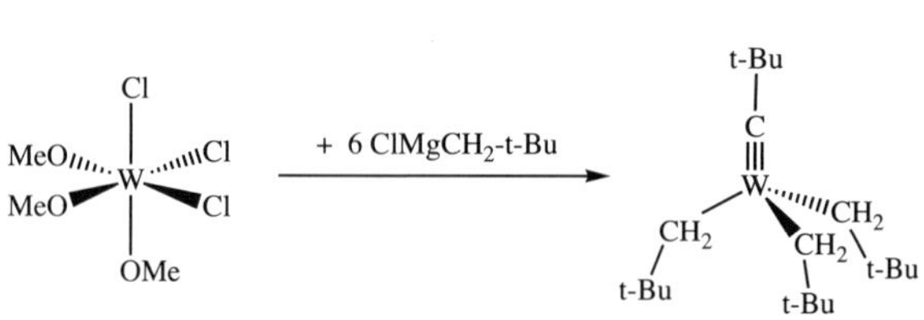

(15)

closely related to $(Me_3CCH_2)_3Ta{=}CHCMe_3$ in having four neopentyl or neopentyl-derived ligands around the metal. As we became aware of the benefits of alkoxides for efficient olefin metathesis, we became curious whether a trialkoxide species such as $(Me_3CO)_3W{\equiv}CCMe_3$ could be prepared and whether it would be an efficient *alkyne metathesis* initiator. We were pleased to find that $(Me_3CO)_3W{\equiv}CCMe_3$ in fact could be prepared readily (Eq 16), and especially pleased to find that alkynes *are* metathesized at room temperature

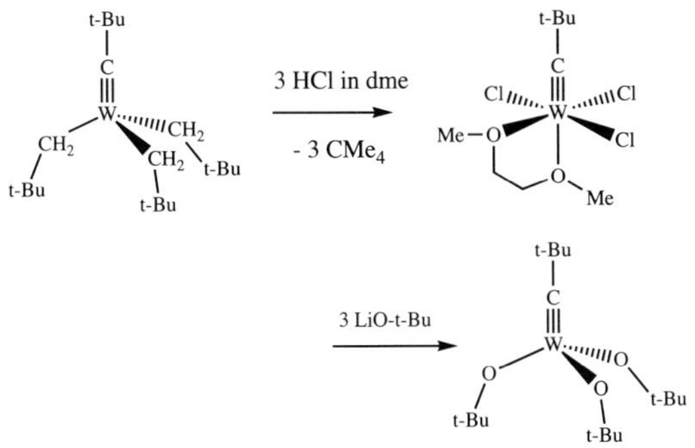

(16)

with $(Me_3CO)_3W{\equiv}CCMe_3$ as the initiator[36] at rates thousands of times faster than "black box" catalysts known in the literature. This finding proved to us that sterically bulky alkoxides were indeed highly beneficial for metathesis activity. In contrast to $(Me_3CO)_3W{\equiv}CCMe_3$, $(dme)Cl_3W{\equiv}CCMe_3$ does not metathesize alkynes, even though a tungstenacyclobutadiene intermediate necessary for that transformation, e.g., $Cl_3W[C(t\text{-}Bu)C(CH_3)C(CH_3)]$, could be isolated and crystallographically characterized![37] Other reactions took precedence.[38-40] $(Me_3CCH_2)_3W{\equiv}CCMe_3$ also did not metathesize internal alkynes; it was simply too unreactive. When the OR group in $(RO)_3W{\equiv}CCMe_3$ was a more electron withdrawing alkoxide or phenoxide, we could isolate and crystallographically characterize intermediate tungstenacyclobutadiene complexes in alkyne metathesis reactions and study them in detail.[41,42]

A large variety of species that contain tungsten-tungsten or molybdenum-molybdenum triple bonds, $X_3M{\equiv}MX_3$ species, were known at that time,[43,44] and in particular hexa-t-butoxide ditungsten, $(Me_3CO)_3W{\equiv}W(OCMe_3)_3$. An interesting question, therefore, was whether $(Me_3CO)_3W{\equiv}W(OCMe_3)_3$ would

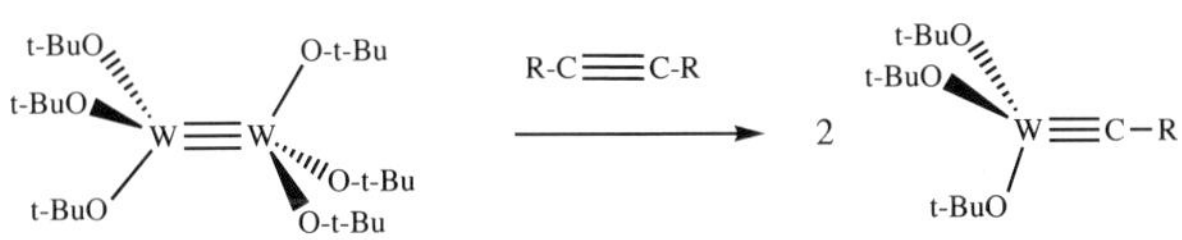

(17)

react with internal alkynes to yield $(Me_3CO)_3W{\equiv}CR$ complexes (Eq 17), which is effectively a metathesis-like reaction? The answer turned out to be yes, and it did so amazingly readily.[45] Although this is not a general reaction for all $X_3M{\equiv}MX_3$ species (M = Mo or W), the fact that it is facile only when X is a relatively sterically bulky alkoxide, had to be more than a coincidence. This discovery opened up a route to $(Me_3CO)_3W{\equiv}CR$ species that did not rely on α hydrogen abstraction reactions,[46] and cemented the relationship between, and interconvertibility of, strong triple bonds between metals, between carbons (in alkynes), and between a metal and a carbon in an alkylidyne complex. It is interesting to note that the C≡N bond in nitriles is also cleaved readily, although the N≡N bond in dinitrogen is not. That spectacular reaction, which requires a (three-coordinate) triamido species, would not be reported until 1995.[47]

But tungsten alkylidynes were important for an additional reason. If a neopentylidyne species could be prepared by removing an α proton from a neopentylidene ligand, could not that reaction be run in the reverse direction, i.e., could not a proton be added to a neopentylidyne species to yield a neopentylidene species? Moreover, could that not be accomplished in a manner that would produce the other desired metal-ligand multiple bond, namely the imido ligand, at the same time? An amido/alkylidyne complex could in fact be prepared readily and the proton then moved from **N** to **C** by a base (Eq 18). A large variety of W(NAr)(CH-t-Bu)$(OR)_2$ complexes then could be

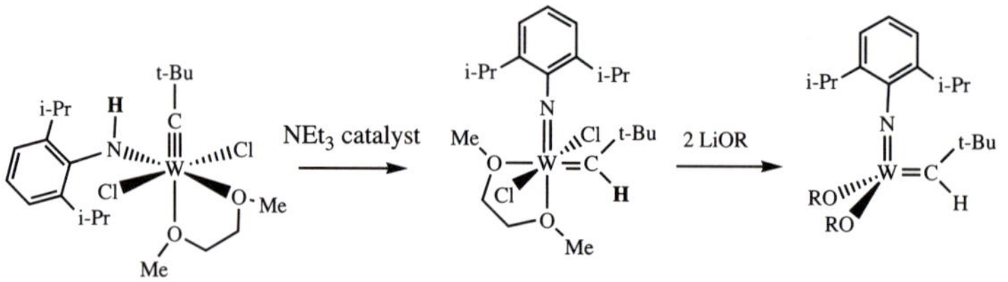

(18)

prepared readily from the dichloride species, as long as OR was large enough to prevent complications that resulted from bimolecular reactions. X-ray structures such as that in Figure 4 showed how the t-butyl group of the neopentylidene ligand points toward the imido ligand (a *syn* orientation), and how the diisopropylphenyl group on nitrogen protects the imido nitrogen and the alkylidene carbon atom against bimolecular reactions.

It soon became clear that W(NAr)(CH-t-Bu)$(OR)_2$ species would metathesize olefins in the expected manner with an activity that correlated roughly with the electron withdrawing power of the OR ligand. Catalysts that contain hexafluoro-t-butoxide ligands appeared to have the highest activities. We found that tungstenacyclobutane intermediates, the very intermediates proposed by Chauvin in the metathesis reaction, actually could be isolated and crystallographically characterized.[48] However, tungstenacyclobutane stabilities proved to be a problem is some circumstances; loss of an olefin from an un-

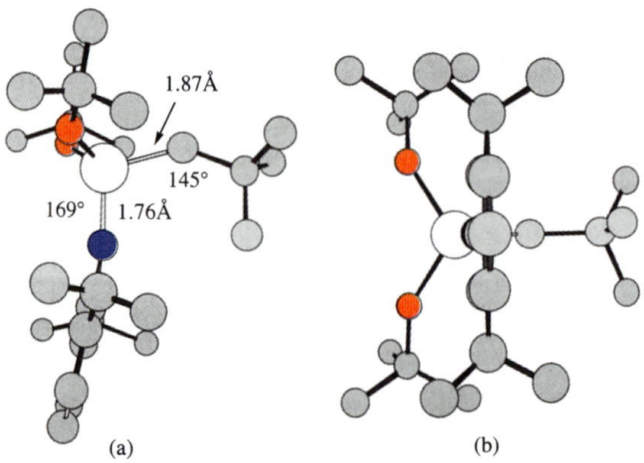

Figure 4. The structure of W(NAr)(CH-t-Bu)$(O\text{-}t\text{-}Bu)_2$; (a) top view; (b) side view.

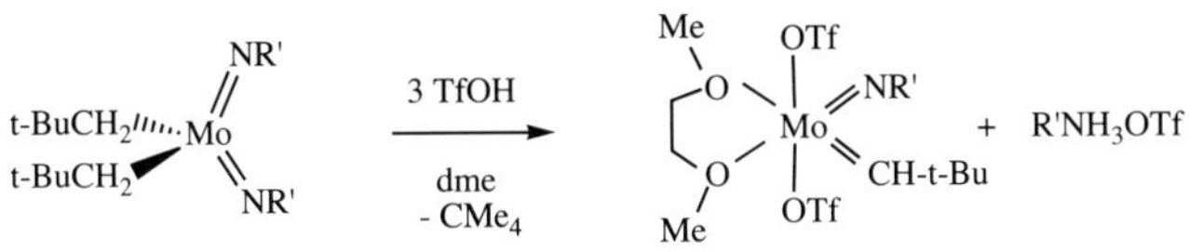

(19)

substituted tungstenacyclobutane intermediate, in particular, could be relatively slow. The metal therefore would remain sequestered in the form of a tungstenacyclobutane. Since Mo-ligand bonds are generally weaker than W-ligand bonds, we felt that a molybdacyclobutane complex might lose an olefin more readily. Therefore, we aimed to synthesize analogous Mo(NAr)(CH-t-Bu)$(OR)_2$ catalysts. A synthesis of bisalkoxides was devised, the key feature of which is the use of an imido "protecting group" that is then removed upon addition of triflic acid (Eq 19), thereby forming in the process a neopentylidene ligand from two neopentyl ligands through an α hydrogen abstraction reaction in a hypothetical Mo(NR')$(CH_2$-t-Bu$)_2$(Triflate$)_2$ intermediate.[49] The imido (NR') and alkoxide groups could be varied widely and a large number of molybdenum catalysts therefore became accessible.

The molybdenum bisalkoxide catalysts are also remarkably active for a wide range of metathesis reactions, again especially when the alkoxide is the highly electron withdrawing OCMe$(CF_3)_2$ group, and molybdacyclobutane intermediates *are* less stable than tungstacyclobutane intermediates toward loss of olefin. The steric bulk of the OCMe$(CF_3)_2$ and Ar (2,6-diisopropylphenyl) groups are also of considerable importance in preventing bimolecular reactions between the intermediate Mo(NAr)(CHR')$(OR)_2$ species in any given reaction. We slowly began to understand the extent to which the metathesis reaction could be *controlled* through systematic variation of the size and the electronic characteristics of the alkoxide and imido groups.

We also discovered that two isomers of any M=CHR catalyst are possible (Eq 20). In the isomer that is usually observed the alkylidene's substituent

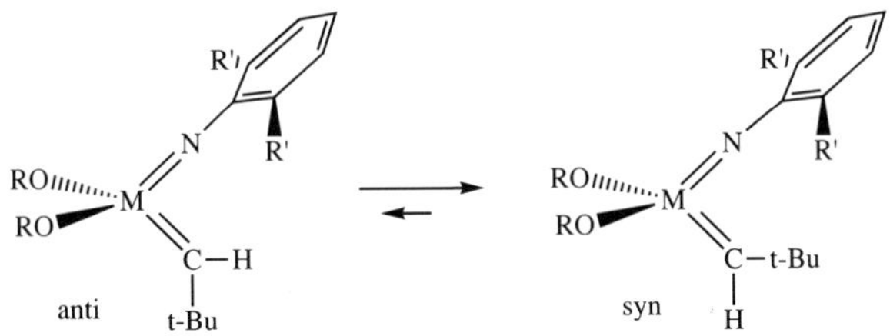

(20)

points toward the imido group (*syn*), while in the other it points away from the imido group (*anti*). *Syn* and *anti* species can be formed during the course of a metathesis reaction, but *syn* and *anti* isomers of any given alkylidene also can interconvert simply through rotation about the Mo=C bond. The rate of rotation of a neopentylidene or neophylidene was found to vary by a factor of at least 10^6, and to depend (largely) on the nature of the alkoxide,[50,51] with rotation being relatively fast in the presence of alkoxide ligands such as t-butoxide and relatively slow in the presence of hexafluoro-t-butoxide ligands. In ring-opening metathesis polymerization (ROMP) studies, one of many variations of the basic olefin metathesis reaction, we proved (for one type of polymerizeable monomer) that the *anti* isomer was the more reactive by a factor of ~10^4. Therefore one, or the other, or both *syn* and *anti* intermediates in a catalytic reaction might be involved, with the outcome depending upon the nature of the imido and alkoxide groups, *and* the reactivity of the olefin involved in the reaction with *syn* and *anti* intermediates, and also the temperature and other variables. This circumstance complicates to a considerable degree a detailed understanding of the metathesis reaction by such catalysts, but at the same time dramatically increases the flexibility of the catalysts and the possibility that a metathesis reaction will proceed efficiently.

Ring-opening metathesis polymerization of cyclic olefins had been studied for many years before well-defined catalysts were developed.[17,18] We naturally asked whether well-defined catalysts behaved as polymerization catalysts for cyclic olefins, and if so, what advantages did they offer? In our only publication together, one that resulted from my sabbatical at Caltech as a Fairchild scholar in 1986, Grubbs and I showed that these new well-defined catalysts did behave in the expected manner, and that the process could be "living" in the right circumstances, i.e., intermediate alkylidenes that contain the growing polymer chain would not decompose.[52] In my laboratory we went on to show that the polymerization process could be controlled in a dramatic fashion to yield polymers with a highly regular repeating structure. Generally polymers with highly regular structures have the most interesting properties and therefore are the most desirable. In many cases we found that the entire polymer structure can be controlled through minute, but critical, changes in the catalyst structure. For example (Eq 21) a highly regular *cis,isotactic* polymer is

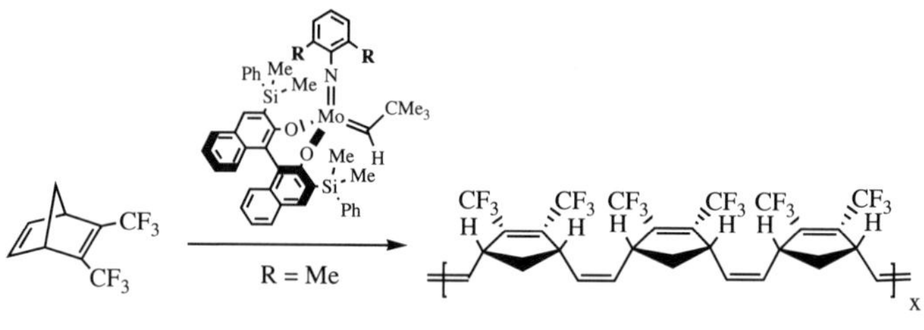

(21)

formed if the catalyst contains a specific binaphtholate ligand and a *dimethyl*phenylimido group (R = CH_3 in the 2,6-disubstituted phenyl group).[53,54] In contrast, if the R group in the imido ligand is *isopropyl*, then the polymer has a *relatively random structure*. The smaller *dimethyl*phenylimido group is believed to allow polymerization to proceed solely via *syn* isomers to give *cis* linkages. Control of the "tacticity," or the way the five-membered rings point in this polymer, is greatest in the presence of a chiral (racemic) binaphtholate ligand, which is a consequence of "enantiomorphic site control" over the polymerization process. In a series of experiments with related monomers that contain two esters made from enantiomerically pure alcohols, we could distinguish between *isotactic* and *syndiotactic* polymers, and therefore could prove the *cis,isotactic* structure, which is formed through enantiomorphic site control, and the *trans,syndiotactic* structure, which is formed through chain-end control, for the first time in ROMP polymers.[55] The degree to which one can control the polymer structure through systematic, tiny variations of the catalyst is of great importance for many ROMP polymers that are made today with high oxidation state alkylidene catalyst initiators.

For some time it had been known that "classical" Mo and W catalysts that would metathesize olefins would also polymerize alkynes. This suggested that metal alkylidenes were responsible.[56-58] We showed that well-defined species do indeed polymerize alkynes[59] or cyclopolymerize 1,6-heptadiynes[60] (Eq 22)

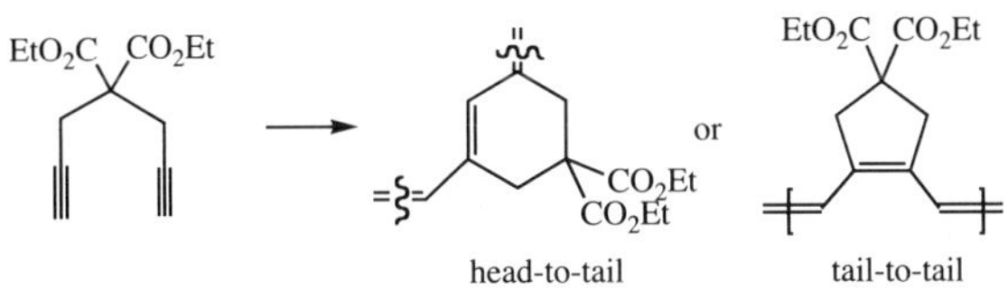

(22)

to yield polymers analogous to those described in the literature. In the case of a cyclopolymerization of 1,6-heptadiynes such as diethyldipropargylmalonate, a six-membered or a five-membered ring can form, depending upon the regiochemistry of thc addition of thc first alkync to thc alkylidcnc. Sincc thc optical properties of such polyenes vary with the polymer's structure and chain length,[61] it is highly desirable to make a polymer with a single structure and to control the molecular weight of that polymer through a polymerization with a well-defined catalyst. Although work is still ongoing in this area, it now appears possible to do so.[62,63]

In 1992 G. C. Fu and R. H. Grubbs published two papers in which they demonstrated how $Mo(NAr)(CHCMe_2Ph)[OCMe(CF_3)_2]_2$ could be employed to make cyclic olefins quickly and efficiently that contain functionalities other than a C=C bond,[64,65] with the only other product being a volatile olefin such as ethylene, propylene, or butene (Fig 5). Rings of many sizes, and even rings that contain tetrasubstituted olefinic bonds, could be prepared. These papers helped awaken organic chemists to the possibility that olefin metathesis by well-defined catalysts had significant potential in organic chemistry. The fact that the Mo and W catalysts in this general class are sen-

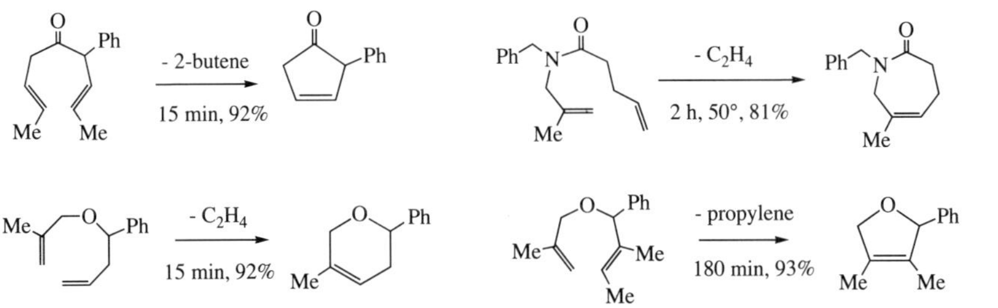

Figure 5. Some ring closing metatheses initiated by $Mo(NAr)(CHCMe_2Ph)[OCMe(CF_3)_2]_2$.

sitive to air, moisture, and some functionalities did not prohibit their use in organic reactions. In 1995 A. H. Hoveyda and his group reported a synthesis of a cyclic natural product, Fluvirucin-B_1, in which one of the steps consisted of a ring closing metathesis (RCM) reaction (Eq 23).[66] This paper demon-

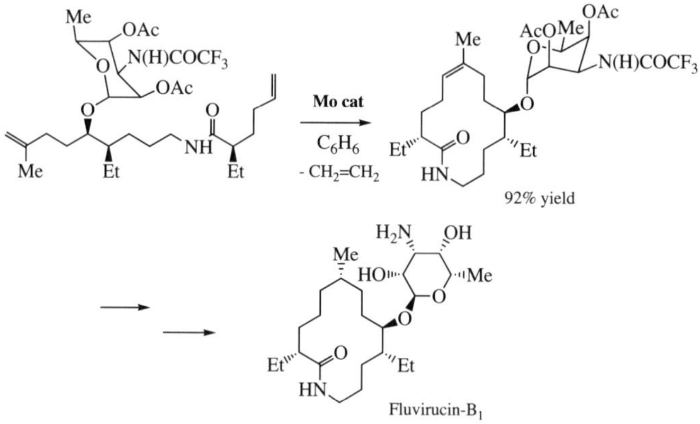

(23)

strated that relatively complex molecules could be prepared via metathesis. These and other applications of metathesis technology, along with the commercial availability of $Mo(NAr)(CHCMe_2Ph)[OCMe(CF_3)_2]_2$, helped fuel applications of Mo metathesis technology to organic chemistry.[67] The possibility that metathesis could be employed routinely in organic chemistry I believe was a significant factor that drove Grubbs to develop ruthenium catalysts in the early 1990's.[20,68,69]

The use of alkyne metathesis catalysts for organic chemistry also was realized in the 1990's, primarily through work by A. Fürstner, who showed that large rings could be prepared through alkyne metathesis with the $(Me_3CO)_3W{\equiv}CCMe_3$ initiator.[70] A cyclic alkyne then could be hydrogenated selectively to yield a *cis* olefin (Lindlar reduction). This alternative to alkene metathesis circumvented a still unsolved problem, formation of largely (~75%) thermodynamically more favored *trans* double bonds in alkene metathesis, instead of *cis* double bonds. An example is the stereoselective synthesis of civetone (Eq 24).[71] Many other reactions that involve formation of a carbon-carbon triple bond through alkyne metathesis as part of a synthetic organic procedure began to appear, such as syntheses of the cytotoxic marine alkaloid motuporamine C,[72] prostaglandin E_2-1,15-lactone,[73] and epothilone

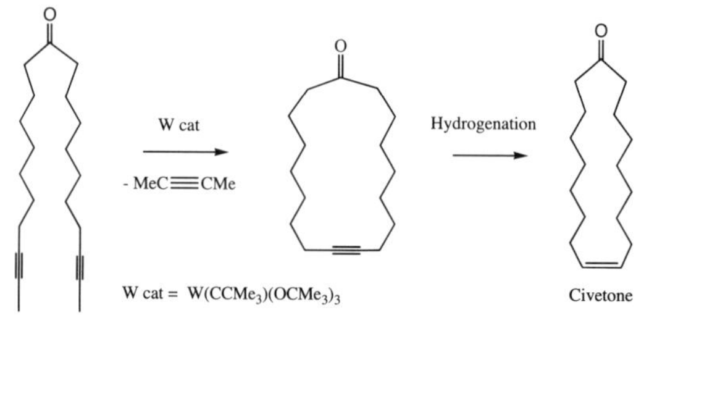

(24)

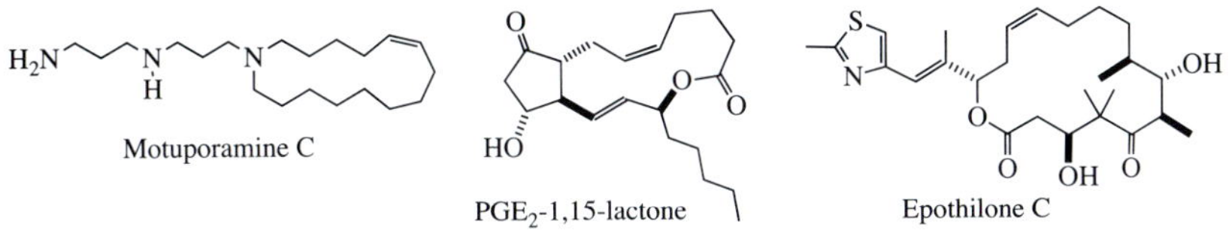

Figure 6. Some molecules whose syntheses involve alkyne metathesis to create the *cis* double bond.

A and C, all of which contain a *cis* olefin within a ring (Fig 6).[74] Alkyne metathesis can be useful in other ways, as shown in the synthesis of S-(+)-citreofuran (Eq 25).[75] Although molybdenum-based alkyne metathesis

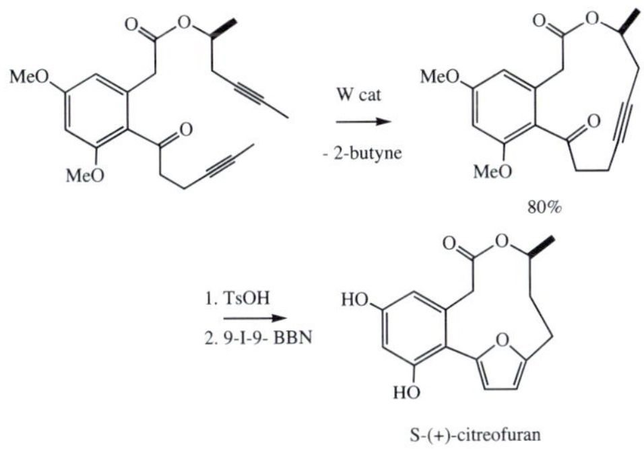

(25)

catalysts were known,[76,77] work by Cummins,[78] Fürstner,[79] and Moore,[80] recently have demonstrated how these relatively inaccessible Mo catalysts can be prepared *in situ* from trisamidoalkylidyne precursors.

With the knowledge that certain sterically protected biphenolates and binaphtholates can be attached to Mo to yield stable neopentylidene or neophylidene ROMP initiators, we turned to the development of enantiomerically pure catalysts for asymmetric metathesis reactions in the mid 1990's. Some preliminary results were obtained with an enantiomerically pure biphenolate catalyst (Fig 7) in 1997. We then began a collaboration with A. H. Hoveyda

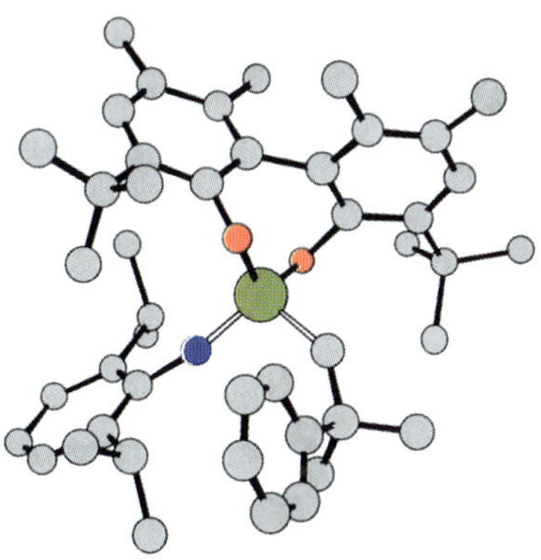

Figure 7. The structure of Mo(NAr)($CHCMe_2Ph$)(3,3'-di-t-butyl-5,5',6,6'-tetramethyl-1,1'-biphenyl-2,2'-diolate).

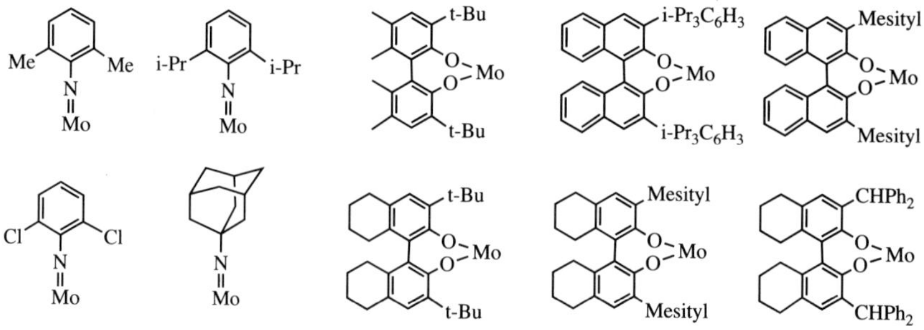

Figure 8. Imido groups and enantiomerically pure diolates (shown in racemic form) that are employed to make asymmetric catalysts..

that was aimed at the application of *asymmetric* metathesis reactions to organic synthesis.[67,81,82] The modularity of the catalysts allowed us to prepare many enantiomerically pure variations that contain one of several different imido ligands and several different diolates (Fig 8). It soon became apparent that with the right catalyst, asymmetric reactions could be efficient in terms of both yield and enantioselectivity, in many cases producing a single enantiomerically pure product in virtually quantitative yield relatively quickly (Eqs 26 and 27).[83] Again, byproducts usually consist of simple olefins such as ethyl-

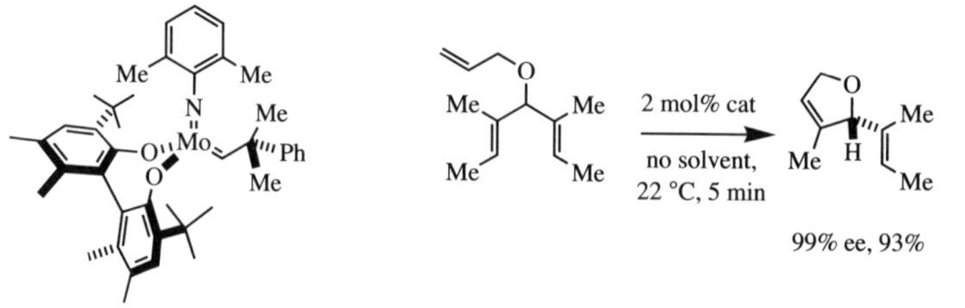

(26)

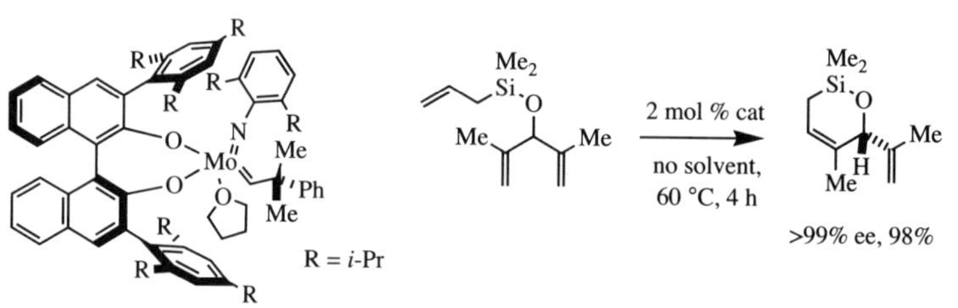

(27)

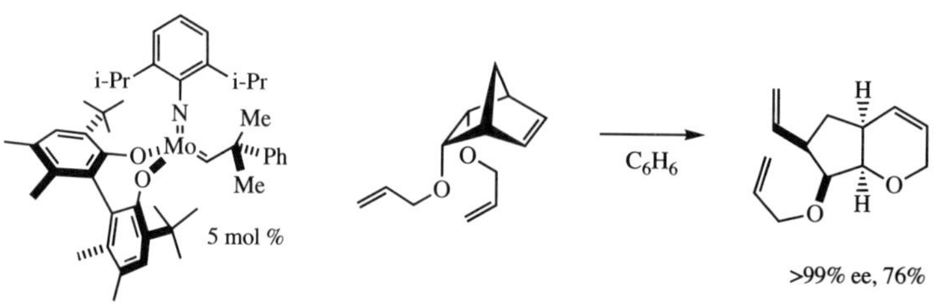

(28)

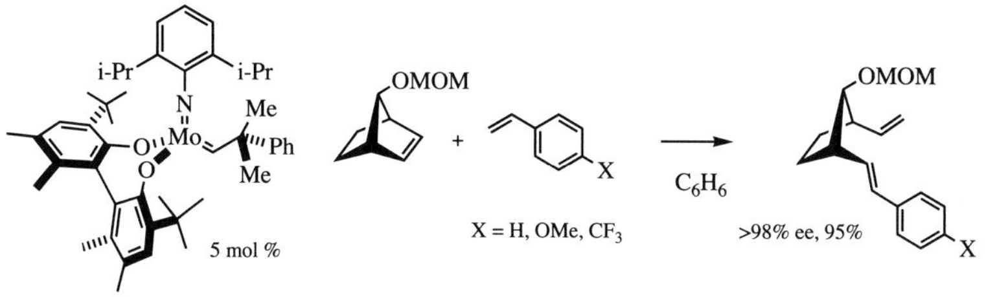

(29)

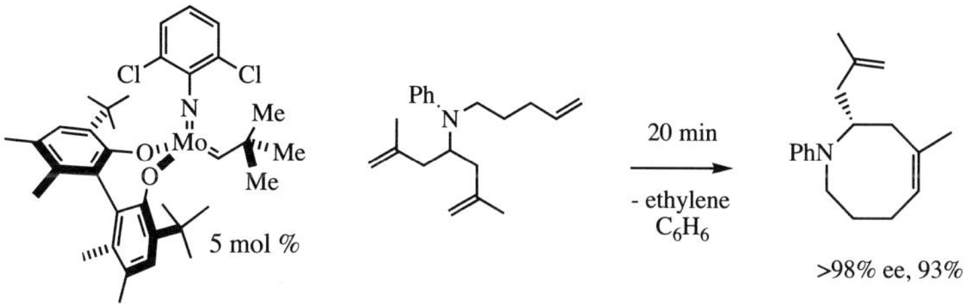

(30)

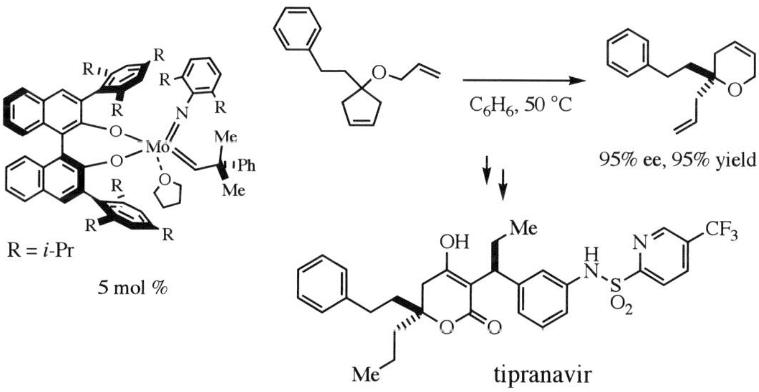

(31)

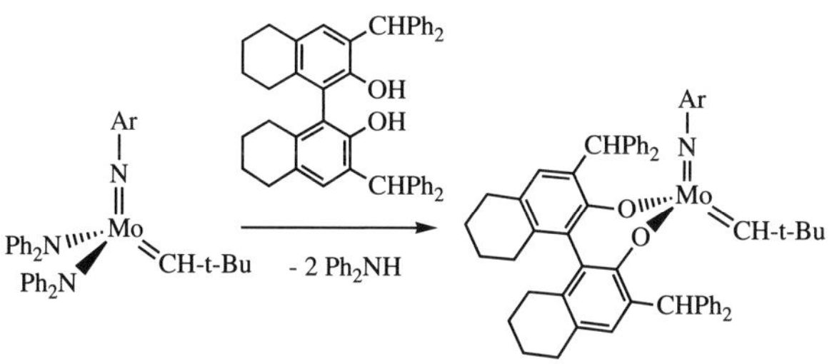

(32)

ene, propylene, or butene. A wide variety of other asymmetric reactions were developed, among them ring-opening/ring-closing metathesis (Eq 28),[84] ring-opening/cross metathesis (Eq 29),[85] and the ring closing of tertiary amines to yield products resembling substructures of various drugs and other natural products (Eq 30).[86] It also was shown that the synthesis of drugs such as tipranavir, an HIV protease inhibitor, could be shortened dramatically through the use of an asymmetric metathesis step to synthesize the problematic enantiomerically pure tertiary ether (Eq 31).[87,88] These are but a few examples of what has been accomplished in the last several years.

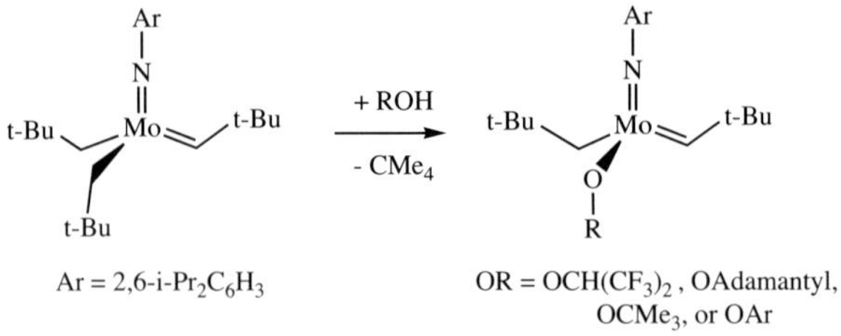

(33)

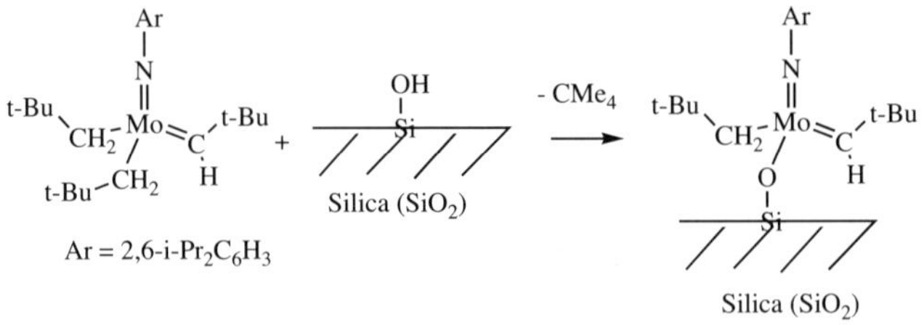

(34)

The isolation and testing of many catalyst variations side by side for one specific asymmetric reaction revealed that the efficiency of the asymmetric catalysts in terms of yield and enantioselectivity can vary dramatically, and that in the long run one could be relatively confident of finding an efficient catalyst. However, dozens of catalysts have to be prepared, isolated, and tested. Therefore, we have begun to explore the possibility of preparing catalysts *in situ* that we know exist as relatively stable species, either through addition of the biphenolate or binaphtholate to the bistriflate precursor (Eq 19), or through addition of the parent biphenol or binaphthol to a bisamido precursor (Eq 32). We have shown that catalysts prepared *in situ* for asymmetric metathesis reactions behave as well as catalysts that have been isolated and purified.[89] Surprisingly, similar approaches using dineopentyl species led to formation of *monoalkoxide* species (Eq 33).[90] Initially we thought these would be relatively poor metathesis catalysts, since dineopentyl species are essentially inactive for metathesis. The presence of one neopentyl and one alkoxide instead of two alkoxides therefore should be detrimental. However, preliminary results suggest that monoalkoxides are highly active catalysts. Recent theoretical calculations in related isoelectronic rhenium systems have suggested why this might be the case.[91] Formation of only monoalkoxides from dineopentyl species recently has allowed the synthesis of "well-defined" and highly active molybdenum-based (and tungsten-based) metathesis catalysts on silica surfaces (Eq 34) that are relatively long-lived because intermediate alkylidenes cannot decompose bimolecularly.[92] A variety of relatively well-defined catalysts now can be prepared on silica surfaces using other Ta(V), Mo(VI), W(VI), and Re(VII) neopentyl precursors (Fig 9).[93] In some cases entirely new reactions can be observed such as "alkane metathesis" by tantalum catalysts.[94] This reaction has now been shown to involve *alkene* metathesis steps.[95]

Rhenium forms metathesis catalysts of the "classical" type. Therefore we

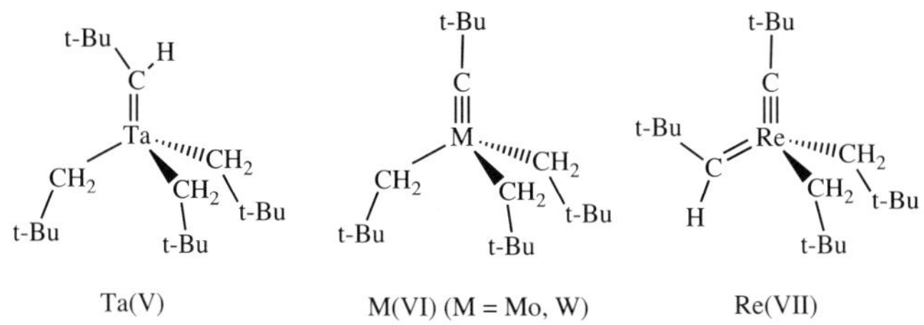

Figure 9. "Clean" sources of catalysts formed upon addition of them to silica surfaces.

felt that we should be able to prepare well-defined catalysts. In order to maintain the four-coordinate geometry around a neutral Re center, a sterically demanding ligand that is triply bound to the metal is required. The only logical choice is a neopentylidyne ligand. In fact, well-defined Re(VII) bisalkoxide catalysts for alkene metathesis can be prepared following the principles learned in W and Mo chemistry,[96,97] since the neopentylidyne ligand does not react with olefins readily (Fig 10). Monoalkoxide mononeopentyl analogs are believed to be even more reactive than bisalkoxides, especially on silica surfaces.[91]

Finally, we have found that M=M double bonds can form where M = Mo,[98] W,[99] or Re[100] as a consequence of decomposition of alkylidene species (Eq 35). Metal-metal double bonds are unusual, especially when the double bond

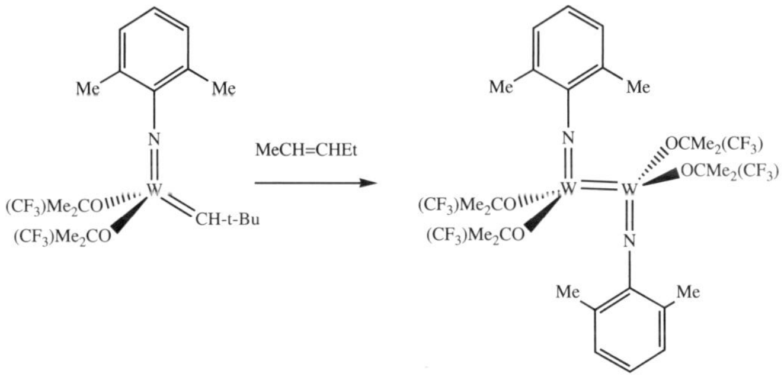

(35)

is not bridged by potentially bridging ligands such as an alkoxide or an imido. The fact that these species themselves will metathesize certain olefins[98] (slowly) raises the possibility that alkylidenes are being reformed from M=M bonds. If this turns out to be the case, it would constitute strong evidence that interconversion of metal-metal, carbon-carbon, and metal-carbon bonds is possible for double bonds, as we know it to be for triple bonds (Eq 17).

We have come an enormous distance in the last 30 years, from "ill-defined" metathesis catalysts to those whose structure and reactivity in solution (and perhaps now on surfaces) we can control with pinpoint precision. Fundamental problems with known catalysts remain, the most important being how to prevent catalyst decomposition and/or how to regenerate catalysts from "clean" decomposition products. I expect these challenges to be met and hope that the consequences of the synthesis and fundamental study of

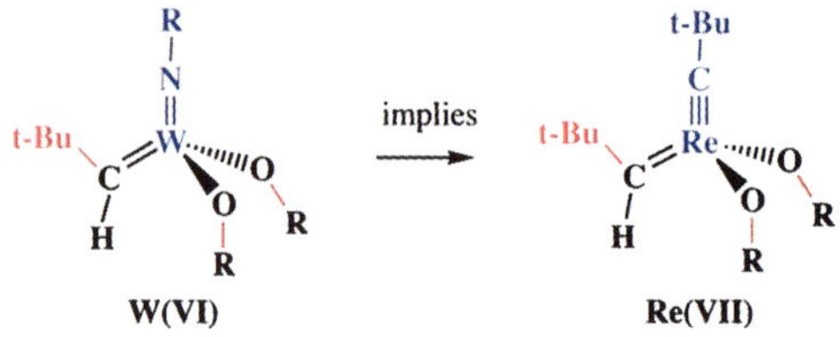

Figure 10. Rhenium neopentylidyne bisalkoxide complexes are olefin metathesis catalysts.

new types of catalysts, and the application of them to a wide variety of problems, will continue to be felt in the coming years.

ACKNOWLEDGEMENT

I am grateful for the freedom given to me in early years at the Central Research Department at duPont, for the opportunity to broaden my efforts and my interests at MIT, and for the support provided by various funding agencies (the National Science Foundation, the National Institutes of Health, the Navy, the Army, and the Department of Energy) over the years for high oxidation state organometallic chemistry and other projects. I am forever indebted to my graduate and postdoctoral students, each of whom in his or her own way has contributed to the massive effort required to turn a few observations and new compounds into chemistry that has had a significant impact on fields outside of inorganic chemistry. Finally, I want to thank my wife Nancy and my family for their love and support.

REFERENCES

1. G. Wilkinson, Nobel Lecture, December 11, 1973.
2. E. O. Fischer, Nobel Lecture, December 11, 1973.
3. Fischer, E. O. *Pure Appl. Chem.* **1970**, *24*, 407.
4. Fischer, E. O. *Pure Appl. Chem.* **1972**, *30*, 353.
5. Fischer, E. O.; Maasbol, A. *Angew. Chem. Int. Ed.* **1964**, *3*, 580.
6. Fischer, E. O.; Kreis, G.; Kreiter, C. G.; Müller, J.; Huttner, G.; Lorenz, H. *Angew. Chem. Int. Ed. Engl.* **1973**, *12*, 564.
7. Schrock, R. R.; Parshall, G. W. *Chem. Rev.* **1976**, *76*, 243.
8. Shortland, A. J.; Wilkinson, G. *J. Chem. Soc., Dalton Trans.* **1973**, 872.
9. "*Note added in proof.* Hexamethylrhenium (K. Mertis and G. Wilkinson) and pentamethyl[t]antalum (R. Schrock, DuPont, Wilmington, private communication) have recently been synthesized;" Geoffrey Wilkinson, Nobel Lecture, December 11, 1973.
10. Juvinall, G. L. *J. Am. Chem. Soc.* **1964**, *86*, 4202.
11. Schrock, R. R.; Meakin, P. *J. Am. Chem. Soc.* **1974**, *96*, 5288.
12. Mowat, W.; Wilkinson, G. *J. Chem. Soc, Dalton Trans.* **1973**, *10*, 1120.
13. Schrock, R. R. *J. Am. Chem. Soc.* **1974**, *96*, 6796.
14. Guggenberger, L. J.; Schrock, R. R. *J. Am. Chem. Soc.* **1975**, *97*, 2935.
15. Schrock, R. R. *J. Am. Chem. Soc.* **1975**, *97*, 6577.
16. Schrock, R. R.; Guggenberger, L. J. *J. Am. Chem. Soc.* **1975**, *97*, 6578.
17. Ivin, K. J. *Olefin Metathesis*; Academic Press: New York, 1983.
18. Ivin, K. J.; Mol, J. C. *Olefin Metathesis and Metathesis Polymerization*; Academic Press: San Diego, 1997.

19. Grubbs, R. H. In *Comprehensive Organometallic Chemistry*; Wilkinson, G., Stone, F. G. A., Abel, E. W., Eds.; Pergamon: New York, 1982; Vol. 8.
20. Grubbs, R. H., Ed. *Handbook of Metathesis. Catalyst Development*; Wiley-VCH: Weinheim, 2003; Vol. 1.
21. Grubbs, R. H.; Burk, P. L.; Carr, D. D. *J. Am. Chem. Soc.* **1975**, *97*, 3265.
22. Grubbs, R. H.; Carr, D. D.; Hoppin, C. R.; Burk, P. L. *J. Am. Chem. Soc.* **1976**, *98*, 3478.
23. Grubbs, R. H.; Hoppin, C. R. *J. Chem. Soc., Chem. Comm.* **1977**, 634.
24. Katz, T. J.; McGinnis, J. *J. Am. Chem. Soc.* **1975**, *97*, 1592.
25. McGinnis, J.; Katz, T. J.; Hurwitz, S. *J. Am. Chem. Soc.* **1976**, *98*, 605.
26. Casey, C. P.; Burkhardt, T. J. *J. Am. Chem. Soc.* **1974**, *96*, 7808.
27. Hérrison, J. L.; Chauvin, Y. *Makromol. Chem.* **1971**, *141*, 161.
28. Pannella, F.; Banks, R. L.; Bailey, G. C. *J. Chem. Soc., Chem. Commun.* **1968**, 1548.
29. Moulijn, J. A.; Reitsma, H. J.; Boelhouwer, C. *J. Catal.* **1972**, *25*, 434.
30. Mortreux, A.; Blanchard, M. *J. Chem. Soc. Chem. Commun.* **1974**, 786.
31. McLain, S. J.; Wood, C. D.; Schrock, R. R. *J. Am. Chem. Soc.* **1979**, *101*, 4558.
32. Schrock, R. R.; Rocklage, S. M.; Wengrovius, J. H.; Rupprecht, G.; Fellmann, J. *J. Molec. Catal.* **1980**, *8*, 73.
33. Wengrovius, J. H.; Schrock, R. R.; Churchill, M. R.; Missert, J. R.; Youngs, W. J. *J. Am. Chem. Soc.* **1980**, *102*, 4515.
34. Clark, D. N.; Schrock, R. R. *J. Am. Chem. Soc.* **1978**, *100*, 6774.
35. Schrock, R. R.; Clark, D. N.; Sancho, J.; Wengrovius, J. H.; Rocklage, S. M.; Pedersen, S. F. *Organometallics* **1982**, *1*, 1645.
36. Wengrovius, J. H.; Sancho, J.; Schrock, R. R. *J. Am. Chem. Soc.* **1981**, *103*, 3932.
37. Churchill, M. R.; Wasserman, H. J. *J. Organomet. Chem.* **1984**, *270*, 201.
38. Churchill, M. R.; Fettinger, J. C.; McCullough, L. G.; Schrock, R. R. *J. Am. Chem. Soc.* **1984**, *106*, 3356.
39. Churchill, M. R.; Ziller, J. W.; Pedersen, S. F.; Schrock, R. R. *J. Chem. Soc., Chem. Commun.* **1984**, 485.
40. Schrock, R. R.; Pedersen, S. F.; Churchill, M. R.; Ziller, J. W. *Organometallics* **1984**, *3*, 1574.
41. Churchill, M. R.; Ziller, J. W.; Freudenberger, J. H., Schrock, R. R. *Organometallics* **1984**, *3*, 1554.
42. Freudenberger, J. H.; Schrock, R. R.; Churchill, M. R.; Rheingold, A. L.; Ziller, J. W. *Organometallics* **1984**, *3*, 1563.
43. Buhro, W. E.; Chisholm, M. H. *Adv. Organometal. Chem.* **1987**, *27*, 311.
44. Chisholm, M. H. *Acc. Chem. Res.* **1990**, *23*, 419.
45. Schrock, R. R.; Listemann, M. L.; Sturgeoff, L. G. *J. Am. Chem. Soc.* **1982**, *104*, 4291.
46. Listemann, M. L.; Schrock, R. R. *Organometallics* **1985**, *4*, 74.
47. Laplaza, C. E.; Cummins, C. C. *Science* **1995**, *268*, 861.
48. Feldman, J.; Schrock, R. R. *Prog. Inorg. Chem.* **1991**, *39*, 1.
49. Schrock, R. R.; Murdzek, J. S.; Bazan, G. C.; Robbins, J.; DiMare, M.; O'Regan, M. *J. Am. Chem. Soc.* **1990**, *112*, 3875.
50. Oskam, J. H.; Schrock, R. R. *J. Am. Chem. Soc.* **1992**, *114*, 7588.
51. Oskam, J. H.; Schrock, R. R. *J. Am. Chem. Soc.* **1993**, *115*, 11831.
52. Schrock, R. R.; Feldman, J.; Cannizzo, L.; Grubbs, R. H. *Macromolecules* **1987**, *20*, 1169.
53. Totland, K. M.; Boyd, T. J.; Lavoie, G. G.; Davis, W. M.; Schrock, R. R. *Macromolecules* **1996**, *29*, 6114.
54. McConville, D. H.; Wolf, J. R.; Schrock, R. R. *J. Am. Chem. Soc.* **1993**, *115*, 4413.
55. O'Dell, R.; McConville, D. H.; Hofmeister, G. E.; Schrock, R. R. *J. Am. Chem. Soc.* **1994**, *116*, 3414.
56. Masuda, T.; Higashimura, T. *Adv. Polym. Sci.* **1987**, *81*, 122.
57. Masuda, T.; Higashimura, T. *Acc. Chem. Res.* **1984**, *17*, 51.
58. Choi, S.-K.; Gal, Y.-S.; Jin, H.-H.; Kim, H.-K. *Chem. Rev.* **2000**, *100*, 1645.
59. Schrock, R. R.; Luo, S.; Lee, J. C. J.; Zanetti, N. C.; Davis, W. M. *J. Am. Chem. Soc.* **1996**, *118*, 3883.

60. Fox, H. H.; Wolf, M. O.; O'Dell, R.; Lin, B. L.; Schrock, R. R.; Wrighton, M. S. *J. Am. Chem. Soc.* **1994**, *116*, 2827.
61. Christensen, R. L.; Faksh, A.; Meyers, J. A.; Samuel, I. D. W.; Wood, P.; Schrock, R. R.; Hultzsch, K. C. *J. Phys. Chem. A* **2004**, *108*, 8229.
62. Schattenmann, F. J.; Schrock, R. R.; Davis, W. M. *J. Am. Chem. Soc.* **1996**, *118*, 3295.
63. Anders, U.; Nuyken, O.; Buchmeiser, M. R. *J. Mol. Catal. A* **2004**, *213*, 89.
64. Fu, G. C.; Grubbs, R. H. *J. Am. Chem. Soc.* **1992**, *114*, 5426.
65. Fu, G. C.; Grubbs, R. H. *J. Am. Chem. Soc.* **1992**, *114*, 7324.
66. Houri, A. F.; Xu, Z. M.; Cogan, D. A.; Hoveyda, A. H. *J. Am. Chem. Soc.* **1995**, *117*, 2943.
67. Schrock, R. R.; Hoveyda, A. H. *Angew. Chem. Int. Ed.* **2003**, *42*, 4592.
68. Grubbs, R. H., Ed. *Handbook of Metathesis. Applications in Organic Synthesis*; Wiley-VCH: Weinheim, 2003; Vol. 2.
69. Grubbs, R. H., Ed. *Handbook of Metathesis. Applications in Polymer Synthesis*; Wiley-VCH: Weinheim, 2003; Vol. 3.
70. Fürstner, A.; Seidel, G. *Angew. Chem. Int. Ed.* **1998**, *37*, 1734.
71. Fürstner, A.; Seidel, G. *J. Organomet. Chem.* **2000**, *606*, 75.
72. Fürstner, A.; Rumbo, A. *J. Org. Chem.* **2000**, *65*, 2608.
73. Fürstner, A.; Grela, K. *Angew. Chem. Int. Ed.* **2000**, *39*, 1234.
74. Fürstner, A.; Mathes, C.; Lehmann, C. W. *Chem. Eur. J.* **2001**, *7*, 5299.
75. Fürstner, A.; Castanet, A.-S.; Radkowski, K.; Lehmann, C. W. *J. Org. Chem.* **2003**, *68*, 1521.
76. McCullough, L. G.; Schrock, R. R. *J. Am. Chem. Soc.* **1984**, *106*, 4067.
77. Murdzek, J. S.; Schrock, R. R. In *Carbyne Complexes*; VCH: New York, 1988.
78. Tsai, Y. C.; Diaconescu, P. L.; Cummins, C. C. *Organometallics* **2000**, *19*, 5260.
79. Fürstner, A.; Mathes, C.; Lehmann, C. W. *J. Am. Chem. Soc.* **1999**, *121*, 9453.
80. Zhang, W.; Kraft, S.; Moore, J. S. *Chem. Commun.* **2003**, 832.
81. Alexander, J. B.; La, D. S.; Cefalo, D. R.; Hoveyda, A.; Schrock, R. R. *J. Am. Chem. Soc.* **1998**, *120*, 4041.
82. Alexander, J. B.; Schrock, R. R.; Davis, W. M.; Hultzsch, K. C.; Hoveyda, A. H.; Houser, J. H. *Organometallics* **2000**, *19*, 3700.
83. La, D. S.; Alexander, J. B.; Cefalo, D. R.; Graf, D. D.; Hoveyda, A. H.; Schrock, R. R. *J. Am. Chem. Soc.* **1998**, *120*, 9720.
84. Weatherhead, G. S.; Ford, J. G.; Alexanian, E. J.; Schrock, R. R.; Hoveyda, A. H. *J. Am. Chem. Soc.* **2000**, *122*, 1828.
85. La, D. S.; Ford, J. G.; Sattely, E. S.; Bonitatebus, P. J.; Schrock, R. R.; Hoveyda, A. H. *J. Am. Chem. Soc.* **1999**, *121*, 11604.
86. Dolman, S. J.; Sattely, E. S.; Hoveyda, A. H.; Schrock, R. R. *J. Am. Chem. Soc.* **2002**, *124*, 6991.
87. Cefalo, D. R.; Kiely, A. F.; Wuchrer, M.; Jamieson, J. Y.; Schrock, R. R.; Hoveyda, A. H. *J. Am. Chem. Soc.* **2001**, *123*, 3139.
88. Kiely, A. F.; Jernelius, J. A.; Schrock, R. R.; Hoveyda, A. H. *J. Am. Chem. Soc.* **2002**, *124*, 2868.
89. Sinha, A.; unpublished results.
90. Sinha, A.; Schrock, R. R. *Organometallics* **2004**, *23*, 1643.
91. Solans-Monfort, X.; Clot, E.; Copéret, C.; Eisenstein, O. *J. Am. Chem. Soc.* **2005**, 14015.
92. Blanc, F.; Copéret, C.; Thivolle-Cazat, J.; Basset, J.-M.; Sinha, A.; Schrock, R. R. *Angew. Chem. Int. Ed.* **2006**, 45, 1216.
93. Copéret, C.; Chabanas, M.; Saint-Arroman, R. P.; Basset, J.-M. *Angew. Chem. Int. Ed.* **2003**, *42*, 156.
94. Taoufik, M.; Schwab, E.; Schultz, M.; Vanoppen, D.; Walter, M.; Thivolle-Cazat, J.; Basset, J.-M. *Chem. Commun.* **2004**, 1434.
95. Soulivong, D.; Copéret, C.; Thivolle-Cazat, J.; Basset, J.-M.; Maunders, B. M.; Pardy, R. B. A.; Sunley, G. J. *Angew. Chem. Int. Ed.* **2004**, *43*, 5366.
96. Toreki, R.; Schrock, R. R. *J. Am. Chem. Soc.* **1992**, *114*, 3367.

97. Schrock, R. R.; Toreki, R.; Vaughan, G. A.; Farooq, A. In *Transition-Metal Carbyne Complexes*; Kreissl, F. R., Ed.; Kluwer Academic: Dordrecht, The Netherlands, 1992.
98. Schrock, R. R.; Lopez, L. P. H.; Hafer, J.; Singh, R.; Sinha, A.; Müller, P. *Organometallics* **2005**, *24*, 5211.
99. Lopez, L. P. H.; Schrock, R. R. *J. Am. Chem. Soc.* **2004**, *126*, 9526.
100. Toreki, R.; Schrock, R. R.; Vale, M. G. *J. Am. Chem. Soc.* **1991**, *113*, 3610.

Portrait photo of Richard R. Schrock by photographer L. B. Hetherington.